KIRK-OTHMER

ENCYCLOPEDIA OF CHEMICAL TECHNOLOGY

THIRD EDITION

INDEX
Volumes 13 to 16

HYDROGEN-ION
ACTIVITY to
PERFUMES

A WILEY-INTERSCIENCE PUBLICATION
John Wiley & Sons
NEW YORK • CHICHESTER • BRISBANE • TORONTO • SINGAPORE

A

Abherents
  in papermaking, 16:808
Abietic acid [514-10-3]
  in rosin size, 16:810
ABI-INFORM information system, 13:328
Abitol 90X, 14:567
Ablative polymer
  from 2,7-naphthalenediol, 15:736
Abrasives
  in magnetic tape, 14:743
ABS. See *Acrylonitrile–butadiene–styrene.*
Absence seizures, 13:137. (See also *Petit mal.*)
Absorptance
  of optical filters, 16:522
Absorption
  in LPG processing, 14:384
  in NO$_x$ abatement, 15:864
  noise, 16:15
Absorption, dielectric, 13:537
Abuscreen kits, 15:88
Acac. See *Acetylacetone.*
Acacia decurrens, 16:951
Acanthoscelides obtectus, 13:465
Acaricides, 13:461
Acar process, 13:745
Acatalasemia
  encapsulated treatment, 15:490
Acceleration tubes
  in ion implantation, 13:709
Accelerator, ion implantation, 13:**706**
Accelerators
  in metal treating, 15:305
  radiation exposure from, 16:217
Acceptance life tests, 15:27
ACCOUNTANTS' INDEX information system,
    13:328
Accurel, 15:104
Acecoline [60-31-1], 15:756. (See also
    Acetylcholine chloride.)
Acenaphthene [83-32-9], 15:717
  1,8-naphthalenedicarboxylic acid anhydride
    from, 15:741
5,6-Acenaphthenedicarboximide [5702-86-3]
  1,4,5,8- naphthalenetetracarboxylic acid from,
    15:744
Acenaphthenequinone [82-86-0], 15:716
Acenaphthylene [208-96-8], 15:717
Acephate [30560-19-1], 13:443
Acetal [1820-50-4]
  formaldehyde reaction, 13:925

Acetaldehyde [75-07-0]
  autoxidation of, 13:23
  aziridine reaction, 13:150
  from ethylene, 16:483, 668
  in latex discoloration, 14:90
  in microbial ephedrine synthesis of, 15:464
  ozone reaction, 16:687
Acetaldehyde cyanohydrin [78-97-7], 13:89. (See
    also *Lactonitrile.*)
Acetaldehyde hemiacetal peracetate [7416-48-0],
    16:687
Acetalimine [20729-41-3], 13:143
4-Acetamido-5-hydroxy-2,7-naphthalenedisulfon-
    ic acid [134-34-9], 15:742
Acetaminomalonic esters, 14:800
Acetaminophen [103-90-2]
  immunologic effects, 13:174
Acetate filters, 16:542
Aceterobacter oryzae, 15:463
Acetic acid [64-19-7], 16:763
  from ethylene, 16:668
  for hydrogen peroxide recovery, 13:21
  from hydroxyacetic acid, 13:91
  in iron reaction, 13:764
  ketene reaction, 13:876
  from methanol, 16:584
  paint removers, 16:763
  production, 15:398
Acetic anhydride [108-24-7]
  isocyanate reaction, 13:798
  novoloid fiber bleaching, 16:131
Acetoacetamide [5977-14-0], 13:887
  ammonia reaction, 13:889
  from diketene, 13:888
Acetoacetanilides, 13:890
Acetoacetic acid [541-50-4]
  blood ketone, 13:605
  derivatives, 13:887
Acetoacetic acid, 2-methyl-3-buten-2-ol ester
    [15973-38-3], 13:888
Acetoacetyl fluoride [2343-88-6]
  from diketene, 13:884
Acetobacter suboxydans, 15:459
Acetobacter xylinum, 15:459
Acetochloro-D-ribofuranose [40554-98-1]
  in adenosine synthesis, 15:756
Acetohexamide [968-81-0], 13:614
Acetomeroctol [584-18-9], 15:170
Acetone [67-64-1]
  aldol condensation, 13:900
  color test for *m*-dinitrobenzene, 15:921
  diketene reaction, 13:884
  ethyl acetate condensation, 13:931
  in isoprene synthesis, 13:831
  from isopropyl alcohol, 13:21
  from kelp, 16:294
  ketene reaction, 13:876
  methacrylates from, 15:357

Acetone (*Continued*)
  nitromethane reaction, *15*:976
  odor-problem chemical, *16*:303
  in paint and varnish removers, *16*:764
  properties, *13*:896
Acetone cyanohydrin [*75-86-5*] (ACN), *15*:889
  methacrylic acid from, *15*:346
  properties, *15*:893
Acetone disulfonic acid [*590-78-8*]
  methyl methacrylate by-product, *15*:359
Acetone phenylhydrazone [*103-02-6*]
  indole synthesis, *13*:218
Acetonitrile [*75-05-8*], *15*:888, 889
  in β-alanine production, *13*:96
  in lithium batteries, *14*:459
  malononitrile preparation, *14*:805
  methacrylonitrile by-product, *15*:363
  properties, *15*:895
Acetonyl methacrylate [*44901-95-3*]
  properties, *15*:354
Acetophenone [*98-86-2*]
  ammonium formate reaction, *13*:901
  from benzoic acid, *13*:934
  properties, *13*:896
Acetophenone lithium enolate [*55905-98-1*], *13·*
    932
o-Acetotoluidide [*120-66-1*], *13*:218
(−)-14-Acetoxycodeine [*76157-64-7*]
  microbial hydrolysis, *15*:462
6-Acetoxy-2,4-dimethyl-*m*-dioxane [*828-00-2*]
  applications of, *13*:248
Acetoxymercuri-2-ethylhexylphenolsulfonate
    [*1301-13-9*], *15*:170
α-Acetoxypropionic acid [*2382-59-4*], *13*:89
β-Acetoxypropionic acid [*2382-59-4*], *13*:96
Acetylacetone [*123-54-6*], *13*:930. (See also *2,4-*
    Pentanedione.)
  catalyst deactivator, *16*:391
  iron complex, *13*:779
  in organolead reactions, *14*:181
  from salicylaldehyde, *13*:78
  uv absorbers, *13*:78
N-Acetyl-H-acid [*134-34-9*], *15*:742. (See also
    4-Acetamido-5-hydroxy-2,7-naphthalenedi-
    sulfonic acid.)
N-Acetylanthranilic acid [*89-52-1*]
  in methaqualone synthesis, *13*:125
Acetylated cedarwood terpene [*32388-55-9*], *16*:
    955
Acetylation
  microbial, *15*:466
1-Acetylaziridine [*460-07-01*], *13*:143, 155
p-Acetylbenzenesulfonamide [*1565-17-9*]
  cyclohexyl isocyanate reaction, *13*:615
Acetylbenzoyl [*579-07-7*]
  properties, *13*:926
Acetylbenzoylmethane [*93-91-4*], *13*:930
Acetylbutyrolactone [*517-23-7*], *13*:888

from ethylene oxide, *13*:888
Acetyl chloride [*75-36-5*]
  ferrocene reaction, *16*:604
  malic acid reaction, *13*:105
Acetylcholine [*51-84-3*], *13*:449
  brain occurrence, *15*:134
  calcium effect on, *15*:584
  formation malfunction of, *14*:266
  as neuroregulator, *15*:756
  nicotine effect on, *13*:424
Acetylcholine bromide [*66-23-9*], *15*:756
Acetylcholinechloride [*60-31-1*], *15*:756
Acetylcholinesterase [*9012-78-6*]
  in insecticide reaction, *13*:452
Acetyl-coenzyme A [*72-89-9*]
  in acetylcholine biosynthesis, *15*:756
Acetyl-*S*-coenzyme A [*72-89-9*]
  in fatty acid synthesis, *15*:587
Acetyl cyclohexanesulfonyl peroxide [*3179-56-4*]
  radical initiator, *13*:358
Acetylene [*74-86-2*]
  health effect, *13*:263
  ICl$_3$ reaction, *13*:664
  Ilosvay's reagent for, *16*:665
  in isoprene synthesis, *13*:831
  in metallic coating, *15*:254
  by N$_2$-fixing catalysts, *15*:958
  nitrogen reaction, *15*:934
  oxalic acid from, *16*:627
Acetylenediurea [*496-46-8*]
  resin from in fluorescent product, *14*:560
Acetylenes
  nitroparaffin reaction, *15*:977
1-Acetylindole [*576-15-8*], *13*:215
Acetyl isoeugenol [*93-29-8*]
  in perfumes, *16*:968
Acetylketene, *13*:879
N-Acetylmethacrylamide [*44810-87-9*]
  properties, *15*:356
N-Acetyl-5-methoxytryptamine [*73-31-4*], *15*:
    771. (See also *Melatonin.*)
1-Acetyl-2-methyl-1,2-epoxypropane [*4478-63-1*]
  from diacetone alcohol, *13*:912
Acetyl nitrate [*591-09-3*]
  nitration reagent, *15*:843
3-(Acetyloxy)-2-cyclobuten-1-one [*38425-52-4*]
  from ketene, *13*:878
(Acetyloxy)triethyllead [*2587-81-7*], *14*:181
Acetylsalicylic acid [*50-78-2*]
  measurement of with electrodes, *13*:725
1-Acetyl-2,4,6,6-tetramethylcyclohexa-1,3-diene
    [*13834-80-5*], *13*:917
ACGIH. See *American Conference of*
    Governmental Industrial Hygienists.
ACh. See *Acetylcholine.*

Acheson process
  for graphite, *15*:172
Achlorhydria
  treatment, *15*:767
Achromobacter obae, *15*:464
Achromobacter tartarogenes, *15*:463
1,2,4-Acid [*116-63-2*], *15*:740. (See also *3-Hydroxy-4-amino-1-naphthalenesulfonic acid.*)
ε-Acid [*117-43-1*], *15*:738, 739. (See also *8-Hydroxy-1,6-naphthalenedisulfonic acid.*)
γ-Acid [*90-51-7*], *15*:740. (See also *4-Hydroxy-6-amino-2-naphthalenesulfonic acid.*)
Acid cresol red [*1733-12-6*], *13*:10
Acidity
  nitroparaffins, *15*:970
Acid lignin, *14*:310
Acid metacresol purple [*2303-01-7*], *13*:10
Acidogen, *15*:767
Acidoride, *15*:767
Acidosis
  in diabetes, *13*:606
Acid rain, *16*:796
Acids
  packaging of, *16*:717
Acid thymol blue [*76-61-9*], *13*:10
Acidulant
  lactic acid in foods, *13*:86
Acidulin, *15*:767
Aciglumin, *15*:767
ACL-56, *13*:230
ACL-59, *13*:230
ACL-60, *13*:230
ACL-66, *13*:230
ACL-85, *13*:230
Aclor, *15*:767
ACN. See *Acetone cyanohydrin.*
Acne
  from chloronaphthalenes, *15*:702
  treatment, *13*:62
Aconitase [*9024-25-3*], *15*:595
Aconitate decarboxylase [*9025-01-8*], *13*:869
Aconitate hydratase [*9024-25-3*], *13*:869
cis-Aconitic acid [*585-84-2*]
  from citric acid, *13*:869
Acorn, *16*:250
Acoustic emission tests
  nondestructive, *16*:68
Acoustic insulation, *13*:513. (See also *Insulation, acoustic.*)
  lead, *14*:157
Acoustic lining
  for sound control, *13*:513
Acrodermatitis enteropathica, *15*:590
  nutrients associated with, *15*:573

Acrolein [*107-02-8*]
  applications of, *13*:248
  kinetics of destruction, *13*:201
Acroleins
  oxo by-products, *16*:646
Acromegaly
  treatment, *15*:780
Acrylamide [*79-06-1*]
  in emulsion polymerization, *14*:87
  in encapsulation, *15*:481
  toxicity, *13*:263
Acrylamide polymers
  in papermaking, *16*:804
Acrylate esters
  organotins from, *16*:578
Acrylic acid [*79-10-7*]
  in cation-exchange resins, *13*:686
  copolymer with ethylene, *16*:420
  in emulsion polymerization, *14*:87
  isoprene reaction, *13*:821
  in papermaking, *16*:804
Acrylic acid, 2,3-epoxypropyl ester [*106-90-1*]
  copolymers from, *15*:205
Acrylic ester copolymers
  in leather coatings, *14*:222
Acrylic inks, *13*:391
Acrylic resins
  tanning with, *14*:216
Acrylics
  odor-problem chemicals, *16*:303
  in paint, *16*:748
Acryloid B-66, *14*:566
Acryloid NAD-10, *14*:567
Acrylonitrile [*107-13-1*], *15*:889
  copolymer, *15*:192
  electrodimerization, *15*:899
  in encapsulation, *15*:480
  as fumigant, *13*:466
  in glutethimide synthesis, *13*:124
  methacrylate copolymerization, *15*:384
  in paint, *16*:760
  polymerization of, *13*:356
  solubility of, *14*:83
  toxicity, *13*:263
Acrylonitrile–butadiene–styrene
  latex technology, *14*:82
  metallic coating of, *15*:266
Acryloxytributylstannane [*13331-52-7*]
  copolymers from, *15*:205
[2-[Acryloyl]ethyl]ferrocene [*34802-13-6*], *15*:187
[2-[Acryloyl]ethyl]ferrocene, polymer with styrene [*35561-01-4*], *15*:189
[[Acryloyl]methyl]ferrocene [*31566-60-6*], *15*:187
1-[[Acryloyl]methyl]ferrocene, polymer with styrene [*32612-76-3*], *15*:189

α-Actalbumin
  in casein, *15*:568
ACTH. See *Adrenocorticotrophic hormone.*
Actin [*9064-37-3*], *14*:419
  calcium effect on, *15*:584
Actinide nitrides, *15*:882
Actinides
  purification and recovery, *16*:182
Actinomycetes
  in N_2 fixation, *15*:948
Actinorhizae
  nitrogen-fixing, *15*:948
Activated-carbon fibers, *16*:136
Activated charcoal
  from pecan shells, *16*:273
Activators
  rare earth, in phosphors, *14*:540
Activity coefficients, *13*:4
Activity measurement
  with electrodes, *13*:729
Acylation
  metallocenes, *16*:604
Acyl cyanides
  nitroparaffin salt reaction, *15*:974
Addison's disease, *13*:173
  nutrients associated with, *15*:573
  treatment, *15*:762
Addition
  initiators used in, *13*:356
Additives
  in lubricants, *14*:477
Adenine [*73-24-5*]
  in adenosine biosynthesis, *15*:757
  in nucleic acid, *15*:587
Adenosine [*58-61-7*]
  as neuroregulator, *15*:756
  properties, *15*:756
Adenosine diphosphate [*58-64-0*]
  nitrogenase inhibitor, *15*:951
Adenosine diphosphate [*58-64-0*] (ADP)
  in CPK assay, *15*:77
Adenosine monophosphate
  ACTH, MSH effect on, *15*:133
Adenosine picrate [*6128-21-8*], *15*:756
Adenosine triphosphate [*56-65-5*]
  in nitrogenase activity, *15*:951
Adenosine triphosphate [*56-65-5*] (ATP)
  in CPK assay, *15*:77
  in insecticide reaction, *13*:429
  in sodium–potassium pump, *15*:574
S-Adenosylhomocysteine [*979-92-0*]
  in adenosine biosynthesis, *15*:757
S-Adenosylmethionine [*29908-03-0*], *15*:587
  in DMT biosynthesis, *15*:763
  in spermine biosynthesis, *15*:777
Adenylate cyclase [*9012-42-4*]
  octopamine effect on, *15*:774

3′-Adenylic acid [*84-21-9*]
  in adenosine biosynthesis, *15*:757
5′-Adenylic acid [*61-19-8*]
  in adenosine biosynthesis, *15*:757
Adenylylation
  microbial, *15*:466
Adhesives, *13*:57
  anaerobic, *15*:491
  aziridine-based, *13*:161
  for nonwoven textiles, *16*:114
  in papermaking, *16*:789
  for plywood, *14*:14
  in plywood mfg, *14*:12
  in wood laminates, *14*:33
Adipic acid [*124-04-9*]
  adiponitrile from, *15*:898
  from cyclohexanone/cyclohexanol, *13*:924
  lubricant from, *14*:496
Adipic dihydrazide [*1071-93-8*]
  in putrescine synthesis, *15*:777
Adipoguanamine [*4341-27-9*]
  from adiponitrile, *15*:901
Adiponitrile [*111-69-3*], *15*:888
  from butadiene, *15*:808
  properties, *15*:897
Adipoyl azide [*25021-15-2*]
  in putrescine synthesis, *15*:777
Adiprene, *13*:815
Admel, *16*:78
Adnephrine, *15*:757
ADP. See *Adenosine diphosphate.*
ADP network services, *13*:327
Adrenal, *15*:757
Adrenal cortex
  liquid crystals in, *14*:418
Adrenaline [*51-43-4*], *13*:219; *15*:757. (See also
    Epinephrine.)
L-Adrenaline [*51-43-4*], *13*:62. (See also
    Epinephrine.)
Adrenaline-bitartrate [*51-42-3*], *15*:757
Adrenaline hydrochloride [*329-63-5*], *15*:757
Adrenalone [*99-45-6*], *13*:62
Adrenamine, *15*:757
Adrenine, *15*:757
Adrenocorticosteroids, *13*:173
Adrenocorticotrophic hormone (ACTH) [*9002-
    60-2*]
  effect on learning and memory, *15*:132
Adrenocorticotropic hormone (ACTH), *15*:758
Adrenocorticotropin [*9002-60-2*]
  zinc effect on, *15*:590
Adrenor, *15*:773
Adrin, *15*:757
Adsorbents
  molecular-sieve zeolites, *15*:659
Adsorption
  in LPG processing, *14*:388

Adsorptive separation
zeolites for, *15*:660
Adult-onset diabetes, *13*:615
Advertisements
fluorescent, *14*:564
Aedes aegypti, *13*:459
Aerial photography
filters for, *16*:535
Áerobacter aerogenes, *15*:466
Aerobes
in N₂ fixation, *15*:949
Aerosol bomb
as insecticide dispersant, *13*:418
Aerosol paint
fluorescent pigments in, *14*:567
Aerosols, *16*:301
incineration, *13*:204
in metallic coating, *15*:270
Aerospace
metallic coatings, *15*:241
Aerospace applications
magnesium in, *14*:611
Aerospace fluids
measurement of level, *14*:434
Aerospace vehicles
lithium alloys for, *14*:458
AEROS SAROAD information system, *13*:301
AEROS SOTDAT information system, *13*:301
Aesclepias syriaca, *13*:422
Aesculus hippocastanum, *16*:251
Aflatoxin
analysis procedure, *16*:264
Aflatoxin B₁ [1162-65-8], *16*:263
Aflatoxin G₁ [1165-39-5], *16*:263
Aflatoxins
in nuts, *16*:263
AFO information system, *13*:328, 334
Afterglow
suppression in matches, *15*:4
Agar [9002-18-0]
substitute, *15*:452
Agarose [9002-18-0]
in CPK measurement, *15*:77
Agar plate test
for milk, *15*:550
Agathophyllum aromaticum, *16*:251
AGDOC
information system, *16*:905
Agglomeration
of milk products, *15*:560
Aging
as an autoimmune phenomenon, *15*:139
immune deficiency diseases, *13*:170
A₂-globulin, *15*:588. (See also *Ceruloplasmin*.)
Agrobacterium radiobacter, *15*:453

Agrobacterium rhizogenes, *15*:453
Agrobacterium tumefaciens, *15*:453, 952
Agroclavine [548-42-5]
microbial synthesis of, *15*:465
AIBN. See *Azobisisobutyronitrile*.
AIHA. See *American Industrial Hygiene Association*.
Air
as insulator, *13*:549
in paint, *16*:746
regeneration, *14*:463
Air conditioner
odors from, *16*:304
Aircraft
fluorinated plastic cable insulation for, *13*:575
lithium alloys in, *14*:457
noise source, *16*:9
oxygen generation systems, *16*:677
oxygen supply in, *16*:671
ozone in, *16*:698
windshields, *13*:979, 988
Aircraft fasteners
ion plating, *15*:267
Aircraft generators
magnetic materials used in, *14*:663
Nomex used in, *13*:556
Aircraft lubricants, *16*:497
Air fresheners, *16*:297, 302
Air Frölich
heat exchanger, *14*:819
Airplane travel
radiation exposure from, *16*:217
Air pollution control
lime and limestone in, *14*:379
Airports
noise source, *16*:10
Air-purge system
for level measurement, *14*:444
Air separation
molecular sieves in, *15*:663
Air-suspension coating
in microencapsulation, *15*:472
Airwick stick-ups, heavy duty, *16*:304
AISI-304 steel
ion implantation, *13*:717
AISI-416 steel
ion implantation, *13*:717
AISI 1018 steel
nitrogen implantation, *15*:269
Aktamin, *15*:773
Ala. See δ-*Aminolevulinic acid*.
ALAD, *14*:198
Alanap, *15*:727
Alanine [302-72-2]
in nuts, *16*:249

β-Alanine [107-95-9], *13*:97.   (See also
    β-*Aminopropionic acid.*)
Alanine transferase [9000-86-6] (SPGT)
    measurement of, *15*:78
N-(N-L-Alanylglycyl)glycine [3146-40-5], *16*:41
Alar, *14*:788
Alarms
    in PM systems, *14*:756
Alberta yellow
    fluorescent dye, *14*:547
Albumin [9048-46-8]
    measurement, *15*:80, 81
Alcaligenes faecalis Var *myxogenes*, *15*:452
Alcantara, *14*:248
Alcohol
    in missile fuel, *14*:518
Alcohol dehydrogenase [9031-79-5]
    zinc in, *15*:590
Alcoholism
    treatment, *13*:133
Alcohols
    autoxidation, *13*:21
    for detergents, *16*:572
    hydrogen peroxide from, *13*:21
    oxo process for, *16*:637
    in varnish and paint removers, *16*:763
Alcohols, detergent-range
    production, *16*:487
Alcohols, polyhydric
    as rust inhibitors, *14*:493
Alcomax, *14*:671
Aldaric acids, *13*:99
Aldehyde-modified resins
    in papermaking, *16*:817
Aldehyde oxidase [9029-07-6]
    Mo content, *15*:693
    molybdenum in, *15*:592
Aldehydes
    ethylene copolymerization, *13*:933
    from α-hydroxy acids, *13*:96
    from nitriles, *15*:891
    from nitroparaffins, *15*:974
    oxo process for, *16*:637
    in perfumes, *16*:967, 968
    in tannages, *14*:217
Aldicarb [116-06-3], *13*:454
Aldolase [9024-52-6]
    zinc in, *15*:590
Aldol reaction, *13*:895
Aldonic acids, *13*:99
Aldosterone [52-39-1]
    sodium control, *15*:579
Aldoximes
    from nitroparaffins, *15*:976
Aldrin [309-00-2], *13*:434
    toxicity, *13*:263

Aleurites fordii, *16*:251
Aleurites moluccana, *16*:251
Aleurites triloba, *16*:251
Aleutian disease (mink)
    treatment, *13*:178
Alfalfa
    lignin content, *14*:297
Alfernol, *14*:656
Alfin process, *16*:557
Algae
    antimicrobial agents used against, *13*:238
    control in swimming pools, *13*:674
    nitrogen-fixing, *15*:948
Alginic acid [9005-32-7]
    bacterial substitute, *15*:452
Aliquat 336, *13*:725
Alizarin [72-48-0], *13*:10, 43
    economic aspects, *13*:53
Alkali disease
    of livestock, *15*:591
Alkaligenes faecalis, *15*:453
Alkali lignins [8068-05-1], *14*:304
Alkali manufacture
    lime in, *14*:380
Alkali metals
    as initiators, *13*:370
Alkaline cupric tartrate, *13*:118.   (See also
    Fehling's solution.)
Alkaline phosphatase [9001-78-9]
    measurement of, *15*:78
    zinc in, *15*:590
Alkaline phosphatases
    magnesium requirement, *15*:579
Alkaloids
    in nuts, *16*:259
Alkanes (C$_5$–C$_8$)
    toxicity, *13*:263
Alkanolamines
    as antimicrobial agents, *13*:241
    from nitro alcohols, *15*:910
Alkanol S, *15*:724
Alkenyl succinic anhydrides
    as sizing agents, *16*:811
*trans-p-n-*Alkoxycinnamic acids
    mesomorphism of, *14*:412
p-Alkoxyphenyl p′-alkylcarbonatobenzoates
    synthesis of, *14*:414
Alkyds
    in paint, *16*:752
Alkyd wire enamel
    thermal conductivity of, *13*:544
Alkylation, *16*:556.   (See also *Organometallics*,
    σ-*bonded alkyls and aryls.*)
    with higher olefin, *16*:484
    by tetraethyllead, *14*:183
    zeolite catalysts in, *15*:652
Alkylbenzene
    limit of radiation tolerance, *14*:519

Alkyl chlorides
  in tetraalkyllead manufacture, *14*:185
4-Alkyl-*N,N*-dialkyl-1-naphthalenecarbox-
  amides, *15*:737
tert-Alkyl hydroperoxides, *13*:363
N-Alkylmaleamic acids, *14*:774
N-Alkylmaleimides, *14*:774
Alkyl mercuric compounds
  toxic hazards, *15*:164
Allene [*463-49-0*]
  carbonylation, *15*:366
Allergens
  in nuts, *16*:259
Allergic dermatitis
  treatment, *13*:60
Allethrin [*584-79-2*], *13*:456, 458
  as insecticide, *13*:425
Allied process
  for oxalic acid, *16*:624
Alloaromadendrene [*25246-27-9*], *16*:313
Allograft reaction
  treatment, *13*:167
Allomaleic acid [*110-17-8*], *14*:770.   (See also
  Fumaric acid.)
D-Allonic-γ-lactone [*29474-78-0*], *13*:100
Alloy 20, *16*:629
Alloy 48, *14*:659
Alloy 600, *16*:152
4750 Alloy, *14*:660
Alloys
  constant modulus, *15*:795
  fusible, *14*:157
  lead, *14*:140
  of lithium, *14*:457
  of magnesium, *14*:592
  magnetic, *14*:646
  manganese, *14*:824
  mercury, *15*:144
  for metal fibers, *15*:221
  molybdenum, *15*:678
  nickel, *15*:791
  niobium, *15*:824
  oxide-dispersion-strengthened, *15*:794
  produced by ion implantation, *13*:716
Alloy steels
  cladding of, *15*:281
Allspice oil [*8006-77-7*], *16*:320
All Weather Wood, *13*:237
Allyl amyl glycolate [*67634-01-9; 67634-00-8*]
  in perfumes, *16*:967
Allyl bromide [*106-95-6*]
  nickel carbonyl reaction, *15*:808
4-Allylcatechol [*1126-61-0*], *13*:45
Allyl chloride [*107-05-1*]
  methanol/mercuric acetate reaction, *15*:164
  toxicity, *13*:263
π-Allylcobalt tricarbonyl [*12144-85-3*], *16*:613

π-Allylic ions
  metal π-complex formation, *16*:592
Allylmagnesium bromide [*1730-25-2*]
  nickel bromide reaction, *15*:808
Allylmanganese pentacarbonyl [*14057-83-1*], *16*:
  597
η^3-Allylmanganese tetracarbonyl [*33307-28-7*],
  16:597
Allyl methacrylate [*96-05-9*]
  properties, *15*:353
DL-2-Allyl-3-methylcyclopent-2-en-4-ol-1-onyl
    DL-*cis,trans*-chrysanthemate [*584-79-7*]
  as insecticide, *13*:456
π-Allylnickel [*12077-85-9*]
  methyl Grignard reaction, *15*:809
Allyloxymethyl methacrylate [*44978-33-8*]
  properties, *15*:353
Allyl urea [*557-11-9*]
  mercuric acetate/methanol reaction, *15*:165
Almond, *16*:250
Almond moth
  control of with pheromones, *13*:481
Almond oil [*8007-69-0*], *16*:321
Alnico 1
  magnetic properties, *14*:671
Alnico 2
  magnetic properties, *14*:671
Alnico 3
  magnetic properties, *14*:671
Alnico 4
  magnetic properties, *14*:671
Alnico 5, *14*:673
  magnetic properties, *14*:671
Alnico 6
  magnetic properties, *14*:671
Alnico 8
  magnetic properties, *14*:671
Alnico 9
  magnetic properties, *14*:671
Alnico 5 Col
  magnetic properties, *14*:671
Alnico 5 DG
  magnetic properties, *14*:671
Alnico 8 HC
  magnetic properties, *14*:671
Alnicos
  properties, *14*:671
Alopecia senilis
  treatment, *13*:60
Alphabet acids, *15*:720
Alpha cellulose [*9004-34-6*]
  in reinforced plastics, *13*:971
Alphaline differential-pressure transmitter
  for level measurement, *14*:446
Alpha particles
  detection of tracks, *16*:826

Alphol, *15*:729
Alum [*10043-01-3*]
  in papermaking, *16*:804
Aluma-Ti, *15*:243
Alumina [*1344-28-1*]
  filler in plastics, *13*:973
  in insulation, *13*:562, 598
  ion exchange in, *13*:678
  in limestone, *14*:350
  in LPG drying, *14*:389
  nitriding, *15*:944
  in spent shale, *16*:351
  Ziegler-Natta catalyst, *16*:461
Alumina (FP) fiber
  in metal laminates, *13*:945
Alumina hydrate [*21645-51-2*]
  in inks, *13*:378
Alumina, hydrated [*21645-51-2*]
  in papermaking, *16*:819
Alumina trihydrate [*21645-51-2*]
  zeolites from, *15*:655
Aluminium oxide trihydrate [*21645-51-2*]
  phosphors from, *14*:541
Aluminizing, *15*:247
Aluminosilicate [*1327-36-2*]
  gel from in zeolite mfg, *15*:654
Aluminosilicate gels
  zeolite from, *15*:654
Aluminosilicates [*1327-36-2*]
  as lubricant, *14*:520
Aluminothermic reduction
  niobium ores, *15*:823
Aluminum [*7429-90-5*]
  as cable covering, *13*:581
  cladding of, *15*:281
  in coatings, *15*:241
  coefficient of friction, *14*:481
  emissivity, *13*:342
  gasket properties, *16*:727
  in ICS, *13*:628
  in integrated circuits, *13*:641
  is neutron absorber, *16*:140
  lithium salts in mfg, *14*:460
  in magnesium alloys, *14*:596
  in N$_2$ fixation, *15*:955
  niobium pentoxide reaction, *15*:826
  from the ocean, *16*:279, 292
  in superhybrid composites, *13*:942
  surface treatment of, *15*:304
  thermal conductivity of, *13*:544
  as thermal insulator, *13*:592
  zinc coatings, *15*:246
Aluminum-27 [*7429-90-5*], *13*:840
Aluminum alkyls, *16*:565
Aluminum alloys
  fiber reinforced, *13*:943
  magnesium in, *14*:610
Aluminum antimonide [*25152-52-7*]
  properties, *14*:286

Aluminum arsenide [*22831-42-1*]
  properties, *14*:286
Aluminum barium magnesium oxide, Eu doped
    [*12254-04-5*]
  phosphor, *14*:527
Aluminum bromide [*7727-15-3*]
  in N$_2$ fixation, *15*:955
Aluminum–calcium compound [*12042-65-8*] (2:1),
    14:604
Aluminum, compd with germanium and
    manganese [*12042-22-7*] (1:1), *14*:694
Aluminum, compd with iron [*12004-62-5*] (3:1)
  in coatings, *15*:243
Aluminum, compd with nickel [*12003-78-0*] (1:1)
  in coatings, *15*:248
Aluminum containers
  manganese alloy in, *14*:841
Aluminum–copper
  phase diagram, *15*:335
Aluminum dichloride iodide, *16*:582
Aluminum etching, *15*:302
Aluminum flakes
  in inks, *13*:379
Aluminum fluoride [*13595-82-9*]
  lithium carbonate reaction, *14*:460
Aluminum foil
  in packings, *16*:737
Aluminum gallium arsenide laser, *14*:59
Aluminum honeycomb filters, *16*:533
Aluminum hydroxide [*21645-51-2*]
  in lithium recovery, *14*:454
  zeolites from, *15*:655
Aluminum isoperoxide [*555-31-7*]
  in grease thickener formulation, *14*:502
Aluminum lactate [*18917-91-4*], *13*:88
Aluminum nitride [*24304-00-5*], *15*:872
  ammonia from, *15*:944
  as grain-growth inhibitor, *14*:655
  properties, *15*:873
Aluminum oxide [*1344-28-1*]
  in anodizing, *15*:309
  in GS process, *13*:849
  in phosphors, *14*:541
Aluminum oxide [*1344-28-1*] (2:3)
  in coatings, *15*:248
Aluminum oxide dihydrate [*1333-84-2*]
  in paper coating, *16*:789
Aluminum oxide trihydrate [*21645-51-2*]
  as filler, *13*:558
  track resistance of, *13*:548
Aluminum permanganate(VII), *14*:860
Aluminum phenoxide [*15086-27-8*]
  in isoprene reaction, *13*:826
Aluminum phosphate [*7784-30-7*], *14*:509
  mica binder, *15*:430
Aluminum phosphide [*20859-73-8*]
  properties, *14*:286

Aluminum resinates
  paper size, *16*:809
Aluminum salts
  in tannages, *14*:215
Aluminum–steel clads, *15*:275
Aluminum sulfate [*10043-01-3*], *16*:778.   (See
  also *Papermakers' alum.*)
Aluminum sulfide [*1302-81-4*]
  in GS process, *13*:849
Aluminum–tantalum clads, *15*:275
Aluminum triacetate [*139-12-8*]
  TEL by-product, *16*:582
Aluminum trichloride [*7446-70-0*]
  as initiator catalyst, *13*:371
Aluminum trioxide [*1344-28-1*]
  in iron-ore processing, *13*:740
Alzheimer's disease
  treatment, *15*:134
Am-350, *16*:741
Amalgams, *15*:144
Amandin, *16*:249
  amino acid composition, *16*:249
Amaranthus retroflexus, *13*:76
Amati process
  for magnesium, *14*:585
Ambergris [*8038-65-1*]
  in perfumes, *16*:949
Amberlite, *13*:696
Amberlite IRA-400
  for iodine adsorption, *13*:655
Amber mica, *15*:420
Amberol resin M82, *14*:567
Ambrein [*473-03-0*]
  in perfumes, *16*:949
Ambrox [*21582-36-5*]
  in perfumes, *16*:968
Ambroxan [*31207-71-3*]
  in perfumes, *16*:969
American Conference of Governmental Industrial
  Hygienists, *13*:261
American Industrial Hygiene Association, *13*:274
American Magnesium Brine Process
  for magnesium, *14*:581
American National Standards Institute, *13*:274
Americium nitride [*12236-36-1*], *15*:872
Ames test
  for carcinogens, *13*:261
  naphthalene, *15*:714
  organic polymers, *13*:548
Amethopterin [*59-05-2*], *13*:172, 472
Amical 48, *13*:231
Amical 77, *13*:231
Amidines
  from nitriles, *15*:889
Amido black [*1064-48-8*]
  in dye-binding test of milk, *15*:552

Amidoximes
  from nitriles, *15*:889
Amikacin [*39831-55-5*]
  microbial stability, *15*:466
Amines
  as antimicrobial agents, *13*:241
  from nitriles, *15*:889
Aminoacetonitrile [*540-61-4*]
  phenol reaction, *15*:774
ω-Aminoacetophenone [*551-93-9*]
  phenylethanolamine from, *15*:775
Amino-ϵ-acid [*129-91-9*], *15*:730.   (See also 8-
  Amino-1,6-naphthalenedisulfonic acid.)
Amino acids
  in nuts, *16*:252
L-Amino acids
  microbial preparation, *15*:459
α-Amino acids
  from nitroparaffins, *15*:977
Amino acid synthesis
  malonates in, *14*:800
Aminoaciduria
  from lead exposure, *14*:198
4-Aminoantipyrine [*83-07-8*]
  cholesterol diagnostic reagent, *15*:82
4-Aminobenzenesulfonamide [*63-74-1*]
  phosgene reaction, *13*:801
1-Amino-4-bromo-2-methyl anthracenedione
  [*81-50-5*]
  diethyl malonate reaction, *14*:800
2-Amino-1-butyl benzoate-benzoic acid complex
  [*7756-96-9*], *13*:143
γ-Aminobutyric acid [*56-12-2*], *15*:767
  as neuroregulator, *15*:766
  properties, *15*:766
  valproic acid effect on, *13*:139
γ-Aminobutyronitrile [*32754-99-7*]
  in spermidine synthesis, *15*:777
DL-Amino-ϵ-caprolactam [*17929-90-7*]
  microbial synthesis of, *15*:463
Aminocarb [*2032-59-9*], *13*:455
2-Amino-4-chlorotoluene [*95-79-4*], *15*:930
2-Amino-6-chlorotoluene [*87-60-5*], *15*:930
3-Aminocrotonamide [*15846-25-0*], *13*:889
7-Aminodeacetoxycephalosporanic acid [*26395-
  99-3*]
  microbial synthesis of, *15*:463
7-Amino-3,4-dimethyloctyl methacrylate [*76392-
  26-2*]
  properties, *15*:354
4-Amino-3,4″-dinitro-*p*-terphenyl [*75802-59-4*],
  14:414
2-Aminoethanethiol [*60-23-2*], *13*:155
2-Aminoethanol [*141-43-5*]
  sulfuric acid reaction, *13*:157
2-(Aminoethyl)-5-chloro-*N*,*N*-diethyl-3-(2-
  fluorophenyl)-1*H*-indole-1-ethylamine
  [*24234-99-9*]
  in flurazepam hydrochloride synthesis, *13*:128

2-Aminoethyl ethyl sulfide [*36489-03-9*], *13*:155

2-Aminoethyl hydrogen sulfate [*969-39-6*]
  aziridine from, *13*:157

3-(2-Aminoethyl)-5-hydroxyindole [*50-67-9*]
  in melatonin biosynthesis, *15*:772

2-(2-Aminoethyl)-2-thiopseudourea [*151-16-6*], *13*:155

Amino G-acid [*86-65-7*], *15*:731.   (See also *7-Amino-1,3-naphthalenedisulfonic acid.*)

1-Aminohexamethyleneimine [*5906-35-4*]
  tolazamide preparation, *13*:617

3-Amino-5-hydrazino pyrazole [*75507-64-1*], *14*: 805

α-Amino-4-hydroxyacetophenone [*51-67-2*]
  octopamine from, *15*:774

3-Amino-5-hydroxy-2,7-naphthalenedisulfonic acid [*90-40-4*], *15*:740

4-Amino-5-hydroxy-1,3-naphthalenedisulfonic acid [*82-47-3*], *15*:740

4-Amino-5-hydroxy-1,7-naphthalenedisulfonic acid [*130-23-4*], *15*:740

4-Amino-5-hydroxy-2,7-naphthalenedisulfonic acid [*90-20-0*], *15*:740

5-Amino-6-hydroxy-2-naphthalenesulfonic acid [*5639-34-9*], *15*:740

Amino (imino) borane [*76082-00-3*], *16*:33

Aminoisobutyronitrile [*96-16-2*], *15*:893

Amino J-acid [*118-33-2*], *15*:731.   (See also *6-Amino-1,3-naphthalenedisulfonic acid.*)

δ-Aminolevulinic acid [*106-60-5*]
  correlation with blood level, *14*:198

δ-Aminolevulinic acid dehydratase [*9036-37-7*]
  blood-lead level affect, *14*:198

Aminomethylation
  metallocenes, *16*:605

7-Amino-4-methylcoumarin [*26093-31-2*]
  as chromogen, *14*:549

4-Aminomethyl-2-mercapto-1,3-imidazole hydrochloride [*53744-46-0*]
  in histamine synthesis, *15*:769

2-Amino-2-methyl-1-propanol [*124-68-5*]
  explosives sensitizer, *15*:983

Aminonaphthalenecarboxylic acids, *15*:744

1-Amino-2,7-naphthalenedisulfonic acid [*76530-15-9*], *15*:730

3-Amino-1,5-naphthalenedisulfonic acid [*131-27-1*], *15*:731

3-Amino-2,7-naphthalenedisulfonic acid [*92-28-4*], *15*:730

4-Amino-1,6-naphthalenedisulfonic acid [*85-75-6*], *15*:730

4-Amino-1,7-naphthalenedisulfonic acid [*85-74-5*], *15*:730

4-Amino-2,6-naphthalenedisulfonic acid [*6362-05-6*], *15*:730

4-Amino-2,7-naphthalenedisulfonic acid [*6521-07-6*], *15*:730

5-Amino-1,3-naphthalenedisulfonic acid [*13306-42-8*], *15*:730

6-Amino-1,3-naphthalenedisulfonic acid [*118-33-2*], *15*:731

7-Amino-1,3-naphthalenedisulfonic acid [*86-65-7*], *15*:731

8-Amino-1,5-naphthalenedisulfonic acid [*117-55-5*], *15*:730

8-Amino-1,6-naphthalenedisulfonic acid [*129-91-9*], *15*:730

Aminonaphthalenesulfonic acids, *15*:728

1-Amino-2-naphthalenesulfonic acid [*81-06-1*], *15*:729

2-Amino-1-naphthalenesulfonic acid [*81-16-3*], *15*:729

4-Amino-1-naphthalenesulfonic acid [*84-86-6*], *15*:729

4-Amino-2-naphthalenesulfonic acid [*134-54-3*], *15*:729

5-Amino-1-naphthalenesulfonic acid [*84-89-9*], *15*:729

5-Amino-2-naphthalenesulfonic acid [*119-79-9*], *15*:729

6-Amino-1-naphthalenesulfonic acid [*81-05-0*], *15*:729

6-Amino-2-naphthalenesulfonic acid [*93-00-5*], *15*:730

7-Amino-1-naphthalenesulfonic acid [*86-60-2*], *15*:730

7-Amino-2-naphthalenesulfonic acid [*92-40-0*], *15*:730

8-Amino-1-naphthalenesulfonic acid [*82-75-7*], *15*:729

8-Amino-2-naphthalenesulfonic acid [*119-28-8*], *15*:729

6-Amino-1,3,5-naphthalenesulfonic acid [*55524-84-0*]
  desulfonation, *15*:728

4-Amino-1,3,5-naphthalenetrisulfonic acid [*76530-15-9*], *15*:731

6-Amino-1,3,5-naphthalenetrisulfonic acid [*55524-84-0*], *15*:731

7-Amino-1,3,6-naphthalenetrisulfonic acid [*118-03-6*], *15*:731

8-Amino-1,3,5-naphthalenetrisulfonic acid [*17894-99-4*], *15*:731

8-Amino-1,3,6-naphthalenetrisulfonic acid [*117-42-0*], *15*:731

5-Amino-1-naphthalenol [*83-55-6*], *15*:740

7-Amino-2-naphthalenol [*118-46-7*], *15*:740

Aminonaphthols, *15*:736

Aminonaphtholsulfonic acids, *15*:736

1-Amino-2-naphthol-4-sulfonic acid [*116-63-2*]
  in phosphorus measurement, *15*:84
cis-2-Amino-5-nitrobenzophenone oxime [*4892-69-7*]
  in nitrozepam synthesis, *13*:129
4-Amino-4″-nitro-*p*-terphenyl [*38190-45-3*], *14*:413
2-Amino-4-nitrotoluene [*99-55-8*], *15*:931
4-Amino-2-nitrotoluene [*119-32-4*], *15*:931
6-Aminonocardicinic acid [*67509-41-5*]
  microbial synthesis of, *15*:463
ω-Aminononanoic acid [*1120-12-3*], *16*:688
6-Aminopenicillanic acid [*551-16-6*]
  microbial preparation, *15*:459
6-Aminopenicillanic acid [*551-16-6*] (6-APA)
  microbial synthesis of, *15*:462
m-Aminophenol [*591-27-5*], *13*:45
o-Aminophenol [*95-55-6*], *13*:45; *15*:924
p-Aminophenol [*123-30-8*], *13*:45; *15*:924
  economic aspects, *13*:53
2-Amino-1-phenylbutane [*34577-88-3*], *15*:976
Amino-2-phenyl-4-delta-1-pyrroline [*22349-30-0*], *14*:802
o-Aminophenylglyoxylic acid [*484-38-8*], *13*:221
2-Amino-5-phenyl-4-oxazolidinone [*2152-34-3*], *15*:136. (See also *Pemoline*.)
Aminopolyamide–epichlorohydrin resins
  as creping adhesives, *16*:808
  in papermaking, *16*:817
Aminopolyamides
  creping adhesive, *16*:808
β-Aminopropionic acid [*107-95-9*], *13*:97. (See also β-*Alanine*.)
β-Aminopropionitrile [*151-18-8*]
  antiaging effect, *15*:139
Amino-R-acid [*92-28-4*], *15*:730. (See also *3-Amino-2,7-naphthalenedisulfonic acid*.)
Amino resins
  in tannages, *14*:216
p-Aminosalicylic acid [*65-49-6*]
  m-aminophenol from, *13*:62
Amino-Schaeffer acid [*5639-34-9*], *15*:740. (See also *5-Amino-6-hydroxy-2-naphthalenesulfonic acid*.)
DL-2-Aminothiazoline-4-carboxylic acid [*16899-18-6*]
  microbial transformation to L-cysteine, *15*:465
2-Amino-1,1,3-tricyano-1-propene [*868-54-2*]
  malononitrile dimer, *14*:804
2-Amino-4,6,6-trimethyl-2,4-cyclohexadiene-1,1,3-tricarbonitrile [*30481-45-9*], *14*:805
5-Amino-1,3,3-trimethylcyclohexane methanamine [*2855-13-2*], *13*:919

Amitraz [*33089-61-1*], *13*:463
Ammine complexes
  of nickel, *15*:806
Ammonia [*7664-41-7*], *15*:872
  absorption by $NiCl_2$, *15*:804
  from aluminum nitride, *15*:944
  from calcium cyanamide, *15*:944
  dihydroxybenzene reactions, *13*:45
  electrode, *13*:720
  hydrogen reaction, *15*:944
  iodine reaction, *13*:659
  iron complex, *13*:765
  isobutylene–air reaction, *15*:363
  in isotope separation, *13*:847
  as leak tracer, *16*:55
  lithium reaction, *14*:459
  as lubricant, *14*:514
  mercuric oxide reaction, *15*:161
  in metal nitriding, *15*:317
  microbial fumaric acid reactions, *15*:464
  nickel recovery, *15*:798
  nitric acid from, *15*:854
  in nuts, *16*:249
  odor-problem chemical, *16*:303
  ozone reaction, *16*:688
  paint removers, *16*:763
  in plasma doping by Si_3N_4, *13*:639
  production from lignite, *14*:328
  toxicity, *13*:263
  use in NO_x abatement, *15*:864
Ammoniacal copper arsenate [*32680-29-8*]
  applications of, *13*:237
Ammonia–epichlorohydrin condensates
  in papermaking, *16*:804
Ammonia synthesis, *15*:944. (See also *Haber-Bosch process*.)
Ammoniated mercury [*10124-48-8*], *15*:170
Ammonium bifluoride [*1341-49-7*]
  in boronizing, *15*:250
  in IC manufacture, *13*:628
Ammonium bisulfite [*10192-30-0*]
  in sulfite pulping, *14*:304
Ammonium chloride [*12125-02-9*]
  in chromizing, *15*:249
  in ISE, *13*:728
  in leak detection, *16*:55
  nitrile reactions, *15*:889
Ammonium dimolybdate [*27546-07-2*], *15*:686
Ammonium disilver(I) hexakiscyanoferrate [*58675-53-9*], *13*:768
Ammonium dithiomolybdate [*16150-60-0*], *15*:690
Ammonium diuranate [*7783-22-4*]
  in fuel reprocessing, *16*:179
Ammonium formate [*540-69-2*]
  acetophenone reaction, *13*:901
Ammonium heptamolybdate [*12027-67-7*], *15*:686
Ammonium hydroxide [*1336-21-6*]
  in insecticide solution, *13*:424

Ammonium iodide [*12027-06-4*]
   in chromizing, *15*:249
Ammonium ions
   removal with molecular sieves, *15*:668
Ammonium magnesium chloride [*39733-35-2*]
   dehydration of, *14*:576
Ammonium molybdate [*12027-67-7*], *15*:690
   calcination, *15*:670
   manufacture, *15*:674
   in phosphorus assay, *15*:84
   toxicity, *15*:679
Ammonium molybdiphosphate [*12026-66-3*]
   in phosphorus measurement, *15*:84
Ammonium nitrate [*6484-52-2*]
   from NH$_4$NO$_3$, *16*:688
   production, *15*:865
   recovery in Higgins contactor, *13*:693
Ammonium oxalate [*1113-38-8*]
   in niobium analysis, *15*:828
   in wool dyeing, *16*:631
Ammonium ozonide [*12161-20-5*]
   ozone reactions, *16*:688
Ammonium permanganate [*13446-10-1*], *14*:845
Ammonium persulfate [*7727-54-0*]
   in polymethacrylate mfg, *15*:388
Ammonium phosphate [*10124-31-9*]
   in matches, *15*:4
Ammonium picrate [*131-74-8*]
   in ISE, *13*:728
Ammonium sulfate [*7783-20-3*]
   as leatherlike material filler, *14*:242
   in leather manufacture, *14*:209
   in pulping, *14*:306
Ammonium sulfide [*12135-76-1*]
   in leak detection, *16*:55
Ammonium tetrathiomolybdate [*15060-55-6*], *15*: 690
Ammonium vanadate [*7803-55-6*]
   toxicity, *15*:574
Ammoxidation
   isobutylene, *15*:363
   nitriles by, *15*:893
Ammunition
   lead alloys for, *14*:146
Amnesia
   treatment, *15*:134
Amniotic fluid
   zinc in, *15*:590
Amoco gas-phase process
   for polyethylene, *16*:394
Amonton's law, *14*:481
Amorphous magnetic alloys, *14*:668
Amoxicillin [*26787-78-0*], *13*:77
AMP. See *Adenosine monophosphate.*
Amphetamine [*300-62-9*]
   measurement, *15*:88
   memory-enhancing agent, *15*:135

Amphetamines
   measurement, *15*:88
Amphiboles
   lithium in, *14*:449
Amphimallon majalis, *13*:481
Ampicillin [*69-53-4*]
   controlled release of, *15*:488
   microbial synthesis, *15*:463
Amyelois transitella, *13*:480
Amygdalase [*51683-43-3*]
   amygdalin splitting, *16*:256
Amygdalin [*29883-15-6*]
   in kernels, *16*:321
   in nuts, *16*:256
Amyl acetate [*628-63-7*]
   in chemical deposition coating, *15*:264
   odor, *16*:297
Amyl alcohols
   oxo products, *16*:650
2-(*tert*-Amyl)anthraquinone [*32588-54-8*]
   autoxidation, *13*:18
Amylase [*9000-92-4*]
   measurement, *15*:79
Amylase-α [*9000-90-2*]
   in distillers malt, *14*:811
   in zanflo preparation, *15*:451
Amyl butyrate [*540-18-1*]
   odor, *16*:301
Amyl cinnamic aldehyde [*122-40-7*], *16*:955
Amyl nitrite [*110-46-3*]
   nitrile nitrosation reaction, *15*:892
Amyloglucosidase [*9032-08-0*]
   as malt replacement, *14*:817
Amyloids
   in nuts, *16*:252
Amylosaccharides, *15*:439
tert-Amyl peroctoate [*686-31-7*]
   half-life of, *13*:360
Amyl salicylate [*2050-08-0*], *16*:955
   in perfumes, *16*:968
Amyris balsamifero L., *16*:321
Amyris oil [*8015-65-4*], *16*:321
Amytryptyline [*549-18-8*]
   controlled release of, *15*:488
Amzoil, *14*:498
Anabaena, *15*:950
Anabaena azollae
   in rice culture, *15*:948
Anabasine [*494-52-0*]
   as insecticide, *13*:424
Anabasis aphylla, *13*:422
Anacardic acid [*18524-94-2*]
   in cashew, *16*:259
   in cashew shells, *16*:271
Anacardium occidentale, *16*:250
Anaerobes
   in N$_2$ fixation, *15*:950

Anagosta kuehniella, 13:481
Analogue signals
 in feedback control, 13:488
Analysis
 membranes in, 15:121
Analytical standards, 15:36
Anaphylaxis
 treatment, 15:762
Anarsia lineatella, 13:473
Anatase [1317-70-0]
 paper pigment, 16:778
Andco-Torrax system
 for incineration, 13:195
Andersen's acid [117-43-1], 15:738. (See also 8-
 Hydroxy-1,6-naphthalenedisulfonic acid.)
Androgens
 as neuroregulators, 15:759
1,4-Androstadiene-3,17-dione [53452-47-4]
 microbial synthesis of, 15:461
5-Androstene [4369-59-9]
 microbial isomerization, 15:468
4-Androstene-3,17-dione [63-05-8]
 microbial reduction, 15:462
 microbial synthesis of, 15:461
Anemia
 nutrients associated with, 15:573
 treatment, 13:88
 treatment of, 13:774
Anemias
 treatment, 15:445
Anesthesia
 from hypnotics overdose, 13:122
 oxygen supply for, 16:670
Anesthetic gases and vapors
 toxicity, 13:274
Anethole [104-46-1]
 in anise oil, 16:321
 as insect attractant, 13:481
Aneurysm
 diagnosis, 15:79
Angiotensin I [484-42-4], 15:760
Angiotensin II [4474-91-3], 15:760
Angiotensin III [13602-53-4], 15:760
Angiotensins, 15:759
Anglesite [7446-14-2], 14:99
Angoumois grain moth
 control of with pheromones, 13:481
Anhydrite [7778-18-9]
 from the ocean, 16:287
Anhydrous ethanol
 in inks, 13:390
Aniba rosaeodora Var. *amazonica*, 16:321
Anilides
 as antimicrobial agents, 13:239
Aniline [62-53-3], 15:917
 in hydroquinone production, 13:48

in hydroquinone reaction, 13:45
 iodination of, 13:653, 664
 isocyanate reaction product, 13:798
 isocyanates from, 13:805
 malic acid reaction, 13:105
 in rocket fuel, 15:856
Aniline blue [39422-21-4]
 complex with (1→3)-linked β-D-glucan, 15:454
Aniline hydrochloride [142-04-1]
 in isocyanate manufacture, 13:805
Aniline sulfate [542-16-5]
 hydroquinone from, 14:877
 in hydroquinone production, 13:51
3-Anilino-2-hydroxypropyl methacrylate [16926-
 84-4], 15:355
C-Anilino-N-phenylsuccinimide [37036-49-3],
 13:105
Animal extractives
 as odor modifier, 16:301
Animal feeds
 malting by-products, 14:822
Animal waxes
 in inks, 13:379
Anionic polymerization, 13:370
Anisaldehyde [123-11-5], 13:73
Anise [8007-70-3], 16:307
Anise oil [8007-70-3], 16:319, 321
Anise, star oil [68952-43-2], 16:321
o-Anisidine [90-04-0], 15:924
p-Anisidine [104-94-9], 15:924
Anisole [100-66-3]
 in boron-10 separation, 16:164
Anisole-boron trifluoride [456-31-5]
 in boron-10 separation, 16:164
Anisomycin [22862-76-6]
 protein synthesis inhibitor, 15:136
Anisotropy
 of wood, 14:2
Anisotropy energy
 in magnetic materials, 14:648
Anisyl acetone [104-20-1]
 as insect attractant, 13:481
Ankylosing spondylitis
 treatment, 13:179
Annealing. See *Metal treatments*.
Anode coatings. See *Metal anodes*.
Anodes
 lead alloys, 14:147
 titanium, 15:173
Anodizing
 of metals, 15:309
 oxalates in, 16:630
Anopheles gambiae, 13:473
Anorexia
 TRH induced, 15:782
Anschutz, Jacob, 13:736
ANSI. See *American National Standards
 Institute*.

Antacid
  magnesium carbonate, *14*:621
Antalka, *15*:767
Antennas
  in laminated-glass windshields, *13*:984
Anthonomus grandis, *13*:420, 482
Anthracene [*120-12-7*]
  chromium hexacarbonyl reaction, *16*:613
π-Anthracenechromium tricarbonyl [*12155-42-9*],
  16:613
Anthranilates
  in perfumes, *16*:968
Anthranilic acid [*118-92-3*], *15*:926
Anthranone [*90-44-8*], *15*:747
Anthraquinone [*84-65-1*], *15*:715
  autoxidation, *13*:16
9,10-Anthraquinone [*84-65-1*]
  as photoinitiator, *13*:368
Anthraquinone autoxidation process
  for hydrogen peroxide, *13*:16
Anthrax, *15*:553
Antiaging drugs, *15*:132.   (See also *Memory-
    enhancing agents and antiaging drugs.*)
Anti-aircraft gun
  analogue control in, *13*:489
Antibodies
  in digoxin measurement, *15*:86
Anticonvulsants, *13*:122.   (See also *Hypnotics,
    sedatives, anticonvulsants.*)
Antidandruff agent
  selenium sulfide, *13*:239
Antidiabetic agents, *13*:605
Antiepileptic drugs, *13*:134
Antiepileptics, *13*:122.   (See also
    Anticonvulsants.)
Antiferromagnetism, *14*:647
Antifoulant
  antimicrobial agent as, *13*:223
Antifouling paints, *16*:579
Antifreeze, *16*:687
Antifriction material
  lead, *14*:157
Antigens
  sources of, *13*:167
Antigorite [*12135-86-3*], *14*:615
Antihistamines, *13*:133.   (See also
    Phenothiazine.)
Anti-inflammatory agent
  diethylammonium hydroquinone sulfonate,
    13:60
Anti-inflammatory agents, *13*:174
Antiknock additives, *16*:579
Antiknock agent
  oxalic acid, *16*:633
Antiknocks
  lead alkyls, *14*:191

Antimicrobial
  salicylaldehyde derivatives, *13*:76
Antimicrobials
  nitro alcohols, *15*:915
Antimitotic
  aziridines, *13*:162
Antimonial lead, *14*:**125**
Antimony [*7440-36-0*], *15*:883
  in coatings, *15*:243
  as grain-growth inhibitor, *14*:655
  lead alloys, *14*:143
  from the ocean, *16*:280
  in silicon doping, *13*:635
  toxicity, *13*:263
Antimony-121 [*14265-72-6*], *13*:841
Antimony-123 [*14119-16-5*], *13*:841
Antimony(III) [*7440-36-0*]
  in phosphors, *14*:534
Antimony lactate [*58164-88-8*], *13*:88
Antimony metal
  from lead refining, *14*:117
Antimony nitride [*12333-57-2*], *15*:872
Antimony oxalate [*16455-98-4*]
  polyester catalyst, *16*:632
Antimony oxide [*1314-60-9*]
  in tartar emetic production, *13*:115
Antimony pentachloride [*7647-18-9*]
  as initiator catalyst, *13*:371
Antimony sulfide [*1345-04-6*]
  in matches, *15*:2
Antimony trioxide [*1309-64-4*]
  in insulation, *13*:548
  in phosphor mfg, *14*:536
Antioxidants
  from dihydroxybenzenes, *13*:58
  in inks, *13*:379
  for LLDPE, *16*:394
  in paper, *16*:794
Antiperspirant
  aluminum lactate as, *13*:88
  zirconium lactate as, *13*:88
Antipsychotic agents, *13*:133.   (See also
    Hypnotics, sedatives, and anticonvulsants.)
Antiseptic
  antimicrobial agent as, *13*:223
Antiseptics
  mercury compounds, *15*:155, 158, 164
Antisera, *13*:167
Antistatic agents
  in magnetic coatings, *14*:742
  for thermoplastics, *15*:724
Antisyphilitics
  mercury compounds, *15*:165
Antiwear agents, *14*:493
Antu, *15*:727
Anxiety
  (−) adrenaline induced, *15*:758

effect of, *13*:122

Anxiolytics, *13*:129.   (See also *Hypnotics, sedatives, anticonvulsants.*)

AOD.   See *Argon–oxygen gassing.*

Aorta walls
  liquid crystals in, *14*:418

6-APA.   See *6-Aminopenicillanic acid.*

Aphid control
  ground mica in, *15*:438

Aphids
  insecticides, *13*:415

Apholate [*52-46-0*], *13*:473

Apios americana, *16*:251

APIPAT
  information service, *16*:902
  information system, *13*:300

Apoferritin [*9013-31-4*], *15*:595

Apollo moon missions
  oxygen generation, *16*:677

Apollo-Soyuz testing program, *16*:845

Apophedrin [*7568-93-6*], *15*:775

Apotransferrin, *15*:595

Apples
  catechol in, *13*:46
  insecticidal protection of, *13*:415

Applications research, *14*:905

APTIC
  information service, *16*:902
  information system, *13*:328

Aquaculture
  oxygen use, *16*:669

Aqua fortis, *15*:853

Aquapel, *16*:779

Aquapels, *13*:885

Aquatetrakis(cyano-*C*)oxomolybdate(2−)
  [*57298-65-4*], *15*:960

AQUATOX information system, *13*:301

L(−) Arabinonamide [*74421-66-2*], *13*:101

L(−) Arabinonic acid [*608-53-7*], *13*:101

DL(−) Arabinonic acid, γ-lactone [*13280-76-7*], *13*:101

L(−) Arabinonic acid, γ-lactone [*51532-86-6*], *13*:101

Arabinose [*147-81-9*]
  in microbial synthesis, *15*:467

D-Arabinose [*31178-68-4*]
  from D-gluconic acid, *13*:99

Arachidic acid [*506-30-9*]
  in nut oils, *16*:252

Arachidonic acid [*506-32-1*]
  in prostaglandin syntheses, *15*:778

Arachin, *16*:249
  amino acid composition, *16*:249

Arachis hypogaea, *16*:251

Arachne, *16*:120

Aragonite [*14791-73-2*]
  in limestone, *14*:344

oölitic, from the ocean, *16*:286

Aramid fibers
  in composites, *13*:972

Araucaria angustifolia, *16*:250

Araucarian nut, *16*:250

Arbutin [*497-76-7*], *13*:46.   (See also
  β-D-*Glucopyranoside.*)

Architectural glass, *13*:990

Architectural treatment
  for sound control, *13*:513

Arc spraying, *15*:254

Area-bonding
  spunbondeds, *16*:86

Areca cathecu, *16*:251

Areca nut, *16*:270

Arecoline [*63-75-2*]
  learning effects, *15*:134

Arecoline [*63-75-2*], *16*:270

Arene π complexes, *16*:592

Arenes
  metal π-complex formation, *16*:592

Arginine [*74-79-3*]
  in nuts, *16*:249, 252

L-Arginine [*74-79-3*]
  microbial transformation to citrulline, *15*:464

Arginine[8]-vasopressin [*11000-17-2*] (AVP)
  avoidance behavior effect, *15*:133

Argon [*7440-37-1*]
  in air, *16*:657
  in arc welding, *14*:608
  ion implantation, *13*:715
  lasers, *14*:44
  as lubricant, *14*:515
  from the ocean, *16*:279

Argon-36 [*13965-95-2*], *13*:840

Argon-38 [*13994-72-4*], *13*:840

Argon-40 [*7440-37-1*], *13*:840
  in isotope separation, *13*:849

Argon laser, *14*:51

Argon–oxygen decarburization, *15*:342

Argon–oxygen decarburizing
  for steelmaking, *16*:667

Argon–oxygen degassing process, *15*:799

Argyrotaenia velutinana, *13*:481

Armco iron, *14*:653

Armco process, *13*:745

Armor plate
  lithium alloy as, *14*:458

Armstrong + Wynne's acid [*3771-14-0*], *15*:738.   (See also *4-Hydroxy-2-naphthalenesulfonic acid.*)

Arndt-Eistert synthesis, *13*:880

Ar nut, *16*:250

Aroclor [*1336-36-3*]
  as insulating fluid, *13*:548

Aroma chemicals, *16*:310, 953

Aromatic polymers
    membrane formation, *15*:104

Arrhenius method
    in predicting thermal life of insulation, *13*:545

Arsenic [*7440-38-2*]
    in coatings, *15*:241
    from the ocean, *16*:279
    in silicon doping, *13*:635
    toxicity, *13*:263

Arsenic-75 [*7440-38-2*], *13*:840

Arsenic acid [*7778-39-4*]
    as insecticide, *13*:420

Arsenicals
    as antimicrobial agents, *13*:238
    as insecticides, *13*:420

Arsenic anhydride [*1327-52-2*]
    nickel carbonate reaction, *15*:806

Arsenic nitride [*26754-98-3*], *15*:872

Arsenic pentoxide [*1303-28-2*]
    as insecticide, *13*:420
    from $Ni_3(AsO_4)_2$, *15*:806

Arsenic trichloride [*7784-34-1*]
    IC dopant, *13*:638
    in LED mfg, *14*:288
    tetraethyllead reaction, *14*:183

Arsenic trioxide [*1327-53-3*]
    as insecticide, *13*:421

Arsenous oxide [*1327-53-3*]
    as insecticide, *13*:420

Arsine [*7784-42-1*]
    from calcium arsenides, *14*:150
    IC dopant, *13*:627
    in LED mfg, *14*:287

Arthritis
    treatment, *13*:111

Artificial-kidney machine, *15*:116

Artinite [*12143-96-3*], *14*:619

Artriona, *15*:761

Asafetida resin
    resorcinol from, *13*:39

Asafetida resins
    resorcinol from, *13*:39

Asahi spunbond, *16*:78

Asarco continuous tapper, *14*:108

Asarone [*5353-15-1*], *16*:320

Asbestos [*1332-21-4*], *14*:615
    fiber, in papermaking, *16*:823
    as grease filler, *14*:503
    as insulation, *13*:597
    in insulation enamels, *13*:580
    in packings, *16*:726
    toxicity, *13*:263

Asbestos composites
    gasket properties, *16*:727

Asbestos–rubber
    gasket properties, *16*:727

ASCATOPICS information system, *13*:295

Ascorbic acid [*50-81-7*]
    ammonium diphosphate reduction, *15*:84
    in niobium analysis, *15*:829
    nitrosamine inhibition, *15*:989

L-Ascorbic acid [*50-81-7*]
    microbial preparation, *15*:459

Ascorbic acid oxidase [*9029-44-1*]
    copper in, *15*:589

Ash
    in cattlehides, *14*:203
    as insecticide carrier, *13*:416

ASI information system, *13*:328, 334

ASLIB information system, *13*:313

Aspara, *15*:760

L-Asparaginase [*9015-68-3*]
    microcapsules, *15*:489

L-Asparagine [*70-47-3*]
    removal of, from tumors, *15*:489

Aspara K, *15*:760

Asparstat, *15*:760

L-Aspartate [*56-84-8*], *15*:76

Aspartateamino transferase [*9000-97-9*] (AST)
    in diagnosis, *15*:75

Aspartate transcarbamylase [*9012-49-1*]
    zinc in, *15*:590

Aspartic acid [*56-84-8*]
    in nuts, *16*:249

DL-Aspartic acid [*617-45-8*], *15*:760

L-Aspartic acid [*56-84-3*], *15*:760

Aspergillus flavus, *13*:106; *16*:263

Aspergillus fumigatus, *13*:784
    in *meso*-tartaric acid synthesis, *13*:114

Aspergillus humicola, *13*:784

Aspergillus melleus, *13*:784

Aspergillus oryzae, *15*:463

Aspergillus parasiticus, *13*:106

Aspergillus terreus, *13*:784, 868

Aspergillus versicolor, *13*:784

Asphalt
    in fabric for insulation, *13*:578
    in insulation, *13*:595

Asphalt emulsions
    paper sizing, *16*:778

Asphalt filler
    limestone, *14*:375

Asphalt fumes
    health effect, *13*:263

Aspidiotus perniciosus, *13*:422, 473

Aspirin [*50-78-2*]
    encapsulation, *15*:476
    immunologic effects, *13*:174
    time-released, *15*:488

Assembly operation
    planning, *16*:515

AST.   See *Aspartate transferase.*

Asthma
treatment, *15*:762
Asthmatic attacks
oxygen-helium cure, *16*:670
Astiban [*1986-66-9*]
sustained-release form, *15*:490
ASTM unified numbering system, *14*:596
Astraphloxine FF [*6320-14-5*], *14*:551
Astrocaryum murumuru, *16*:251
Astrocaryum rulgare, *16*:251
Astronomical filters, *16*:552
Asymmetric membranes, *15*:106
Atherosclerosis
liquid crystals implicated in, *14*:418
nutrients associated with, *15*:573
Atlantis Deep deposits, *16*:290
Atmosphere
ozone in, *16*:696
Atmosphere regeneration
lithium hydroxide for, *14*:463
Atomic weight
standardization of, *15*:38
Atomizing burners, *13*:196
ATP.   See *Adenosine triphosphate.*
ATPase [*9000-83-3*]
calcium effect on, *15*:584
Atropine [*51-55-8*]
microbial hydrolysis, *15*:462
Attalea funifera, *16*:251
Attapulgite
as insecticide carrier, *13*:416
Attenuation
sound, *13*:516
Attractants
in insect control, *13*:419
in insect control technology, *13*:480
Audio cassette, *14*:733
Auditory response.   See *Noise pollution.*
Auger recombination, *14*:274
Auramine O [*2465-27-2*], *14*:546
Aureobasidium pullulans, *15*:455; *16*:748
Aurothioglucose [*12192-57-3*]
in rheumatoid arthritis therapy, *13*:174
Aurothiomalic acid [*24145-43-5*], *13*:111
Auslegeschrift, *16*:891
Austenite, *15*:338
Autographa californica, *13*:480
Autoimmune disease, *15*:139
Autoimmune response
suppression, *13*:172
Autoimmunity
treatment, *13*:167
Automatic control, *13*:485.   (See also *Feedback control.*)
Automatic Subject Citation Alert, *13*:295
Automation
in nuclear reactor design, *16*:171

Automobiles
lubricants for, *14*:485
Autoxidation
initiators used in, *13*:356
Avalanche diodes
microwave power output of, *15*:504
Aviation
lubricants in, *14*:485
Avoidance behavior
effect of neuropeptides, *15*:133
Avoidance responses
effect of acth on, *15*:133
AVP.   See *Arginine*[8]*-vasopressin.*
Axtar, *16*:78
Azaferrocene [*11077-12-6*], *16*:593
8-Azaguanine [*134-58-7*]
encapsulation, *15*:477
Azathioprine [*446-86-6*]
antiaging effect, *15*:139
in neoplasm therapy, *13*:172
Azauridine [*39455-15-7*]
in neoplasm therapy, *13*:172
Azelaic acid [*123-99-9*], *16*:686
lubricant from, *14*:496
Azelon, *16*:688
Azeotropes
of ketones, *13*:895
Azetidines, *13*:142
Azinphos ethyl [*2642-71-9*], *13*:441
Azinphosmethyl [*86-50-0*], *13*:437, 441
Aziridine [*151-56-4*]
polymerization, *13*:155
properties, *13*:143
Aziridine-aluminum chloride complex [*74411-30-6*]
benzene reaction, *13*:151
Aziridine dimer [*27233-25-6*], *13*:155
Aziridine, lithium salt [*33734-99-5*], *13*:148
1-Aziridine-methoxytetrafluoroboric acid complex [*75431-52-6*]
methanol reaction, *13*:151
Aziridine, potassium salt [*29521-21-9*]
bromobenzene reaction, *13*:148
Aziridines, *13*:143.   (See also *Ethylenimines.*)
manufacture, *13*:157
1-Aziridine-tetrafluoroboric acid complex [*17409-60-8*]
phenol reaction, *13*:151
Aziridine tetramer [*39539-24-0*], *13*:157
Aziridine trimer [*39539-24-0*], *13*:155
Aziridinium sulfate [*29086-57-5*], *13*:143
Aziridinium tetrafluoroborate [*17409-60-8*], *13*:143
(1-Aziridinyl)amine [*1721-30-8*], *13*:148
α-Aziridinylbenzyl methyl ketone [*6713-46-8*], *13*:148

1-(1-Aziridinyl)-1-dimethylaminoethane [*5498-98-6*], *13*:155

2-(1-Aziridinyl)ethanol [*1072-52-2*], *13*:151

5-(1-Aziridinyl)-3-pentenonitrile [*74411-28-2*], *13*:151

(1-Aziridinyl)thiocarboxanilide [*19116-37-1*], *13*:155

3-(1-Aziridyl)-1-phenylprop-2-en-1-one [*22048-28-8*], *13*:151

Azobenzene [*103-33-3*], *15*:917
    isocyanate reaction product, *13*:795
    reduction of, *13*:22

Azobis(2-amidinopropane) dihydrochloride [*2997-92-4*]
    as polymerization initiator, *14*:93

Azobisbutyronitrile [*78-67-1*]
    as polymerization initiator, *14*:93

1,1-Azobiscyclohexanecarbonitrile [*2094-98-6*]
    half-life of, *13*:366

2,2′-Azobis[2,4-dimethyl]pentanenitrile [*4419-11-8*]
    half-life of, *13*:366

2,2′-Azobis(2,4-dimethylvaleronitrile) [*4419-11-8*], *15*:889
    properties, *15*:901

Azobisisobutyronitrile [*78-67-1*], *15*:893
    initiator, *15*:190
    methacrylate ester reaction, *15*:350

2,2′-Azobis(isobutyronitrile) [*78-67-1*], *15*:889
    properties, *15*:901

2,2′-Azobis(2-methylpropionitrile) [*78-67-1*]
    half-life of, *13*:366

Azodicarbonamide [*123-77-3*]
    in spunbondeds mfg, *16*:94

Azo initiators, *13*:365

Azolla nilotica, *15*:948

Azonitriles, *13*:366

Azoospermia
    from 1,2-dibromo-2-chloropropane, *13*:469

Azosol Brilliant Yellow 6GF [*2478-20-8*], *14*:546

Azospirillum brasilense, *15*:949

Azospirillum lipoferum, *15*:949

Azotobacter, *15*:949

Azotobacter indicus, *15*:452. (See also *Beijerinckia indica*.)

Azotobacter paspali, *15*:949

Azotobacter sp, *15*:946

Azotobacter vinelandii, *15*:452, 950

Azotobacter vinelandii UW45, *15*:952

p-Azoxyanisole [*1562-94-3*]
    liquid crystalline range of, *14*:399

Azoxybenzene [*495-48-7*], *15*:917

Aztran, *14*:235

Azurol, *15*:735

B

B-1900 [*12773-54-5*], *15*:789

Babassu nut, *16*:251

Babcock test, *15*:552

Baby foods
    nitrates and nitrites in, *15*:62

B-acid [*17894-99-4*], *15*:731. (See also *8-Amino-1,3,5-naphthalenetrisulfonic acid.*)

Bacillus, *15*:950

Bacillus anthracis, *15*:553

Bacillus calmette-guerin (BCG)
    as immunotherapeutic agent, *13*:176

Bacillus megatherium
    disinfection by ozone, *16*:705

Bacillus popillae, *13*:471

Bacillus sphaericus, *13*:471; *15*:466

Bacillus subtilus, *15*:551, 553

Bacillus thuringiensis, *13*:471

Bacon
    nitrosamines in, *15*:993

Bacteria
    effect on cattlehides, *14*:208
    nitrogen-fixing, *15*:946
    in polysaccharide preparation, *15*:439
    sterilization by ozone, *16*:703

Bactericide
    chlormethoxypropylmercuric acetate, *15*:164

Bactericides
    in leather manufacture, *14*:225
    in paint, *16*:750

Bactofugation
    of milk, *15*:527

Badische acid [*86-60-2*], *15*:730. (See also *7-Amino-1-naphthalenesulfonic acid.*)

Baeyer's acid [*132-57-0*], *15*:738. (See also *7-Hydroxy-1-naphthalenesulfonic acid.*)

Bagasse
    lignin content, *14*:297
    in paper, *16*:771

Bag inks, *13*:382

Bags
    as packaging, *16*:720

Bahnmetall, *14*:457

Bainite, *15*:341

Baits
    in insecticide formulations, *13*:417

Bakery products
    lactates in, *13*:86

BAL. See *British antilewisite.*

Balata, *13*:827

Balke process, *15*:826

Ball-point inks, *13*:395

Balsam tolu

odor compensation of, *16*:303

Bambarra groundnut, *16*:250

Bamberger's reaction, *13*:52

Bamboo
lignin content, *14*:297
paper from, *16*:768

Banbury mixer, *15*:633

Band viscometer, *13*:376

B and W fiber
in metal laminates, *13*:953

Barbital [*57-44-3*], *15*:907

Barbiturates, *13*:122. (See also *Hypnotics.*)
measurement, *15*:88

Barbituric acid [*67-52-7*], *13*:122; *14*:800, 803

Bardac 22, *13*:234

Bardac LF, *13*:234

Bardet process,
for mica paper, *15*:431

Barex, *14*:248

Barges
for LPG transportation, *14*:390

Barita [*7727-43-7*]
in inks, *13*:378

Barite [*13462-86-7*]
from the ocean, *16*:290

Barium [*7440-39-3*]
as magnetic insulator, *14*:694
from the ocean, *16*:279, 292

Barium-130 [*15055-15-9*], *13*:841

Barium-132 [*15065-85-7*], *13*:841

Barium-134 [*15193-77-8*], *13*:841

Barium-135 [*14698-58-9*], *13*:841

Barium-136 [*15125-64-1*], *13*:841; *16*:188

Barium-137 [*13981-97-0*], *13*:841
fission product, *16*:188

Barium-138 [*15010-01-2*], *13*:841

Barium carbonate [*513-77-9*]
in case hardening, *15*:315
in ferrite mfg, *14*:674
nitrogen reaction, *15*:934

Barium chloride fluoride, Eu doped [*68876-91-5*]
x-ray phosphor, *14*:544

Barium dipotassium hexakiscyanoferrate [*31389-21-6*], *13*:767

Barium dithiophosphate
lubricant antioxidant, *14*:492

Barium ethyl sulfate [*6509-22-4*]
KCN reaction, *15*:892

Barium ferrate [*13773-23-4*], *13*:777

Barium ferrite [*12009-00-6*], *13*:776
as magnetic insulator, *14*:694

Barium fluorosilicate [*17125-80-3*]
as insecticide, *13*:421

Barium gluconate [*22561-74-6*]
in iron(II) gluconate synthesis, *13*:774

Barium iodate [*10567-69-8*], *13*:666
for ^{131}I disposal, *16*:180

Barium iodide [*13718-50-8*], *13*:661

Barium lithol red [*1103-38-4*]
in inks, *13*:392

Barium magnesium aluminum oxide [*55134-50-4*]
(1:1:16:27)
in phosphors, *14*:542

Barium magnesium silicate [*68201-15-0*]
phosphor, *14*:543

Barium manganate(V) [*12231-83-3*], *14*:845, 857

Barium manganate(VI) [*7787-35-1*], *14*:845

Barium metaborate [*13701-59-2*]
applications of, *13*:237

Barium molybdate [*7787-37-3*], *15*:687

Barium nitride [*12047-79-9*], *15*:872

Barium oxalate [*516-02-9*]
in tripotassium tris(oxalato)ferrate trihydrate synthesis, *13*:780

Barium oxide [*1304-28-5*]
barium peroxide from, *13*:23
in glass electrode, *13*:5

Barium perferrite [*12230-58-9*], *13*:776

Barium permanganate [*7787-36-2*], *14*:845
ethanol reduction, *14*:857

Barium peroxide [*1304-29-6*]
acid reaction, *13*:16
hydrogen peroxide from, *13*:23
in niobium processing, *15*:823
oxygen-generation systems, *16*:675

Barium sodium niobate [*12323-03-4*]
in nonlinear optics, *14*:64

Barium sulfate [*7727-43-7*]
color standard, *14*:563
electric insulation filler, *13*:565
in inks, *13*:378
ion exchange in, *13*:678
paper pigment, *16*:789

Bark phenolic acids, *14*:308

Barley, *14*:810. (See also *Malts and malting.*)
breeding, *14*:816
gramine in, *13*:220

Barnacles
antimicrobial agents used against, *13*:238

Barquat MB 50, *13*:234

Barrels
as packaging, *16*:719

Barrier coatings
on paper, *16*:791

Barrier polymers, *15*:92. (See also *Membrane technology.*)

Barriers
to noise, *16*:17

Barringtonite [*5145-48-2*], *14*:619

Barthrin [*70-43-9*], *13*:457

Barton process
for lead monoxide, *14*:164

Basalt
  iron in, *13*:735
Base-stock policy
  in operations planning, *16*:515
Basic impulse level (BIL), *13*:542
Basic lead carbonate [*1319-46-6*]
  properties, *14*:171
Basic oxygen furnace
  for steelmaking, *16*:666
Basic oxygen process
  for steelmaking, *16*:666
Basil oil [*8015-73-4*], *16*:321
Bate
  in leather manufacture, *14*:209
Bathophenanthroline [*1662-01-7*]
  in nitrate electrode, *13*:725
Bat repellents, *16*:579
Batteries
  maintenance-free, *14*:145
  membranes for, *15*:124
  mercury, *15*:154
  mercury compounds in, *15*:162
  MnO_2 for, *14*:864
Battery grids
  lead alloys for, *14*:144
Battery scrap
  recycling, *14*:129
Bauhinia reticulata
  tartaric acid in, *13*:113
Bauxite [*1318-16-7*]
  in LPG drying, *14*:389
Bayer-Villiger oxidation reaction, *13*:924
Bay leaf oil [*8006-78-8*], *16*:321
Bay oil [*8006-78-8*], *16*:319
BCD. See *Binary-coded decimal output.*
BCG. See *Bacillus calmette-guerin.*
B_4C on B fiber
  in metal laminates, *13*:953
BD catalysts, *15*:248
BDP. See *Bundle-drawing process.*
Beaches
  minerals from, *16*:289
Beacon, *14*:811
Beads
  in paint, *16*:749
Beam-splitter cubes, *16*:533
Bean weevil
  fumigant activity on, *13*:465
Bearings
  indium in, *13*:210
  lead alloys, *14*:142
  lead for antifriction in, *14*:157
  lubricants for, *14*:516
  lubrication, *14*:483
Beauveria bassiana, *13*:471
Beaverboard, *14*:4

Béchamp process, *15*:917
Beckenbach kilns, *14*:365
Bedding
  nonwoven, *16*:123
Beechnut, *16*:250
Beef, *15*:62.   (See also *Meat products.*)
Beer
  nitrosamines in, *15*:993
Beer's law, *13*:159; *16*:523
Beeswax
  in packings, *16*:734
Beet armyworm
  control of with pheromones, *13*:480
Beetles
  mesomorphism in exocuticle of, *14*:420
Beet sugar
  catechol in, *13*:46
Beijerinckia indica, *15*:452.   (See also
    Azotobacter indicus.)
Beilstein's Handbuch der Organischen Chemie,
    13:287
Bel Baie II, *16*:782
Bel bond inverform process, *16*:784
Bell test
  environmental stress-crack resistance, *16*:426
Belt separators
  magnetic, *14*:719
BEM. See *n-Butyl(ethyl)magnesium.*
Bemegride [*64-65-3*]
  CNS stimulant, *15*:135
Bementite [*66733-93-5*], *14*:825
Bem-liese, *16*:78
Bendiocarb [*22781-23-3*], *13*:453
Beneficiation
  of iron ore, *13*:738
Ben nut, *16*:251
Bentone
  in packings, *16*:738
Bentone 38, *14*:567
Bentone thickener, *14*:504
Bentonite [*1302-78-9*]
  as grease thickener, *14*:502
  in iron-ore processing, *13*:739
  mica substitute, *15*:432
  as solid lubricant, *14*:508
Bentonites
  as insecticide carrier, *13*:416
Benzaldehyde [*100-52-7*], *13*:917; *16*:312
  from almond, *16*:321
  from amygdalin, *16*:256
  aziridine reactions, *13*:150
  in microbial ephedrine synthesis of, *15*:464
  in perfumes, *16*:954
Benzalethylamine [*6852-54-6*]
  phenyl isocyanate reaction,
    13:796

Benzalkonium chloride [68391-01-5]
  applications of, 13:234
Benzamide [55-21-0]
  dehydration, 15:905
7-Benzamido-4-hydroxy-2-naphthalenesulfonic
    acid [132-87-6], 15:742
4-Benzamido-5-hydroxy-1,7-naphthalenedisulfo-
    nic acid [6361-49-5], 15:742
Benzene [71-43-2]
  disulfonation, 13:48
  iodination of, 13:671
  kinetics of destruction, 13:201
  mercuration, 15:163
  mercuric acetate reaction, 16:587
  in mineral oils, 13:551
  nitric acid reaction, 15:916
  odor-problem chemical, 16:303
  oxidation to maleic anhydride, 14:781
  in paint, 16:760
  propionyl chloride reaction, 13:934
  propylene tetramer reaction, 16:484
  toxicity, 13:264
π-Benzene-π-benzoylbenzenechromium [52445-
    45-1], 16:607
Benzenechromium tricarbonyl [12082-08-5], 16:
    599
π-Benzene-π-cyclopentadienyl chromium
    [12093-16-2], 16:607
π-Benzene-π-cyclopentadienylmolybdenum
    [12153-25-2], 16:607
Benzenediazonium chloride [100-34-5]
  2-hydroxy-3(N-phenyl)propyl methacrylate
    reaction, 15:356
1,2-Benzenediol [120-80-9], 16:625
π-Benzene-π-diphenylchromium(0) [12095-46-4],
    16:614
m-Benzenedisulfonic acid [78-48-6]
  mononitrotoluene synthesis, 15:929
  in resorcinol production, 13:48
m-Benzenedisulfonic acid, disodium salt [831-
    59-4]
  resorcinol from, 13:48
Benzene hexachloride [58-89-9]
  as insecticide, 13:432
Benzene–maleic anhydride adduct [1846-38-4],
    14:776
π-Benzene-π-methoxycarbonylbenzenechromium
    [77430-30-9], 16:607
Benzene propanal [104-53-0]
  in perfumes, 16:968
Benzenesulfonyl chloride [98-09-9], 15:905
  aziridine reaction, 13:143
Benzidine [92-87-5]
  skin absorption, 13:260
Benzil [134-81-6]
  in isoprene reaction, 13:823
  mfg, 13:927
  in phenytoin synthesis, 13:135

  as photoinitiator, 13:368
  as repellent, 13:476
1,2-Benzisothiazoline-3-one [2634-33-5]
  applications of, 13:245
Benzoate
  preservative, 13:247
Benzodiazepines
  measurement, 15:88
1,3-Benzodioxole [274-09-9], 13:43.   (See also
    Methylenedioxybenzene.)
Benzoguanamine [91-76-9], 15:905
Benzoic acid [65-85-0], 13:247
  in antiaging drug, 15:140
  decarboxylation, 13:934
  in grease thickener formulation, 14:502
  lecithin preservative, 14:254
  microbial synthesis of, 15:465
  naphthalene azeotrope, 15:699
  in perfumes, 16:950, 954
Benzoic sulfimide [81-07-2]
  in adhesive, 15:491
Benzoin [119-53-9]
  odor compensation of, 16:303
  oxidation of, 13:927
  in perfumes, 16:949
Benzoin ethers
  as photoinitiator, 13:369
Benzonitrile [100-47-0], 15:889
  olefin reaction, 16:614
  properties, 15:904
Benzophenone [119-61-9]
  as photoinitiator, 13:368
  properties, 13:896
Benzopurpurine 4B [992-59-6], 13:10
Benzo[a]pyrene [50-32-8]
  in oils, 16:352
p-Benzoquinone [106-51-4], 13:52
  in adhesive, 15:491
  in Nenitzescu synthesis, 13:219
Benzoylacetone [93-91-4]
  in perfume formulations, 13:927
N-Benzoyl-acid [6361-49-5], 15:742.   (See also
    *4-Benzamido-5-hydroxy-1,7-naphthalenedi-
    sulfonic acid*.)
Benzoyl chloride [98-88-4]
  in liquid crystal synthesis, 14:414
  resorcinol reaction, 13:45
N-Benzoylindole [1496-76-0]
  halogenation, 13:215
N-Benzoyl J-acid [132-87-6].   See *7-Benzamido-
    4-hydroxy-2-naphthalenesulfonic acid*.
Benzoyl nitrate [6786-32-9]
  in indole nitration, 13:215
p-Benzoyloxybenzoates
  synthesis of, 14:414
Benzoyl peroxide [94-36-0], 13:29
  as bleach for lecithin, 14:259

Benzoyl peroxide (*Continued*)
  in polymethacrylate mfg, *15*:389
  toxicity, *13*:264
N-Benzoyl-*N*-phenylhydroxylamine [*304-88-1*]
  in niobium analysis, *15*:829
Benzoyl resorcinol [*131-56-6*], *13*:45. (See also
  2,4-Dihydroxybenzophenone.)
Benzyl acetate [*140-11-4*], *16*:955
  in perfumes, *16*:968
Benzyl alcohol [*100-51-6*]
  naphthalene azeotrope, *15*:699
Benzyl benzoate [*120-51-4*]
  as repellent, *13*:476
Benzyl bromoacetate [*5437-45-6*]
  applications of, *13*:231
Benzyl chloride [*100-44-7*]
  2-dimethylaminoethyl methacrylate reaction,
    15:356
  toxicity, *13*:264
Benzyl-*p*-chlorophenol [*120-32-1*]
  applications of, *13*:226
Benzyl cyanide [*140-29-4*]
  in phenylethylamine synthesis, *15*:775
Benzyldichlorosilane [*18173-99-4*]
  isoprene reaction, *13*:826
2-(*N*-Benzyl-*N*,*N*-dimethyl)amminium-ethyl
  methacrylate chloride, *15*:356
5-Benzyl-3-furylmethyl DL-*cis,trans*-
  chrysanthemate [*10453-86-8*]
  as insecticide, *13*:457
N-Benzylhistidineprolamide [*32926-29-7*]
  in TRH synthesis, *15*:782
1-[2-(Benzylideneamino)ethyl]aziridine [*4025-
  38-1*], *13*:150
Benzyllithium–*N*,*N*,*N*′,*N*′-tetramethylethylened-
  iamine complex [*20471-78-7*]
  polymerization reagent, *14*:469
Benzyl methacrylate [*2495-37-6*]
  properties, *15*:353
5-Benzyloxyindole [*1215-59-4*]
  in serotonin synthesis, *15*:770
5-Benzyloxyindole-3-glyoxylamide [*22424-62-0*]
  in serotonin synthesis, *15*:770
5-Benzyloxyindole-3-glyoxylchloride [*22424-61-9*]
  in serotonin synthesis, *15*:770
Benzyloxymethyl methacrylate [*76392-18-2*]
  properties, *15*:353
Benzylpenicillin [*61-33-6*]
  microbial hydrolysis of, *15*:462
Benzyl phenyl ketone [*451-40-1*]
  oxidation of, *13*:927
Benzyl salicylate [*118-58-1*], *16*:955
Benzylsodium [*1121-53-5*], *16*:557
N-Benzyl,*N*-tertbutoxycarbonylhistidine
  [*20898-44-6*]
  in TRH synthesis, *15*:782

Bergamot oil [*8007-75-8*], *16*:319, 321
Bergapteu [*484-20-8*], *16*:316
Bergwachs, *14*:340
Berkelium nitride [*56509-31-0*], *15*:872
Berlin green [*14433-93-3*]
  from prussian blue oxidation, *13*:771
Berlock dermatitis, *16*:320
Bertholletia excelsa, *16*:250
Beryllia [*1304-56-9*]
  in insulation, *13*:562
Beryllium [*7440-41-7*]
  from the ocean, *16*:280
  toxicity, *13*:264
Beryllium-9 [*7440-41-7*], *13*:840
Beryllium fiber
  in metal laminates, *13*:945
Beryllium nitride [*1304-54-7*], *15*:872
Beryllium oxide [*1304-56-9*]
  use in lasers, *14*:53
Bessel function, *14*:945
Bessel's equation, *14*:945
Bessemer converter
  in steelmaking, *16*:667
Bessemer-furnace iron, *13*:740
Bessemer process, *14*:824
Betamax, *14*:750
Betamethasone [*378-44-9*], *13*:173
Beta (Mg–Al) [*12254-22-7*], *14*:603
Beta-scission
  of alkoxy radicals, *13*:360
Betel nut, *16*:251
BET measurements
  of anode surface area, *15*:177
BET method
  for zeolite areas, *15*:659
Betol, *15*:734
Betterton-Kroll process, *14*:123
Betts process
  electrorefining lead, *14*:117
Betula lenta L., *16*:321
Betzes/Shabet, *14*:811
BHA. See *Butylated hydroxy anisole.*
BHT. See *Butylated hydroxytoluene.*
Biacetyl [*431-03-8*], *13*:905. (See also *2,3-
  Butanedione.*)
  mfg, *13*:927
  properties, *13*:898
Biarsine [*15942-63-9*], *16*:39
Biazzi process, *13*:804
Bibenzoyl [*134-81-6*]
  properties, *13*:926
Bicarbonate/carbonate
  in liquid membranes, *15*:119
Bicarbonates
  in leather manufacture, *14*:213

[(2,3,4,5,6-η)-Bicyclo[5.1.0]octa-3,5-diene-2-yl]tri-
   carbonyliron [*41853-19-4*], *16*:602
Bidim, *16*:78
Biguanidines
   as antidiabetic agents, *13*:613
BIL.   See *Basic impulse level.*
Bile
   in *S*(−)-malic acid biosynthesis, *13*:107
Bilharzia
   molluscicide use, *16*:579
Bilirubin [*635-65-4*]
   measurement, *15*:80, 83
Billboards
   fluorescent, *14*:564
Binapacryl [*485-31-4*], *13*:428
Binary-coded decimal (BCD) output
   in pH meters, *13*:9
Binderine, *14*:125
Binders
   in chlorate candles, *16*:675
   for nonwoven textiles, *16*:114
   in papermaking, *16*:822
   in spunbondeds mfg, *16*:89
Binodrenal, *15*:773
Binomial distribution
   for failure calculations, *15*:26
Biochemical nomenclature, *16*:41
Biocide N-521, *13*:244
Biocide N-948, *13*:243
Biocide N-1386, *13*:243
Biocides, *13*:415.   (See also *Insecticides.*)
   organoleads, *14*:193
   organophosphorus insecticides as, *13*:453
   organotin compounds, *16*:578
Biomedical instrumentation, *15*:75.   (See also
   Medical diagnostic reagents.)
Biomet 204, *13*:236
Biomethylation
   of mercury, *15*:166
Biopal VRO 20, *13*:230
Biopan P-1487, *13*:242
Biopermethrin [*28434-01-7*], *13*:458
Biopolymer CS, *15*:455
Bio-Quat 50-24, *13*:234
Bioresmethrin [*28434-01-7*], *13*:458
BIOSIS PREVIEWS information system, *13*:303
Biotin [*58-85-5*]
   in nuts, *16*:253
   sulfur in, *15*:587
Biotite [*1302-27-8*], *15*:416
   lithium in, *14*:449
BIPA information system, *13*:328
4,4′-Biphenylenebis[dimethyl]silanol [*4852-15-7*],
   15:200

Bipolar junction transistor
   in ICS, *13*:621
2,2′-Bipyridine [*366-18-7*]
   as iron chelate ligand, *13*:778
Birch bark oil [*68917-50-0*], *16*:321
Birefringence
   of liquid crystals, *14*:422
Birkeland-Eyde process
   for dinitrogen oxidation, *15*:943
Birnessite [*1244-32-5*], *14*:855
Birthmarks
   laser removal, *14*:70
Bis-η^3-allylnickel [*12077-85-9*], *16*:592
Bis(π-allyl)nickel [*12077-85-9*], *15*:808
Bis[1-(2-aminoethyl)aziridine]-cupric chloride
   complex [*73214-90-1*], *13*:151
Bis(2-aminoethyl) sulfide [*871-76-1*], *13*:155
Bis(4-aminophenyl)methane [*101-77-9*] (MDA)
   in isocyanate manufacture, *13*:805
2,4-Bis[(4-aminophenyl)methyl]aniline [*25834-
   80-4*]
   in isocyanate manufacture, *13*:805
1,1′-Bis[(3-aminopropyl)dimethylsilyl]ferrocene
   [*12190-11-3*], *15*:202
Bis(1-aziridinyl)phenylarsine [*20825-66-5*], *13*:150
[Bis(benzenethiolato)ferrate]di-μ-thioxodithioxo-
   molybdate(2−) [*72089-28-2*], *15*:962
Bis(η^6-biphenyl)chromium [*12099-15-9*], *16*:608
Bis(2,2′-bipyridine)iron(II)dihydroxy bis(2,2′-
   bipyridine), *13*:778
Bis[bis(trimethylsilyl)methyl]tetrakis(η^5-2,4-cycl-
   o- pentadien-1-yl)[μ-(dinitrogen-*N*,*N′*)]-
   dizirconium [*66970-17-0*], *15*:956
Bis(1,4-bromoacetoxy)-2-butene [*20679-58-7*]
   applications of, *13*:231
2,3,4,5-Bis(butyl 2-ene)tetrahydrofurfural [*126-
   15-8*]
   as repellent, *13*:476
Bischler synthesis
   of indoles, *13*:218
1,1′-Bis(chlorodimethylsilyl)ferrocene [*12319-30-
   1*], *15*:202
N,*N*-Bis(2-chloroethyl)succinamide [*6265-83-4*],
   13:148
Bis-chloromethyl ether [*542-88-1*]
   ion-exchange contaminant, *13*:695
1,1-Bis(*p*-chlorophenyl)ethylene [*10605-50-2*],
   13:461
Bis(*p*-chlorophenyl) sulfone [*29315-80-8*], *13*:462
   as insecticide, *13*:429
1,1-Bis(*p*-chlorophenyl)-2,2,2-trichloroethanol
   [*54532-36-4*]
   in insect control, *13*:461
Bischofite [*13778-96-6*], *14*:622

Bis(cyclopentadienyl)nickel [*1271-28-9*], *15*:807

Bisdehydroconiferyl alcohol
in lignin, *14*:301

1,1'-Bis[3-(dichlorophenyl)-1,1,3,3-tetramethyldi-
siloxanyl-]ferrocene [*12321-03-8*], *15*:200

N,N-Bis(2-diethylaminoethyl)methacrylamide
[*76392-05-7*]
properties, *15*:356

4,4'-Bis(*N,N'*-dimethylamino)benzophenone
[*90-94-8*]
as photoinitiator, *13*:368

1,1'-Bis[(dimethylamino)dimethylsilyl]ferrocene
[*35625-88-8*], *15*:202

Bis(dimethylamino)phosphoryl fluoride [*115-26-
4*]
as insecticide, *13*:445

1,1'-Bis[dimethyl[3-(oxiranylmethoxy)propyl]si-
lyl]ferrocene [*1295-93-8*], *15*:202

1,1'-Bis[dimethyl[oxo-3-oxiranylmethoxy)-
propyl]silyl]ferrocene [*1295-93-8*], *15*:202

1,1'-Bis[dimethyl[3-[(trimethylsilyl)oxy]propyl]-
silyl]ferrocene [*1295-94-9*], *15*:202

Bis(dinitrogen)bis[1,2-ethanediylbis[diphenyl-
phosphine]-*P,P'*]molybdenum [*41700-58-7*]
HBr reaction, *15*:956

Bis(dinitrogen)tetrakis(dimethylphenylphosphine)molybdenum [*55925-48-9*], *15*:956

Bis(dinitrogen)tetrakis(methyldiphenylphosphine)molybdenum [*33248-03-2*], *15*:956

π-Bisdiphenylchromium [*33154-48-2*], *16*:614

1,2-Bis(diphenylphosphino)ethane [*1663-45-2*]
in N$_2$-containing complexes, *15*:956

Bis[2-(diphenylphosphinyl)ethyl]phenyl
phosphine
in metal-nitrogen complexes, *15*:957

2,2-Bis[*p*-(2,3-epoxypropoxy)phenyl]propane
[*1675-54-3*]
copolymers from, *15*:205

Bis[1,2-ethanediylbis[diphenylphosphine]-*P,P'*]-
fluoro- [hydrazinato(2−)-*N*]molybdenum(1
ketone reaction, *15*:957

Bis[μ-(ethanethiolato)]-μ-thioxohexa-μ³-thioxob-
is [tris(ethanethiolato)-μ³-thioxotriferrate]
dimolybdate(3-) [*67583-54-4*], *15*:962, 963

1,1'-Bis(ethenyldimethylsilyl)ferrocene [*1295-65-
4*], *15*:202

1,1'-Bis(ethoxydimethylsilyl)ferrocene [*12320-
47-7*]
cyclization, *15*:202

Bis(2-ethyl butanoic acid), triethyleneglycol ester
[*95-08-9*], *13*:981

Bisglycine monohydrochloride [*13059-60-4*], *15*:
769

Bis-(4'-*n*-heptyloxybenzylidene)-1,4-phenylene-

diamine [*24679-01-4*]
polymorphism of, *14*:410

Bishexamethylenetriamine [*143-23-7*]
in papermaking, *16*:817

N,N'-Bis(2-hydroxy benzyl)ethylenediamine-
N,N'-diacetic acid [*35998-29-9*], *13*:786

N-Bis(2-hydroxyethyl)lactamide [*64058-34-0*],
13:90

1,1'-Bis(hydroxymethyl)ferrocene [*1291-48-1*]
cross-linking agent, *15*:196

1,1'-Bis[(3-hydroxypropyl)dimethylsilyl]ferrocene
[*1295-84-7*], *15*:202

Bis(π-indenyl)nickel [*52409-46-8*], *15*:807

Bis(4-isocyanatocyclohexyl)methane [*5124-30-1*],
13:808. (See also *4,4'-Dicyclohexylmethane
diisocyanate.*)

Bis(4-isocyanatomethyl)cyclohexane [*10347-54-
3*], *13*:806

Bis(mesityl) nickel [*77509-07-0*], *16*:583

Bis(methacryloyloxyethyl) sulfide [*35411-32-6*]
properties, *15*:355

Bis(2-methoxyethyl mercuric) silicate, *15*:170

Bis[(1,2,3,4,5-η)-1-methyl-2,4-cyclopentadien-
1-yl]cobalt [*12146-91-7*]
oxidation, *16*:604

Bis(3-methylphenoxy)phosphinous chloride
[*21719-86-8*]
nickel complex formation, *15*:808

Bis(4-methylphenoxy)phosphinous chloride
[*20003-41-2*]
nickel complex formation, *15*:808

Bismuth [*7440-69-9*]
in magnesium alloys, *14*:596
in metallic coating, *15*:268
recovery in lead refining, *14*:124

Bismuth-209 [*7440-69-9*], *13*:842

Bismuth, compd with manganese [*12010-50-3*]
(1:1), *14*:694

Bismuth nitride [*12232-97-2*], *15*:872

Bis(pentachloro-2,4-cyclopentadien-1-yl) [*2227-
17-0*]
as insecticide, *13*:463

Bis[(1,2,3,4,5-η)-1,2,3,4,5-pentamethyl-2,4-cyclop-
entadien-1-yl]titanium [*11136-37-1*], *15*:955

1,1'-Bis(pentamethyldisiloxanyl)ferrocene
[*38903-97-8*], *15*:200

1,1'-Bis(1,1,3,3,3-pentamethyldisiloxanyl)ferro-
cene [*12243-60-6*], *15*:200

Bisphenol A [*80-05-7*], *13*:900
epoxy resins from, *13*:973
from etching polycarbonate, *16*:834
in hydroquinone production, *13*:52

trans-1,2-Bis(*n*-propylsulfonyl)ethylene [*1113-14-0*]
mildewcide in paints, *16*:754

Bis(pyridine)iodine(1+) nitrate [*39001-42-8*], *13*:
665

N,N'-Bis(4-sulfonamidophenyl)urea [*1773-43-9*], *13*:801

Bis(2,2′,2″-terpyridine)iron(2+) [*17455-70-8*], *13*:781

1,1′-Bis(1,1,3,3-tetramethyl-3-phenyldisiloxanyl)-ferrocene [*12321-04-9*], *15*:200

Bis(tri-*n*-butylphosphine)nickel dibromide [*15242-92-9*], *15*:809

Bis(trichloromethyl)sulfone [*3064-70-8*] applications of, *13*:243

Bis(tricyclohexylphosphine)hydridonickel chloride [*25703-57-5*], *15*:809

α,α-Bis(trifluoromethyl)-1-aziridinylmethanol [*773-23-9*], *13*:150

Bis(trifluoromethyl)ketene [*684-22-0*] properties, *13*:375

Bis(trifluoromethyl)thioketene [*7445-60-5*] properties, *13*:875

3,3-Bis(trifluoromethyl)-4-[2.2.2-trifluoro-1-(trifluoromethyl)ethylidene]-2-oxetanone [*10436-05-2*], *13*:883

2,4-Bis[2.2.2-trifluoro-1-(trifluoromethyl)ethylidene]-1,3-dithietane [*7445-61-6*], *13*:883 from bis(trifluoromethyl)ketene, *13*:878 thioketene from, *13*:882

Bis(trimethylphosphine)dimethylnickel [*60802-48-4*], *15*:809

Bis(triphenylphosphine)nickel nitrosyl bromide [*14586-72-2*], *15*:809

Bis(triphenylphosphine)phenylnickel chloride [*38415-93-3*], *15*:809

Bis(trityl)nickel [*7544-48-1*], *15*:809

Bisulfite [*15181-46-1*] in emulsion polymerization, *14*:85

Bithermal process, *13*:849

Bitter almond. See *Almond oil.*

Bitter almond oil [*8013-76-1*], *16*:320

Bitterfeld cell, *14*:871

Bitter magnet, *14*:681

Bitterns from seawater, *16*:281

Bitumen, *16*:333 pyrolysis, *16*:337

Biuret of hexamethylene diisocyanate [*4035-89-6*] properties, *13*:789

Blackbody radiation, *13*:337

Black walnut, *16*:248 development, *16*:260

Bladan [*107-49-3*], *13*:437

Bladder nut, *16*:251

Blades high-speed composite, *13*:947

BLAISE information system, *13*:292

Blaise reaction, *15*:892

Blanc fixe [*7727-43-7*]

in inks, *13*:378

Blast furnace lead ore processing, *14*:104

Blast-furnace process, *13*:739

Blasting in metal cleaning, *15*:302

Bleaches fragrances, *16*:970

Bleaching by ozone, *16*:683

Bleaching agent hydrogen peroxide, *13*:27

Bleeding detection, *15*:81

Bleeding ulcers treatment, *15*:780

Blenders, *15*:619. (See also *Mixing and blending.*)

Blending. See *also Mixing and blending.* operations planning of, *16*:518

Bleomycin [*11056-06-7*] encapsulation of, *15*:477

Blind staggers of livestock, *15*:591

Blister blight nickel fungicide for, *15*:814

Bloch walls, *14*:688

Blocking filters, *16*:535

Block mica, *15*:421

Bloedite [*15083-77-9*], *14*:637

Blood analysis of lead content in, *14*:192 mineral nutrient constituents, *15*:574 oxygenation, *15*:490 soluble dried, meat by-product, *15*:72

Blood–air distribution coefficient, *13*:259

Blood glucose adrenaline effect on, *15*:758

Blood ions electrodes for, *13*:732

Blood meal meat by-product, *15*:72

Blood oxygen, *16*:669

Blood plasma from animals, *15*:72

Blood-serum albumin in casein, *15*:568

Blood urea nitrogen (BUN) measurement, *15*:80

Blood-volume expander dextran, *15*:441

Bloom test for lead-calcium analysis, *14*:149

Blowdowns noise control, *16*:23

Blow molding HDPE, *16*:447

Blown fibers
    novolac, *16*:130
Blue grass seed
    rust prevention in, *15*:814
Blue phase
    as exhibited in cholesterics, *14*:410
Blueprinting
    oxalates in, *16*:633
B-lymphocytes, *13*:167
    in vivo separation, *15*:490
Board, corrugated, *16*:800
Boat hulls
    of reinforced plastics, *13*:975
Bodansky unit
    phosphatase activity, *15*:80
Bode method, *13*:501
Body odor
    in perfumes, *16*:970
Boeing 747
    windshield, *13*:990
Boeniger acid [*116-63-2*], *15*:740. (See also *3-Hydroxy-4-amino-1-naphthalenesulfonic acid.*)
Bofors nitration process, *15*:847
Boilermakers' ear, *16*:1
Boiling-water reactor (BWR), *16*:150
Boiling-water reactors
    fuel assembly, *16*:165
Bois de rose oil [*8015-77-8*], *16*:319, 321. (See also *Rosewood oil.*)
Boletic acid [*110-17-8*], *14*:770. (See also *Fumaric acid.*)
Boliden process
    for lead smelting, *14*:130
Bolometers, *13*:340
Bolt lubricants, *14*:512
BOM. See *n-Butyl-n-octylmagnesium.*
Bomah nut, *16*:251
Bomb shelters
    oxygen generation, *16*:673
BON. See *3-Hydroxy-2-naphthalenecarboxylic acid.*
BONA. See *3-Hydroxy-2-naphthalenecarboxylic acid.*
Bonding
    nonwovens, *16*:114
Bonding, ionic, *13*:536
Bonding, metallic, *13*:534
Bonduc nut, *16*:251
Bone meal
    in feeds, *14*:378
    meat by-product, *15*:72
Bones
    hydroxyapatite forms, *15*:584
Book cover, *16*:772
Book covers
    nonwoven, *16*:94

Book inks, *13*:382
Book matches, *15*:2
Book preservation, *16*:586, 811
Boolean logic, *13*:279
Boophilus decolaratus, *13*:473
Boranes, *16*:562
Boranetriamine [*14939-14-1*], *16*:33
Borax [*1303-96-4*], *13*:422
    in leather manufacture, *14*:213
    in metallic coating, *15*:252
    in metal treating, *15*:310
Borazon, *15*:883
Bordeaux mixture, *13*:475
Borescope, *15*:542
Boric acid [*1303-96-4*], *13*:422
    in matches, *15*:4
    in metallic coating, *15*:252
    in nuclear reactors, *16*:150
    pH of system with LiOH, *16*:150
    in phosphor mfg, *14*:542
Boric acid, anhydrous [*13460-50-9*], *16*:33
Boric oxide [*1303-86-2*]
    in semiconductor manufacture, *14*:287
Borides
    in case hardening, *15*:320
Borneol [*507-70-0*]
    from pine, *16*:327
Bornyl acetate [*76-49-3*]
    from pine, *16*:327
Borofused wire dies, *15*:250
Borohydrides
    in magnetic tape mfg, *14*:740
Boron [*7440-42-8*]
    ion-exchange resins for, *13*:688
    ion implantation, *13*:717
    isotope separation processes, *13*:853
    as metallic coating, *15*:247
    from the ocean, *16*:279
    in silicon doping, *13*:635
    in superhybrid composites, *13*:942
Boron-10 [*14798-12-0*], *13*:840
    in nuclear reactors, *16*:161
    separation, *16*:163
Boron-11 [*14798-13-1*], *13*:840
Boron alkyls, *16*:562
Boron carbide [*12069-32-8*]
    in case hardening, *15*:320
    in metallic coating, *15*:252
Boron–carbon fiber
    in metal laminates, *13*:945
Boron chloride [*12069-32-8*]
    (4:1)
    in coatings, *15*:241
Boron deficiency
    in nuts, *16*:253
Boron fiber
    in metal composites, *13*:943

Boronizing, *15*:250, 319
Boron nitride [*10043-11-5*], *15*:872
  properties, *15*:874
Boron nitride (cubic) [*10043-11-5*]
  in metallic coating, *15*:266
Boron oxide [*1303-86-2*]
  in nitride formation, *15*:877
Boron oxide [*1303-86-2*] (2:3)
  niobium boride from, *15*:834
Boron tribromide [*10294-33-4*]
  IC dopant, *13*:638
Boron trichloride [*10294-34-5*]
  trialkylaluminum reaction, *16*:564
Boron trifluoride [*7637-07-2*]
  in boron-10 separation, *16*:164
  Grignard reaction, *16*:564
  health effect, *13*:264
  as initiator catalyst, *13*:371
  in ion implantation, *13*:708
  in malic acid reaction, *13*:105
  nitration reagents, *15*:841
Boron trifluoride-dimethyl ether [*353-42-4*]
  boron-10 from, *16*:163
Boron trifluoride etherate [*109-63-7*]
  LiAlH₄ reaction, *16*:564
  NaBH₄ reaction, *16*:564
Boron–tungsten fiber
  in metal laminates, *13*:945
Borosilicate glass
  in nuclear waste disposal, *16*:210
Borosilicates
  as lubricant, *14*:520
Borsic fiber
  in metal laminates, *13*:945, 953
Bosch process, *16*:681
Botanical oils. See *Oils, essential.*
Bottles, *16*:722
Bouguer-Lambert law, *16*:523
Bovine ACTH [*39319-42-1*], *15*:758
Bovine neurotensin [*55508-42-4*], *15*:772
Boxes, *16*:798
  as packaging, *16*:722
Bragg diffraction
  in liquid crystals, *14*:409
Bragg peak, *16*:830
Brain acetylcholine neurotransmitter
  formation of by lecithin, *14*:261
Brake fluid
  diacetone alcohol, *13*:915
Brake fluids, hydraulic
  lubricants for, *14*:499
Brake lining
  oxalic acid use in, *16*:632
Bran
  attractant, *13*:417
Branched olefins, *16*:480
Branding
  as scar source on leather, *14*:207

Brass [*12597-71-6*]
  emissivity, *13*:342
  in packings, *16*:734
Brass flakes
  in inks, *13*:379
Brattleboro rats, *15*:133
Braunite, *14*:825
Brazilian mint [*68917-18-0*], *16*:309
Brazil nut, *16*:250
Brazing alloys
  lithium in, *14*:458
Breadnut, *16*:250, 252
Breast-cancer detection
  microwave techniques for, *15*:514
Breathing apparatus, *16*:677
Breeder reactor
  fuel elements, *16*:165
Breeder-reactor fuel
  reprocessing of, *16*:173
Breeder reactors
  boron-10 used in, *16*:163
Breeding ratio
  nuclear, *16*:188
Brevibacterium ammoniagenes, *13*:107; *15*:465
Breweries
  use of ozone in, *16*:709
Brewster's angle, *14*:52
Brick
  emissivity, *13*:342
  as insulation, *13*:597
  sound coefficients of, *13*:513
Bricks
  magnesia in, *14*:636
Brigham Filter No. 75, *16*:530
Brighteners
  dihydroxybenzenes used in, *13*:58
Brightening, *15*:303
Brilliant green [*633-03-4*], *14*:551, 552
Brilliant sulfoflavine FF [*2391-30-2*], *14*:546, 553
Brilliant yellow [*6441-64-1*], *13*:10
Brines
  deep-sea, minerals from, *16*:290
  lithium in, *14*:449
Briquet oxygéné, *15*:2
Briquetting
  iron ore, *13*:739
  lignite and brown coal, *14*:330
British antilewisite [*59-52-9*]
  as arsenic antidote, *13*:421
Brodie purifier, *15*:708
Broenner's acid [*93-00-5*], *15*:730. (See also *6-Amino-2-naphthalenesulfonic acid.*)
Broken-back syndrome
  from toxaphene poisoning, *13*:433

Broken hill furnace
  lead processing, *14*:110
Bromacil [*314-40-9*], *13*:888
Bromat, *13*:234
Bromelia, *15*:734
Bromination
  initiators used in, *13*:356
Bromine [*7726-95-6*]
  antimicrobial activity of, *13*:232
  discovery, *16*:283
  mercury reaction, *15*:160
  from the ocean, *16*:279
Bromine-79 [*14336-94-80*], *13*:840
Bromine-81 [*14380-59-7*], *13*:840
Bromine (Br₂) [*7726-95-6*]
  toxicosis from, *15*:598
Bromine nitride [*15162-90-0*], *15*:872
Bromine pentafluoride [*7789-30-2*]
  uranium reaction, *16*:182
Bromine trifluoride [*7787-71-5*]
  uranium reaction, *16*:182
Bromism
  carbromal induced, *13*:125
Bromisovalum [*496-67-3*]
  properties of, *13*:122
Bromoacetyl chloride [*22118-09-8*]
  in quazepam synthesis, *13*:130
Bromobenzene [*108-86-1*]
  in phenyllithium synthesis, *14*:471
α-Bromobenzyl methyl ketone [*23022-83-5*], *13*: 148
Bromobis[1,2-ethanediylbis [diphenylphosphine]-*P,P'*] [hydrazido(2−)-*N*]molybdenum(1, *15*: 957
Bromobis[1,2-ethanediylbis[diphenylphosphine]-*P,P'*][1,1-dimethylhydrazido(2-)-*N*]tungsten bromide [*56178-22-4*], *15*:958
Bromo(*n*-butyldiazenido)bis[1,2-ethanediylbis[diphenyl- phosphine]-*P,P'*]molybdenum [*68022-28-6*], *15*:957
Bromochlorodimethylhydantoin [*126-06-7*]
  applications of, *13*:231
Bromocresol green [*76-60-8*], *13*:10
  in albumin, *15*:81
Bromocresol purple [*115-40-2*], *13*:10
  in leak detection, *16*:55
3-Bromo-1-cyclohexyloxypropane [*22498-19-7*]
  in spermine synthesis, *15*:777
Bromo(diethyl) aluminum [*760-19-0*]
  propylene polymerization cocatalyst, *16*:459
β-Bromoethyl acetate [*927-68-4*]
  in acetylcholine bromide synthesis, *15*:756
β-Bromoethylamine [*107-09-5*]
  sodium sulfite reaction, *15*:781
2-Bromoethyl methacrylate [*4513-56-8*]
  properties, *15*:355
3-Bromoindole [*1484-27-1*], *13*:215

1-Bromo-2-isocyanatobenzene [*1592-00-3*], *13*:798
1-Bromo-3-isocyanatobenzene [*23138-55-8*], *13*: 798
1-Bromo-4-isocyanatobenzene [*2493-02-9*], *13*:798
3-Bromo-2-methyl-1-butene [*51872-48-1*]
  isoprene reaction product, *13*:824
2-Bromo-3-methylbutyryl chloride [*52574-82-0*]
  in bromisovalum synthesis, *13*:125
2-Bromo-2-nitropropane-1,3-diol [*52-51-7*]
  applications of, *13*:231
Bromophenol blue [*115-39-9*], *13*:10
  in sodium lactate analysis, *13*:88
4-Bromophenyl methacrylate [*36889-09-5*]
  properties, *15*:355
Bromophos [*2104-96-3*], *13*:440
Bromopicrin [*464-10-8*], *15*:975
Bromosuccinic acid [*923-06-8*]
  thiomalic acid from, *13*:110
N-Bromosuccinimide [*128-08-5*]
  in indole halogenation, *13*:215
Bromothymol blue [*76-59-5*], *13*:10
Bromotrichloromethane [*75-62-7*]
  isoprene reaction, *13*:825
Bromoxynil [*1689-84-5*], *13*:76
Brønsted acid
  zeolites as, *15*:650
Brønsted acids
  as initiator catalysts, *13*:371
Bronze [*12597-90-5*]
  as cable covering, *13*:582
  in coatings, *15*:241
  coefficient of friction, *14*:481
  in packings, *16*:734
Bronze foil
  in packings, *16*:737
Brookfield Torsion viscometer, *13*:376
Brookfield viscometer, *16*:757
Brosimum alicastrum, *16*:250
Brown coal, *14*:313
Browning
  by microwaves, *15*:502
Brown ring test, *13*:777; *15*:867
Brown rot, *13*:225
Browns Ferry nuclear power station
  fire at, *16*:232
BRS
  information-retrieval system, *16*:901
Brucella abortus, *15*:553
Brucellosis, *15*:553
  treatment, *13*:61
Brucine [*357-57-3*]
  in lactic acid resolution, *13*:81
  in phloionic acid reaction, *13*:118
  in poison nut, *16*:257

Brucite [*1317-43-7*], *14*:615
Brugia malayi, *13*:413
BTC-100, *13*:234
BTC-1010, *13*:234
Buanine [*73-40-5*]
  in nucleic acid, *15*:587
Bubbler system
  for level measurement, *14*:444
Bubbles, magnetic, *14*:694
Bucherer amination, *15*:728
Buffer
  malic acid as, *13*:108
Buffers
  and pH, *13*:1
Bufotenine [*487-93-4*], *13*:220
Building materials
  from lignite ash, *14*:340
  limestone, *14*:376
Building paints, *16*:752
Built-up mica, *15*:429
Bulking compound
  magnesium carbonate, *14*:621
Bulk-molding compound, *13*:975
Bulk separation, *15*:663
Bullets
  lead alloys for, *14*:146
Bull's eye
  in laminated glass, *13*:981
Bumium bulbocastanum, *16*:251
BUN. See *Blood urea nitrogen.*
Buna N. See *Butyl rubber.*
Buna-S, *15*:894
Bundle-drawing process
  for metal fibers, *15*:227
Bune
  lignin content, *14*:297
Bunium flexuosum, *16*:250
Buoyancy devices
  lithium hydride in, *14*:463
Buoyancy method
  for polymer density, *16*:437
Buriti nut, *16*:250
Burlap bags, *16*:720
Burners
  for incineration, *13*:197
Burn-in
  product testing, *15*:17
Burning profile
  coals, *14*:331
Burnup
  nuclear fuel, *16*:184
Burrus diode, *14*:277
Bursa equivalent, *13*:168
Busan 11-M1, *13*:237
Business data bases, *13*:326
Business form inks, *13*:381

Business forms
  carbonless paper, *15*:487
Business information, *13*:305
Butadiene [*106-99-0*]
  adiponitrile from, *15*:899
  aziridine reaction, *13*:151
  chlorination, *15*:899
  copper(II) halide reaction, *16*:611
  in emulsion form, *14*:93
  ethylene reaction, *15*:808
  HCN reaction, *15*:899
  maleic anhydride reaction, *14*:774
  methacrylate copolymerization, *15*:384
  polymerization of, *13*:356
  solubility of, *14*:83
1,3-Butadiene [*106-99-0*]
  from butane, *13*:672
  copolymer, *15*:192
  methacrylate ester reaction, *15*:349
  methyl methacrylate reaction, *15*:350
η^4-Butadieneiron tricarbonyl [*12078-32-9*], *16*:593
1,3-Butadienylferrocene [*12126-86-2*], *15*:186
Butane [*106-97-8*]
  dehydrogenation, *13*:672
  nitration, *15*:848
  oxidation to maleic anhydride, *14*:782
  storage, *14*:393
n-Butane [*106-97-8*]
  in LPG, *14*:384
  properties, *14*:384
1,4-Butanediamine [*110-60-1*]
  in spermine synthesis, *15*:777
1,4-Butanediol [*110-63-4*]
  in antimicrobial agents, *13*:238
  from formaldehyde, *15*:412
  from maleic anhydride, *14*:780
2,3-Butanediol [*513-85-9*]
  oxidation of, *13*:925
1,3-Butanediol methacrylate [*1189-08-8*]
  properties, *15*:354
1,4-Butanediol methacrylate [*2082-81-7*]
  properties, *15*:354
2,3-Butanedione [*431-03-8*], *13*:905
1-Butanol [*71-36-3*]
  from oxo process, *16*:637
  oxo product, *16*:650
2-Butanol [*78-92-2*]
  dehydrogenation of, *13*:905
tert-Butanol [*75-65-0*]
  hydroquinone reaction, *13*:43
2-Butanone, *13*:903. (See also *Methyl ethyl ketone.*)
[(1,2,3-η)-2-Butanyl]tricarbonylchloroiron [*12287-49-9*], *16*:615
1-Butene [*106-98-9*]
  copolymer with ethylene, *16*:388
  diborane reaction, *16*:564

1-Butene (*Continued*)
  from 3-hexene and methacrylate, *15*:352
  isomerization, *15*:808
  in LPG, *14*:384
  lube oils from, *14*:499
  polymerization, *16*:476
  properties, *14*:384; *16*:475
2-Butene [*107-01-7*]
  from 1-butene, *15*:808
  lube oils from, *14*:499
cis-2-Butene [*590-18-1*]
  in LPG, *14*:384
  properties, *14*:384
trans-2-Butene [*624-64-6*]
  in LPG, *14*:384
  properties, *14*:384
2-Butenedioic acid [*110-17-8*], *13*:772
3-Buten-2-one [*79-94-4*]
  isoprene reaction, *13*:821
1-(2-Butenyl)-aziridine [*41104-25-0*], *13*:151
(1,2,3-η)-2-Butenyl iodonickel [*33728-90-4*]
  in isoprene polymerization, *13*:829
Butonate [*126-22-7*], *13*:441
2-(2-Butoxyethyl)-ethylthiocyanate [*112-56-1*]
  as insecticide, *13*:429
2-Butoxyethyl methacrylate [*13532-94-0*]
  properties, *15*:353
Butoxypolypropylene glycol [*9003-13-8*]
  as repellent, *13*:476
1-Butoxypropyl methacrylate [*76392-23-9*]
  properties, *15*:353
Butter, *15*:561
Buttermilk, *15*:561
Butternut, *16*:250
Butter oil, *15*:560
N-Butylacetanilide [*91-49-6*]
  as repellent, *13*:476
Butyl acetate [*123-86-4*]
  buoyancy immersion liquid, *16*:437
  in paint and varnish removers, *16*:764
Butyl acrylate [*141-32-2*]
  solubility of, *14*:83
Butyl alcohol [*71-36-3*]
  as odorant, *16*:298
tert-Butyl alcohol [*75-65-0*], *13*:22
  methacrylates from, *15*:357
Butylamine [*109-73-9*]
  odor-problem chemical, *16*:303
sec-Butylamine [*13952-84-6*], *13*:904
tert-Butylamine [*75-64-9*]
  ozone reaction, *16*:687
2-*t*-Butylaminoethyl methacrylate [*3775-90-4*]
  properties, *15*:347
2-*tert*-Butylanthraquinone [*84-47-9*]
  autoxidation, *13*:18
Butylated hydroxyanisole [*25013-16-5*]
  antioxidant, *13*:58
  economic aspects, *13*:53

  in pecan butter, *16*:266
Butylated hydroxy anisole
  (BHA) [*121-00-6*], *13*:45
Butylated hydroxytoluene [*128-37-0*]
  (BHT)
  antiaging effects, *15*:138
2-(*tert*-Butylazo)-2-methylpropionitrile [*25149-46-6*]
  half-life of, *13*:366
n-Butyl benzoate [*136-60-7*]
  hardening agent, *15*:479
n-Butyl-*sec*-butylmagnesium [*39881-32-8*]
  (DBM), *16*:560
t-Butylcatechol [*27213-78-1*]
  isoprene inhibitor, *13*:833
4-*tert*-Butylcatechol [*98-29-3*], *13*:45
n-Butyl chloride [*109-69-3*]
  lithium reaction, *14*:467
  magnesium reaction, *16*:560
sec-Butyl chloride [*78-86-4*]
  lithium reaction, *14*:469
tert-Butyl chloride [*507-20-0*]
  lithium reaction, *14*:470
tert-Butylcyanoketene [*29342-22-1*]
  properties, *13*:875
4-*tert*-Butylcyclohexanone [*98-53-3*] (4-TBC)
  hydroboration, *16*:565
cis-*p*-*tert*-Butylcyclohexyl acetate [*10411-92-4*], *16*:955
trans-*p*-*tert*-Butylcyclohexyl acetate [*1900-69-2*], *16*:955
N-*t*-Butyl-*N*-(diethylphosphono)
  methacrylamide [*76392-07-9*]
  properties, *15*:356
n-Butyl-6,6-dimethyl-5,6-dihydro-1,4-pyrone-2-carboxylate [*532-34-3*]
  as repellent, *13*:476
1,3-Butylene dimethacrylate [*1189-08-8*]
  properties, *15*:347
2,3-Butylene methacryloyloxyethyl borate
  [*76329-36-4*]
  properties, *15*:355
Butylene oxide [*26249-20-7*]
  HCl scavenger, *16*:763
  isomerization of, *13*:906
Butylenes
  dimerization, *16*:493
  heptenes from, *16*:493
  olefins from, *16*:480
  oxidation to maleic anhydride, *14*:783
n-Butyl(ethyl)magnesium [*62202-86-2*] (BEM), *14*:617, *16*:559
Butyl glycolate [*7397-62-8*], *13*:94
tert-Butyl hydroperoxide [*75-91-0*], *13*:363
  formation of, *13*:22
  as polymerization initiator, *14*:93

tert-Butylhydroquinone [*1948-33-0*], *13*:45, 60
n-Butyl isocyanate [*111-36-4*]
  properties, *13*:789
n-Butyllactamide [*3328-88-9*], *13*:90
Butyl lactate [*138-22-7*]
  properties, *13*:89
n-Butyl lactate [*138-22-7*], *13*:87, 89
Butyllithium [*109-72-8*]
  as initiator, *15*:193
  in isoprene polymerization, *13*:828
n-Butyllithium [*109-72-8*]
  in batteries, *14*:459
  mfg and properties, *16*:556
  properties, *14*:466
  sodium *tert*-butoxide reaction, *16*:558
sec-Butyllithium [*598-30-1*]
  mfg and properties, *16*:556
  properties, *14*:469
tert-Butyllithium [*594-19-4*]
  properties, *14*:470
n-Butyllithium–*N*,*N*,*N'*,*N'*-tetramethylethylene-
    diamine complex [*26880-57-9*]
  toluene reaction, *14*:469
Butyl methacrylate [*97-88-1*]
  in emulsion polymerization, *14*:84
n-Butyl methacrylate [*97-88-1*]
  properties, *15*:347
sec-Butyl methacrylate [*2998-18-7*]
  properties, *15*:353
tert-Butyl methacrylate [*585-07-9*]
  properties, *15*:353
2-*tert*-Butyl-4-methoxyphenol [*121-00-6*], *13*:45
3-*tert*-Butyl-4-methoxyphenol [*88-32-4*], *13*:45
t-Butyl 2-methyl-4-chlorocyclohexanecarboxylate
    [*12002-53-6*]
  as insect attractant, *13*:481
4-*n*-Butyl-2-methylenebutanedioate [*6439-57-2*]
  properties, *13*:871
n-Butyl-*n*-octylmagnesium [*69929-18-6*] (BOM),
    16:560
OO-tert-Butyl *O*-isopropyl monoperoxycarbonate
    [*2372-21-6*], *13*:361
4-Butyloxybenzal-4-ethylaniline [*29743-15-5*]
  liquid crystalline range of, *14*:399
tert-Butyl peracetate [*107-71-1*]
  half-life of, *13*:360
tert-Butyl perbenzoate [*614-45-9*]
  half-life of, *13*:360
tert-Butyl permanganate(VII), *14*:859
tert-Butyl peroxide [*110-05-4*], *13*:22. (See also
    Di-tert-butyl peroxide.)
tert-Butyl peroxyacetate [*107-71-1*], *13*:29
tert-Butyl peroxy-2-ethylhexanoate [*3006-82-4*]

half-life of, *13*:360
tert-Butyl peroxyneodecanoate [*748-41-4*]
  half-life of, *13*:360
tert-Butyl peroxypivalate [*927-07-1*]
  radical initiator, *13*:357
2-(*p-tert*-Butylphenoxy)cyclohexyl 2-propynyl
    sulfite [*2312-35-8*]
  as insecticide, *13*:463
o-sec-Butylphenyl methylcarbamate [*3766-81-2*],
    13:452
Butyl rubber [*9010-85-9*]
  gas permeability, *15*:118
  from isoprene, *13*:819
  weather resistance, *13*:547
Butyl rubber [*9003-18-3*] (Buna N, NBR)
  gasket properties, *16*:727
n-Butylsodium [*3525-44-8*], *16*:558
Butyl thiolmethacrylate [*54667-21-9*]
  properties, *15*:355
n-Butyltin trichloride [*1118-46-3*]
  properties, *16*:573
n-Butyraldehyde [*123-72-8*]
  aldol reaction, *16*:637
  oxo intermediate, *16*:639
γ-Butyrolactone [*96-48-0*], *13*:97; *14*:780
  in lithium batteries, *14*:459
  NaOH reaction, *15*:770
BWR. See *Boiling-water reactor*.
BX, *13*:584

C

C-103, *15*:830
CAB
  information service, *16*:902
Cabbage looper
  control of with pheromones, *13*:480
Cable
  insulation coverings for, *13*:564
  insulation of, *13*:554
Cable coverings
  lead alloys, *14*:142
Cables
  polyethylene, *16*:401
Cable sheathing
  lead alloys for, *14*:146
Cables, shipboard
  insulation for, *13*:571
Cabosil M-s, *14*:566
CABP. See *Calcium-binding protein*.
Cacodyl [*471-35-2*]
  iodoarsine synthesis, *13*:668
  methyl iodide reaction, *13*:669

δ-Cadinene [*483-76-1*], *16*:313
γ-Cadinene [*483-74-9*], *16*:313
10-α-Cadinol [*481-34-5*], *16*:313
Cadmium [*7440-43-9*]
  in coatings, *15*:241
  electrode, *13*:720
  in fuel reprocessing, *16*:182
  inducement of immunodeficiency, *13*:170
  in magnesium alloys, *14*:596
  from the ocean, *16*:280
  surface treatment of, *15*:304
  toxicity, *13*:264
Cadmium-106 [*14378-39-3*], *13*:841
Cadmium-108 [*14191-62-9*], *13*:841
Cadmium-110 [*14191-63-0*], *13*:841
Cadmium-111 [*14336-64-2*], *13*:841
Cadmium-112 [*14336-65-4*], *13*:841
Cadmium-113 [*14336-66-4*], *13*:841
Cadmium-114 [*14041-58-8*], *13*:841
Cadmium-116 [*14390-59-1*], *13*:841
Cadmium chloride [*10108-64-2*]
  lecithin reaction, *14*:256
  as solid lubricant, *14*:508
Cadmium(II) iodide [*7790-80-9*]
  geometry of, *13*:774
Cadmium iodide [*7790-80-9*]
  as solid lubricant, *14*:508
Cadmium molybdate [*13972-68-4*], *15*:687
Cadmium nitride [*12380-95-9*], *15*:872
Cadmium orange [*1306-23-6*]
  in inks, *13*:378
Cadmium red [*58339-34-7*]
  in inks, *13*:378
Cadmium salts
  olefin fiber dyes, *16*:361
Cadmium sulfide [*1306-23-6*]
  HDPE stabilizer, *16*:446
Cadmium yellow [*1306-23-6*]
  in inks, *13*:378
Cadra cautella, *13*:481
Cadra figulilella, *13*:481
Caffeine [*58-08-2*], *15*:907
CAIN information system, *13*:290
Calamenene [*22339-23-7*], *16*:313
Calcimatic kiln, *14*:366
Calcination
  of limestone, *14*:360
  nuclear wastes, *16*:210
Calciretard, *15*:760
Calcite [*13397-26-7*]
  in limestone, *14*:344
  from the ocean, *16*:286
  thermal decomposition, *16*:337
Calcitonin [*21215-62-3*]
  in blood-Ca²⁺ *15*:584
  in calcium homeostasis, *15*:586

Calcium [*7440-70-2*]
  in coatings, *15*:252
  electrode, *13*:720
  lead alloys, *14*:148
  in limestone, *14*:344
  measurement, *15*:80
  nutrient properties, *15*:571
  from the ocean, *16*:279, 280, 292
  titration procedure, *13*:731
  in walnuts, *16*:261
Calcium-40 [*14092-94-5*], *13*:840
Calcium-42 [*14333-05-2*], *13*:840
Calcium-43 [*14333-06-3*], *13*:840
Calcium-44 [*14255-03-9*], *13*:840
Calcium-46 [*13981-77-6*], *13*:840
Calcium-48 [*13981-76-5*], *13*:840
Calcium acetate [*62-54-4*]
  in grease thickeners, *14*:502
Calcium acetate monohydrate [*5743-26-0*]
  toxicity, *15*:573
Calcium alginate [*9000-35-0*]
  encapsulation with, *15*:474
  in pancreas cell encapsulation, *15*:490
Calcium aluminum oxide [*12005-50-4*] (1:12:19)
  in phosphors, *14*:541
Calcium antimonide [*60649-94-7*] (1:2), *14*:151
Calcium antimonide [*53095-92-4*] (2:1), *14*:151
Calcium antimonide (3:1), *14*:149
Calcium antimonide [*11062-30-9*] (3:2), *14*:151
Calcium antimonide [*52037-42-0*] (5:3), *14*:151
Calcium arsenate [*7784-44-1*]
  carrier of pesticides, *14*:380
  as insecticide, *13*:414, 421
  in lead refining, *14*:117
Calcium arsenate, basic [*74966-85-1*]
  as insecticide, *13*:421
Calcium arsenide [*39372-67-3*] (1:1), *14*:151
Calcium arsenide [*59763-73-4*] (1:3), *14*:151
Calcium arsenide [*54065-14-4*] (2:1), *14*:151
Calcium arsenide [*12255-53-7*] (3:2), *14*:151
Calcium arsenide [*59094-39-4*] (5:3), *14*:151
Calcium aspartate [*10389-09-0*], *15*:760
Calcium-binding protein, *15*:585
Calcium bisulfite [*13780-03-5*]
  in sulfite pulping, *14*:304
Calcium carbide [*75-20-7*]
  from lime, *15*:944
  lime in mfg, *14*:380
Calcium carbonate [*13397-26-7*]
  electric insulation filler, *13*:565
Calcium carbonate [*471-34-1*]
  in blast furnace flux, *13*:740
  from calcium cyanamide, *15*:944
  in case hardening, *15*:315
  as filler, *13*:558
  filler in plastics, *13*:973
  in inks, *13*:378
  in limestone, *14*:344

in nondestructive testing, *16*:56
paper filler, *16*:812
paper pigment, *16*:777
removal with hydroxyacetic acid, *13*:94
from seawater, *16*:280
from shells, *16*:285
toxicity, *15*:573
Calcium caseinate, *15*:568
lime in refining of, *14*:380
Calcium chloride [*10043-52-4*]
in lithium recovery, *14*:453
as retention aid, *16*:806
toxicity, *15*:573
Calcium compd with magnesium [*12133-32-3*] (1:
2), *14*:603
Calcium cyanamide [*156-62-7*]
from calcium carbide, *15*:943
Calcium dicesium hexakiscyanoferrate [*15415-35-7*], *13*:767
Calcium difluoride [*7789-75-5*]
as solid lubricant *14*:508
Calcium dipotassium hexakiscyanoferrate
[*20219-00-5*], *13*:767
Calcium dithiophosphate
lubricant antioxidant, *14*:492
Calcium ferrate [*35764-67-1*], *13*:777
Calcium ferrite [*12013-33-1*], *13*:776
Calcium fluoride [*7789-75-5*]
filler in gaskets, *16*:732
optical-filter coating, *16*:526
toxicity, *15*:573
Calcium gluconate [*299-28-5*]
in iron(II) gluconate synthesis, *13*:774
oxalic acid poisoning treatment, *16*:629
Calcium l-glutamate [*19238-49-4*], *15*:767
Calcium hexakiscyanoferrate [*13821-08-4*], *13*:767
Calcium hydroxide [*1305-62-0*]
in hair removal from cowhide, *14*:209
from lime, *14*:344
lithium carbonate reaction, *14*:463
pH standard, *13*:3
Calcium hypochlorite [*7778-54-3*], *13*:229
lead dross treatment, *14*:157
odor control with, *16*:303
preparation from lime, *14*:380
Calcium iodate [*7789-80-2*]
feed additive, *13*:673
Calcium lactate [*814-80-2*], *13*:85, 88
lime in refining of, *14*:380
oxalic acid poisoning treatment, *16*:629
Calcium–magnesium–bismuth compound
[*12233-18-0*] (1:2:2), *14*:124
Calcium malonate [*19455-76-6*]
purification of malonic acid, *14*:795
Calcium molybdate [*7789-82-4*], *15*:687
toxicity, *15*:573, 679

Calcium monosilicide [*12049-73-9*]
polysilane from, *13*:408
Calcium naphthenate [*61789-36-4*], *15*:752
Calcium nitride [*12013-82-0*], *15*:872
Calcium nitride silicide [*12356-60-4*], *15*:935
Calcium oxalate [*563-72-4*]
in body tissues, *16*:630
stone formation, *15*:583
Calcium oxide [*1305-78-8*]
in glass electrode, *13*:5
iodine reaction, *13*:653
in iron-ore processing, *13*:740
in lignitic coal, *14*:334
lime, *14*:344
in spent shale, *16*:351
Calcium paraperiodate [*13763-58-1*], $Ca_5(IO_6)_2$,
13:653, 667
Calcium pentasulfide [*1332-68-9*]
in insecticide, *13*:422
Calcium perferrite [*12524-96-8*], *13*:776
Calcium permanganate tetrahydrate [*7789-81-3*],
14:845
Calcium peroxide [*1305-79-9*], *14*:356
Calcium peroxide, cmpd with hydrogen peroxide
[*55396-26-4*]
dehydration, *16*:680
oxygen generation, *16*:680
Calcium potassium hexakiscyanoferrate [*20219-00-5*] (1:2:1), *13*:767
Calcium propionate [*4075-81-4*]
preservative, *13*:247
Calcium silicate [*1344-96-3*]
building material, *14*:376
filler in plastics, *13*:973
as insulation, *13*:597
as thermal insulator, *13*:592
Calcium silicide [*12013-56-8*]
nitrogen reaction, *15*:934
Calcium silicon nitride [*12213-72-8*], *15*:934
Calcium stannate [*12013-46-6*]
in lead refining, *14*:117
Calcium stearate [*1592-23-0*]
lime in refining of, *14*:380
LLDPE stabilizer, *16*:395
in papermaking, *16*:823
Calcium stearoyl-2-lactylate [*5793-94-2*], *13*:86
Calcium sulfate [*7778-18-9*]
in control of nitration ratios, *15*:929
in lithium recovery, *14*:453
paper pigment, *16*:789
in pulping, *14*:306
Calcium sulfide [*20548-54-3*]
in in iron ore processing, *13*:742
Calcium sulfite [*10257-55-3*]
in insecticide, *13*:422
from pulping liquor, *14*:306

Calcium sulfoaluminate.   See *Satin white.*
Calcium sulfonates
    as rust inhibitors, *14*:493
Calcium superoxide [*12133-35-6*]
    oxygen generation, *16*:679
Calcium tartrate [*15808-04-5*]
    solubility, *13*:113
    in tartaric acid manufacture, *13*:114
    from wine fermentation, *13*:114
Calcium tartrate tetrahydrate [*5892-21-7*]
    in tartaric acid production, *13*:118
Calcium tetrasulfide [*1332-67-8*]
    in insecticide, *13*:422
Calcium thiosulfate [*10124-41-1*]
    in insecticide, *13*:422
Calcium tungstate [*7790-75-2*]
    x-ray phosphor, *14*:544
Calculators
    ICS in, *13*:646
Calgon pickle
    hides, *14*:213
Californium nitride [*70420-93-8*], *15*:872
Callitroga hominovorax, *13*:471
Calomel [*10112-91-1*], *13*:422; *15*:144, 159.   (See
    also *Mercurous chloride.*)
Calomel electrode, *13*:6
Calrod
    heating elements, *14*:636
Calutrons, *13*:844
Cambric
    in electric insulation, *13*:578
CAMP.   See *Cyclic adenosine monophosphate.*
Camphene [*79-92-5*], *16*:312
    toxaphene from, *13*:433
Camphor [*76-22-2*], *16*:329
    from basil oil, *16*:321
Camphor oil [*8008-51-3*], *16*:321
Canada deuterium–uranium reactor.   See
    CANDU reactor.
Cananga odorata Hook f., *16*:322
Cananga odorata Hook f. et Thomson, *16*:329
Canarium buconicum, *16*:250
Canarium commune, *16*:250
Cancer
    immunotherapeutic agents for, *13*:167
    from *N*-nitrosamines, *15*:988
    nutrients associated with, *15*:573
    and radiation exposure, *16*:220
Cancer therapy
    microwave techniques for, *15*:514
Candlenut, *16*:251
Candu-PHW reactors
    cladding, *15*:830
CANDU reactor, *16*:162
Canizzaro reaction
    salicylaldehyde, *13*:71

Canning
    of meat products, *15*:69
CAN–OLE information system, *13*:292
Cans
    solders for, *14*:154
Capacitance
    in level measurement, *14*:431
Capacitors
    mica films in, *15*:418
    nondestructive testing of, *16*:57
Capacitor tissue, *16*:772
CAPD.   See *Continuous ambulatory peritoneal
    dialysis.*
Capilair, *14*:248
Capri system
    for patent information, *16*:922
Caproic acid [*142-62-1*]
    as insect attractant, *13*:481
Caprolactam [*105-60-2*]
    from toluene, *13*:876
ε-Caprolactam [*105-60-2*]
    from nitrocyclohexane, *15*:979
Caprolactone [*502-44-3*]
    from cyclohexanone, *13*:922
Caproyl chloride [*142-61-0*]
    resorcinol reaction, *13*:44
Capsicum [*8023-77-6*], *16*:314
Captan [*133-06-2*], *14*:788
Car accessories
    from polypropylene, *16*:468
Caraway oil [*8000-42-8*], *16*:322
Carbamates
    as insecticides, *13*:453
Carbamazipine [*294-46-4*]
    properties of, *13*:135
Carbamoyl chloride [*433-72-9*]
    in ethinamate synthesis, *13*:126
Carbanilide [*102-07-8*]
    isocyanate reaction product, *13*:798
Carbaryl [*63-25-2*], *13*:453; *15*:729
    as DDT replacement, *15*:714
    toxicity, *13*:264
Carbenicillin [*4697-36-3*]
    microbial synthesis, *15*:463
Carbide [*12070-13-2*], *14*:515
Carbide lime, *14*:347
Carbide tips
    nitride coating of, *15*:878
Carbitol [*111-90-0*]
    in lecithin manufacture, *14*:260
Carbodiimides
    from isocyanates, *13*:797
Carbofuran [*1563-66-2*], *13*:62, 453
Carbohydrates
    in cattlehide, *14*:204
    in nuts, *16*:249, 252

Carbon [*7440-44-0*]
  affect on iron magnetism, *13*:738
  emissivity, *13*:342
  encapsulation, *15*:478
  fiber properties, *13*:970
  ion implantation, *13*:717
  in lignite, *14*:327
  niobium pentoxide reaction, *15*:826
  from the ocean, *16*:279
  use in detoxification, *15*:489
Carbon-12 [*7440-44-0*], *13*:840
Carbon-13 [*14762-74-4*], *13*:840
Carbon-14 [*14762-75-5*]
  disposal of, *16*:208
  radiation exposure from, *16*:217
Carbon black [*1333-86-4*]
  electric insulation filler, *13*:565
  filler in plastics, *13*:973
  as grease thickener, *14*:502
  in HDPE, *16*:428
  HDPE stabilizer, *16*:446
  in inks, *13*:377
  in paper, *16*:779
  toxicity, *13*:264
  uv stabilizer for olefin fibers, *16*:361
Carbon blacks
  in magnetic tape, *14*:742
Carbon dioxide [*124-38-9*]
  absorption, *16*:680
  in air, *16*:657
  from carbon monoxide, *13*:665
  electrode, *13*:720
  in fumigation, *13*:466
  as insulator, *13*:549
  lasers, *14*:44; *16*:62
  as leak tracer, *16*:55
  lithium hydroxide reaction, *14*:463
  as lubricant, *14*:514
  metabolic transport, *15*:588
  permeability in polymers, *15*:118
  removal by molecular sieves, *15*:663
  toxicity, *13*:264
Carbon dioxide laser, *14*:51
Carbon dioxide lasers
  in fusion research, *14*:77
Carbon dioxide scrubbers, *16*:679
Carbon disulfide [*75-15-0*]
  aziridine reaction, *13*:155
  as fumigant, *13*:466
  fumigation of nuts with, *16*:263
  nitro alcohol reaction, *15*:913
  odor-problem chemical, *16*:303
Carbon fibers
  from novoloids, *16*:136
Carbon–graphite
  coefficient of friction, *14*:481
Carbonic acid [*463-79-6*]
  in ion-exchange removal, *13*:687

Carbonic anhydrase [*901-03-0*]
  zinc in, *15*:590
Carbon-in-gelatin filters, *16*:528
Carbonitrides, *15*:877
Carbonitriding, *15*:316
Carbonium ion salts
  as initiator catalysts, *13*:371
Carbonization
  of brown coal, *14*:334
Carbonless copy paper, *15*:477, 480
Carbon monoxide [*630-08-0*]
  copolymer with ethylene, *16*:420
  in DR processes, *13*:754
  ethylene copolymer, *13*:933
  in iron-ore processing, *13*:741
  isotope separation of, *13*:847
  metallocene insertion, *16*:610
  metallocene reaction, *16*:613
  methanol from, *15*:403
  nitro group reaction, *13*:807
  oxalic acid from, *16*:627
  oxidation with iodine pentoxide, *13*:665
  toxicity, *13*:265
Carbon-13 monoxide [*1641-69-6*]
  in isotope separation, *13*:849
Carbon nitride [*12069-92-0*], *15*:872
1008 Carbon steel
  in explosive cladding, *15*:278
Carbon steels
  cladding of, *15*:281
Carbon suboxide [*504-64-3*]
  from malonic acid, *14*:794
  polymerization, *13*:879
  properties, *13*:875
Carbon subsulfide [*627-34-9*]
  properties, *13*:875
Carbon tetraboride [*558-13-4*]
  niobium boride from, *15*:834
Carbon tetrabromide [*558-13-4*]
  isoprene reaction, *13*:825
Carbon tetrachloride [*56-23-5*]
  in etch gas, *13*:639
  as fumigant, *13*:466
  in initiator systems, *13*:367
  in isoprene reaction, *13*:825
  toxicity, *13*:265
Carbonylation
  of methanol, *16*:584
Carbonyl-1,1'-bis(aziridine) [*1192-75-2*], *13*:143
Carbonyl difluoride [*353-50-4*]
  trifluoroacetonitrile reaction, *13*:800
Carbonylhydrobis(triphenylphosphine)platinum,
  trichlorostannate [*56237-13-9*]
  oxo process catalyst, *16*:639, 642
Carbonyl sulfide [*463-58-1*]
  removal of from LPG, *14*:389

Carbonyltris(phosphorustrifluoride)nickel
    [*14219-40-0*], *15*:807
Carbonyltris(trimethyphosphite)nickel [*17084-
    87-6*], *15*:807
Carbophenothion [*786-19-6*], *13*:443
Carbophenothion methyl [*953-17-3*]
    as insecticide, *13*:443
1,7-Carborane [*16986-24-6*], *13*:404
Carboxbenzylpyroglutamate-*N*-benzylhistidine
    prolamide [*32926-30-0*]
    in TRH synthesis, *15*:782
Carboxbenzylpyroglutamic acid [*1155-62-0*]
    in TRH synthesis, *15*:782
Carboxyesterase [*9016-18-6*], *13*:474
2-Carboxyethyl methacrylate [*13318-10-0*]
    properties, *15*:354
Carboxyhemoglobin
    methylene chloride-induced, *16*:767
Carboxylic acids
    nitriles from, *15*:893
Carboxymethyl cellulose [*9000-11-7, 9004-32-4*]
    in paper processing, *16*:780
2-(*N*-Carboxymethyl-*N*,*N*-dimethylamino)ethyl
    methacrylate chloride [*45156-26-1*], *15*:352
Carboxymethyl methacrylate [*76391-99-6*]
    properties, *15*:354
α-Carboxymethyl β-sulfosuccinic acid [*41344-35-
    8*]
    as soap builder, *13*:92
Carboxypeptidase A [*11075-17-5*]
    zinc in, *15*:590
Carboxypeptidase b [*9025-24-5*]
    zinc in, *15*:590
Carboys, *16*:722
Carbrital, *13*:122
Carbromal [*77-65-6*]
    properties of, *13*:122
Carburizing
    in case hardening, *15*:313
Carbutamide [*339-43-5*]
    antidiabetic agent, *13*:614
Carcinogen
    catechol as, *13*:56
Carcinogenicity
    of cyclodienes, *13*:437
    of fumigants, *13*:469
    nitrosamines, *15*:991
Carcinogens
    aflatoxins, *16*:263
    Ames test, *13*:261
    NIOSH list, *13*:261
    N-nitrosamines, *15*:988
    paint and varnish removers, *16*:767
    shale oil products, *16*:352
Carcinoma
    diagnosis, *15*:78

Cardamom oil [*8000-66-6*], *16*:322
Cardanol [*37330-39-5*]
    in cashew shells, *16*:271
Cardiac enzymes
    diagnostic tests, *15*:75
Carding, *16*:111
Cardol [*16737-82-9*]
    in cashew, *16*:259
δ-3-Carene [*13466-78-9*], *16*:313
Cargo loading
    queuing theory applied to, *16*:520
Carius method
    iodine determination, *13*:659
Carnallite [*1318-27-0*], *14*:615
    dehydration of, *14*:576
    magnesium from, *14*:582
    magnesium ore, *14*:570
    from the ocean, *16*:281
Carnauba wax
    in packings, *16*:734
β-Carotene [*7235-40-7*], *14*:471
    in palm oil, *16*:253
Carpet
    production from polypropylene, *16*:383
Carpet backing
    latexes for, *13*:871
Carpeting
    as acoustic material, *13*:519
Carpets
    fibers for, *16*:376
    nonwoven, *16*:123
Carrageenin salts
    added to milk, *15*:554
Carriers
    for odor modifiers, *16*:301
CARS. See *Coherent anti-Stokes Raman
    spectroscopy.*
Cars, electric
    lithium batteries for, *14*:450
Cartons, folding, *16*:798
Cartridge, recording, *14*:733
Carum carvi L., *16*:322
Carus cell, *14*:874
Carvacrol [*499-75-2*], *16*:325, 328
Carvone [*99-49-0*], *16*:312
d-Carvone [*2244-16-8*]
    from caraway, *16*:322
l-Carvone [*6485-40-1*]
    from spearmint, *16*:328
Carya illinoensis, *16*:250
Caryophyllene [*87-44-5*], *16*:313
Cascade theory
    in isotope separation, *13*:839
CA SEARCH information system, *13*:303
Case hardening, *15*:313

Casein [*9000-71-9*], *15*:568
  paper binder, *16*:789
  in papermaking, *16*:822
  paper sizing, *16*:778
Casein glue
  in laminated wood, *14*:34
Cashew, *16*:250
Cashew nutshell oil, *16*:271
Cassella acid [*131-27-1*], *15*:731.   (See also *3-*
  Amino-1,5-naphthalenedisulfonic acid.)
Cassettes
  magnetic tape in, *14*:732
Cassia [*8007-80-5*], *16*:307
Cassia oil [*8007-80-5*], *16*:322
Cassia, redistilled oil [*8007-80-5*], *16*:320
Cassiterite [*1317-45-9*], *16*:285.   (See also *Tin*
  oxide.)
  from beach sand, *16*:289
  columbites with, *15*:822
  from the ocean, *16*:285
Castanea crenata, *16*:250
Castanea dentata, *16*:250
Castanea mollissima, *16*:250
Castanea pumila, *16*:250
Castanea sativa, *16*:250
Castanin, *16*:249
Cast epoxy
  thermal conductivity of, *13*:544
Casting
  lead–antimony alloys for, *14*:148
  liquid metal for fibers, *15*:229
  methacrylate sheets, *15*:385
  plastic pipe, *13*:976
Castings
  of magnesium alloys, *14*:596
Cast iron, *13*:736
Castoreum [*8023-83-4*]
  catechol in, *13*:46
  in perfumes, *16*:950
Castor oil [*8001-79-4*]
  in grease thickeners, *14*:502
  in packings, *16*:734
  in varnishes, *13*:554
  as vehicle, *13*:395
Castung 403UV, *14*:567
Catalan furnace, *13*:735
Catalase [*9001-05-2*]
  iron in, *15*:595
Catalysis
  by iodides, *13*:671
  by ion-exchange resins, *13*:702
  by metal complexes, *16*:610
  shape-selective, *15*:651
Catalysts, *13*:355.   (See also *Initiators.*)
  1-alkylaziridine, *13*:162
  for butane oxidation, *14*:783
  for isobutylene oxidation, *15*:362

  isoprene polymerization, *13*:829
  mercury, *15*:154
  molybdenum compounds, *15*:693
  molybdenum for denitrification, *15*:694
  naphthenates, *15*:752
  nickel-containing, *15*:812
  nitration, *15*:844
  for nitrogen reduction, *15*:955
  organotins as, *16*:579
  organotransition metals, *16*:583
  for oxo process, *16*:637
  zeolites, *15*:650, 664
Catalytic combustion
  in NO_x abatement, *15*:861
Catalytic incineration, *13*:203
Catechin [*120-80-9*]
  catechol from, *13*:39
Catechol [*120-80-9*], *13*:39.   (See also
  Hydroquinone, resorcinol, and catechol.)
  in (−) adrenaline synthesis, *15*:757
  economic aspects, *13*:53
  microbial transformation to L-DOPA, *15*:464
  from phenol, *13*:29
  properties, *13*:40
  vanadium complex for N_2 reduction, *15*:959
  vegetable tannins derived from, *14*:214
Catecholamine [*51-43-4*]
  in memory enhancement, *15*:133
Catechol phosphonate [*934-35-0*]
  plant growth stimulant, *13*:63
1,2-Catechol sulfate [*4074-55-9*]
  in urine, *13*:46
Catechu [*8001-48-7*]
  catechol from, *13*:39
Cathartic
  magnesium sulfate as, *14*:640
Cation-exchange membranes
  in chlor-alkali production, *15*:120
Cationic polymerization, *13*:371
Caustic soda [*1310-73-2*]
  paint and varnish stripping, *16*:762
Cauterization
  by lasers, *14*:71
CB.   See *Citizen's radio bands.*
CBD.   See *Dichlorotetramethylcyclobutene.*
CBPI information system, *13*:328
CC.   See *Cholecalciferol; Vitamin D₃.*
CCD.   See *Charge-coupled devices.*
CDB-90, *13*:230
CDT.   See *Cyclododecatriene.*
CECE.   See *Combined electrolysis and chemical*
  exchange.
Cedarwood
  odor compensation of, *16*:303
Cedarwood oil texas [*68990-83-0*], *16*:322
Cedrone [*3477-88-3*], *13*:930
CEE information system, *13*:328

Celcon, *16*:377
  in electrical insulation, *13*:547
Celestite [*14291-02-2*]
  in ferrite mfg, *14*:675
Celestolide [*13171-00-1*]
  in perfumes, *16*:968
Celestra, *16*:78
CELEX information system, *13*:328
Celguard, *15*:104
Celite 545, *15*:929
Cell-mediated immune system, *13*:168
Cell membrane
  liquid crystallinity of, *14*:419
  smecticlike vesicles in, *14*:406
Cellobiose [*528-50-7*]
  in microbial syntheses saccharides, *15*:468
Cellophane [*9005-81-6*]
  lactic acid used for, *13*:87
  in magnetic tape, *14*:734
Cells, living
  ion exchange in, *13*:678
Cellulose [*9004-34-6*]
  analysis in paper, *16*:793
  biodegradation of, *13*:225
  enzymatic conversion of, *14*:309
  methacrylate grafting, *15*:390
  nitric acid reaction, *15*:856
  nonwovens from, *16*:106
  in nuts, *16*:252
Cellulose acetate [*9004-35-7*]
  in CPK measurement, *15*:77
  fiber membranes, *15*:108
  gas permeability, *15*:118
  insulating properties of, *13*:556
  as ion-exchange support, *13*:725
  in laminated glass, *13*:978
  in magnetic tape, *14*:733
  moisture regain, *16*:358
  nonwoven fabrics from, *16*:108
  properties, *16*:358
Cellulose acetate butyrate [*9004-36-8*]
  in contact lenses, *15*:126
  in encapsulation, *15*:478
  stability, *15*:382
Cellulose diacetate [*9035-69-2*]
  asymmetric membranes of, *15*:106
Cellulose esters
  membrane formation, *15*:104
Cellulose fiber
  acoustic performance of, *13*:518
  in composites, *13*:971
Cellulose fibers
  in composites, *13*:970
Cellulose nitrate [*9004-70-0*]
  effect of malic acid on, *13*:110
  etching of, *16*:834
  in laminated glass, *13*:978
  as magnetic tape binders, *14*:741
  as track-etch detector, *16*:827

Cellulose tricarbanilate
  in paper analysis, *16*:793
Cellulose xanthate [*9032-37-5*]
  in bonding nonwovens, *16*:119
Celontin, *13*:135
Celotex, *14*:4
Cement
  insulating, *13*:597
  retarder, *13*:93
Cementation coatings, *15*:247
Cementite [*12169-32-3*], *15*:340
Cements
  magnesium oxychloride, *14*:624
  oxysulfate, *14*:638
Cementstone, *14*:344
Central nervous system (CNS)
  neuroregulators in, *15*:754
Central Patents Index
  information system, *16*:905
Centrifugal extrusion
  in microencapsulation, *15*:473
Centrifugal mixers, *15*:629
Centrifuge
  in fuel reprocessing, *16*:179
Centrophenoxine [*51-68-3*], *15*:138.   (See also
      Meclofenoxate(III).)
α-Cephalin [*5681-36-7*], *14*:406.   (See also
      Phosphatidylethanolamine.)
  in commercial lecithin, *14*:250
  in nuts, *16*:253
Cephalosporins
  microbial hydrolysis of, *15*:463
Ceramic 1
  magnetic properties, *14*:671
Ceramic 2
  magnetic properties, *14*:671
Ceramic 3
  magnetic properties, *14*:671
Ceramic 4
  magnetic properties, *14*:671
Ceramic 5
  magnetic properties, *14*:671
Ceramic 6
  magnetic properties, *14*:671
Ceramic 7
  magnetic properties, *14*:671
Ceramic 8
  magnetic properties, *14*:671
Ceramic fibers
  gasket properties, *16*:727
Ceramic magnets, *14*:673
Ceramics
  glass as lubricant, *14*:520
  lithium salts in, *14*:460
  nickel compounds in, *15*:813
  nondestructive testing of, *16*:52
  novoloids fibers in, *16*:135

Ceratocystis ulmi, 13:482

Cereal rusts
control by NiSO₄, 15:814

Cerebrocuprein
copper in, 15:588

Cerex, 16:76

Ceric sulfate [24670-27-7]
reduction by hydrogen peroxide, 13:16
resorcinol reaction, 13:41

Cerium-136 [15758-67-5], 13:841

Cerium-138 [15758-26-6], 13:841

Cerium-140 [14191-73-2], 13:841

Cerium-142 [14119-20-1], 13:841

Cerium carbide [12012-32-7]
nitrogen reaction, 15:934

Cerium, compd with cobalt [12214-13-0] (1:5), 14:676

Cerium, compd with cobalt [12014-88-9] (2:17), 14:676

Cerium fluoride [13400-13-0]
in fluoride electrode, 13:727

Cerium(III) [18923-26-7]
phosphor activator, 14:540

Cerium(IV) oxide [1306-38-3]
phosphors from, 14:541

Cerium magnesium aluminum oxide [55070-88-7] (1:1:11:19)
in phosphors, 14:541

Cerium nitride [25764-08-3], 15:872, 934

Cerium terbium magnesium aluminate (0.65:0.35:1:11:19)
phosphor, 14:527

Ceruloplasmin [9031-37-2], 15:595
copper in, 15:588

Cerussite [14476-15-4], 14:99

Cesium [7440-46-2]
amalgamation with mercury, 15:147
as lubricant, 14:515
oxidation, 16:656

Cesium-133 [7440-46-2], 13:841

Cesium-137 [10045-97-3]
in level measurement, 14:437
in nuclear waste, 16:209

Cesium fluoride [13400-13-0]
in isocyanate synthesis, 13:800

Cesium manganate(VI) [25583-22-6], 14:859

Cesium molybdate [13597-64-3], 15:687

Cesium nitride [12134-29-1], 15:872

Cesium oxide [20281-00-9]
in glass electrode, 13:5

Cesium ozonide [12053-67-7], 16:688

Cesium permanganate [13456-28-5], 14:845

Cetyl alcohol [36653-82-4]
solvent purification, 15:983

Cetylpyridinium chloride (CPC) [123-03-5]
applications of, 13:234

Cetyltrimethyl–ammonium bromide [57-09-0]
applications of, 13:234

Cevadilline [1415-76-5], 13:427

Cevadine [62-59-9], 13:427

Cevine [124-98-1]
insecticides from, 13:427

CFM. See *Dichlorodifluoromethane.*

CFR. See *Code of Federal Regulations.*

CFT. See *Crystal-field theory.*

Chabazite [12251-32-0], 15:639

Chagas' disease, 13:413

Chain-transfer constants
methacrylate in various solvents, 15:387

Chalcopyrite [1308-56-1], 15:796

Chalk, 14:344

Chamois
oil tanning of, 14:217

Chamosite
as iron source, 13:748

Chance-Pilkington filters, 16:535

Change-can mixers, 15:631

Chanoclavine [2390-99-0]
microbial synthesis of, 15:465

Chapman-Enskog kinetic theory, 14:924

Char
in DR processes, 13:754

Charcoal, 15:85
in case hardening, 15:315
mercury adsorbent, 15:168

Charge-coupled devices (CCD), 14:686

Charge-transfer complexes
of iodine, 13:650
as lubricants, 13:672

Chariots
noise of, 16:2

Chase econometrics, 13:326

Chavicol [501-92-8]
from bay oil, 16:321

Chavicol methyl ether [140-67-0]
from basil, 16:321

ChE. See *Cholinesterase.*

Check weighing, 16:55

Cheese, 15:562
nitrosamines in, 15:993

Chelafrin, 15:757

Chelating agent
malic acid, 13:106
tartaric acid, 13:113

CHEMCON information systems, 13:291

CHEMDEX information system, 13:304

CHEMDOC
information system, 16:905

Chemetals process
for manganese dioxide, 14:862

Chemfets, 13:729

Chemical Abstracts Service, 13:278, 292
information system, 16:903

Chemical Abstracts Service (*Continued*)
  registry numbers, *13*:280
Chemical bright dips, *15*:308
Chemical conversion coatings, *14*:509
Chemical Information System (NIH–EPA), *13*:
  300
Chemical lasers, *14*:60
Chemical literature.   See *Information retrieval.*
Chemical nomenclature, *16*:28
Chemical reactions
  laser-induced, *14*:73
Chemicals from brine, *16*:277.   (See also *Ocean
  raw materials.*)
Chemical toilets
  odor control agent, *15*:915
Chemical vapor deposition
  metallic coatings, *15*:262
  nitrides, *15*:877
Chemical vapor deposition (CVD)
  in IC manufacture, *13*:636
  in LED and semiconductor laser manufacture,
    14:287
  in thin-film fabrication, *14*:693
Chemiluminescence
  of phosphors, *14*:527
Chemithermomechanical pulp (CTMP), *16*:770
CHEMLINE information system, *13*:304
CHEMNAME information system, *13*:304
Chemosterilants
  in insect control, *13*:472
Chemosterilization
  of insects with aziridines, *13*:162
Chemotherapeutics, *13*:162
Chemotherapy
  immunotherapeutic agents, *13*:179
Cherry-Crandall unit
  lipase activity, *15*:80
Chestnut, *16*:250
  as vegetable tannin, *14*:214
Chicago acid [*82-47-3*], *15*:740.   (See also *4-
  Amino-5-hydroxy-1,3-
  naphthalenedisulfonic acid.*)
China clay
  paper uses, *16*:777
China woodnut, *16*:251
Chinawood oil [*8007-00-9*]
  in inks, *13*:381
Chinese chestnut
  protein content, *16*:249
Chinquapin nut, *16*:250
Chipboard, *14*:17
Chipcore, *14*:17
Chirauli nut, *16*:252
Chisel test, *15*:286
Chitin [*1398-61-4*]
  diflubenzuron effect on, *13*:459

Chitin synthesis inhibitors, *13*:459
Chitin synthetase [*9030-18-6*]
  diflubenzuron effect on, *13*:459
Chitosan [*9012-76-4*]
  in papermaking, *16*:818
Chloral hydrate [*302-17-0*]
  properties of, *13*:122
Chlorapatite [*12345-16-3*], *14*:857
Chlorate candle, *16*:673
Chlorate cells
  metal anodes for, *15*:180
Chlorates
  oxygen generation from, *16*:673
Chlorazepate dipotassium [*57109-90-7*]
  properties of, *13*:122
Chlorbenside [*103-17-3*], *13*:463
β-Chlordane [*12789-03-6*], *13*:434
Chlordecone [*143-50-0*]
  from mirex, *13*:436
Chlordene [*3734-48-3*], *13*:434
Chlordiazepoxide [*58-25-3*], *13*:128
Chlordimeform [*6164-98-3*], *13*:463
Chlorendic anhydride [*115-27-5*], *14*:788
Chlorfenvinphos [*470-90-6*], *13*:443
Chloride of lime [*7778-54-3*], *16*:265.   (See also
  Calcium hypochlorite.)
  in bleaching nuts, *16*:265
Chloride shift
  in CO_2 transport, *15*:588
Chlorinated hydrocarbons
  in paint and varnish removers, *16*:762
Chlorinated polyethylene
  in cable coverings, *13*:585
Chlorinated rubber
  as binder, *13*:393
  corrosion resistant paint, *16*:755
Chlorinated TSP [*56802-99-4*], *13*:229
Chlorination
  initiators used in, *13*:356
Chlorination–dehydrochlorination process
  paraffins to olefins, *16*:492
Chlorine [*7782-50-5*], *13*:229
  from chlorate, *16*:675
  diaphragm cell production of, *15*:120
  mercury reaction, *15*:159
  metabolic functions, *15*:587
  metal anodes in mfg, *15*:173
  nutrient properties, *15*:571
  from the ocean, *16*:279
  pulp bleaching, *16*:770
  reduction by hydrogen
    peroxide, *13*:16
  toxicity, *13*:265
Chlorine-35 [*13981-72-1*], *13*:840
Chlorine-37]13981-73-2], *13*:840
Chlorine (Cl_2) [*7782-50-5*]
  toxicosis from, *15*:598
Chlorine compounds
  in lubricants, *14*:479

Chlorine dioxide [*10049-04-4*]
    from chlorate, *16*:675
    odor control with, *16*:303
    pulp bleaching, *16*:770
    in water purification, *13*:229
Chlorine oxide [*10049-04-4*] (1:2), *16*:675
Chlorine trifluoride [*7790-91-2*], *16*:732
    uranium reaction, *16*:181
Chloroacetic acid [*79-11-8*]
    2-dimethylaminoethyl methacrylate reaction,
        15:356
    NaCN reaction, *15*:907
γ-Chloroacetoacetyl chloride [*41295-64-1*]
    from diketene, *13*:884
Chloroacetone [*78-95-5*]
    o-aminophenylglyoxylic acid reaction, *13*:221
    aziridine reaction, *13*:148
4-Chloroacetylcatechol [*99-40-1*], *13*:45
    in (−) adrenaline synthesis, *15*:757
Chloroacetyl chloride [*79-04-9*]
    in (−)-adrenaline synthesis, *15*:757
    catechol reaction, *13*:44
    in diazepam synthesis, *13*:127
    from ketene, *13*:876
1-(3-Chloroallyl)-3,5,7-triaza-1-azoniaadaman-
        tane chloride [*4080-31-3*]
    applications of, *13*:248
Chloroamine-T, *13*:230
o-Chloroaniline [*95-51-2*], *15*:924
p-Chloroaniline [*106-47-8*], *15*:924
2-Chloro-*p*-anisidine [*29242-84-0*], *15*:924
Chlorobenzene [*108-90-7*]
    nitration, *15*:924
    in phenyllithium synthesis, *14*:471
p-Chlorobenzene-*N*-*N*-dibutylsulfoxamide
        [*127-59-3*]
    as insecticide, *13*:474
p-Chlorobenzenesulfonamide [*98-64-6*]
    chlorpropamide preparation, *13*:616
Chlorobenzilate [*510-15-6*], *13*:461
o-Chlorobenzyl chloride [*611-19-8*], *15*:926
p-Chlorobenzyl *p*-chlorophenyl sulfide [*103-17-
        3*], *13*:463
    as insecticide, *13*:463
Chlorobis(η⁵-2,4-cyclopentadien-1-yl)titanium
        [*60955-54-6*], *15*:957
Chloro-bis[(1,2,3,4,5-η)-2,4-cyclopentadienyl]
        [methacryloyl]titanium, *15*:187
γ-Chlorobutyronitrile [*628-20-6*]
    potassium phthalimide reaction, *15*:766
7-Chloro-5-(2-chlorophenyl)-1,3-dihydro-2*H*-1,4-
        benzodiazepin-2-hydrazone [*40070-48-2*]
    in triazolam synthesis, *13*:131, 132
Chlorocyclohexane [*542-18-7*]
    microbial transformation of, *15*:463
3-Chlorocyclopentene [*96-40-2*]
    isoprene reaction, *13*:824

2-Chloro-1-(2,4-dichlorophenyl)vinyl diethyl
        phosphate [*470-90-6*]
    as insecticide, *13*:443
Chloro(diethyl) aluminum [*96-10-6*]
    propylene polymerization cocatalyst, *16*:459
7-Chloro-1-(2-(diethylamine)ethyl)-1,3-dihydro-
        5-phenyl-2*H*-1,4-benzodiazepin-2-one [*1101-
        71-9*]
    sedative, *13*:126
7-Chloro-1-(2-(diethylamino)ethyl)-5-(2-fluoro-
        phenyl)-1,3-dihydro-2*H*-1,4-benzodiazepin-
        2-one [*17617-23-1*]
    sedative, *13*:126
7-Chloro-1,3-dihydro-1-methyl-5-phenyl-2*H*-1,4-
        benzodiazepin-2-one 4-oxide [*2888-64-4*]
    in diazepam synthesis, *13*:127
7-Chloro-1,3-dihydro-5-phenyl-2*H*-1,4-benzodia-
        zepine-2-thione [*4547-02-8*]
    in estazolam synthesis, *13*:130
7-Chloro-2,3-dihydro-5-phenyl-1*H*-2,4-benzodia-
        zepin-1-one 4-oxide [*58112-81-5*]
    in lorazepam synthesis, *13*:129
4-Chloro-7-dimethylamino-1-methyl-quinolone-
        2-aldehyde [*31136-92-2*], *14*:803
1-Chloro-3,5-dimethyl-2,6-octadiene [*23446-46-0*]
    isoprene reaction product, *13*:825
1-Chloro-2,4-dinitrobenzene [*97-00-7*]
    aziridine reaction, *13*:148
2-Chloroethylamine hydrochloride [*870-24-6*]
    from aziridine, *13*:143, 158
N-(2-Chloroethyl)benzenesulfonamide [*6453-86-
        7*], *13*:143
Chloroethylene bisthiocyanate [*24689-89-2*]
    applications of, *13*:243
Chloroethylmercury [*107-27-7*], *14*:183
N-(2-Chloroethyl)-*N*-methylacetamide [*17225-
        69-3*], *13*:143
2-(2-Chloroethyl)-*N*-methylpiperazine [*50846-
        01-0*]
    in mesoridazine synthesis, *13*:133
1-β-Chloroethylthio-3-methyl-3-phospholene-
        1-oxide [*20098-58-2*]
    isoprene reaction product, *13*:826
Chlorofenethol [*80-08-6*], *13*:461
Chlorofluorocarbons
    release into atmosphere, *16*:697
7-Chloro-5-(2-fluorophenyl)-1,3-dihydro-1-me-
        thyl-2*H*-1,4-benzodiazepin-2-one [*3900-31-0*]
    in quazepam synthesis, *13*:130
    sedative, *13*:126
Chloroform [*67-66-3*]
    dihydroxybenzene reactions, *13*:45
    exposure limit, *13*:257
    health effect, *13*:265
    methyl iodide reaction, *13*:670

Chloroform (Continued)
from organic pollutants, 13:229
1-Chlorohexane [544-10-5]
magnesium addition, 16:561
7-Chloro-2-hydrazino-5-phenyl-3H-1,4-benzodiazepine [18091-89-9]
in estazolam synthesis, 13:131
2-Chlorohydroquinone [615-67-8]
developer, 13:60
Chlorolignin [8068-02-8]
in ore flotation, 14:309
Chloromagnesium hydroxide [10233-03-1]
Ziegler-Natta catalyst, 16:460
Chloromaleic anhydride [96-02-6], 14:778
isoprene reaction, 13:821
Chloromalonic acid [600-33-9], 14:795
o-Chloromercuricphenol [90-03-9], 15:169
2-Chloro-1-methoxy-2-methyl-3-butene [57513-11-8]
isoprene reaction product, 13:824
(E)4-Chloro-1-methoxy-2-methyl-2-butene [57513-12-9]
isoprene reaction product, 13:824
3-Chloro-2-methoxypropylmercuric acetate, 15:169
cis-5-Chloro-2-(methylamino)benzophenone oxime [6997-78-0]
in diazepam synthesis, 13:128
2-Chloromethyl-1,3-butadiene [4075-28-9]
isoprene reaction product, 13:824
2-Chloro-3-methyl-1,3-butadiene [1809-02-5]
methyl methacrylate reaction, 15:349
1-Chloro-3-methyl-2-butene [503-60-6]
isoprene reaction product, 13:824
2-Chloro-2-methyl-3-butene [2190-48-9]
isoprene reaction product, 13:824
1-Chloro-3-methyl-4-(2-cyclopentenyl)-2-butene [1781-59-5]
isoprene reaction product, 13:824
1-Chloromethyl-3,5-dimethoxybenzene [6652-32-0]
sodium diethyl malonate reaction, 14:797
2-Chloro-3,4-methylenedioxybenzyl DL-cis,trans-chrysanthemate [70-43-9]
as insecticide, 13:457
4-Chloromethylimidazole [23785-22-0]
in histamine synthesis, 15:769
5-Chloro-2-methyl-4-isothiazoline-3-one [26530-03-0]
in antimicrobial agent, 13:247
Chloromethyl methacrylate [27550-73-8]
properties, 15:355
α-Chloromethyl methyl ether [107-30-2]
isoprene reaction, 13:824
Chloromethyl methyl ether (CMME) [107-30-2]
in ion-exchange resin mfg, 13:687
1-Chloromethylnaphthalene [86-52-2], 15:703
nitrile from, 15:716

1-Chloronaphthalene [90-13-1], 15:700
in iron(II) phthalocyanine synthesis, 13:782
Chloronaphthalenes
preparation, 15:702
4-Chloro-1-naphthalenol [604-44-4]
from 1-naphthalenol, 15:733
Chloroneb [2675-97-6], 13:59
2-Chloro-5-nitrobenzene sulfonic acid [96-73-1], 15:924
2-Chloro-1-nitroethane [625-47-8], 15:975
Chloronitroparaffins, 15:913
4-Chloro-2-nitro-N-phenylbenzenamine [16611-15-7]
in clobazam synthesis, 13:131
O-2-Chloro-4-nitrophenyl O,O-dimethyl phosphorothionate [2463-84-5]
as insecticide, 13:440
O-3-Chloro-4-nitrophenyl O,O-dimethyl phosphorothionate [500-28-8]
as insecticide, 13:440
1-Chloro-2-nitropropane [2425-66-3]
soil fungicide, 15:984
2-Chloro-1-nitropropane [503-76-4], 15:975
3-Chloro-1-nitropropane [16694-52-3], 15:975
4-Chloro-4-oxo-2-butenoic acid pentyl ester [26367-51-1], 14:773
2-Chloro-3-pentene [1458-99-7]
isoprene reaction, 13:824
o-Chlorophenol [95-57-8]
catechol from, 13:47
hydrolysis, 13:48
p-Chlorophenol [106-48-9]
naphthalene azeotrope, 15:699
Chlorophenol red [4430-20-0], 13:10
N-(2-Chlorophenyl) acetoacetamide [93-70-9], 13:887
p-Chlorophenyl benzenesulfonate [80-38-6]
as insecticide, 13:462
8-Chloro-1-phenyl-1H-1,5-benzodiazepine-2,4(3H,5H)-dione [22316-55-8]
in clobazam synthesis, 13:130
p-Chlorophenyl p-chlorobenzenesulfonate [80-33-17]
as insecticide, 13:462
p-Chlorophenyl diethyl phosphate [5076-63-1], 13:451
1-(4-Chlorophenyl)-3-(2,6-difluorobenzoyl)urea [25367-38-5]
in insect control, 13:459
p-Chlorophenyl diiodomethyl sulfone [20018-12-6]
applications of, 13:231
o-Chlorophenyl isocyanate [3320-83-0], 13:801
p-Chlorophenyl isocyanate [104-12-1]
properties, 13:789
p-Chlorophenylmalononitrile [32122-64-8]
antifouling agent, 14:806

m-Chlorophenyl *N*-methylcarbamate [*4090-00-0*], 13:452

o-Chlorophenyl *N*-methylcarbamate [*3942-54-9*], 13:452

Chlorophenyl methyl silicone
 lubricating oils of, 14:500

p-Chlorophenyl phenyl sulfone [*80-00-2*]
 in insect control, 13:461

3-Chloro-1-phenylprop-2-en-1-one [*3306-07-08*]
 aziridine reaction, 13:151

p-Chlorophenyl sulfone [*80-07-9*], 13:463

p-Chlorophenyl sulfoxide [*7047-28-1*], 13:462

1-(*o*-Chlorophenyl)-2,2,2-trichloroethanol
 [*10291-39-1*]
 as insecticide, 13:429

S-(2-Chloro-1-phthalimidoethyl) *O,O*-diethyl
 phosphorothionate [*10311-84-1*]
 as insecticide, 13:442

Chlorophyll [*1406-65-1*], 14:615
 as photosensitizer, 13:370

Chloropicrin [*76-06-2*]
 as fumigant, 13:466
 from nitromethane, 15:975

Chloroprene [*126-99-8*]
 emulsion polymerization of, 14:93
 toxicity, 13:265

Chloroprene rubber [*126-99-8*] (CR), 16:
 727. (See also *Neoprene*.)

β-Chloropyruvaldoxime [*14337-41-8*]
 from γ-chloroacetoactyl chloride, 13:890

3-Chloroquinoline [*612-59-9*], 13:216

Chlorosulfonated polyethylene [*9008-08-6*]
 gasket properties, 16:727
 insulation properties of, 13:575

Chlorosulfonic acid [*7790-94-5*]
 in ion exchange, 13:686
 naphthalene reaction, 15:720

2-Chlorothiaxanthone [*86-39-5*]
 as photoinitiator, 13:368

o-Chlorotoluene [*95-49-8*], 15:926
 isoprene reaction, 13:825

N'-(4-Chloro-*o*-tolyl)-*N,N*-dimethylformamidine
 [*6164-98-3*]
 as insecticide, 13:463

2-Chloro-1-(2,4,5-trichlorophenyl)vinyl dimethyl
 phosphate [*22248-79-9*]
 as insecticide, 13:443

Chlorotriethyllead [*1067-14-7*], 14:181

Chlorotrifluoride [*7790-91-2*]
 in iron(III) fluoride synthesis, 13:772

Chlorotrifluoromethane [*75-72-9*]
 in etch gas, 13:639

(5-Chloro-2-((2,2,2-trifluoromethyl)amino)phe-
 nyl)(2-fluorophenyl) methanone [*50939-39-4*]
 in quazepam synthesis, 13:130

Chlorotrimethylsilane [*75-77-4*]
 ketene derivative, 13:882

Chlorovinyl iododichloride [*18964-25-5*], 13:664

Chloro(4-vinylphenyl)bis(tributylphosphine)pal-
 ladium [*54407-93-1*], 15:187

Chloro(4-vinylphenyl)bis(triphenylphosphine)-
 platinum [*76095-57-3*], 15:187

Chlorpropamide [*94-20-2*], 13:616

Chlorpyrifos [*2921-88-2*], 13:440

Chlorpyrifos methyl [*5598-13-0*]
 as insecticide, 13:440

Chlorthion [*500-28-8*], 13:439

Chocolate
 attractant, 13:417

Cholecalciferol [*67-97-0*]
 in calcium homeostasis, 15:586

Cholecystitis
 diagnosis, 15:79

Choledocholithiasis
 diagnosis, 15:79

Cholest-4-en-3-one [*601-57-0*]
 in cholesterol detection, 15:82

Cholesteric liquid crystals, 14:409

Cholesterol [*57-88-5*]
 in corticosteroid biosynthesis, 15:762
 in encapsulation, 15:476
 liquid crystals from, 14:408
 measurement, 15:80
 from mevalonic acid, 13:99
 mobilization by dilinoleyl phosphatidylcholine,
  14:251
 reduction of with ion-exchange, 13:704

Cholesterol ester hydrolase [*9026-00-0*]
 diagnostic reagent, 15:82

Cholesterol-ester storage disease
 liquid crystals in, 14:421

Cholesterol oxidase [*9028-76-6*]
 diagnostic reagent, 15:82

Cholesteryl myristate [*1989-52-2*]
 blue phase from, 14:410

Cholesteryl nonanoate [*1182-66-7*]
 liquid crystalline range of, 14:399

Cholesteryl propionate [*663-31-8*]
 blue phase of, 14:410

Cholestyramine [*11041-12-6*]
 in cholesterol treatment, 13:704

Choline [*62-49-7*]
 learning effects, 15:134
 in peanuts, 16:255

Choline acetylase [*9012-78-6*]
 in acetylcholine biosynthesis, 15:756

Cholinergic mechanisms
 in learning, 15:134

Cholinesterase [*51-84-3*]
 insecticide effect on, 13:449

Chondroitin sulfate [*9007-28-7*]
 manganese requirement, 15:594
 sulfur in, 15:587

Choristoneura fumiferana, *13*:480

Chromated copper arsenate [*37337-13-6*]
    applications of, *13*:237

Chromates
    ion-exchange recovery, *13*:702

Chromating
    of metals, *15*:307

Chromatium, *15*:950

Chromatography
    in diagnostic testing, *15*:88

Chromatography, hydrodynamic
    in particle-size analysis, *14*:89

Chrome orange [*1344-37-2*]
    in inks, *13*:378

Chrome sulfate, *14*:210

Chrome yellow [*1344-37-2*]
    in inks, *13*:378

Chromic acid [*1308-14-1*]
    affect on ion-exchange matrix, *13*:684
    in metal treating, *15*:308
    paint stripper, *16*:765
    toxicity, *13*:265
    toxicosis from, *15*:598

Chromic fluoride [*7788-97-8*]
    dye mordant, *16*:631

Chromindur, *14*:679

Chromite [*1308-31-3*]
    from beach sand, *16*:289

Chromium [*7440-47-3*]
    in coatings, *15*:241
    ion implantation, *13*:707
    lasers, *14*:54
    in magnesium alloys, *14*:596
    in magnetic alloys, *14*:694
    as magnetic material, *14*:647
    metabolic functions, *15*:592
    nickel alloys, *15*:792
    nutrient properties, *15*:571
    from the ocean, *16*:280

Chromium-50 [*14304-94-0*], *13*:840

Chromium-51 [*14392-02-0*]
    reactor corrosion product, *16*:155

Chromium-52 [*14092-98-9*], *13*:840

Chromium-53 [*13981-78-7*], *13*:840

Chromium-54 [*14304-97-3*], *13*:840

Chromium bromide [*10031-25-1*]
    in magnetooptics, *14*:705

Chromium bromide [*10031-25-1*] (1:3)
    Faraday effect, *14*:705

Chromium carbide [*12105-81-6*]
    in chromizing, *15*:249
    nitride form, *15*:877

Chromium carbide [*12552-55-5*] (7:3), *15*:249

Chromium carbide [*12105-81-6*] ($Cr_{23}C_6$), *15*:252

Chromium chloride [*10060-12-5*]
    in GTT, *15*:592

Chromium–cobalt–iron alloys
    as magnetic materials, *14*:678

Chromium dichloride [*10049-05-5*]
    in metallic coating, *15*:252

Chromium dioxide [*12018-01-8*]
    in magnetic tape, *14*:737
    in magnetic tape mfg, *14*:738

Chromium hexacarbonyl [*13007-92-6*]
    cyclopentadiene reaction, *16*:613

Chromium, hexavalent [*18540-29-9*]
    in metal treating, *15*:308

Chromium(III) chloride [*10025-73-7*]
    toxicity, *15*:573

Chromium(III) chloride hexahydrate [*10060-12-5*]
    toxicity, *15*:573

Chromium nitride [*24094-93-7*]
    properties, *15*:874

Chromium nitride [*24094-93-7*] (1:1), *15*:872

Chromium nitride [*12053-27-9*] (2:1), *15*:872

Chromium oxide [*1308-39-9*]
    in coatings, *15*:241
    in magnetic tape, *14*:701
    paint pigment, *16*:745

Chromium oxide [*12018-01-8*] (1:2)
    as magnetic insulator, *14*:694

Chromium sulfate [*10101-53-8*]
    in leather manufacture, *14*:210

Chromium trichloride [*10025-73-7*]
    in metallic coating, *15*:252

Chromium trioxide [*1333-82-0*]
    in anodizing, *15*:310
    silanol reaction, *16*:443

Chromium(VI)
    toxicity, *13*:265

Chromizing
    metallic coatings, *15*:248

Chromogenic materials
    encapsulated, *15*:491

Chromotropic acid [*148-25-4*], *15*:739.   (See also
    *4,5-Dihydroxy-2,7-naphthalenedisulfonic
    acid.*)

Chronic brain syndrome
    treatment, *13*:133

Chronodrug, *15*:488

Chronology
    standards, *15*:36

D-*trans*-Chrysanthemic acid [*28434-01-7*]
    in bioresmethrin, *13*:457

Chrysanthemum cinerariaefolium, *13*:424

Chrysanthemum coccineum, *13*:424

Chrysene [*218-01-9*]
    health effect, *13*:265

Chrysotile [*12007-29-5*]
    in gaskets, *16*:727

Chufa nut, *16*:250

Chymotrypsin [*9004-07-3*]
    effect on LHRH, *15*:771

Chypre
    in perfumes, *16*:962

CI 13400 [*10214-21-8*], *15*:740
CI 15585 [*2092-56-0*]
  in inks, *13*:378
CI 15865 [*3564-21-4*]
  in inks, *13*:378
CI Acid Black 7 [*8004-59-9*], *15*:730
CI Acid Black 24 [*3071-73-6*], *15*:731
CI Acid Black 35 [*6527-60-2*], *15*:731
CI Acid Blue 42 [*6656-04-8*], *15*:740
CI Acid Blue 58 [*71798-74-8*], *15*:738
CI Acid Blue 61 [*6837-41-4*], *15*:731
CI Acid Blue 70 [*1323-94-0*], *15*:740
CI Acid Blue 113 [*3351-05-1*], *15*:729
CI Acid Blue 169 [*6370-10-1*], *15*:739
CI Acid Brown 14 [*5850-16-8*], *15*:731
CI Acid Green 12 [*10241-21-1*], *15*:729
CI Acid Orange 12 [*1934-20-9*], *15*:738
CI Acid Red 14 [*3567-69-9*], *15*:738
CI Acid Red 26 [*3761-53-3*], *15*:739
CI Acid Red 41 [*5850-44-2*], *15*:739
CI Acid Red 52 [*3520-42-1*], *14*:547
CI Acid Red 70 [*6226-74-0*], *15*:738
CI Acid Red 73 [*5413-75-2*], *15*:739
CI Acid Red 88 [*1658-56-6*], *15*:731
CI Acid Red 99 [*3701-40-4*], *15*:738
CI Acid Red 115 [*6226-80-8*], *15*:739
CI Acid Red 133 [*6417-36-4*], *15*:742
CI Acid Red 186 [*52677-14-8*], *15*:740
CI Acid Violet 3 [*1681-60-3*], *15*:739
CI Acid Violet 7 [*4321-69-1*], *15*:742
CI Acid Yellow 1 [*846-70-8*], *15*:739
CI Acid Yellow 7 [*2391-30-2*], *14*:546
CI Basic Green [*633-03-4*], *14*:551
CI Basic Red 1 [*989-38-8*], *14*:547
CI Basic Red 12 [*6370-14-5*], *14*:551
CI Basic Violet 10 [*81-88-9*], *14*:547
CI Basic Yellow 40 [*12221-86-2*], *14*:547
CI Direct Black 19 [*6428-31-5*], *15*:740
CI Direct Black 22 [*6473-13-8*], *15*:740
CI Direct Blue 15 [*2429-74-5*], *15*:740
CI Direct Blue 26 [*7082-31-7*], *15*:738
CI Direct Blue 27 [*6420-15-1*], *15*:739
CI Direct Blue 71 [*4399-55-7*], *15*:740
CI Direct Blue 98 [*6656-03-7*], *15*:738
CI Direct Blue 120 [*3626-40-2*], *15*:731
CI Direct Blue 127 [*6411-51-4*], *15*:738
CI Direct Blue 128 [*6226-72-8*], *15*:738
CI Direct Brown 31 [*2429-51-4*], *15*:740
CI Direct Brown 62 [*8003-56-3*], *15*:731
CI Direct Green 33 [*6248-18-8*], *15*:730
CI Direct Green 42 [*5938-86-3*], *15*:740
CI Direct Green 51 [*6428-19-9*], *15*:731
CI Direct Orange 13 [*6470-22-0*], *15*:730
CI Direct Orange 26 [*3626-36-6*], *15*:742

CI Direct Orange 49 [*6420-32-2*], *15*:730
CI Direct Orange 69 [*8004-79-3*], *15*:730
CI Direct Orange 74 [*6104-56-9*], *15*:731
CI Direct Red 4 [*6240-41-3*], *15*:730
CI Direct Red 15 [*5413-69-4*], *15*:731
CI Direct Red 16 [*6227-02-7*], *15*:740
CI Direct Red 22 [*6448-80-2*], *15*:730
CI Direct Red 81 [*2610-11-9*], *15*:742
CI Direct Red 149 [*6420-39-9*], *15*:742
CI Direct Violet 7 [*6227-10-7*], *15*:742
CI Direct Violet 11 [*6227-19-6*], *15*:731
Cie color diagram, *14*:535
Cigarette odors, *16*:305
Cigarettes
  self-lighting, *15*:5
Cigarette smoke
  catechol in, *13*:46
CIIT information system, *13*:301
CI Mordant Brown 1 [*3564-15-6*], *15*:731
CI Mordant Brown 40 [*6369-32-0*], *15*:742
CI Mordant Brown 65 [*5852-26-6*], *15*:740
CI Mordant Red 7 [*2868-75-9*], *15*:740
Cine
  magnetic material for, *14*:751
Cinemoid plastic filters, *16*:535
Cineole [*470-82-6*], *16*:322.   (See also
    Eucalyptol.)
  from basil oil, *16*:321
1,4-Cineole [*470-67-7*]
  from lime, *16*:324
1,8-Cineole [*470-82-6*], *16*:313
  from lime, *16*:324
Cinerin I [*25402-06-6*]
  as insecticide, *13*:424
Cinerin II [*121-20-0*]
  as insecticide, *13*:424
CIN information system, *13*:328
Cinnabar [*19122-79-3*], *15*:143, 162.   (See also
    Mercuric sulfide.)
Cinnamaldehyde [*104-55-2*], *16*:312
  in perfumes, *16*:954
Cinnamic acid [*621-82-9*]
  microbial synthesis of, *15*:465
  in perfumes, *16*:949, 954
Cinnamic aldehyde [*104-55-2*]
  from cassia oil, *16*:322
  from cinnamon, *16*:322
Cinnamomum cassia Blume, *16*:322
Cinnamomum zeylanicum Nees, *16*:322
Cinnamon [*8007-80-5*], *16*:307
Cinnamon bark oil [*8007-80-5*], *16*:322
Cinnamon ceylon oil [*8006-79-9*], *16*:319
Cinnamonitrile [*4360-47-8*], *15*:907
Cinnamonum camphorae Sieb, *16*:321
Cinnamyl alcohol [*104-54-1*]
  in perfumes, *16*:954

CI Pigment Red 48 [*3564-21-4*]
   in inks, *13*:378
CI Pigment Red 49 [*1248-18-6*], *15*:729
CI Pigment Red 53 [*2092-56-0*]
   in inks, *13*:378
CI Pigment Red 54 [*6373-10-0*], *15*:738
Circuit breaker oils, *13*:551
Circuits, integrated, *13*:621.   (See also
   Integrated circuits.)
Circulators
   microwave ferrites in, *14*:667
CIS INDEX data base, *13*:334
CIS information system, *13*:300
CI Solvent Green 5 [*2744-50-5*], *14*:553
CI Solvent Yellow 43 [*12226-96-9*], *14*:549
CI Solvent Yellow 44 [*2478-20-8*], *14*:546, 547
CI Solvent Yellow 135, *14*:547
CI Solvent Yellow 160:1, *14*:547
Cistus ladaniferus, *16*:323, 950
Cistus oil.   See *Labdanum oil.*
Citizen's radio bands, *15*:497
Citraconic acid [*498-23-7*]
   properties, *13*:871
Citraconic anhydride [*616-02-4*]
   properties, *13*:871
Citral [*5392-40-5*]
   in lemon oil, *16*:324
   odor modification with, *16*:304
Citralva [*5146-66-7*], *16*:969
Citric acid [*77-92-9*]
   encapsulation, *15*:476
   in iron citrate, *13*:765
   itaconic acid from, *13*:869
   from ketene, *13*:876
   lime-refined, *14*:380
   in niobium analysis, *15*:828
   oil emulsion of, *15*:490
Citric acid cycle
   in L-glutamate synthesis, *15*:767
Citrobacter, *15*:949
Citronellal [*106-23-0*]
   from citronella oil, *16*:322
Citronella oil [*8000-29-1*], *16*:322
Citronellol [*106-22-9*], *16*:312, 955
   from citronella oil, *16*:322
   fragrance of, *16*:959
1-Citronellol [*6812-78-8*], *16*:327.   (See also
   Rhodinol.)
Citronellyl acetate [*150-84-5*], *16*:312
L-Citrulline [*372-75-8*]
   microbial synthesis, *15*:464
Citrullus vulgatis, *16*:251
Citrus aurantifolia Swingle, *16*:324
Citrus aurantium L. subsp *arara*, *16*:326
Citrus aurantium L. subsp *bergamia*, *16*:321
Citrus aurantium L. subsp *amara* L., *16*:324, 325

Citrus aurantium Var. *dulcis*, *16*:325
Citrus limonum, *16*:324
Citrus oils
   odor modification with, *16*:303
Citrus paradisi Macfayden, *16*:323
Civet [*68916-26-7*], *16*:314
   in perfumes, *16*:950
Civetone [*542-46-1*], *16*:950
   as civet cat odor, *13*:922
Clabazam [*22316-47-8*]
   properties of, *13*:122
Cladding, *15*:275, 277.   (See also *Explosively
   clad metals.*)
Cladene, *15*:769
Cladosporium resinae, *13*:249
Clad vessels
   metallic coating for, *15*:258
Claflin refiner, *16*:777
Claims
   in patent specifications, *16*:866
Claims files
   information system, *16*:913
CLAIMS information system, *13*:299
Claisen-Schmidt condensation, *13*:73, 895
Clarifiers
   for milk processing, *15*:527
Clarino, *14*:234, 235
Clay
   electric insulation filler, *13*:565
   filler in plastics, *13*:973
   in inks, *13*:378
   as insecticide carrier, *13*:416
   as ion-exchange material, *13*:685
   as thermal insulator, *13*:592
Clays
   lithium in, *14*:449
   in paper coatings, *16*:788
Clean Air Act, *13*:254; *16*:760
Cleaning, *15*:296.   (See also *Metal surface
   treatments.*)
Cleaning systems
   in dairy plants, *15*:543
Clean room
   in laminated-glass mfg, *13*:982
Clearfield mixer, *15*:629
Clearing nut, *16*:251
Clearon, *13*:230
Clementine, *16*:185
Cleve's acid, *15*:723
1,6-Cleve's acid [*119-79-9*], *15*:723, 729.   (See
   also *5-Amino-2-naphthalenesulfonic acid.*)
1,7-Cleve's acid [*119-28-8*], *15*:723, 729.   (See
   also *8-Amino-2-naphthalenesulfonic acid.*)
Clinch River breeder reactor (CRBR), *16*:191

Clinical chemistry. See *Medical diagnostic reagents.*
Clinoptilolite [*12173-10-3*], *15*:639
  as ion-exchange material, *13*:685
Clinton formation
  iron ore from, *13*:752
Clocks, geologic
  isotopes as, *13*:860
Clonazepam [*1622-61-3*]
  properties of, *13*:135
Clonopin, *13*:135, 138
Cloran TM [*1782-06-5*], *14*:788
Clorox, *15*:906
Clostridium pasteurianum, *15*:950
Clostridium sp, *15*:946
Cloth
  cutting by lasers, *14*:66
Clothing
  fibers for, *16*:376
  fluorescent, *14*:564
Cloud point
  of nonionic surfactants, *14*:88
Cloud seeding
  silver iodide for, *13*:672
Clove bud oil [*8000-34-8*], *16*:322
Clove leaf oil [*8015-97-2*]
  in perfumes, *16*:950
Clover
  lignin content, *14*:297
Cloxacillin [*61-72-3*]
  microbial synthesis, *15*:463
Clusius thermal-diffusion column, *13*:844
CMC. See *Carboxymethyl cellulose.*
CMD. See *Manganese dioxide, chemical.*
CME. See *Crucible melt extraction.*
CMME. See *Chloromethyl methyl ether.*
CNS. See *Central nervous system.*
CO_2 acceptor process, *14*:338
Coacervation
  in encapsulation, *15*:477
Coagulant, blood
  calcium lactate as, *13*:88
Coagulation
  latices, *14*:85
Coal
  in DR processes, *13*:754
  in ion exchange, *13*:678
  lignitic, *14*:313
  magnetic purification, *14*:729
  microwave desulfurization, *15*:516
  mine dusting, *14*:381
  risk as fuel vs nuclear, *16*:220
  subbituminous, *14*:313
Coalescers
  in LPG processing, *14*:389
Coal gasification plants
  health effect, *13*:265

Coal liquefaction
  molybdenum catalysts for, *15*:694
Coal Mine Health and Safety Act, *13*:254
Coals
  classification, *14*:315
Coal tar
  naphthalene from, *15*:706
Coal-tar distillation
  catechol from, *13*:48
Coal tar products
  toxicity, *13*:265
Coated fabrics
  as leather substitutes, *14*:232
Coaters
  for paper, *16*:786
Coating
  magnetic tape processes, *14*:744
  of particles, *15*:472
Coating colors, *16*:786
Coating processes
  for leather substitutes, *14*:237
Coatings, *15*:241. (See also *Metallic coatings.*)
  decorative, *15*:266
  in latices, *14*:82
  lead alloys, *14*:155
  for leather, *14*:225
  nondestructive testing, *16*:57
  nondestructive testing of, *16*:52
  for paper, *16*:785, 821
  from polyethylene, *16*:431
  removal, *16*:762
  sprayed, *15*:253
  wear-resistant carbides, *15*:249
Coatings, industrial, *16*:742. (See also *Paint.*)
Coatings, powder
  as insulation, *13*:554
Cobalt [*7440-48-4*]
  affect on iron magnetism, *13*:738
  in coatings, *15*:242
  in driers, *13*:379
  in magnetic alloys, *14*:694
  as magnetic material, *14*:647
  metabolic functions, *15*:597
  nutrient properties, *15*:571
  from the ocean, *16*:280, 292
  oxygen reaction in chlorate candle, *16*:675
Cobalt-58 [*13981-38-9*]
  reactor corrosion product, *16*:155
Cobalt-59 [*7440-48-4*], *13*:840
Cobalt-60 [*10198-40-0*]
  in level measurement, *14*:434
  radiographic use, *16*:64
  reactor corrosion product, *16*:155
Cobalt acetate [*71-48-7*]
  as dye fastener, *15*:309
Cobalt alloy [*59299-14-8*] (CoCrAlY)

Cobalt alloy (*Continued*)
  as metallic coating, *15*:266
Cobalt carbonyl [*10210-68-1*]
  Ni(CO)$_4$ coproduct, *15*:807
  oxo process catalyst, *16*:611
Cobalt carbonylhydrotris(triphenylphosphine)
    rhodium [*17185-29-4*]
  from oxo process, *16*:640
Cobalt chloride [*7647-79-9*] (1:2)
  sodium diphenylphosphinate reaction, *15*:217
Cobalt, compd with dysprosium [*12017-58-2*] (5:
    1), *14*:676
Cobalt [1,1′,1″,1‴-(η^4-1,3-cyclobutadiene-1,2,3,4-
    tetrayl)- tetrakis[benzene]-
    η^5-2,4-cyclopentadien-1-yl] [*1278-02-0*], *16*:
    615
Cobalt 2-ethylhexanoate [*13586-82-8*]
  from oxo catalysts, *16*:646
Cobalt ferrite [*12052-28-7*]
  as magnetic insulator, *14*:694
  in video tape, *14*:701
Cobalt hexakiscyanoferrate [*15415-49-3*] (2:1),
    13:771
Cobalt hydroxide [*21041-93-0*]
  in oxygen-generation systems, *16*:675
Cobaltic hexamminetriiodide [*13841-85-5*], *13*:660
Cobalticinium-1,1′-dicarboxylate salts
  titanocene copolymers, *15*:210
Cobalt(III) hydroxide [*1307-86-4*]
  from CeSO$_4$ and O$_3$, *16*:688
Cobalt(III) nitrate [*15520-84-0*]
  in anode coating mfg, *15*:178
Cobalt(II) iodide [*15238-00-3*], *13*:661
Cobalt molybdate [*13762-14-6*], *15*:687
Cobalt naphthenate [*61789-51-3*], *15*:752
  as catalyst, *15*:752
  from oxo catalysts, *16*:646
  as promoter, *13*:364
Cobalt nitride [*12432-98-3*], *15*:872
Cobalt nitride [*12259-10-8*] (2:1), *15*:872
Cobaltocene [*1277-43-6*]
  tetrafluoroethylene reaction, *16*:604
Cobaltocenium ion
  reduction, *16*:604
Cobalt oleate [*14666-94-5*]
  from oxo catalysts, *16*:646
Cobalt oxide [*1308-06-1*]
  in anode coatings, *15*:173
Cobalt oxide [*1308-06-1*] (3:4)
  spinel, in anode coatings, *15*:177
Cob nut, *16*:250
Cocaine [*50-36-2*]
  measurement, *15*:88
Cockroach
  insecticide for, *13*:421
Cocoa butter
  in lecithin manufacture, *14*:257

Coco nitrile [*61789-53-5*]
  properties, *15*:908
Coconut, *16*:248, 250
Coconut oil
  in lecithin manufacture, *14*:257
Cocos nucifera, *16*:250
N-Coco-trimethylenediamene [*61791-63-7*]
  applications of, *13*:242
COD. See *Cyclooctadiene.*
Code of Federal Regulations (CFR), *16*:714
Codling moth
  control of with pheromones, *13*:480
COE. See *Cube-on-edge.*
Coenzyme B$_{12}$ [*13870-90-1*], *15*:597
  function, *15*:597
Coenzyme Q$_{10}$ [*303-98-0*]
  vanadium effect on, *15*:598
Coextrusion
  metal-laminate fabrication, *13*:948
Coherence
  of laser light, *14*:47
Coherent anti-Stokes Raman spectroscopy
    (CARS), *14*:72
Cohobation
  of essential oils, *16*:314
Coins
  from explosion-clad metal, *15*:293
Coke
  in case hardening, *15*:315
  in DR processes, *13*:754
  in iron mfg, *13*:740
  in iron-ore processing, *13*:739
  as molecular sieve, *15*:937
Coke in-line mixer, *15*:623
Coke oven emissions
  toxicity, *13*:266
Colback, *16*:78
Colbond, *16*:78
COLD
  information service, *16*:902
Cold welding
  metal coatings, *15*:260
Cold-working
  of metals, *15*:329
COLEX information system, *13*:291
Collagen [*9059-25-0*]
  in cattlehides, *14*:203
  cross-linking in, *15*:588
  uses, *14*:227
Collodion [*9004-70-0*]
  microcapsule wall, *15*:479
Colloids
  consolute temperature, *14*:88
Collotype inks, *13*:395
Color
  in paint, *16*:757

Colorados, *14*:225
Colorants
 for paper, *16*:779
 for plastics, *14*:566
Color filters, *16*:534
Color formation
 by fluorescent dyes, *14*:558
Coloring
 metals, *15*:310
Color lakes
 in paper, *16*:779
Color rendition index, *14*:537
Color temperature, *13*:339
Columbite [*1310-23-2*], *15*:821
 from beach sand, *16*:289
Columbium [*7440-03-1*]. See *Niobium*.
Columns
 continuous, *14*:957
Coma
 from hypnotics overdose, *13*:122
 nutrients associated with, *15*:573
Comalloy, *14*:680
Combined electrolysis and chemical-exchange process
 in isotope separation, *13*:852
Combustion
 lignites, *14*:331
 oxygen role in, *16*:656
Combustors, *13*:183. (See also *Incinerators*.)
Comics
 ink for, *13*:393
Cominco process
 for lead smelting, *14*:130
Commercial inks, *13*:381
Comminution
 of meat products, *15*:67
Commiphora erythraea, *16*:952
Commiphora myrrha, *16*:951
Commodity chemicals, *14*:904
Communication
 with lasers, *14*:71
 noise interference, *16*:6
Communication systems
 diodes for, *14*:279
Comol, *14*:680
Compaction
 nuclear wastes, *16*:211
Complexometric titrations, *13*:732
Composite mechanics, *13*:952
Composite membranes, *15*:102, 106
Composites
 hybrid, *13*:943
 novoloid fibers in, *16*:132
 wood derived, *14*:1

Composite wood
 lignins for, *14*:311
Composition board, *13*:532. (See also *Particle board*.)
Compression
 in LPG processing, *14*:387
Compressors
 noise control, *16*:23
COMPUSTAT information system, *13*:327, 328
Computerized information sources, *13*:326
Computer memories, *14*:667
Computers
 ICS in, *13*:645
 in information retrieval, *13*:279
 in instrumentation and control, *13*:488
 use in operations planning, *16*:507
Con A. See *Concanavalin*.
Conarachin, *16*:249
 amino acid composition, *16*:249
Concanavalin A (Con A) [*11028-71-0*]
 in lymphocyte-blast transformation, *13*:175
Concentration
 standardization of, *15*:40
Concrete
 emissivity, *13*:342
 limestone in, *14*:374
 nuclear waste storage, *16*:211
 as radiation shield, *16*:240
 sound coefficients of, *13*:513
Condensed milk, *15*:555
Conduction
 in insulators, *13*:592
Conductivity
 in level measurement, *14*:431
 in nondestructive testing, *16*:61
 in polymeric systems, *14*:90
Conductors, *13*:534. (See also *Insulation, electric*.)
Cone-and-screw mixer, *15*:632
Cone-drop test, *16*:99
Confidence limits
 reliability tests, *15*:22
Confused flour beetle
 fumigant activity on, *13*:465
Congressional information service, *13*:334
Coniferyl alcohol [*485-35-5*]
 dehydrogenation of, *14*:301
 enzymatic reactions, *14*:301
 lignin from, *14*:295
Connecticut Yankee, *16*:155
Conquest/Bonanza, *14*:811
Conrad-Limpach reaction, *13*:890
Construction
 limestone uses, *14*:372
Consumer products
 nondestructive testing of, *16*:47
Consumer Product Safety Commission, *13*:260

Contac, *15*:488

Contact angle, *16*:808

Contact lenses, *15*:92. (See also *Membrane technology*.)
  permeability, *15*:126

Contactors
  for phase separation, *14*:**912**

Contact poisons
  as insecticides, *13*:418

Containers, *16*:714. (See also *Packaging materials, industrial*.)
  paper, *16*:798
  from polyethylene, *16*:431, 451
  from polypropylene, *16*:468

Containers, bulk, *16*:715

Contamination control
  in IC manufacture, *13*:624

Continental shelves, *16*:284

Continuous ambulatory peritoneal dialysis, *15*: 116

Control, automatic, *13*:485

Control, feedback, *13*:485

Controlled-release, *15*:92. (See also *Membrane technology*.)

Controlled-release membranes, *15*:123

Controlled-release systems
  microencapsulated, *15*:485

Conversion coatings
  chromate, *15*:307

Conversion rolling
  explosion bonding application, *15*:291

Convulsions
  nutrients associated with, *15*:573
  treatment, *13*:122

Conwed, *16*:78

Cookeite [*1302-92-7*], *14*:450

Cooking
  microwave, *15*:500

Cooking odors, *16*:305

Coolanol 45, *14*:496

Coolant
  for nuclear reactors, *16*:152

Coolants
  in nuclear reactors, *16*:140

Cooling towers
  noise control, *16*:22

Coordination complexes, *16*:592. (See also *Organometallics—metal π complexes*.)

Coordination polymers, *15*:212

Copaene [*3856-25-5*], *16*:313

Copolymers
  of polypropylene, *16*:464

Copper [*7440-50-8*]
  alloys with nickel, *15*:792
  as cable covering, *13*:582
  cladding, *15*:281
  in coatings, *15*:241

electrode for, *13*:726

emissivity, *13*:342

gasket properties, *16*:727

lead alloys of, *14*:142

lead by-product, *14*:107

in magnesium alloys, *14*:596

as magnetic material, *14*:647

metabolic functions, *15*:588, 593

in metal treating, *15*:305

nutrient properties, *15*:571

in nuts, *16*:260

from the ocean, *16*:292

oxidation, *16*:668

in packings, *16*:737

recovery by ion exchange, *13*:702

from sulfide ores, *16*:668

thermal conductivity of, *13*:544

wear coefficient, *14*:482

Copper-63 [*14191-84-5*], *13*:840

Copper-65 [*14119-06-3*], *13*:840

Copper acetoarsenite [*12540-34-0*]
  as insecticide, *13*:421

Copper alloys
  cladding of, *15*:281

Copper–aluminum [*12004-15-8*] (1:2)
  by precipitation hardening, *15*:337

Copper–aluminum clads, *15*:275

Copper arsenate, basic [*12774-48-0*]
  as insecticide, *13*:421

Copper arsenide [*12005-75-3*], *15*:241

Copperas [*7720-78-7*], *13*:777.
  (See also *Iron(II)*.)

Copper chromite [*11104-65-7*]
  oxo catalyst use, *16*:646

Copper citrate [*866-82-0*]
  as algicide, *13*:238

Copper compound with gold [*12006-52-9*] (3:1)
  magnetic behavior, *14*:657

Copper compound with manganese [*12272-98-9*] (1:1), *14*:648

COPPER DEVELOPMENT ASSOCIATION
  information system, *13*:328

COPPER DEVELOPMENT DATA BASE, *13*: 327

Copper drossing
  in lead processing, *14*:109

Copper ethylenediaminetetraacetic acid (EDTA) [*54453-03-1*]
  as algicide, *13*:238

Copper extraction
  DRI used in, *13*:762

Copper flakes
  in inks, *13*:379

Copper(I) acetylide [*16753-36-9*]
  in acetylene analysis, *16*:665

Copper(II) chloride [*7447-39-4*]
  toxicity, *15*:573
  Wacker-process catalyst, *16*:611

Copper iodomercurate(II) [13876-85-2], 13:661
Copper–maraging steels
  cladding of, 15:276
Copper molybdate [13767-34-5], 15:687
Copper naphthenate [1338-02-9]
  applications of, 13:237
Copper–nickel–cobalt alloys
  as magnetic materials, 14:679
Copper–nickel–iron alloys
  as magnetic materials, 14:679
Copper nitride [1308-80-1], 15:872
Copper oleate [1120-44-1]
  as preservative, 13:238
Copper oxychloride [1332-40-7]
  oxygen generation catalyst, 16:679
Copper phthalocyanine [147-14-8]
  as grease thickener, 14:519
Copper 8-quinolinate [10380-28-6]
  applications of, 13:236
Copper refining
  lithium in, 14:458
Copper sulfate [7758-99-8]
  in bordeaux mixture preparation, 13:476
  in flotation, 14:102
  in teart treatment, 15:593
  toxicity, 15:573
Copper sulfate pentahydrate [7758-99-8]
  toxicity, 15:573
Copper tartrate [52327-55-6]
  in biuret reagent, 15:81
Copper tetraammine carbonate, 16:668
Coproporphyrin, 14:199
Copying paper
  stabilization with malic acid, 13:110
Copyright, 16:851
Copy system
  carbonless, 15:477, 480
Coquilla nut, 16:251
Corderoite [53237-82-4], 15:144
Cordley, 14:248
Cordovan, 14:201
Corfam, 14:234
Coriander oil [8008-52-4], 16:320, 322
Coriandrum sativum L., 16:322
Coridothymus capitatus Reichb., 16:325
Corium
  in cattlehide, 14:203
Cork
  gasket properties, 16:727
  in gaskets, 16:732
  as thermal insulator, 13:592
Cork–rubber
  gasket properties, 16:727
Corncobs
  lignin content, 14:297
Corn earworm
  control of with pheromones, 13:480

Corning filters, 16:535
Corn oil [8001-30-7]
  lecithin from, 14:250
Corn sugar, 15:440. (See also *Dextrose.*)
  in leather manufacture, 14:213
Corn syrup
  meat-curing agent, 15:62
Corona
  in solid insulation systems, 13:542
Corona discharge
  in cable insulation, 13:569
  in ozone generation, 16:690
Coronadite [12414-82-3], 14:855
Corona tests, 16:57
Corovin PP-S, 16:76
Corrosion
  in light-water reactors, 16:150
Corrosion damage
  detection of, 16:48
Corrosion inhibitors
  molybdates, 15:694
  naphthenates as, 15:752
Corrosion resistance
  enhancement of by ion implantation, 13:716
  of lead alloys, 14:159
Corrosive sublimate of mercury. See *Mercuric chloride.*
Corrugated board, 16:799
Corrugated inks, 13:381
Cortelan, 15:761
Corticosteroids
  as neuroregulators, 15:761
Corticosterone [50-22-6], 13:173
  ACTH effect on, 15:759
  properties, 15:761
Cortisol [50-23-7], 15:761. (See also *Hydrocortisone.*)
  ACTH effect on, 15:759
Cortisone [53-06-5], 13:173
  microbial synthesis, 15:460
  properties, 15:761
Cortistab, 15:761
Cortisyl, 15:761
Corylin, 16:249
Corylus americana, 16:250
Corylus avellana, 16:250
Corynebacterium, 15:949
Corynebacterium diptheriae, 15:553
Corynebacterium granulosum
  in cancer chemotherapy, 13:176
Corynebacterium parvum
  in cancer chemotherapy, 13:176
Cosan P, 13:245
Cosmic rays
  tracks of, 16:840
  ultraheavy, 16:842

Cost effectiveness, *14*:753.   (See also
    Maintenance.)
COT.   See *Cyclooctatetraene.*
Cotin 300, *13*:236
Cotinus nitida, *13*:481
Cottage cheese, *15*:563
Cotton
    moisture regain, *16*:358
    nonwoven fabrics from, *16*:107
    in packings, *16*:734
Cotton bags, *16*:720
Cotton dust
    toxicity, *13*:266
Cotton leafworm
    control of with pheromones, *13*:481
Cotton linters
    in paper, *16*:771
Cotton seed
    bacteriostat, *16*:583
Cottonseed oil [*8001-29-4*]
    in packings, *16*:734
Coumaphos [*56-72-4*], *13*:437, 449
Coumarin [*91-64-5*], *13*:73; *16*:955
    derivatives of, *13*:46
    in perfumes, *16*:951, 954
Coumarin 7 [*27425-55-4*], *14*:550
Coumarone [*271-89-6*]
    from naphthalene mfg, *15*:707
Council of the International Association of
    Chemical Societies, *16*:29
Court of Customs and Patent Appeals, *16*:863
Coxiella burneti, *15*:553
CPCS CHEMRIC information system, *13*:301
CPE.   See *Polyethylene, chlorinated.*
CPK.   See *Creatine phosphokinase.*
CPK isoenzyme [*9001-15-4*], *15*:89
CPM.   See *Critical path method.*
CPS.   See *Critical path scheduling.*
CPSC.   See *Consumer Product Safety
    Commission.*
CPVC.   See *Critical pigment volume
    concentration.*
CR.   See *Chloroprene rubber.*
CR-39 [*142-22-3*]
    as track-etch detector, *16*:827
Crabgrass
    control, *16*:587
Cracking
    in isoprene synthesis, *13*:830
    zeolite catalysts for, *15*:651
Crane 425A, *14*:513
Crates
    as packaging, *16*:722
CRBR.   See *Clinch River breeder reactor.*
Cream, *15*:560.   (See also *Milk products.*)
Cream of tartar [*868-14-4*], *13*:111, 114.   (See
    also *Vinaccia.*)

manufacture of, *13*:114
Creatine [*57-00-1*]
    in CPK assay, *15*:77
Creatine phosphate [*67-07-2*]
    in CPK assay, *15*:77
Creatine phosphokinase [*9001-15-4*] (CPK), *15*:76
    in diagnosis, *15*:75
Creatinine [*60-27-5*]
    assay, *15*:81
    carbon adsorption, *15*:489
    in digoxin assay, *15*:86
Creatinine sulfate [*31377-28-3*]
    serotonin complex, *15*:770
CRECORD information system, *13*:328, 334
Creosote [*8021-39-4*]
    in glulam treatment, *14*:38
    in wood preservation, *13*:225
Creping aids, *16*:808, 821
Cresol [*93-51-6*]
    toxicity, *13*:266
Cresol carbonate [*617-09-4*]
    salicylaldehyde from, *13*:75
p-Cresol esters
    in perfumes, *16*:968
Cresol phosphate [*78-30-8*]
    salicylaldehyde from, *13*:75
o-Cresolphthalein [*596-27-0*]
    in calcium, *15*:83
Cresol red [*1733-12-6*], *13*:10
Cresylic acid [*1319-77-3*]
    froth stabilizer, *14*:102
    in paint and varnish removers, *16*:764
Cretinism
    nutrients associated with, *15*:573
CRI.   See *Color rendition index.*
Criegee zwitterion, *16*:684
Cristobalite [*14464-46-1*], *15*:658
Critical path method (CPM), *16*:516
Critical path scheduling (CPS), *16*:516
Critical pigment volume concentration (CPVC),
    16:745
Crocein acid [*132-57-0*], *15*:738.   (See also *7-
    Hydroxy-1-naphthalenesulfonic acid.*)
CRONOS information system, *13*:328
Crossed-aldol condensation, *13*:73
Cross-linking
    in insulation materials, *13*:545
Cross-tie walls, *14*:689
Crotonaldehyde [*4170-30-3*]
    ketene reaction, *13*:877
Crotoxyphos [*7700-17-6*], *13*:449
Crown ethers
    in liquid membranes, *15*:119
Crowntex, *16*:78

Crucible melt extraction, *15*:229
Crucibles
  for mica melts, *15*:425
Crude oil
  naphthenic acid content, *15*:751
Crufomate [*299-86-5*], *13*:449
Crushers, *14*:358
Cryocomminution
  with N$_2$, *15*:940
Cryoforming
  metals, *15*:940
Cryogenic crushing
  nuclear wastes, *16*:211
Cryogenic fluids
  in bearing lubrication, *14*:517
Cryogrinding
  with N$_2$, *15*:940
Cryolite [*15096-52-3*]
  as insecticide, *13*:414, 421
Cryopulverizing
  with N$_2$, *15*:940
Cryptococcus laurentii, *15*:464
Cryptococcus melibiosum, *13*:784
Cryptomelane [*52014-08-1*], *14*:825, 855
Crystal-field theory (CFT), *16*:594
Crystal violet [*548-62-9*], *13*:10
Crystal violet lactone [*1552-42-7*]
  in carbonless copy paper, *15*:487
CS acid [*82-75-7*], *15*:739. (See also *4-Hydroxy-1,5-naphthalenedisulfonic acid.*)
CSM. See *Chlorosulfonated polyethylene.*
CSPE. See *Chlorosulfonated polyethylene.*
440C stainless steel
  ion implantation, *13*:717
CTMP. See *Chemithermomechanical pulp.*
CUADRA associates, *13*:304
α-Cubebene [*37571-83-8*]
  as insect attractant, *13*:482
Cube-on-edge (COE), *14*:653
Culex fatigans, *13*:413
Culex quinquefasciatus, *13*:456, 458
Cumara nut, *16*:251
Cumene [*98-82-8*]
  in DIPB production, *13*:49
Cumene hydroperoxide [*80-15-9*], *13*:363. (See also *tert-Cumyl hydroperoxide.*)
  in adhesive, *15*:491
Cumic acid [*536-66-3*]
  microbial preparation, *15*:460
Cumilate, *13*:236
Cumin [*8014-13-9*], *16*:307
Cumyldimethylcarbinol [*3445-42-9*]
  dihydroxybenzene reaction, *13*:50
tert-Cumyl hydroperoxide [*80-15-9*], *13*:363
Cumylmethylketone [*645-13-6*]
  dihydroxybenzene reaction, *13*:50

Cunico, *14*:679
Cunico I, *14*:694
Cunico II, *14*:694
Cunife, *14*:679
Cunife I, *14*:694
  magnetic properties, *14*:671
Cunife II, *14*:694
Cunninghamella blakesleeana, *15*:465
Cupellation
  in lead refining, *14*:121
Cupferron [*135-20-6*]
  in niobium analysis, *15*:829
Cupric chloride [*7447-39-4*]
  oxidation catalyst, *13*:53
Cupric hexakiscyanoferrate [*13601-13-3*] (2:1), *13*:771
Cupric lactate [*814-81-3*], *13*:89
Cupric oxalate [*814-91-5*]
  in photography, *16*:633
Cupric sulfate [*7758-98-7*]
  in initiator systems, *13*:367
Cupric sulfate pentahydrate [*7758-99-8*]
  in Fehling's solution, *13*:118
Cupric sulfide [*1317-40-4*]
  in copper electrode, *13*:726
Cupric tetrammineiodide [*74685-94-2*], *13*:660
Cupronickel [*11114-42-4*], *15*:258
Cupro–nickel [*11114-43-5*] (70:30)
  oxalic acid storage in, *16*:629
Cupro–nickel [*11114-42-4*] (90:10)
  oxalic acid storage in, *16*:629
Cupronickel (70% copper)
  cladding, *15*:293
Cupronickel (90% copper)
  in metal cladding, *15*:294
Cuprous chloride [*7758-89-6*]
  oxidation catalyst, *13*:53
Cuprous chloride [*12258-96-7*] (Cu$_2$Cl$_2$)
  in initiator systems, *13*:367
Cuprous cyanide [*544-92-3*]
  as insecticide, *13*:422
Cuprous iodide [*7681-65-4*], *13*:661
Cuprous iodomercurate [*13876-85-2*], *15*:169
Cuprous oxide [*1317-39-1*]
  applications of, *13*:237
Curdlan [*54724-00-4*], *15*:453
Cured meats, *15*:62
Curie-point writing, *14*:694
Curing
  of meat products, *15*:62
Curing agents
  ketone peroxides for resins, *13*:895
Curium nitride [*56509-28-5*], *15*:872
Curl control
  of paper, *16*:821

Curtius rearrangement
  in isocyanate synthesis, *13*:799
Curvularia falcata, *15*:461
Cushing's disease
  nutrients associated with, *15*:573
Cushing's syndrome, *13*:173
Cutting
  with lasers, *14*:66
Cuyuna range
  iron ore from, *13*:751
CVD.   See *Chemical vapor deposition.*
Cyanic acid [*420-05-3*]
  lactic acid reaction, *13*:84
Cyanide
  electrode for, *13*:726
  ozone reaction, *16*:688
Cyanides
  toxicity, *13*:266
Cyanide wastes
  ozone oxidation of, *16*:710
Cyaniding, *15*:316
Cyanine dyes
  as fluorescent pigments, *14*:551
Cyanoacetate esters, *15*:907
Cyanoacetic acid [*372-09-8*], *14*:801; *15*:889
  benzaldehyde reaction, *15*:907
  properties, *15*:906
Cyanoacetic acid nitrile [*109-77-3*], *14*:794.   (See
    also *Malononitrile.*)
α-Cyanoacrylate [*25067-29-2*]
  nitromethane solvent for, *15*:982
Cyanobacteria
  nitrogen-fixing, *15*:948
m-Cyanobenzylamine [*10406-24-3*], *15*:906
Cyanobiphenyls
  as liquid crystals, *14*:414
Cyanocobalamin [*68-19-9*]
  as dietary supplement, *15*:597
N-(2-Cyanoethyl)methacrylamide [*24854-94-2*]
  properties, *15*:356
2-Cyanoethyl methacrylate [*4513-53-5*]
  properties, *15*:354
N-(2-Cyanoethyl)-*N*-methylacrylamide [*76392-
    04-6*]
  properties, *15*:356
3-Cyanoethyl-5-methylindole [*65881-14-3*]
  in melatonin synthesis, *15*:772
Cyanogen [*2074-87-5*], *15*:883
Cyanogen chloride [*506-77-4*]
  malononitrile preparation, *14*:805
Cyanoguanidine [*461-58-5*], *15*:901.   (See also
    Dicyandiamide.)
  benzonitrile reaction, *15*:905
Cyanol [*118-46-7*], *15*:740.   (See also *7-Amino-
    2-naphthalenol.*)
Cyanomer A-370, *15*:389
Cyanomethyl methacrylate [*7726-87-6*]

  properties, *15*:354
8-Cyano-1-naphthalenesulfonic acid [*83-20-5*]
  8-amino-1-naphthalenecarboxylic acid from,
    15:747
Cyanoolefins
  from nitriles, *15*:892
p-Cyanophenol [*767-00-0*], *13*:72
α-Cyano-3-phenoxybenzyl alcohol [*39515-47-4*]
  in fenvalerate synthesis, *13*:457
α-Cyano-3-phenoxybenzyl DL-*cis*-3-(2,2- .
    dibromovinyl)-2,2-dimethylcyclopropanecar-
    boxylate [*52820-00-5*]
  as insecticide, *13*:457
p-Cyanophenyl diethyl phosphate [*6132-16-7*],
    13:451
Cyanosis
  oxygen supply in, *16*:670
Cyanothioacetamide [*7357-70-2*], *14*:805
Cyanuric chloride [*108-77-0*]
  in bonding nonwovens, *16*:120
Cybotactic groups
  in nematic liquid crystals, *14*:407
Cycads
  in N_2-fixing associations, *15*:948
Cyclamal [*103-95-7*]
  in perfumes, *16*:967
Cyclic adenosine monophosphate [*60-92-4*]
  (CAMP), *15*:587
Cyclic ketones.   See *Ketones.*
Cyclic polyenes
  as chromogens, *14*:552
Cyclobutadieneiron tricarbonyl [*12078-17-0*], *16*:
    596
  salts from, *16*:602
1,3-Cyclobutanedione [*15506-53-3*], *13*:878, 883,
    887
Cyclobutanes
  from maleic anhydride, *14*:776
Cyclobutanone [*1191-95-3*]
  properties, *13*:923
Cyclobutenes
  from maleic anhydride, *14*:776
Cyclocosanone [*6907-39-7*] .
  properties, *13*:923
Cyclodecanone [*1502-06-3*]
  properties, *13*:923
Cyclodienes
  as insecticides, *13*:434
Cyclododecane [*294-62-2*], *16*:602
Cyclododecanone [*830-13-7*]
  properties, *13*:923
trans,trans,cis-Cyclododecatriene [*39881-80-6*]
  isoprene reaction product, *13*:823
1,5,9-Cyclododecatriene [*4904-61-4*] (CDT)
  nickel acetylacetonate reaction, *15*:808
1,5,9-Cyclododecatriene-centro-nickel(0) [*12126-
    69-1*], *16*:602

Cyclododecatriene nickel [*39330-67-1*], *15*:806
[(1,2,5,6,9,10-η)-1,5,9-Cyclododecatriene]nickel,
  stereoisomer [*12126-69-1*], *15*:808
Cycloheptadecanone [*3661-77-6*]
  properties, *13*:923
Cycloheptanone [*502-42-1*]
  properties, *13*:923
π-Cycloheptatrieneiron tricarbonyl [*36343-88-1*],
  16:602
Cyclohexadecanone [*2550-52-9*]
  properties, *13*:923
2,4-Cyclohexadiene-6-diazo-1-one [*4024-72-0*]
  rearrangement, *13*:881
2,5-Cyclohexadiene-1,4-dione [*106-51-4*], *13*:41
3,5-Cyclohexadiene-1,2-dione [*583-63-1*], *13*:41
Cyclohexane [*110-83-8*]
  liquid-phase nitration, *15*:979
  oxidation of, *13*:924
1,2-Cyclohexanedione [*765-87-7*]
  in catechol production, *13*:51
  properties, *13*:926
1,3-Cyclohexanedione [*504-02-9*], *13*:46
  properties, *13*:928
  in resorcinol production, *13*:52
1,4-Cyclohexanedione [*637-88-7*]
  properties, *13*:931
Cyclohexanol [*108-93-0*]
  from cyclohexane, *13*:924
Cyclohexanone [*108-94-1*]
  catechol from, *13*:51
  in ethinamate synthesis, *13*:126
  peracid oxidation, *13*:922
  properties, *13*:897
  properties, mfg, *13*:924
Cyclohexanone oxime [*100-64-1*]
  from nitrocyclohexanone, *15*:979
Cyclohexanone peroxide [*12262-58-7*], *13*:364
Cycloheximide[*66-81-9*]
  protein synthesis inhibitor, *15*:136
3-Cyclohexyl dihydro-2,5-furandione [*5962-96-9*],
  14:777
(1,2-Cyclohexylenedinitrilo)tetraacetic acid [*482-54-2*]
  in water treatment, *13*:730
Cyclohexyl isocyanate [*3173-53-3*]
  p-acetylbenzene sulfonamide reaction, *13*:615
  properties, *13*:789
Cyclohexyl methacrylate [*101-43-9*]
  properties, *15*:353
4-Cyclohexyl-4-methyl-2-pentanone [*4927-39-3*]
  odor compensation with, *16*:304
Cyclohexyloxymethyl methacrylate [*76392-12-3*]
  properties, *15*:353
Cyclohexylphosphonic dichloride [*1005-22-7*]
  resorcinol reaction, *13*:61
Cyclononadecanone [*6907-38-6*]
  properties, *13*:923

Cyclononanone [*3350-30-9*]
  properties, *13*:923
Cyclooctadecanone [*6907-37-5*]
  properties, *13*:923
Cyclooctadiene [*111-78-4*]
  copper(II) halide reaction, *16*:611
1,5-Cyclooctadiene [*111-78-4*] (COD)
  Ni complex reaction, *15*:808
Cyclooctanone [*502-49-8*]
  properties, *13*:923
Cyclooctatetraene [*629-20-9*]
  nickel halide reaction, *15*:806
Cyclooctatetraenechromium tricarbonyl [*12093-03-7*], *16*:600
Cyclooctatetraene [*629-20-9*] (COT)
  metal complexes, *16*:600
Cyclooctatetraeneiron tricarbonyl [*12093-05-9*],
  16:599
Cyclooctatetraenemolybdenum tricarbonyl
  [*12108-93-9*], *16*:599
Cyclooxygenase [*39391-18-9*]
  in prostaglandin syntheses, *15*:780
Cyclopentadecanone [*502-72-7*]
  properties, *13*:923
Cyclopentadiene [*542-92-7*]
  dichloroketene reaction, *13*:877
  ferrocene synthesis, *16*:611
  isoprene reaction, *13*:821
  maleic anhydride reaction, *14*:775
  methacrylate ester reaction, *15*:349
  sodium reaction, *16*:558
2,4-Cyclopentadienecarboxylic acid [*45520-93-2*],
  13:881
2,4-Cyclopentadieneketene [*4727-22-4*], *13*:881
π-Cyclopentadienylbenzenechromium [*12093-16-2*], *16*:599
π-Cyclopentadienylcobalt cyclopentadiene
  [*33032-03-0*], *16*:604
π-Cyclopentadienylcobalt 1,5-cyclooctadiene
  [*12184-35-9*], *16*:602
π-Cyclopentadienylcobalt 1,3,5,7-
  cyclooctatetraene [*12110-49-5*], *16*:602
π-Cyclopentadienyl complexes
  reactions, *16*:603
η⁵-Cyclopentadienyl-η⁴-cyclobutadienecobalt
  [*12306-95-5*], *16*:593
η⁵-Cyclopentadienyl-η⁶-cycloheptatrienylchromi-
  um [*43138-04-1*], *16*:593
π-Cyclopentadienyl-σ-cyclopentadienyliron
  dicarbonyl [*12247-96-0*], *16*:601
Cyclopentadienyl dicarbonyl cobalt [*12078-23-8*]
  polymer-supported catalyst, *15*:194
Cyclopentadienyldimethylsilane [*1295-15-4*]
  ferrocene derivative, *15*:202
Cyclopentadienyl ions
  metal π-complex formation, *16*:592

Cyclopentadienylium-h^3-cycloheptatrienylium, *16*:592

2,4-Cyclopentadienylmagnesium bromide [*34766-86-4*]
  ferric chloride reaction, *16*:612

Cyclo-2,4-pentadienyl methacrylate [*76741-96-3*]
  properties, *15*:353

(η^5-2,4-Cyclopentadien-1-yl)[(1,2,3,4,5-η)-1-[2-meth- acryloylethyl]-2,4-cyclopentadien-1-yl]iron, polymer with methyl acrylate [*35443-63-1*], *15*:189

Cyclopentadienylnickel carbonyl dimer [*12170-92-2*], *15*:808

Cyclopentadienylnickel nitrosyl [*12071-73-7*], *15*:807

η^5-Cyclopentadienylnickel nitrosyl [*12071-73-7*], *16*:593

π-Cyclopentadienylrhenium tricarbonyl [*12079-73-1*]
  sulfonation, *16*:606

π-Cyclopentadienylrhodium cyclopentadiene [*1293-77-2*], *16*:604

Cyclopentadienylsodium [*4984-82-1*], *16*:557

1,2-Cyclopentanedione [*3008-40-0*]
  properties, *13*:926

1,3-Cyclopentanedione [*3859-41-4*]
  properties, *13*:928

Cyclopentanone [*120-92-3*], *15*:892
  properties, *13*:923

(*Z*)-Cyclopenten-1-one-2-(2-pentenyl) [*41031-88-3*]
  sodium dimethyl malonate reaction, *14*:799

π-Cyclopentenyl-π-cyclopentadienylnickel [*31811-17-3*], *16*:604

Cyclophosphamide [*6055-19-2*]
  antiaging effect, *15*:139
  in neoplastic therapy, *13*:172

Cyclopropane [*75-19-4*]
  from 1,3-dichloropropane, *13*:672

Cyclopropanone [*5009-27-8*]
  from ketene, *13*:877
  properties, *13*:923

Cyclotetradecanone [*3603-99-4*]
  properties, *13*:923

Cyclotridecanone [*832-10-0*]
  properties, *13*:923

Cyclotrimethylenetrinitramine [*121-82-4*] (RDX), *15*:844

Cyclotron
  in isotope detection, *13*:859

Cyclotron-resonance tubes, *15*:503

Cycloundecanone [*878-13-7*]
  properties, *13*:923

Cylinder stock
  in crude oil distillation, *14*:485

Cymbopogon martini Stapf var. *motia*, *16*:326

Cymbopogon winterianus Jowitt, *16*:322

p-Cymene [*99-87-6*], *16*:313
  microbial oxidation, *15*:460

p-Cymen-8-ol [*1197-01-9*], *16*:313

Cyperus esculentus, *16*:250

Cyperus papyrus, *16*:768

Cyprus vitriol
  nitric acid from, *15*:853

Cysteine [*52-90-4*]
  antiaging effects, *15*:138

L-Cysteine [*52-90-4*]
  microbial synthesis of, *15*:465

Cysteine sulfinic acid [*2381-08-0*]
  in taurine biosynthesis, *15*:781

Cystic fibrosis
  screening of with ISES, *13*:732

Cystine [*56-89-3*]
  in cattle hair, *14*:209
  in nuts, *16*:249

Cytochrome b [*9035-37-4*]
  sulfur in, *15*:587

Cytochrome c oxidase [*9001-16-5*]
  copper in, *15*:589

Cytochrome P-450 [*9035-50-1*]
  antiepileptic drug reaction, *13*:135

Cytochromes, *13*:782

Cytoplasmic membrane
  quaternary ammonium agents' effect on, *13*:223

Cytosine [*71-30-7*]
  in nucleic acid, *15*:587

Cytox 3522, *13*:243

Cytox 3711, *13*:243

Cytox 3810, *13*:243

Cytoxan, *13*:172

Czochralski crystal growth
  in IC manufacture, *13*:625

Czochralski technique
  for semiconductor preparation, *14*:286

D

DAA.   See *Diacetone alcohol.*

D-acid [*6361-37-1*], *15*:739.   (See also *4-Hydroxy-1,6-naphthalenedisulfonic acid.*)

Dacron
  in electrical insulation, *13*:556

Dacus cucurbitae, *13*:481

Dacus dorsalis, *13*:481

DAFSA-LES LIAISONS FINANCIERES
  information system, *13*:328

DAFSA LIAISONS FINANCIERES information
  system, *13*:334

Daguerre's process, *13*:649

Dahl's acid [*81-05-5*], *15*:729, 739. (See also *6-Amino-1-naphthalenesulfonic acid.*)

Dahl's acid II [*85-74-5*], *15*:730. (See also *4-Amino-1,7-naphthalenedisulfonic acid.*)

Dahl's acid III [*85-75-6*], *15*:730. (See also *4-Amino-1,6-naphthalenedisulfonic acid.*)

Dairy products
  lactic acid in, *13*:85

Dalmane, *13*:122

β-Damascenone [*23726-93-4*]
  in perfumes, *16*:953

*trans-*β-Damascenone [*23726-93-4*], *16*:312

α-Damascone [*24720-09-0*]
  in perfumes, *16*:967

β-Damascone [*23726-91-2*]
  in perfumes, *16*:967

Dammar gum [*9000-16-2*]
  in matches, *15*:5

Damping
  in vibration control, *13*:526

Dandruff
  treatment, *13*:60

DAP. See α,α'-*Diaminopimelic acid.*

Daphnia, *13*:55

Darcy's law, *15*:99, 224

Darzens-Claisen condensation, *13*:895

Dashpots, *13*:532

Data
  reference standards, *15*:36

Data-base management, *13*:509

Data bases. See *Information retrieval.*
  business, *13*:326
  for patent information, *16*:901

Data compilations, *13*:285

Data resources, *13*:326

Dating
  fission track, *16*:843

Dating, isotope, *13*:860

Davison 714, *15*:660

Dawsonite [*12011-76-6*], *16*:335

Day-glo blaze orange, *14*:558

Daylight fluorescent pigments
  in inks, *13*:379

Daylight pigments, *14*:546

DBCP. See *1,2-Dibromo-2-chloropropane.*

DBM. See *n-Butyl-sec-butylmagnesium.*

DDA. See *4,4'-Dichlorodiphenylacetic acid.*

DDCN. See *4,4'-Dichlorodiphenylacetonitrile.*

DDD. See *1,1-Dichloro-2,2-bis(p-chlorophenyl)-ethane.*

DDE. See *1,1-Dichloro-2,2-bis(p-chlorophenyl)-ethylene.*

DDT. See *1,1,1-Trichloro-2,2-bis(p-chlorophenyl)ethane.*

DDT [*50-29-3*]
  toxicity, *13*:266

DDT dehydrochlorinase [*50-29-3*], *13*:474

DDVP. See *Dichlorvas.*

DEAC. See *Diethylaluminum chloride.*

Deacidite, *13*:696

DEAE cellulose [*9013-34-7*]
  insulin purification, *13*:611

DEAE-Sephadex, *15*:77

Deanol [*108-01-0*]
  learning effects, *15*:134

Death
  from hypnotics overdose, *13*:122

Debutanizer
  in LPG processing, *14*:385

Debye-Hückel theory, *13*:3

Decahydronaphthalene [*91-17-8*], *15*:704. (See also *Decalin.*)

*cis-*Decahydronaphthalene [*493-01-6*], *15*:704
  properties, *15*:705

*trans-*Decahydronaphthalene [*493-02-7*], *15*:704

Decalcomania inks, *13*:396

Decalin [*91-17-8*], *15*:705. (See also *Decahydronaphthalene.*)
  properties, *15*:705

Decalone [*1579-21-1*]
  microbial reduction, *15*:461

Decamethrin [*52820-00-5*], *13*:458

Decamethyl triphosphoramide [*3952-76-9*], *13*:445

Decanal [*112-31-2*], *16*:312

1,10-Decanediol methacrylate [*6701-13-9*]
  properties, *15*:354

2,5-Decanedione [*41368-32-5*]
  properties, *13*:931

Decanol [*112-30-1*]
  as electrode solvent, *13*:724

Decanonitrile [*1975-78-6*], *15*:888

Decarboxylation
  microbial, *15*:466

2-Decenal [*3913-71-1*]
  in orange oil, *16*:325

1-Decene [*872-05-9*]
  disproportionation, *16*:484
  properties, *16*:481

1-*n*-Decylaziridine [*46237-34-7*], *13*:148

n-Decyl chloride [*1002-69-3*]
  aziridine reaction, *13*:148

n-Decyl methacrylate [*3179-47-3*]
  properties, *15*:353

*cis-*2-Decyl-3-(5-methylhexyl)oxirane [*2904-22-6*]
  encapsulation, *15*:487

4-Decyloxy-3-benzoic acid [*5519-23-3*]
  mesomorphism, *14*:413
4,4-Decyloxy-3-bromobenzoic acid [*5519-23-3*],
  14:413
Decyloxyresorcinol [*25934-35-4*], *13*:43
5-Decyne [*1942-46-7*]
  nitromethane reaction, *15*:977
Deep-sea nodules, *16*:291. (See also
  Ferromanganese nodules.)
  metal content, *14*:825
Deer hides
  for leather, *14*:224
Deethanizer
  in LPG processing, *14*:385
Defects
  nondestructive analyses, *16*:50
Defoamers
  in lube oils, *14*:496
  in papermaking, *16*:791, 806
Deformation
  of metals, *15*:327
Deguelin [*522-17-8*]
  insecticide, *13*:426
Dehumidifier
  lithium chloride as, *14*:461
Dehydration
  of meat, *15*:70
  zeolite catalysts in, *15*:652
Dehydroacetic acid [*520-45-6*]
  from diketene, *13*:879
Dehydrodiconiferyl alcohol [*4263-87-0*]
  in lignin, *14*:301
Dehydrogenation
  isobutyrates, *15*:365
  paraffins, *16*:492
  zeolite catalysts in, *15*:652
Dehydrorotenone [*4440-77-1*], *13*:426
Delnet, *16*:78
Delphi technique, *14*:901
Delrin
  in electrical insulation, *13*:547
Deltamax, *14*:651, 659
Delta rays
  from fission particles, *16*:831
Delustering agents, *16*:111
Delweve, *16*:78
Demethanizer
  in LPG processing, *14*:386
Demethylation
  microbial, *15*:465
N-Demethylvindoline [*54759-53-4*]
  microbial synthesis of, *15*:465
Demeton [*8065-48-3*], *13*:438, 445
Demeton methyl [*8022-00-2*], *13*:446
Demineralization
  by ion exchange, *13*:700
Demodex
  effect on cattlehide, *14*:208

Denitrification
  in biological–cycle, *15*:941
Dental caries
  effect of fluoride, *15*:593
Dental cements
  itaconic–acrylic acid copolymers for, *13*:871
Dental materials
  methacrylates, *15*:393
Denton filters, *16*:537
Deodorants and antiperspirants
  fragrances, *16*:970
Deodorants, industrial, *16*:297
Deodorant soaps
  TCC in, *13*:239
Deoxyribonucleic acid (DNA)
  cholesteric properties, *14*:419
2-Deoxy-D-ribose [*533-67-5*]
  in nucleic acids, *15*:587
Depakene, *13*:135
Depigmentation
  agents for, *13*:63
Depilatories
  for cattlehides, *14*:209
Depolymerization
  in insulation materials, *13*:545
Depository libraries
  for patents, *16*:900
Depot devices, *15*:124
Derris elliptica, *13*:425
Derris malaccinsis, *13*:426
Derwent, *13*:299
  information systems, *16*:905
Desalination, *15*:92. (See also *Membrane
  technology.*)
  economics of, *15*:114
  energetics, *15*:112
  energy consumption, *15*:114
  ion-exchange resins for, *13*:688
Desalination membranes, *15*:114
Desalination of water
  membrane, *15*:106
Desalination plants
  table, *16*:282
Desferal, *13*:786
Desglycinamide[9]-arginine[8]-vasopressin [*37552-
  33-3*] (DG-AVP)
  avoidance behavior effect, *15*:133
Desglycinamide[9]-lysine[8]-vasopressin [*37552-33-3*]
  (DG-LVP)
  avoidance behavior effect, *15*:133
Desiccants
  ion exchange resins as, *13*:703
Design
  reliability of materials, *15*:9
Design FMA (DFMA), *15*:12
Design patents, *16*:891

N-Desmethyldiazepam [*1088-11-5*]
  as hypnotic-sedative metabolite, *13*:132
Desmodur N, *13*:792
Desmodur W, *13*:801, 807
Desmodur Z-4370, *13*:807
11-Desoxycorticosterone [*64-85-7*], *13*:173
11-Desoxycortisol [*152-58-9*]
  in corticosteroid biosynthesis, *15*:762
Desulfovibrio, *15*:950
Desulfurization
  microwave techniques, *15*:516
Detector
  infrared, *13*:340
Detergent alcohols
  oxo products, *16*:649
Detergent builders
  lactic acid in, *13*:88
  zeolites, *15*:668
Detergents
  in automotive oils, *14*:488
  fragrances, *16*:968
  in lube oils, *14*:494
  α-olefins for, *16*:572
  oxo alcohols for, *16*:638
Detonation gun, *15*:254
Detoxification systems
  in the human body, *15*:587
Deuterio-oxalic acid [*21481-13-0*], *16*:621
Deuterium [*7782-39-0*]
  in nuclear reactors, *16*:161
  separation, *16*:162
  separation processes, *13*:847
Deuterium fluoride lasers, *14*:61
Deuterium nitride [*13550-49-7*]
  in isotope separation, *13*:849
Deuterium oxide [*7789-20-0*]
  in isotope separation, *13*:849
Devarda's alloy [*8049-11-4*]
  in nitric acid analysis, *15*:867
Developer 8, *15*:745
Developers
  dihydroxybenzenes, *13*:57
Devonian-Mississippian shales, *16*:333
Devrinol, *15*:729
Dexamethasone [*50-02-2*], *13*:173
Dextran [*9004-54-0*]
  methylation, *15*:442
  in sugar, *15*:440
Dextranase [*9025-70-1*]
  in sugar processing, *15*:440
Dextran hydrogen sulfate [*9042-14-2*]
  antilipemic effect, *15*:445
Dextransucrase [*9032-14-8*], *15*:440
Dextrin [*9004-53-9*]
  in nuts, *16*:252
  oxalic acid from, *16*:621

Dextrose [*50-99-7*], *15*:440.   (See also *Corn sugar*.)
  meat-curing agent, *15*:62
  in soybean lecithin, *14*:251
Dezincing
  lead, *14*:122
DFP. See *Diisopropyl fluorophosphate.*
DG-AVP. See *Desglycinamide9-arginine8-vasopressin.*
DG-LVP. See *Desglycinamide9-lysine8-vasopressin.*
D-gun. See *Detonation gun.*
DHCC. See *Dihydroxycholecalciferol.*
Diabetes
  from iron toxicity, *15*:595
Diabetes mellitus, *13*:605
  treatment, *13*:606; *15*:780
Diabinese, *13*:616
Diabrotica, *13*:473
Diacetone alcohol [*123-42-2*], *13*:908
  in paint and varnish removers, *16*:764
  properties, *13*:898, 912
Diacetone-2-keto-L-gulonic acid [*26996-98-5, 28300-95-0*]
  microbial oxidation, *15*:459
Diacetone-L-sorbose [*53452-77-0*]
  microbial oxidation, *15*:459
Diacetyl, 2,3-butanedione [*431-03-8*]
  in bonding nonwovens, *16*:115
1,1'-Diacetylferrocene [*1273-94-5*], *16*:604
1,2-Diacetylferrocene [*12766-56-2*], *16*:604
1,3-Diacetylindole [*17537-64-3*], *13*:215
Diacetyl peroxide [*110-22-5*], *13*:359
  half-life of, *13*:359
Diacetyltartaric acid [*1826-99-9*], *13*:112
Diacetyltartaric anhydride [*74113-55-6*], *13*:112
  thermolysis to carbon suboxide, *13*:882
1,2-Diacylglycero-3-phosphorylcholine [*8002-43-5*]
  in commercial lecithin, *14*:250
Diacyl peroxides
  as initiators, *13*:359
Diagnostic chemicals.   See *Medical diagnostic reagents.*
Diagnostic reagent systems, *15*:75
Diaion, *13*:696
Dialdehydes
  from oxo process, *16*:644
Dialifor [*10311-84-9*], *13*:443
Dialkyldimethyl–ammonium chloride [*5538-94-3*]
  applications of, *13*:234
Dialkyl peroxides
  as initiators, *13*:361
Dialkyl peroxydicarbonates, *13*:362
Diallyl disulfide [*2179-57-9*]
  in garlic, *16*:311

Diallyl phthalate [131-17-9]
  in resin insulators, 13:558
DIALOG
  information systems, 16:901
DIALOG information system, 13:290
Dialysis, 15:92, 115. (See also *Membrane technology*.)
  membranes for, 15:116
DIAM 21, 13:242
Diamagnetism, 14:646
Diaminoacetone [23645-02-5]
  in histamine synthesis, 15:769
2,4-Diamino-3,5-dicyanothiophene [38189-94-5], 14:805
2,2-Diaminoethyldisulfide [51-85-4]
  antiaging effects, 15:138
1,6-Diaminohexane. See *Hexamethylenediamine.*
1,3-Di(aminomethyl)benzene [1477-55-0], 15:906
1,3-Di(aminomethyl)cyclohexane [2579-20-6], 15:906
2,4-Diamino-6-phenyl-1,3,5-triazine [91-76-9], 15:905
meso-α,α′-Diaminopimelic acid [3829-23-0] (DAP)
  microbial decarboxylation of, 15:466
2,3-Diaminotoluene [2687-25-4]
  effect on TDI yield, 13:804
2,4-Diaminotoluene [95-80-7], 15:931
3,4-Diaminotoluene [496-72-0]
  effect on TDI yield, 13:804
Diammonium barium hexakiscyanoferrate [60700-20-1], 13:766
Diammonium calcium hexakiscyanoferrate(II) [60674-40-0], 13:766
Diammonium magnesium hexakiscyanoferrate [66139-49-9], 13:766
Diamonds
  from the ocean, 16:285
DIANE information system, 13:292
Dianisidine diisocyanate [91-93-0]
  properties, 13:789
Diano hardy II
  spectrophotometer, 14:563
Diaper covers
  nonwoven, 16:123
Diaphragm cell
  membranes, 15:120
Diarylide yellow [6358-85-6]
  in inks, 13:379
Diathermy
  microwave techniques for, 15:514
Diatomaceous earth [7631-86-9], 13:595. (See also *Diatomite*.)
  filler in plastics, 13:973
Diatomite [61790-53-2]
  as insulation, 13:597
  in matches, 15:5

thermal conductivity of, 13:595
zeolites from, 15:655
1,4-Diazabicyclooctane peroxyhydrate [39810-25-8], 13:15
Diazene [3618-05-1]
  in nitrogen reduction, 15:958
Diazepam [439-14-5]
  in ketazolam synthesis, 13:132
  properties of, 13:122
Diazepam-*N*-oxide [2888-64-4]
  in temazepam synthesis, 13:130
Diazinon [333-41-5], 13:441
Diazoketones
  ketenes from, 13:881
Diazomethane [334-88-3]
  ketene reaction, 13:877
  lactic acid alkylation, 13:84
Diazotized *o*-chloro-*m*-toluidene-*p*-sulfonic acid
  diazotized, in nickel pigment formation, 15:814
Diazotype process, 13:881
Diazoxide USP [364-98-7]
  hyperglycemic agent, 13:619
Dibactol, 13:234
Dibarium hexakiscyanoferrate [13821-06-2], 13:766
  in ferrocyanic acid synthesis, 13:766
Dibasic lead phosphite [12141-20-7], 14:168
Dibasic lead phthalate [17976-43-1], 14:172
Dibasic lead sulfate [12036-76-9], 14:167
Dibekacin [34493-98-6]
  microbial stability, 15:466
Dibenzenechromium [1271-54-1], 16:598, 607
Dibenzenerhenium(+1) cation [11077-50-2]
  hydride addition, 16:607
Dibenzenetitanium [52462-43-8], 16:613
Dibenzo-18-crown-6-polyether [14605-55-1], 13:43
Dibenzonitrile palladium dichloride [14220-64-5], 16:614
1,2-Dibenzoylethane [495-71-6]
  properties, 13:931
trans-Dibenzoylethylene [959-28-4], 14:773
Dibenzoylmethane [120-46-7], 13:930
Dibenzoyl peroxide [94-36-0]
  half-life of, 13:359
2,5-Di(benzoylperoxy)-2,5-dimethylhexane [618-77-1]
  half-life of, 13:360
Dibenzoylquinone dioxime [120-52-5]
  in electric insulation, 13:565
Dibenzylchlorosilane [18407-39-1]
  isoprene reaction, 13:826
DIBK. See *Diisobutyl ketone.*
Diborane [19287-45-7], 16:564
  IC dopant, 13:627

isoprene reaction, *13*:826
laser excitation, *14*:73
thermal decomposition, *14*:73
Dibromalonic acid [*595-45-9*], *14*:795
Dibromoacetonitrile [*3252-43-5*]
isoprene reaction, *13*:825
Di-μ-bromobis(η³-2-propenyl)dinickel [*12012-90-7*], *15*:808
Dibromochloropropane [*96-12-8*]
toxicity, *13*:266
1,2-Dibromo-2-chloropropane [*96-12-8*]
as fumigant, *13*:466
1,2-Dibromo-3-chloropropane (DBCP) [*96-12-8*]
as fumigant, *13*:466
1,2-Dibromo-2,4-dicyanobutane [*35691-65-7*]
applications of, *13*:231
1,2-Dibromoethane (EDB) [*106-93-4*]
in TEL, *14*:189
1,1'-Dibromoferrocene [*1293-65-8*]
Ullman coupling, *15*:198
Dibromofluorescein [*596-03-2*]
mercuric acetate reaction, *15*:164
Dibromo[hydrazido(2−)- *N*]-tris[dimethylphenylphosphine]tungsten [*58081-44-0*], *15*:957
Dibromomaleic anhydride [*1122-12-9*], *14*:777
trans-1,4-Dibromo-2-methyl-2-butene [*16526-19-5*]
isoprene reaction product, *13*:824
2,2'-Dibromo-3-nitrilopropionamide [*10222-01-2*]
applications of, *13*:231
2,3-Dibromopropyl methacrylate [*3066-70-4*]
properties, *15*:355
Dibutyl adipate [*105-99-7*]
as repellent, *13*:476
N,N-Dibutylaminoethyl methacrylate [*2397-75-3*]
properties, *15*:354
N-(3-Dibutylaminopropyl)methacrylamide [*76392-03-5*]
properties, *15*:356
Di-*n*-butylborane [*34266-14-3*]
properties, *16*:562
Di-*n*-butylchloroborane [*1730-69-4*]
properties, *16*:562
Di-*tert*-butyl-*p*-cresol [*728-37-0*]
as insulator antioxidant, *13*:551
lubricant antioxidant, *14*:492
Di-*tert*-butyl diperoxyazelate [*16580-06-6*]
half-life of, *13*:360
Dibutyl fumarate [*105-75-9*], *14*:777
2,5-Di-*tert*-butylhydroquinone [*1322-72-1*], *13*:60
(3,5-Di-*t*-butyl-4-hydroxyphenyl)methylene malononitrile [*10537-47-0*]
antifouling agent, *14*:806
Di-*n*-butyl itaconate [*2155-60-4*]
for food packaging materials, *13*:870

properties, *13*:871
Di-*tert*-butylketene [*59578-99-3*], *13*:877
Dibutylmagnesium [*39881-32-8*], *16*:560. (See also *n-Butyl-sec-butylmagnesium.*)
Di-*n*-butylmagnesium [*1191-47-5*], *16*:555
Di-*sec*-butylmagnesium [*17589-14-9*]
properties, *16*:559
Di-*tert*-butylmagnesium [*14627-81-7*]
properties, *16*:559
Di-*n*-butylmagnesium [*1191-47-5*] (DNBM)
mfg and properties, *16*:559
Dibutyl maleate [*105-76-0*], *14*:777
2,6-Di-*tert*-butyl-4-methylphenol [*128-37-0*]
LDPE antioxidant, *16*:414
LLDPE antioxidant, *16*:395
olefin antioxidant, *16*:497
Di-*tert*-butyl peroxide [*110-05-4*], *13*:22. (See also *tert-Butyl peroxide.*)
half-life of, *13*:362
initiator for ferrocenes, *15*:198
olefin reaction, *16*:483
in polymethacrylate mfg, *15*:387
2,2-Di(*tert*-butylperoxy)butane [*2167-23-9*]
half-life of, *13*:362
1,1-Di(*tert*-butylperoxy)cyclohexane [*3006-86-8*]
half-life of, *13*:362
2,5-Di(*tert*-butylperoxy)-2,5-dimethylhexane [*78-63-7*]
half-life of, *13*:362
2,5-Di(*tert*-butylperoxy)-2,5-dimethylhexyne-3 [*68-27-5*]
half-life of, *13*:362
Di(*tert*-butylperoxy)ketals
as initiators, *13*:361
1,1-Di(*tert*-butylperoxy)-3,3,5-trimethylcyclohexane [*6731-36-8*]
half-life of, *13*:362
N,N'-Di-*sec*-butyl-*p*-phenylenediamine [*101-96-2*]
olefin antioxidant, *16*:497
Di-*sec*-butyl phenylphosphonate [*38299-61-5*]
in fuel processing, *16*:181
Dibutyl phosphate [*107-66-4*], *16*:180
by irradiation of tributyl phosphate, *16*:240
2-(Dibutylphosphono)ethyl methacrylate [*3729-11-1*]
properties, *15*:355
Dibutyl phthalate [*84-74-2*]
in inks, *13*:390
as repellent, *13*:476
Dibutyl sebacate [*109-43-3*]
as glass laminate plasticizer, *13*:981
as ion-exchanger solvent, *13*:725
Di-*n*-butylstannic dichloride [*638-18-1*]
immunodeficiency effect, *13*:170

Di-*n*-butyl succinate [*141-03-7*]
  as repellent, *13*:476
Di-*tert*-butylthioketene [*16797-75-4*]
  properties, *13*:875
Di-*n*-butyltin bis(isooctyl mercaptoacetate)
  [*25168-24-5*]
  properties, *16*:573
Di-*n*-butyltin bis(laurylmercaptide) [*1185-81-5*]
  properties, *16*:573
Di-*n*-butyltin diacetate [*1067-33-0*]
  properties, *16*:573
Di-*n*-butyltin dichloride [*683-18-1*]
  properties, *16*:573
Di-*n*-butyltin dilaurate [*77-58-7*]
  properties, *16*:573
Di-*n*-butyltin oxide [*818-08-6*]
  toxicity, *16*:577
Dicalcium ferrite [*12013-62-6*], *14*:356
Dicalcium hexakiscyanoferrate [*13821-08-4*], *13*:
  767
Dicalcium phosphate [*7757-93-9*]
  in feeds, *14*:378
  from lime, *14*:380
Dicalcium silicate [*10034-77-2*]
  in cement, *14*:376
  in magnesium mfg, *14*:585
Dicapthon [*2463-84-5*], *13*:439
1,7-Dicarba-*closo*-dodecacarborane-12 [*16986-
  24-6*], *13*:402
Dicarbonylbis(*o*-phenyldimethylphosphine)nickel
  [*76404-14-3*], *15*:807
Dicarbonylbis(phosphorustrifluoride)nickel
  [*13859-78-4*], *15*:807
Dicarbonylbis(trimethylphosphite)nickel [*16787-
  28-3*], *15*:807
Dicarbonylbis(triphenylarsine)nickel [*15709-52-
  1*], *15*:807
Dicarbonylbis(triphenylstibine)nickel [*28042-59-
  3*], *15*:807
Dicarbonyl(η^5-2,4-cyclopentadien-1-yl)(dinitrogen)manganese [*38600-58-7*]
  in N_2 fixation, *15*:958
Dicarbonyl(η^5-2,4-cyclopentadien-1-yl)phenyldia-
  zenidomanganate(−1) lithium [*64682-78-6*]
  in N_2 fixation, *15*:958
Dicarbonyl(η^5-2,4-cyclopentadien-1-yl)(dimethyl-
  diazene-*N*) manganese [*65878-98-0*]
  in N_2 fixation, *15*:958
Dicarbonyl(η^5-2,4-cyclopentadien-1-yl)(methyldi-
  azenido *N'*)manganate(−1) [*65878-97-9*]
  in N_2 fixation, *15*:958
Dicarbonyl[μ-(dinitrogen-*N,N'*)]-
  tetrakis[(1,2,3,4,5-η)- 1,2,3,4,5-pentamethyl-
  2,4-cyclopentadien-1-yl]di-, zirconium
  [*61396-33-6*], *15*:955

Dicarbonylhydrobis(triphenylphosphine)
  ruthenium [*21029-29-8*]
  oxo process catalyst, *16*:639, 642
Dicarbonyl[(1,2,3,4,5-η)-1-vinyl-2,4-cyclopentadi-
  en-1-yl] cobalt [*73231-00-2*], *15*:186
1,1'-Dicarboxycobaltocenium
  hexafluorophosphate [*33638-89-0*], *16*:604
Dicatechol borate, di-*o*-tolylguanidine salt
  [*16971-82-9*], *13*:63
Dicers
  noise control, *16*:24
Dicetyl peroxydicarbonate [*26322-14-5*]
  half-life of, *13*:363
Di-μ-chloride[(1,2,3-η)-2-methyl-2-palladium]
  [*12081-18-4*], *16*:603
2,8-Dichloroadenine [*2914-09-2*]
  in adenosine synthesis, *15*:756
2,8-Dichloroadenosine [*75802-60-7*]
  in adenosine synthesis, *15*:756
Dichloroaluminum hydride etherate
  in isoprene polymerization, *13*:829
Dichloroamine-T, *13*:230
o-Dichlorobenzene [*95-50-1*]
  in isocyanate manufacture, *13*:803
  nitration, *15*:925
  wood preservative, *13*:480
p-Dichlorobenzene [*106-46-7*]
  as fumigant, *13*:467
  nitration, *15*:925
3,3'-Dichlorobenzidine [*91-94-1*]
  in inks, *13*:378
p,p'-Dichlorobenzilic acid [*23851-46-9*], *13*:462
  in insect control, *13*:461
p,p'-Dichlorobenzophenone [*90-98-2*], *13*:461, 462
  in insecticide, *13*:429
7,7-Dichlorobicyclo[3.2.0]hept-2-en-6-one [*5307-
  99-3*]
  tropolone from, *13*:877
1,1-Dichloro-2,2-bis(*p*-chlorophenyl)ethane
  (DDD; TDE) [*72-54-8*]
  as insecticide, *13*:429
  in insecticide, *13*:429
Dichlorobis(η^5-2,4-cyclopentadien-1-yl)titanium
  [*1271-19-8*]
  N_2 reduction, *15*:954
1,1-Dichloro-2,2-bis-(*p*-ethylphenyl)ethane [*72-
  56-0*]
  as insecticide, *13*:431
Dichlorobis[(1,2,3,4,5-η)-1,2,3,4,5-pentamethyl-
  2,4-cyclopentadien-1-yl]titanium [*11136-36-
  0*], *15*:955, 956
Dichlorobis(2-propanolato)titanium [*762-99-2*],
  15:955
Di-μ-chlorobis(η^3-2-propenyl)dinickel [*12145-00-
  5*]

acetylene coreaction, *16*:610
Dichlorobis(triphenylphosphine) platinum
[*10199-34-5*]
oxo process catalyst, *16*:640
3,4-Dichloro-1-butene [*760-23-6*], *15*:899
cis-1,4-Dichloro-2-butene [*1476-11-5*]
NaCN reaction, *15*:899
trans-1,4-Dichloro-2-butene [*110-57-6*]
NaCN reaction, *15*:899
Dichlorocarbene [*1605-72-7*]
isoprene reaction, *13*:824
Di-*μ*-chlorodichlorobis(*η²*-ethylene)diplatinum
[*12073-36-8*], *16*:614
Dichlorodicyanobenzoquinone [*84-58-2*]
oxidizing agent, *15*:185
2,3-Dichloro-5,6-dicyanobenzoquinone [*84-58-2*]
lignin oxidation, *14*:299
β,β′-Dichlorodiethyl ether [*111-44-4*]
as fumigant, *13*:466
Dichlorodiethyllead [*13231-90-8*], *14*:184
1,1-Dichloro-2,2-difluoroethylene [*79-35-6*]
isoprene reaction, *13*:824
1,1-Dichloro-2,2-difluoro-3-isopropenylcyclobu-
tane [*74869-67-3*]
isoprene reaction product, *13*:824
Dichlorodifluoromethane [*75-71-8*] (CFM)
in atmosphere, *16*:697
1,1-Dichloro-2,2-difluoro-3-methyl-3-vinylcyclob-
utane [*74631-06-4*]
isoprene reaction product, *13*:824
Dichlorodimethylhydantoin [*118-52-5*]
applications of, *13*:230
Dichlorodimethylsilane [*75-78-5*]
dimethyl polysilanes from, *13*:408
Dichlorodioxochromium [*14977-61-8*]
in nitride formation, *15*:877
4,4′-Dichlorodiphenylacetic acid [*83-05-6*] (DDA)
DDT metabolite, *13*:431
4,4′-Dichlorodiphenylacetonitrile [*20968-04-1*]
(DDCN)
DDT metabolite, *13*:431
Dichlorodiphenylsilane [*80-10-4*], *13*:408
1,2-Dichloroethane [*1300-21-6*]
ammonia reaction, *13*:158
1,2-Dichloroethane (EDC) [*107-06-2*]
in TEL, *14*:189
Dichloroethylaluminum [*563-43-9*], *14*:183
Di(2-chloroethyl) ether [*111-44-4*]
catechol reaction, *13*:43
Dichloroketene [*4591-28-0*]
cyclopentadiene reaction, *13*:877
properties, *13*:875
Dichloromaleic anhydride [*1122-17-4*], *14*:777
1,4-Dichloro-2-methyl-2-butene [*29843-58-1*]
isoprene reaction product, *13*:824

3,4-Dichloro-2-methyl-1-butene [*53920-89-1*]
isoprene reaction product, *13*:824
1,2-Dichloro-4-nitrobenzene [*99-54-7*], *15*:924
properties, *15*:925
1,4-Dichloro-2-nitrobenzene [*89-61-2*]
properties, *15*:925
1,1-Dichloro-1-nitroethane [*594-72-9*]
as fumigant, *13*:466
in nitroethane determination, *15*:981
Dichlorophene [*97-23-4*]
as antimicrobial agents, *13*:226
2,4-Dichlorophenyl benzenesulfonate [*97-16-5*]
as insecticide, *13*:462
2,4-Dichlorophenyldiazonium chloride [*13617-
98-6*]
in bilirubin detection, *15*:83
3,4-Dichlorophenyl isocyanate [*102-36-3*]
properties, *13*:789
Di-(*p*-chlorophenyl)methylcarbinol [*80-06-8*]
in insect control, *13*:461
1,2-Dichloropropane [*78-87-5*]
as fumigant, *13*:466
1,3-Dichloropropane [*142-28-9*]
zinc reaction, *13*:672
trans-1,3-Dichloropropene [*542-75-6*]
as fumigant, *13*:466
1,3-Dichloro-2-propyl methacrylate [*44978-88-3*]
properties, *15*:355
2,6-Dichloroquinone chloroimide [*101-38-2*]
phenol reaction, *15*:551
Dichlorosilane [*4109-96-0*]
in Si epitaxy, *13*:626
meso-2,3-Dichlorosuccinic acid [*1114-09-6*], *14*:
778
3,4-Dichlorotetramethylcyclobutene [*1194-30-5*]
(CBD)
nickel carbonyl reaction, *15*:808
Dichloro(triphenylphosphine)ruthenium(II)
[*32010-88-1*]
in isoprene reaction, *13*:825
DL-*cis,trans*-3-(2,2-Dichlorovinyl)-2,2-dimethyl-
cyclopropanecarboxylate [*52645-53-1*]
as insecticide, *13*:457
Dichlorvos [*62-73-7*] (DDVP), *13*:442
Dichroic filters, *16*:549
Dicobalt hexacarbonyl bis(tri-*n*-butylphosphine)
[*14911-28-5*]
oxo process catalyst, *16*:639
Dicobalt octacarbonyl [*10210-68-1*]
oxo process catalyst, *16*:638
Dicofol [*54532-36-4*], *13*:461, 474
1,1′-Dicopperferrocene [*76082-12-7*]
coupling, *15*:198
Dicrotophos [*141-66-2*], *13*:446, 890

Dictionaries
    technical, *13*:289
Dicumyl peroxide [*80-43-3*]
    in electric insulation, *13*:565
    half-life of, *13*:362
Dicyandiamide [*461-68-5*]
    adiponitrile reaction, *15*:901
    benzonitrile reaction, *15*:905
    in leather manufacture, *14*:216
    nitrile reactions, *15*:892
    in paper coatings, *16*:791
    phenethylamine reaction, *13*:619
1,3-Dicyanobenzene [*626-17-5*], *15*:906
cis-1,4-Dicyano-1-butene [*2141-58-4*]
    reduction, *15*:899
cis-1,4-Dicyano-2-butene [*1119-85-3*]
    reduction, *15*:899
trans-1,4-Dicyano-1-butene [*2141-59-5*]
    reduction, *15*:899
trans-1,4-Dicyano-2-butene [*18715-38-3*]
    reduction, *15*:899
Dicyclohexylamine [*101-83-7*]
    as rust inhibitor, *14*:493
Dicyclohexylammonium nitrite [*3129-91-7*]
    N-nitrosamines from, *15*:993
Dicyclohexylcarbodiimide [*538-75-0*], *15*:784
4,4′-Dicyclohexylmethane diisocyanate [*5124-30-1*]
    properties, *13*:789
Dicyclohexyl peroxydicarbonate [*1561-49-5*]
    half-life of, *13*:363
Dicyclopentadiene [*77-73-6*]
    copper(II) halide reaction, *16*:611
    in lactic acid reaction, *13*:87
Di-η^5-cyclopentadienyl-η^3-allyltitanium(III) [*12110-59-7*], *16*:593
Di-η^5-cyclopentadienyliron [*102-54-5*], *16*:593. (See also *Ferrocene*.)
Di(2,4-dichlorobenzoyl) peroxide [*133-14-2*]
    half-life of, *13*:359
Didymium
    in Mg alloys, *14*:596
    in optical filters, *16*:554
Die castings
    magnesium alloys for, *14*:598
Dieckmann cyclization, *13*:929
Dieldrin [*60-57-1*], *13*:435
    toxicity, *13*:263
Dielectric constant
    various materials, *13*:537
Dielectric heating, *15*:497
Dielectric properties
    materials, foods, *15*:500
Dielectrics, *13*:536. (See also *Insulation, electric.*)
Diels-Alder reaction
    isoprene, *13*:820
    of maleic anhydride, *14*:774

Die-molding processes
    for plastic composites, *13*:975
Diesel engines
    lubricants for, *14*:488
Diesel fuel
    nitroparaffins in, *15*:984
Diester MIL-L-6085 oil
    limit of radiation tolerance, *14*:519
Diet
    antigen source, *13*:167
    relation to diabetes, *13*:606
Diethanolamine [*111-42-4*]
    in lactamide preparation, *13*:90
    in LPG purification, *14*:389
α,α-Diethyacetophenone [*6175-45-7*]
    as photoinitiator, *13*:369
Diethyl acetamidomalonate [*1068-90-2*]
    in L-glutamic acid synthesis, *15*:767
Diethyl adipate [*141-28-6*]
    in putrescine synthesis, *15*:777
Diethylaluminum chloride [*96-10-6*], *16*:586
    1-butene polymerization with, *16*:476
    in isoprene reaction, *13*:823
    properties, *16*:565
Diethylaluminum ethoxide [*1586-92-1*]
    properties, *16*:565
Diethylaluminum hydride [*871-27-2*], *16*:567
Diethylaluminum iodide [*2040-00-8*]
    1-butene polymerization with, *16*:476
    properties, *16*:565
Diethylamine [*109-87-7*]
    from 1-ethylaziridine, *13*:151
    isoprene reaction, *13*:826
1-(2-(Diethylamine)ethyl)-1,3-dihydro-7-nitro-5-phenyl-2*H*-1,4-benzodiazepin-2-one dihydrochloride [*30195-52-9*]
    sedative, *13*:126
7-Diethylaminocoumarin [*20571-42-0*], *14*:550
Diethylaminoethyl methacrylate [*105-16-8*]
    in papermaking, *16*:804
2-Diethylaminoethyl methacrylate [*105-16-8*]
    in emulsion polymerization, *14*:87
    properties, *15*:354
O,O-Diethyl O-*p*-aminophenyl phosphorothioate [*3735-01-1*]
    from parathion, *13*:439
3-Diethylaminopropyl methacrylate [*17577-32-1*]
    properties, *15*:354
Diethylammonium hydroquinonesulfonate
    inflammation inhibitor, *13*:60
Diethyl barbituric acid [*57-44-3*], *13*:122
Diethylbenzene [*25340-17-4*]
    in divinylbenzene, *13*:688
Diethyl benzyloxymalonate [*5778-35-8*]
    in tartronic acid preparation, *13*:118
Diethylbis(ethylsulfonyl)lead [*14595-72-3*], *14*:183

Diethylborane [*5518-25-2*]
  properties, *16*:562
Diethyl butylidene-malonate [*13937-11-6*], *14*:799
Diethyl butylmalonate [*133-08-4*]
  in phenyl butazone synthesis, *14*:803
Diethylchloroborane [*5314-83-0*]
  properties, *16*:562
O,O-Diethyl *O*-(3-chloro-4-methyl-2-oxo-2*H*-1-
    benzopyran-7-yl) phosphorothionate [*56-72-
    4*]
  as insecticide, *13*:448
O,O-Diethyl *S-p*-chlorophenylthiomethyl
    phosphorodithioate [*786-19-6*]
  as insecticide, *13*:442
Diethyl 2-cyanoethylmalonate [*17216-62-5*]
  sodium dimethyl malonate reaction, *14*:799
2,4-Diethyl-1,3-cyclobutanedione [*75379-05-4*],
    13:883
Diethyl (3,5-dimethoxybenzyl)malonate [*5859-
    68-7*], *14*:797
Diethyldimethyllead [*1762-27-2*], *14*:181, 184
Diethyl dimethylmalonate [*1619-62-1*]
  diethyl malonate reaction, *14*:799
Diethylene glycol [*111-46-6*]
  diglycolic acid from, *13*:94
  in inks, *13*:382
  naphthalene azeotrope, *15*:699
Diethylene glycol methacrylate [*2358-84-1*]
  properties, *15*:354
Diethylenetriamine [*111-40-0*]
  metallic copolymers from, *15*:205
  in missile fuel, *14*:518
  in papermaking, *16*:817
Diethylenetriamine–adipic acid polyamide
    [*25085-20-5*]
  in papermaking, *16*:804
Diethylethoxyaluminum [*1586-92-1*]
  nickel acetylacetonate reaction, *15*:808
O,O-Diethyl *O*-2-(ethylthio)ethyl
    phosphorothionate [*8065-48-3*]
  in insecticide, *13*:445
O,O-Diethyl *S*-2-(ethylthio)ethyl
    phosphorothiolate [*126-75-0*]
  as insecticide, *13*:447
  in insecticide, *13*:445
O,O-Diethyl *S*-(ethylthio)methyl
    phosphorodithioate [*298-02-2*]
  as insecticide, *13*:447
2,5-Di(2-ethylhexanoylperoxy)-2,5-dimethylhex-
    ane [*13052-09-0*]
  half-life of, *13*:360
Diethyl hexyl phosphoric acid [*298-07-7*]
  indium extraction, *13*:208
Di-2-ethylhexyl phthalate [*117-81-7*]
  insulation properties of, *13*:551
Di(2-ethylhexyl) sebacate [*122-62-3*]
  in grease formulation, *14*:504

  in lubricants, *14*:496
3,3-Diethyl-5-hydroxymethyl-1,2,3,4-tetrahydro-
    pyridine-2,4-dione [*20096-03-1*]
  in methyprylon synthesis, *13*:124
Diethyl [(4-hydroxyphenyl)methylene] malonate
    [*17315-86-5*], *14*:804
Diethyl(iodo) aluminum [*2040-00-8*]
  propylene polymerization cocatalyst, *16*:458
O,O-Diethyl *O*-2-isopropyl-4-methylpyrimidyl-
    6-phosphorothionate [*333-41-5*]
  as insecticide, *13*:440
Diethyl itaconate [*2409-52-1*]
  properties, *13*:871
Diethyl ketone [*96-22-0*]
  properties, *13*:897
Diethyllead [*35557-72-3*], *14*:181
Diethyllead dichloride [*13231-90-8*], *16*:587
Diethylmagnesium [*557-18-6*]
  properties, *16*:559
Diethyl maleate [*141-05-9*], *14*:777
Diethyl malonate [*105-53-3*], *14*:794
  aziridine reaction, *13*:150
  in L-glutamic acid synthesis, *15*:767
  mesityl oxide reaction, *13*:917
  properties, *14*:797
Diethylmercury [*627-44-1*], *16*:586
N,N-Diethylmethacrylamide [*5441-99-6*]
  properties, *15*:356
Diethyl methacryloylphosphonate [*76392-33-1*]
  properties, *15*:355
Diethyl (α-nitromethylbenzyl) malonate [*22975-
    21-9*]
  in amino-2-phenyl-4Δ-1 pyrroline synthesis,
    14:803
Diethyl *p*-nitrophenyl phosphate [*311-45-5*]
  as insecticide, *13*:439
O,O-Diethyl *O-p*-nitrophenyl phosphorothionate
    [*56-38-2*]
  as insecticide, *13*:438
O,S-Diethyl *O-p*-nitrophenyl phosphorothioate
    [*597-88-6*]
  as insecticide, *13*:439
Diethyl oxalate [*95-92-1*], *16*:632
Diethyl oximinomalonate [*6829-41-0*]
  in L-glutamic acid synthesis, *15*:767
Diethylphosphatoethyl methacrylate [*814-35-7*]
  properties, *15*:355
N-(Diethylphospho)methacrylamide [*76392-06-8*]
  properties, *15*:356
Diethyl phosphonocyanidate [*33326-12-4*]
  dialkyl malonate reaction, *14*:798
Diethyl-*o*-phosphoric acid [*598-02-7*]
  as insecticide, *13*:438
Diethyl phosphorofluoridate [*358-74-7*]
  as insecticide, *13*:437
Diethylphosphorothioic acid [*2465-65-8*]
  from parathion, *13*:439

Diethyl sodium phthalimidomalonate [29097-77-6]
  ethyl chloroacetate reaction, 15:760
Diethylstannic dichloride [866-55-7]
  immunodeficiency effect, 13:170
Diethyl *p*-terphenyl-*p*-*p*″-carboxylate [37527-56-3]
  liquid crystalline range of, 14:399
3,3-Diethyl-1,2,3,4-tetrahydropyridine-2,4-dione [77-04-3]
  in methyprylon synthesis, 13:124
8,13-Diethyl-3,7,12,17-tetramethylporphyrin-2-propionic acid [644-00-8], 16:41
Diethyltin oxide [3682-12-0]
  toxicity, 16:577
N-*N*-Diethyl-*m*-toluamide [134-62-3]
  as repellent, 13:477
O,*O*-Diethyl *O*-(3,5,6-trichloro-2-pyridyl) phosphorothionate [2921-88-2]
  as insecticide, 13:440
Diethylzinc [557-20-0], 16:555, 585
Differential-pressure transmitters, 14:446
Diffucap, 15:488
Diffusion
  in ion exchange, 13:683
  in metallic coating, 15:242
  molecular transfer by, 14:912
  steady-state, 14:913
  thermal, 13:602
  unsteady-state, 14:917
Diffusion-bonding
  metal-laminate fabrication, 13:948
Diffusion, gaseous
  in isotope separation, 13:855
Diffusion separation, 15:92.   (See also *Membrane technology.*)
Diffusion, thermal
  in isotope separation, 13:845
Diflubenzuron [25367-38-9], 13:459
(Difluoroamino)pentafluorosulfur [13693-10-2], 14:73
Difolatan [2425-06-1], 14:788
Digestion coefficients
  of nuts, 16:260
Digital audio tape, 14:750.   (See also *Tape, digital audio.*)
Digital displays, 14:395.   (See also *Liquid crystals.*)
Digital image analysis, 16:55
Digital signals
  in feedback control, 13:488
Digitaria, 15:949
Diglycerides
  in ice cream, 15:565
Diglycolamine [929-06-6]
  in LPG purification, 14:389
Diglycolic acid [110-99-6], 13:94.   (See also *Oxydiacetic acid.*)

Diglycolide [502-97-6], 13:91.   (See also *Glycolide.*)
Digoxin [20830-75-5]
  measurement, 15:86
Dihalo Sticks, 13:231
Di-*n*-hexyl adipate [110-33-8]
  as glass laminate plasticizer, 13:981
Dihexylamine [143-16-8]
  nitrosation, 15:989
Di-*n*-hexylmagnesium [37509-99-2], 14:618
Di-*n*-hexylmagnesium [37509-99-2] (DNHM)
  properties, 16:559
Dihycon, 13:135
Dihydroanthracene hydroperoxide [18300-71-5]
  formation of, 13:21
2,3-Dihydro-2,2-dimethyl-7-benzofuranyl *N*-methylcarbamate [1563-66-2]
  as insecticide, 13:453
Dihydrogen. See *Hydrogen.*
β-Dihydroheptachlor [14168-01-5]
  as insecticide, 13:434
4,7-Dihydroindole [26686-10-2], 13:218
1,3-Dihydro-4-methyl-2*H*-benzimidazol-2-one [19190-68-2]
  in isocyanate manufacture, 13:805
1,3-Dihydro-1-methyl-7-nitro-5-phenyl-2*H*-1,4-benzodiazepin-2-one [2011-67-8]
  sedative, 13:126
2,5-Dihydro-3-methyl-1-oxide-2-phospholene [1193-10-8]
  isoprene reaction product, 13:822
Dihydromyrcenol [18479-58-8]
  fragrance compound, 16:969
2,5-Dihydroperoxy-2,5-dimethylhexane [3025-88-5], 13:363
Dihydrophenazine [613-32-1]
  oxidation of, 13:22
4,5-Dihydro-*N*-phenol-2-thiazolamine-tetrafluoroboric acid complex [74411-27-1], 13:151
4,5-Dihydro-*N*-phenyl-2-oxazolamine [27151-01-5], 13:151
Dihydro-3-(phenylthio)-2,5-furandione [57242-92-9], 14:779
Dihydrositosterol [83-45-4]
  soybean phosphatides associated with, 14:252
Dihydroxyacetone [96-26-4]
  properties, 13:896
2,3-Dihydroxybenzaldehyde [24677-78-9], 13:46, 70.   (See also *Pyrocatechuic aldehyde.*)
2,4-Dihydroxybenzaldehyde [95-01-2], 13:70.   (See also *β-Resorcylic aldehyde.*)
2,5-Dihydroxybenzaldehyde [1194-98-5], 13:70.   (See also *Gentisaldehyde.*)
2,6-Dihydroxybenzaldehyde [387-46-2], 13:70.   (See also *γ-Resorcylic aldehyde.*)

3,4-Dihydroxybenzaldehyde [*139-85-5*], *13*:46, 70. (See also *Protocatechuic aldehyde.*)

3,5-Dihydroxybenzaldehyde [*26153-38-8*], *13*: 70. (See also *α-Resorcylic aldehyde.*)

1,2-Dihydroxybenzene [*120-80-0*], *13*:40. (See also *Catechol.*)

1,3-Dihydroxybenzene [*108-46-3*], *13*:39. (See also *Resorcinol.*)

1,4-Dihydroxybenzene [*123-31-9*], *13*:39. (See also *Hydroquinone.*)

2,4-Dihydroxybenzoic acid [*89-86-1*], *13*:46. (See also *β-Resorcylic acid.*)

2,5-Dihydroxybenzoic acid [*490-79-9*], *13*: 46. (See also *Gentisic acid.*)

2,6-Dihydroxybenzoic acid [*303-07-1*], *13*: 46. (See also *γ-Resorcylic acid.*)

2,4-Dihydroxybenzophenone [*131-56-6*], *13*:44, 45. (See also *Benzoyl resorcinol.*)

1,4-Dihydroxy-2-butene [*110-64-5*] in endosulfan synthesis, *13*:435

3,4-Dihydroxybutyl methacrylate [*62180-57-8*] properties, *15*:353

Dihydroxycholecalciferol in calcium homeostasis, *15*:586

1,25-Dihydroxycholecalciferol [*32222-06-3*] (1,25-DHCC), *15*:585 in blood-Ca²⁻ *15*:585

24,25-Dihydroxycholecalciferol [*40013-87-4*] (24,25-DHCC) in blood-Ca²⁻ *15*:585

4,6-Dihydroxy-4,6-dimethylheptan-2-one [*74631-04-2*], *13*:914

Dihydroxydimethylsiloxane α-hydrogen [*31692-79-2*], *13*:410

2-2′-Dihydroxydiphenyl ether [*1965-09-9*] in o-chlorophenol hydrolysis, *13*:48

4,5-Dihydroxy-4,5-diphenylimidazolidin-2-one [*5157-13-1*] in phenyltoin synthesis, *13*:136

1,4-Di(2-hydroxyethoxy)benzene [*104-38-1*], *13*:43

Dihydroxymaleic acid [*526-84-1*], *13*:112

4,5-Dihydroxy-2,7-naphthalenedisulfonic acid [*148-25-4*], *15*:739

4,5-Dihydroxy-1-naphthalenesulfonic acid [*83-65-8*], *15*:738

6,7-Dihydroxy-2-naphthalenesulfonic acid [*92-27-3*], *15*:738

3-(3,4-Dihydroxyphenyl)alanine [*587-45-1*] in (−) adrenaline biosynthesis, *15*:758

L-3(3,4-Dihydroxyphenyl)alanine [*59-92-7*] (L-DOPA) microbial synthesis of, *15*:464

2,4-Dihydroxyphenylarsonic acid. See *Resorcinolarsonic acid.*

Dihydroxytartaric acid [*617-48-1*], *13*:112

2,5-Dihydroxyterephthalic acid [*610-92-4*], *13*:45

2,2-Diindol-3-yl-3-indolinone [*17646-95-6*], *13*: 217

1,1′-Diiodoferrocene [*12145-93-6*] Ullman coupling, *15*:198

4′,5′-Diiodofluorescein [*38577-97-8*], *13*:671

Diiodogermylene [*13573-08-5*] in chemical deposition coating, *15*:263

Di(iodomercuric)methane [*27329-48-2*], *13*:670

Diisobutylaluminum chloride [*1779-25-5*] properties, *16*:565

Diisobutylaluminum hydride [*1191-15-7*] properties, *16*:565

Diisobutyl carbinol [*108-82-7*], *13*:916

Diisobutylchloroborane [*13317-64-1*] properties, *16*:562

Diisobutyl ketone [*108-83-8*] manufacture, *13*:915 properties, *13*:898

Diisobutyl ketone [*108-83-8*] (DIBK) in niobium processing, *15*:825

Diisobutylmagnesium [*17589-14-9*] properties, *16*:559

Diisobutyryl peroxide [*3437-84-1*] half-life of, *13*:359

Diisocyanates toxicity, *13*:267

1,4-Diisocyanatobutane [*4538-37-8*] in putrescine synthesis, *15*:777

Diisopropenylferrocene [*12289-78-0*], *15*:186

Diisopropylamine [*108-18-9*] color stabilizer, *13*:916

m-Diisopropylbenzene [*99-62-7*] hydroperoxidation, *13*:48 in resorcinol production, *13*:48

Diisopropylbenzene monohydroperoxide [*98-49-7*], *13*:363

Diisopropylbenzenes (DIPB) from cumene, *13*:50

Diisopropyl fluorophosphate [*55-91-4*] (DFP) learning effects, *15*:134

2,4-Diisopropylidene-1,3-dithietane [*10181-64-3*], *13*:883

Diisopropyl peroxydicarbonate [*105-64-6*] half-life of, *13*:363

Di-J-acid [*87-03-6*], *15*:742. (See also *7,7′-Imino-bis-4-hydroxy-2-naphthalenesulfonic acid.*)

Diketene [*674-82-8*] ammonia reaction, *13*:889 in ketazolam synthesis, *13*:132 ozonolysis, *14*:795

Diketene dimer. See *Dehydroacetic acid.*

Dilantin, *13*:135

Dilauroyl peroxide [*105-74-8*] half-life of, *13*:359

Dilauryl 3,3′-thiobispropionate [123-28-4]
  in LLDPE, 16:395
Dilauryl thiodiproprionate [123-28-4]
  as olefin fiber antioxidant, 16:360
Dilead hexakiscyanoferrate [14402-61-0], 13:767
  in ferrocyanic acid synthesis, 13:766
Dilex 101, 15:253
1,1′-Dilithiumferrocene [33272-09-2]
  duodoferrocene complex, 15:198
1,1′-Dilithiumferrocene-TMEDA complex
    [32677-77-3]
  duodoferrocene complex, 15:198
Dimanganese decacarbonyl [10170-69-1], 14:845
  reduction by sodium, 14:851
DIMDI information system, 13:292
Dimefax [115-26-4], 13:445
Dimensionally stable anodes, 15:172, 173. (See
    also *Metal anodes.*)
Dimension stone, 14:360
1,1′-Di(mercury chloride)ferrocene [37328-05-5]
  coupling, 15:198
Dimersol process, 16:485
Dimesitylchromium tetrahydrofuranate [14689-
    01-1], 16:615
Dimesoperiodic acid [14922-00-0], 13:666
Dimethoate [60-51-5], 13:415, 438, 446
3,4-Dimethoxybenzaldehyde [120-14-9]
  in dopamine synthesis, 15:764
1,2-Dimethoxybenzene [91-16-7], 13:44. (See
    also *Veratrole.*)
Dimethoxyethane [110-71-4]
  nickel halide reaction, 15:804
3,4-Dimethoxy β-nitrostyrene [4230-93-7]
  in dopamine synthesis, 15:764
α,α-Dimethoxy-α-phenylacetophenone [24650-
    42-8]
  as photoinitiator, 13:369
Di-4-methoxyphenyl-*trans*-1,4-cyclohexanedicar-
    boxylate [24707-00-4]
  liquid crystalline range of, 14:399
3-(Dimethoxyphosphinyloxy)-*N,N*-dimethylcro-
    tonamide [141-66-2]
  as insecticide, 13:446
2,5-Dimethoxyterephthalic acid [21004-11-5]
  fluorescent brightener, 13:59
Dimethrin [70-38-2], 13:458
N,N-Dimethylacetamide [127-19-5]
  polyamide solvent, 14:417
N,N-Dimethylacetoacetamide [2044-64-6], 13:
    887, 889
Dimethylaluminum chloride [1184-58-3], 16:586
Dimethylamine [124-40-3]
  in isoprene reaction, 13:826
  in lactamide preparation, 13:90
Dimethylamine–epichlorohydrin condensates
  in papermaking, 16:804

p-Dimethylaminobenzaldehyde [100-10-7]
  phenyl isocyanate reaction, 13:795
1-(Dimethylamino)-3-chloro-2-methylpropane
    [23349-86-2]
  in methotrimetrazine synthesis, 13:134
2-(Dimethylamino)-5,6-dimethyl-4-pyrimidinyl
    N,N-dimethylcarbamate [23103-98-2]
  as insecticide, 13:454
2-Dimethylaminoethanol [108-01-0], 15:
    134. (See also *Deanol.*)
2-(*N,N′*-Dimethylamino)ethanol [108-01-0]
  in paint stripping, 16:763
Dimethylaminoethanol [018-01-0] (DMAE)
  antiaging effect, 15:138
3-Dimethylaminoethyleneiminophenyl *N*-
    methylcarbamate hydrochloride [22299-30-9]
  as insecticide, 13:455
(2-Dimethylaminoethyl)ferrocene [33042-18-1],
    16:606
2-Dimethylaminoethyl methacrylate [2867-47-2],
    15:352
  in emulsion polymerization, 14:87
  properties, 15:347
N-(Dimethylaminoethyl)methacrylamide
    [13081-44-2]
  properties, 15:356
(Dimethylaminomethyl)ferrocene [1271-86-9],
    15:196; 16:605
3-(*N,N*-Dimethylaminomethyl)-5-methylindole
    [30218-58-7]
  in melatonin synthesis, 15:772
(Dimethylaminomethyl)ruthenocene [33293-45-
    7], 15:196
trans-Dimethylaminonitrostilbene [4584-57-0]
  in liquid crystal analysis, 14:397
N-(3-Dimethylaminopropyl)methacrylamide
    [5205-93-6]
  properties, 15:356
Dimethylaminostyrene [10448-42-7], 15:191
p-*N,N*-Dimethylaminostyrene [2039-80-7]
  copolymer, 15:192
4-Dimethylamino-3-tolyl *N*-methylcarbamate
    [2032-59-9]
  as insecticide, 13:455
4-Dimethylamino-3,5-xylyl *N*-methylcarbamate
    [315-18-4]
  as insecticide, 13:454
2,3-Dimethylaniline [87-59-2]
  1-naphthalenesulfonic acid reaction, 15:723
Dimethylaniline (DMA) [121-69-7]
  in promoter system, 13:364
Dimethylarsinous iodide [676-75-5], 13:669
2,2-Dimethylaziridine [2658-24-4]
  properties, 13:143
2,2-Dimethyl-1,3-benzodioxol-4-yl *N*-
    methylcarbamate [22781-23-3]
  as insecticide, 13:454

Dimethylbenzylcarbinol [100-86-7]
in perfumes, 16:968
2,4-Dimethylbenzyl dL-*cis*,*trans*-chrysanthemate
[70-38-2]
as insecticide, 13:457
Dimethylborane [7216-97-9]
properties, 16:562
2,3-Dimethylbutadiene [513-81-5]
mesityl oxide reaction, 13:917
2,3-Dimethyl-2-butene [563-79-1]
hydroformylation reactivity, 16:645
3,3-Dimethyl-1-butene [558-37-2], 15:187
2-(*N*,*N*-Dimethylcarbamoyl)-3-methylpyrazol-
5-yl *N*,*N*-dimethylcarbamate [644-64-4]
as insecticide, 13:454
Dimethyl carbate (*cis*-dimethyl bicyclo[2.2.1]-5-
heptene-2,3-dicarboxylate [39589-98-5]
as repellent, 13:477
Dimethyl 2-carbomethoxy-1-methylvinyl
phosphate [7786-34-7]
as insecticide, 13:446
Dimethylchloroborane [1803-36-7]
properties, 16:562
Dimethyl 2-chloro-2-diethylcarbamoyl-1-
methylvinylphosphate [13171-4-6]
as insecticide, 13:446
2,4-Dimethyl-1,3-cyclobutanedione [20117-50-4],
13:883
cis-3,3-Dimethyl-δ′,α-cyclohexaneacetaldehyde
[24393-78-0]
as insect attractant, 13:482
trans-3,3-Dimethyl-δ′,α-cyclohexaneacetaldehyd-
e [24398-89-8]
as insect attractant, 13:482
5,5-Dimethyl-1,3-cyclohexanedione [126-81-8],
13:917
properties, 13:928
cis-3,3-Dimethyl-δ′,β-cyclohexaneethanol
[26532-23-0]
as insect attractant, 13:482
1,5-Dimethyl-1,5-cyclooctadiene [3760-14-3]
isoprene reaction product, 13:822
1,6-Dimethyl-1,5-cyclooctadiene [3760-13-2]
isoprene reaction product, 13:822
Dimethyldiallylammonium chloride [7398-69-8]
in papermaking, 16:804
Dimethyl 1,2-dibromo-2,2-dichloroethyl
phosphate [300-76-5]
as insecticide, 13:442
O,*O*-Dimethyl *S*-(1,2-dicarbethoxyethyl)-
phosphorodithioate [121-75-5]
as insecticide, 13:441
O,*O*-Dimethyl-*O*-2,5-dichloro-4-bromophenyl
phosphorothionate [2104-96-3]
as insecticide, 13:440
O,*O*-Dimethyl *O*-2,5-dichloro-4-iodophenyl
phosphorothionate [18181-70-9]

as insecticide, 13:440
2,3-Dimethyl-2,3-dinitrobutane [3964-18-9], 15:
977
2,2-Dimethyl-1,3-dinitropropane [762-98-1], 15:
976
4,4-Dimethyl-1,3-dioxane [766-15-4]
in isoprene synthesis, 13:831
4,4-Dimethyl-*m*-dioxane, 13:831
3,6-Dimethyl-*p*-dioxane-2,5-dione. See *Lactide.*
Dimethyl 1,1-diphenyl ethane [719-79-9]
as insulator, 13:551
Dimethyl disulfide [624-92-0]
from pulp mills, 16:795
Dimethylethanolamine [108-01-0]
in ion-exchange resins, 13:687
Dimethyl ether [115-10-6], 15:399
from methanol, 15:668
methyl methacrylate by-product, 15:359
1-[(1,1-Dimethylethyl)azo]cyclohexanecarboni-
trile [25149-47-7]
half-life of, 13:366
2-[(1,1-Dimethylethyl)azo]-2,4-dimethyl
pentanenitrile [55912-18-0]
half-life of, 13:366
2-[(1,1-Dimethylethyl)azo]-4-methoxy-2,4-dimet-
hyl pentanenitrile [55912-17-9]
half-life of, 13:366
2-[(1,1-Dimethylethyl)azo]-2-methylbutanenitrile
[52235-20-8]
half-life of, 13:366
2,4-Dimethyl-5-ethyl-6,8-dioxabicyclo[3.2.1]oct-
ane [54832-20-1]
as insect attractant, 13:482
O,*O*-Dimethyl *S*-2-(ethylthio)ethyl
phosphorothiolate [8022-00-2]
as insecticide, 13:445, 447
Dimethylformamide [68-12-2]
in paint and varnish removers, 16:764
phenyl isocyanate reaction, 13:795
2,5-Dimethylfuran [625-86-5]
2,5-hexanedione from, 13:931
2,4-Dimethyl-2,4-heptadiene-6-one [29179-02-0],
13:914
2,6-Dimethyl-3,5-heptanedione [18362-64-6]
properties, 13:928
2,6-Dimethyl-4-heptanol [108-82-7], 13:915
2,6-Dimethyl-4-heptanone [108-83-8]
solvent, 15:722
2,6-Dimethyl-4-heptanone 108-83-8], 13:
915. (See also *Diisobutyl ketone.*)
2,5-Dimethyl-1,6-hexanediol methacrylate
[76392-00-2]
properties, 15:354
2,5-Dimethyl-3,4-hexanedione [4388-87-8]
properties, 13:926
3,4-Dimethyl-2,5-hexanedione [25234-79-1]
properties, 13:931

5,5-Dimethyl-2,4-hexanedione [*29284-62-6*]
  properties, *13*:928
Dimethylhydrazine [*57-14-7*]
  in missile fuel, *14*:518
1,1-Dimethylhydrazine [*57-14-7*]
  1-butene polymerization with, *16*:476
Dimethylhydrogensilyl-1,1′-ferrocenyl-dimethysi-
  lylpoly[dimethylenedimethylsilyl-1,1′-ferro-
  cenyldimethylsilane] [*76082-28-5*], *15*:200
Dimethylhydrogensilylmethylene-1,1′-ferrocenyl-
  methylenedimethylsilylpoly[dimethylenedi-
  methylsilylmethylene-1,1′-ferrocenylmethyl-
  enedimethylsilane] [*76082-29-6*], *15*:200
Dimethyl 1-hydroxy-2-trichloroethylphosphonate
  [*52-68-6*]
  as insecticide, *13*:441
2,3-Dimethylindole [*91-55-4*], *13*:215
3,3-Dimethylindolenine [*1504-14-9*], *13*:215
Dimethyl itaconate [*617-52-7*]
  properties, *13*:871
Dimethylketene [*598-26-5*]
  polymerization, *13*:878
  properties, *13*:875
Dimethylmagnesium [*2999-74-8*]
  properties, *16*:559
Dimethyl malonate [*108-59-8*]
  properties, *14*:797
Dimethylmalonic acid [*595-46-0*]
  dimethylketene from, *13*:880
Dimethylmercury [*593-74-8*], *13*:668; *16*:586
N,N-Dimethylmethacrylamide [*6976-91-6*]
  properties, *15*:356
Dimethyl(3-methyl-2-butenyl)amine [*2588-78-6*]
  isoprene reaction product, *13*:826
O,O-Dimethyl *S*-(*N*-methylcarbamoyl)methyl
  phosphorodithioate [*60-51-5*]
  as insecticide, *13*:445
5,5-Dimethyl-2(1-methylethylidene)-1,3-dioxane-
  4,6-dione [*4858-67-7*], *13*:880
3,3-Dimethyl-4-(1-methylethylidene)-2-oxetanon-
  e [*3173-79-3*], *13*:883
3,3-Dimethyl-4-(1-methylethylidene)-2-thietane-
  thione [*10181-61-0*], *13*:883
O,O-Dimethyl *O*-3-methyl-4-methylthiophenyl
  phosphorothionate [*55-38-9*]
  as insecticide, *13*:440
O,O-Dimethyl *O*-(3-methyl-4-nitrophenyl)
  phosphorothionate [*122-14-5*]
  as insecticide, *13*:440
O,O-Dimethyl *O*-(1-methyl-2-(1-
  phenylcarbethoxy)vinyl) phosphate [*7700-
  17-6*]
  as insecticide, *13*:448
2,6-Dimethylnaphthalene [*28804-88-8*], *15*:716
Dimethylnickel [*54836-89-4*], *15*:809
O,O-Dimethyl *O-p*-nitrophenyl
  phosphorothionate [*298-00-0*]

  as insecticide, *13*:439
2,6-Dimethyl-1,3,7-octatriene [*6876-07-9*]
  isoprene reaction product, *13*:823
1,3-Dimethylol-5,5-dimethylhydantoin [*6440-58-
  0*]
  preservative, cosmetic, *13*:242
Dimethylolurea [*140-95-4*]
  in encapsulation, *15*:480
3,5-Dimethyl 2*H*-1,3,5 oxadiazine-2,4,6(3*H*,5*H*)
  trione [*36209-52-6*], *13*:798
5,5-Dimethyl-2,4-oxazolidinedione [*695-53-4*]
  in trimethadione synthesis, *13*:137
O,O-Dimethyl *S*-4-oxo,1,2,3-benzotriazin-3-(4*H*)-
  ylmethyl phosphorodithioate [*86-50-0*]
  as insecticide, *13*:441
3,4-Dimethylphenol [*95-65-8*]
  naphthalene azeotrope, *15*:699
N-(2,4-Dimethylphenyl)acetoacetamide [*97-36-
  9*], *13*:887
Dimethyl(phenyl)chlorosilane [*768-33-2*]
  in methyllithium analysis, *14*:470
N,N-Dimethyl-4-[(phenylimino)methyl]aniline
  [*889-37-2*]
  isocyanate reaction, *13*:795
Dimethylphenylphosphine [*672-66-2*]
  in N$_2$-containing complex, *15*:956
2-(Dimethylphosphato)propyl methacrylate
  [*76392-35-3*]
  properties, *15*:355
Dimethylphosphinomethyl methacrylate [*41392-
  09-0*]
  properties, *15*:355
Dimethylphosphonoethyl methacrylate [*22432-
  83-3*]
  properties, *15*:355
O,S-Dimethyl phosphoramidothioate [*10265-92-
  6*]
  as insecticide, *13*:443
Dimethyl phthalate [*131-11-3*]
  as repellent, *13*:477
O,O-Dimethyl *S*-phthalimidomethyl
  phosphorodithioate [*732-11-6*]
  as insecticide, *13*:442
1-[(1,1-Dimethylpropyl)azo]cyclohexanecarbo-
  nitrile [*55912-19-1*]
  half-life of, *13*:366
(2,2-Dimethylpropyl)trimethyllead [*17544-69-3*],
  14:184
2,6-Dimethyl-4-pyrone [*1004-36-0*]
  from diketene, *13*:884
Dimethyl silicone
  limit of radiation tolerance, *14*:519
Dimethyl silicone oil [*9016-00-6*]
  as defoamer in grease formulation, *14*:503
α,p-Dimethyl styrene [*1195-32-0*], *16*:313
Dimethyl sulfate [*77-78-1*]
  iodine reaction, *13*:668

lactic acid alkylation, *13*:84
monomer quaternization, *14*:93
Dimethyl sulfide [*75-18-3*]
  from kraft lignin, *14*:311
  from lignin, *14*:296
  methyl iodide reaction, *13*:669
  oxidation, *16*:686
  from peppermint oil, *16*:309
  polymer reaction, *14*:93
  from pulp mills, *16*:795
Dimethyl sulfoxide [*67-68-5*], *16*:686
  methyl iodide reaction, *13*:669
2,2-Dimethylthioazolidine [*19351-18-9*], *13*:150
Dimethylthioketene [*44158-23-8*], *13*:881
Dimethyltin dichloride [*753-73-1*], *16*:576
  properties, *16*:573
N,N-Dimethyl-*p*-toluidine [*99-17-1*]
  in adhesive, *15*:491
4′-(Dimethyltriazeno)acetanilide [*1933-50-0*], *13*:
  476
O,O-Dimethyl *O*-2,4,5-trichlorophenyl
  phosphorothionate [*299-84-3*]
  as insecticide, *13*:440
Dimethyl 3,5,6-trichloro-2-pyridyl phosphate
  [*5598-52-7*]
  as insecticide, *13*:440
N,N-Dimethyltryptamine [*61-50-7*] (DMT)
  properties, *15*:763
N,N-Dimethyltryptamine methiodide [*13558-34-4*], *15*:763
N,N-Dimethyltryptamine picrate [*20203-37-6*]
  as neuroregulator, *15*:763
Dimethylzinc [*544-97-8*], *16*:585
Dimetilan [*644-64-4*], *13*:455
Dimolybdenum oxyoctachloride [*77727-64-1*], *15*:
  683
Dimpylate [*333-41-5*], *13*:884
Dineopentylmagnesium [*19978-31-5*]
  properties, *16*:559
Diniobium carbide [*12011-99-3*], *15*:834
Diniobium nitride [*12033-43-1*], *15*:834
Dinitriles, *15*:892
Dinitro alcohols, *15*:913
Dinitroaniline [*25321-14-6*]
  in isocyanate mfg, *13*:804
Dinitrobenzene [*25154-54-5*]
  in nitrobenzene mfg, *15*:920
m-Dinitrobenzene [*99-65-0*]
  nitrobenzene analysis, *15*:921
  steroid diagnostic reagent, *15*:85
2,4-Dinitrobenzoic acid [*610-30-0*], *15*:931
4,6-Dinitro-*o-sec*-butylphenol [*88-85-7*]
  as insecticide, *13*:429
2,4-Dinitrochlorobenzene [*97-00-7*]
  properties, *15*:925
2,4-Dinitro-1-chlorobenzene [*97-00-7*], *15*:923,
  924

2,6-Dinitro-1-chlorobenzene [*606-21-3*], *15*:923
Dinitro-*o*-cresol [*534-52-1*]
  toxicity, *13*:267
4,6-Dinitro-*o*-cresol [*534-52-1*]
  as insecticide, *13*:428
Dinitrocresol (DNOC) [*534-52-1*]
  as insecticide, *13*:428
1,1-Dinitrocyclohexane [*4028-15-3*]
  properties, *15*:970
4,4′-Dinitrodiphenyl disulfide [*100-32-3*], *15*:924
1,1-Dinitroethane [*600-40-8*]
  properties, *15*:970
1,2-Dinitroethane [*7570-26-5*]
  properties, *15*:970
aci-1,1-Dinitroethane. See *1-Isonitro-1-
  nitroethane.*
Dinitrogen. See *Nitrogen.*
[μ-(Dinitrogen-*N,N′*)bis(dinitrogen)]tetrakis-
  [(1,2,3,4,5-η)-1,2,3,4,5-pentamethyl-2,4-cyclo-
  pentadien-1-yl] dititanium [*67062-55-9*], *15*:
  955
Dinitrogen fixation. See *Nitrogen fixation.*
Dinitrogen molybdenum [*43060-93-1*]
  nitrogenase intermediate, *15*:693
Dinitrogen nitride [*10544-72-6*], *15*:883
[μ-(Dinitrogen-*N,N′*)]tetrakis[(1,2,3,4,5-
  η)-1,2,3,4,5-penta- methyl-2,4-
  cyclopentadien-1-yl] dititanium [*67062-55-9*],
  15:955
Dinitrogen tetroxide [*10544-72-6*]
  propylene oxidation to oxalic acid, *16*:625
Dinitrogen trioxide [*10544-73-7*]
  methyl methacrylate reaction, *15*:351
Dinitromethane [*625-76-3*]
  properties, *15*:970
1,2-Dinitronaphthalene [*24934-47-2*], *15*:725
1,3-Dinitronaphthalene [*606-37-1*], *15*:725
  nomenclature, *15*:701
1,4-Dinitronaphthalene [*6921-26-2*], *15*:725
1,5-Dinitronaphthalene [*605-71-0*], *15*:725
1,6-Dinitronaphthalene [*607-46-5*], *15*:725
1,7-Dinitronaphthalene [*24824-25-7*], *15*:725
1,8-Dinitronaphthalene [*602-38-0*], *15*:725
2,3-Dinitronaphthalene [*1875-63-4*], *15*:725
2,6-Dinitronaphthalene [*24824-26-8*], *15*:725
2,7-Dinitronaphthalene [*24824-27-9*], *15*:725
gem-Dinitroparaffins, *15*:979
Dinitrophenols
  as insecticides, *13*:428
1-(2,4-Dinitrophenyl)aziridine [*27141-65-7*], *13*:
  148
2,2-Dinitropropane [*595-49-3*]
  by liquid-phase nitration, *15*:979
  by oxidation, *15*:978
  from propane, *15*:851

2,2-Dinitropropane (*Continued*)
  properties, *15*:970
4,4′-Dinitrostilbene [*2501-02-2*], *15*:928
Dinitrotoluene [*25321-62-5*]
  isocyanate from, *13*:804
2,3-Dinitrotoluene [*602-01-7*], *15*:927
2,4-Dinitrotoluene [*121-14-2*]
  properties, *15*:931
2,5-Dinitrotoluene [*121-14-2*], *15*:926, 927
2,6-Dinitrotoluene [*606-20-2*], *15*:926
3,4-Dinitrotoluene [*610-39-9*], *15*:927
Dinitrotoluenes (DNTS), *15*:931
  production, *15*:846
Dinocap [*39300-45-3*], *13*:428
Dinoseb [*88-85-7*], *13*:428
Dioctadecyldimethylammonium chloride [*107-64-2*]
  in encapsulation, *15*:477
Di-*n*-octyl phenylphosphonate [*7664-38-2*]
  as electrode solvent, *13*:724
Dioctylphosphinic acid [*683-19-2*]
  zinc acetate reaction, *15*:217
Di-*n*-octylstannic dichloride [*3542-36-7*]
  immunodeficiency effect, *13*:170
Di-*n*-octyltin bis(isooctyl mercaptoacetate) [*26401-97-8*]
  properties, *16*:573
Di-*n*-octyltin oxide [*870-08-6*]
  toxicity, *16*:577
Diodes
  edge emitting, *14*:279
Diodes, light-emitting, *14*:269
Diode sputtering, *15*:267
Diolefins
  metal π-complex formation, *16*:592
Diosphenol [*490-03-9*], *13*:925
Dioxacarb [*6988-21-2*], *13*:76, 453
Dioxane [*123-91-1*]
  toxicity, *13*:267
Dioxane dibromide [*21992-70-1*]
  in indole halogenation, *13*:215
2,3-*p*-Dioxanedithiol *S,S*-bis(*O,O*-diethylphosphorodithioate) [*78-34-2*]
  as insecticide, *13*:442
Dioxathion [*78-34-2*], *13*:443
Dioxazine violet [*6358-30-1*]
  in inks, *13*:379
1,3-Dioxolane [*646-06-0*]
  in lithium batteries, *14*:459
[2(1,3-Dioxolan-2-yl)phenyl-*N*-methylcarbamate] [*6988-21-2*]
  insecticide, *13*:76
O-(1,3-Dioxolan-2-yl)-phenyl *N*-methylcarbamate [*6988-21-2*]
  as insecticide, *13*:453
3,3′-Dioxybis[3,5,5-trimethyl]-1,2-dioxolane [*4507-98-6*], *13*:916

(Dioxygen)(1-methylimidazole)[(21*H*,23*H*-porphine-5,10,15,20-tetracyltetra-2,1-phenylene)-tetrabis (2,2-dimethylpropanamidato)]iron [*52629-15-9*], *13*:783
Dioxyphenylalanine [*59-92-7*], *13*:46.   (See also *L-DOPA*.)
Dioxy R acid [*92-27-3*], *15*:738.   (See also *6,7-Dihydroxy-2-naphthalenesulfonic acid*.)
Dioxy S acid [*83-65-8*], *15*:738.   (See also *4,5-Dihydroxy-1-naphthalenesulfonic acid*.)
DIPB.   See *Diisopropylbenzene.*
Diphenylacetyl chloride [*1871-76-7*]
  diphenylketene from, *13*:880
Diphenylacetylene [*501-65-5*]
  metal π complex reaction, *16*:615
Diphenylacetylenedicobalt hexacarbonyl [*14515-69-6*], *16*:593
Diphenyl-ε-acid [*129-93-1*], *15*:731.   (See also *6,8-Di(phenylamino)-1-naphthalenesulfonic acid*.)
Diphenylamine [*122-39-4*]
  nitrobenzene analysis, *15*:921
Diphenylaminesulfonic acid [*101-57-5*]
  in iron analysis, *13*:787
6,8-Di(phenylamino)-1-naphthalenesulfonic [*129-93-1*], *15*:731
1,4-Diphenyl-1,4-butanedione [*495-71-6*], *13*:932
  properties, *13*:931
N,N′-Diphenylcarbodiimide [*622-16-2*]
  aziridine reaction, *13*:155
Diphenylketene [*525-06-4*]
  properties, *13*:875
N,N′-Diphenylmalamide [*621-10-3*], *13*:105
Diphenylmercury [*587-85-9*]
  lithium reaction, *14*:471
Di(phenylmercury)dodecenyl succinate [*27236-65-3*]
  applications of, *13*:236
4,4′-Diphenylmethane diisocyanate [*101-68-8*], *13*:789, 795.   (See also *4,4′-Methylenebis(phenyl isocyanate)*.)
  HCN reaction, *13*:797
  properties, *13*:789
4-(Diphenylmethylene)-3,3-diphenyl-2-oxetanone [*6925-23-1*], *13*:883
4,7-Diphenyl-1,10-phenanthroline [*1662-01-7*]
  in iron analysis, *13*:787
N,N′-Diphenyl-*p*-phenylenediamine [*74-31-7*], *13*:45
2-(Diphenylphosphino)ethylamine potassium salt, *13*:150
1,3-Diphenyl-1,3-propanedione [*120-46-7*]
  properties, *13*:928
Diphenylsilandiol [*947-42-2*], *15*:202

Diphenylstannic dichloride [*1135-99-5*]
  immunodeficiency effect, *13*:170
Diphenyl-*p*-styrylphosphine [*40538-11-2*]
  polymer as Rh-ligand oxo catalyst, *16*:640
Diphenyl sulfone [*127-63-9*], *13*:462
Diphenylthiocarbazone [*60-10-6*]
  lead reagent, *14*:137
1,3-Diphenyluretidone [*74631-12-2*]
  phenyl isocyanate reaction, *13*:796
Diphtheria, *15*:553
Dipicrylamine [*131-73-7*]
  for potassium separation, *16*:284
Dipivaloylmethane [*1118-71-4*]
  in lithium recovery, *14*:454
Dipotassium cobalt(II) hexakiscyanoferrate
  [*13821-10-8*], *13*:767
Dipotassium copper(II) hexakiscyanoferrate
  [*14481-39-1*], *13*:767
Dipotassium 7-hydroxy-1,3-
  naphthalenedisulfonate [*842-18-2*]
  Bucherer amination, *15*:728
Dipotassium manganese(II) hexakiscyanoferrate
  [*15631-19-3*], *13*:767
Dipotassium nickel(II) hexacyanoferrate [*13601-
  16-6*], *13*:767
Dipotassium sodium hexakiscyanoferrate [*31940-
  93-9*], *13*:769
Dipropyl disulfide [*629-19-6*]
  in onions, *16*:311
Dipropylene glycol [*25265-71-8*]
  in inks, *13*:382
  naphthalene azeotrope, *15*:699
Di-*n*-propyl isocinchomeronate [*136-45-8*]
  as repellent, *13*:477
Dipropyl methacryloyl phosphate [*76392-34-2*]
  properties, *15*:355
Di-*n*-propyl peroxydicarbonate [*16066-38-9*]
  half-life of, *13*:363
Di-*n*-propylstannic dichloride [*867-36-7*]
  immunodeficiency effect, *13*:170
Di-*n*-propylzinc [*628-91-1*], *16*:585
Dipryl, *16*:78
Dipslide technique, *13*:224
Dipterocarpus apitong, *14*:5
Dipterocarpus kerwing, *14*:5
Dipteryx odorata, *16*:251
4,4'-Dipyridyl [*553-23-4*]
  in isoprene reaction, *13*:823
Direct-flame incinerators, *13*:202
Direct-level measurements, *14*:428
Direct-reduced iron, *13*:754
Direct reduction, iron by, *13*:754
N,*N*-Disalicylidene-1,2-propanediamine [*94-91-7*]
  chelating agent, *13*:72
Discharge lamps
  phosphors for, *14*:544

Disclosure
  in patent specifications, *16*:864
DISCLOSURE information system, *13*:328
Disco, *14*:231
Discount factor
  in operations planning, *16*:506
Disease
  control of with insecticides, *13*:414
  liquid crystals implicated in, *14*:395
Disinfectant
  antimicrobial agent as, *13*:223
Disinfectants
  iodophors, *13*:673
Disinfection
  by ozone, *16*:702
Disks
  magnetic, *14*:732
Disodium disilanolate
  metallo-siloxanes from, *13*:409
Disodium ethylenebisdithiocarbamate [*142-59-6*]
  applications of, *13*:243
  as repellent, *13*:476
Disodium hydrogen phosphate [*7558-79-4*]
  pH standard, *13*:3
Disodium itaconate [*5363-69-9*]
  toxicity test, *13*:870
Disodium 1,5-naphthalenedisulfonate dihydrate
  [*76758-30-0*], *15*:723
  1,5-naphthalenediol from, *15*:735
Disodium *p*-nitrophenyl phosphate [*4624-83-9*]
  phosphatase reagent, *15*:80
Disodium phenyl phosphate [*3279-54-7*]
  in milk, *15*:551
  phosphatase reagent, *15*:80
Disodium phosphate [*7558-79-4*]
  added to milk, *15*:554
  in metal treating, *15*:306
Disodium thiosalicylate [*19296-18-5*]
  ethylmercuric chloride reaction, *15*:164
Disparlur
  encapsulation, *15*:488
Dispersants
  in lube oils, *14*:494
  in papermaking, *16*:806, 822
Dispersion
  of liquids, *15*:619
Displacer instruments
  for level measurement, *14*:439
Disposable nonwovens
  nonwoven, *16*:123
Disposable textiles
  nonwovens, *16*:99
Disproportionation
  organolead compounds, *14*:184
Dissipation factor, *13*:540
Dissociation
  laser-induced, *14*:74

Distance
  laser measurement, *14*:67
Distearyl thiodiproprionate [*693-36-7*]
  as olefin fiber antioxidant, *16*:360
Distillation
  extractive, ketones in, *13*:903
  in isotope separation, *13*:846
Distribution
  operations planning for, *16*:509
Distribution curve
  of ions in ion implantation, *13*:711
Disulfides
  in enzymes, *15*:587
Disulfoton [*298-04-4*], *13*:447
11,15-(Ditetrahydropiranyl)prostaglandin $F_{2\alpha}$
    [*37786-09-7*]
  in PGf$_2$ and PGf$_{2\alpha}$ syntheses, *15*:780
Dithane D 14, *13*:243
Dithiocarbamate [*148-18-5*]
  nickel-poisoning treatment, *15*:812
Dithioethylene phosphorochloridite [*35437-33-3*]
  isoprene reaction, *13*:826
Dithiomalonamide [*6944-34-9*], *14*:805
Dithionite.   See *Sodium hydrosulfite.*
Dithizone [*60-10-6*]
  for mercury determination, *15*:153
Ditoluenechromium [*12087-58-0*]
  metalation, *16*:607
Diuretics
  mercury compounds, *15*:165
Diuretic therapy
  potassium treatment, *15*:588
Divinylbenzene [*1321-74-0*]
  in ion exchange, *13*:680
Divinylferrocene
  polymerization of, *15*:194
1,1'-Divinylferrocene [*1291-66-3*], *15*:186
Di-*p*-xylylene [*1633-22-3*]
  in encapsulation, *15*:482
Dizziness
  (−) adrenaline induced, *15*:758
DMA.   See *Dimethylaniline.*
DMAE.   See *Dimethylaminoethanol.*
DMT.   See *N,N-Dimethyltryptamine.*
DMTT.   See *Tetrahydro-3,5-dimethyl-2H-
    1,3,5-thiadiazine-2-thione.*
DNA.   See *Deoxyribonucleic acid.*
DNBM.   See *Di-n-butylmagnesium.*
DNHM.   See *Di-n-hexylmagnesium.*
DNOC.   See *Dinitrocresol.*
Docan process, *16*:87
Doctrine of equivalents, *16*:879
Document preservation, *16*:811
Document reproduction inks, *13*:381
Dodecacarbonyltriruthenium [*15243-33-1*]
  oxo process catalyst, *16*:639

1,2,3,4,5,5,6,7,8,9,10,10-Dodecachlorooctahydro-
    1,3,4-metheno-2*H*-cyclobuta-[*c,d*]-pentalene
    [*2385-85-5*]
  as insecticide, *13*:436
trans-8,*trans*-10-Dodecadienol [*57002-06-9*]
  as insect pheromone, *13*:480
Dodecane [*112-40-3*]
  naphthalene azeotrope, *15*:699
2,5-Dodecanedione [*32781-66-1*]
  properties, *13*:931
Dodecanethiol [*1322-36-7*]
  from propylene tetramer, *16*:482
2-Dodecenal [*4826-62-4*]
  in orange oil, *16*:325
1-Dodecene [*112-41-4*]
  properties, *16*:481
Dodecenes
  production, *16*:494
cis-7-Dodecenyl acetate [*14959-86-5*]
  as insect pheromone, *13*:480
cis-8-Dodecenyl acetate [*28079-04-1*]
  as insect pheromone, *13*:480
cis-9-Dodecenyl acetate [*16974-11-1*]
  as insect pheromone, *13*:480
Dodecenylsuccinic anhydride [*25377-73-5*]
  from propylene tetramer, *16*:484
Dodecyl mercaptan [*112-55-0*]
  in initiator systems, *13*:365
t-Dodecyl mercaptan [*25103-58-6*]
  in polymethacrylate mfg, *15*:387
Dodecylphenol [*27193-86-8*], *16*:497
2-Dodecylthioethyl methacrylate [*14216-26-3*]
  properties, *15*:355
DOE–RECON information system, *13*:291
Dole system, *15*:549
Dolime [*50933-69-2*]
  in magnesium manufacture, *14*:577
Dolomite [*16389-88-1*], *14*:343, 615.   (See also
    Lime and limestone.)
  in iron-ore processing, *13*:739
  in limestone, *14*:344
  magnesium ore, *14*:570
  thermal decomposition, *16*:337
Dolomite, calcined, *14*:621
Dolomite, half-calcined, *14*:621
Dolomites
  lithium in, *14*:449
Domalt, *14*:816
Donnan dialysis, *15*:101
Donnan potential, *15*:101
L-DOPA [*59-92-7*], *13*:46
  memory-enhancing agent, *15*:135
  in Parkinsonism treatment, *15*:764
Dopamine [*51-61-6*], *13*:63
  in (−) adrenaline biosynthesis, *15*:758
  GHB effect on, *13*:98
  in memory enhancement, *15*:136

as neuroregulator, *15*:763
properties, *15*:763
Dopamine hydrobromide [*645-31-8*]
properties, *15*:764
Dopamine hydrochloride [*62-31-7*]
properties, *15*:764
Dopamine β hydroxylase [*9013-38-1*]
copper in, *15*:589
in (−) noradrenaline biosynthesis, *15*:774
in octopamine biosynthesis, *15*:774
Dopamine picrate [*75802-62-9*]
properties, *15*:764
Doping
of silicon, *13*:633
Doppler effect
in LMFBRS, *16*:199
Doré metal
from lead refining, *14*:121
Doriden, *13*:122
Dorr thickeners, *14*:577
Doryl
thermal conductivity of, *13*:544
Dose–response relationship
radiation, *16*:220
Dosimeter
sound, *16*:5
Double-arm kneading mixers, *15*:631
Doubling time
in breeder reactors, *16*:188
Dough conditioner
stearyl lactylates as, *13*:85
Douglas pine
catechol in, *13*:46
Dounreay fast reactor, *16*:201
Dowcide 1, *13*:227
Dowcide 2, *13*:226
Dowcide 7, *13*:226
Dowcide A, *13*:227
Dowcide B, *13*:226
Dowcide G-ST, *13*:226
Dowex, *13*:696
Dowex HCR-W2, *13*:679
Dowex SBR, *13*:679
Dowex 50X8, *13*:679
Dowicil A-40, *13*:244
Dowicil S-13, *13*:244
Downtime
minimization of, *14*:757
DOW Seawater Process
for magnesium, *14*:577
DPS detergent test, *15*:552
DR. See *Direct reduction.*
Drainage aids
in papermaking, *16*:806
Drapery
as acoustic material, *13*:519

Drawing
metals, *15*:333
DRI. See *Direct-reduced iron.*
DRI CAPSULE DATA BASE information
system, *13*:328
DRI CENTRAL DATA BANK information
system, *13*:328
Dried meat, *15*:70
Driers
naphthenate salts as, *15*:752
Drilling
with lasers, *14*:65
Drossing
in lead processing, *14*:109
Drugs
clinical measurement, *15*:85
Drugstore beetle
fumigant activity on, *13*:465
Drum dryers
for milk, *15*:557
Drum memories, *14*:701
Drums
as packaging, *16*:716
Drum separators
magnetic, *14*:720
Dry cells
mercury in, *15*:162
Drying
gases, by molecular sieves, *15*:663
leather, *14*:220
microwave, *15*:515
paper, *16*:785
Dry milk, *15*:556
Dry offset inks, *13*:379
Dry-strength additives
for paper, *16*:814
DSA. See *Dimensionally stable anode.*
DSN processes. See *Nitric acid.*
Duboisia hopwoodii, *13*:422
Ductility
of metals, *15*:326
Duolite, *13*:696
Duon, *16*:78
Duplex bonding
spunbondeds, *16*:91
Duplex melting, *15*:342
Duplicator inks, *13*:381
Dupont filters, *16*:544
Duranickel Alloy 301, *15*:787
Durotex, *13*:236
Dusts
in insecticide formulations, *13*:416
Dutch elm disease, *13*:482
Dwarfism
nutrients associated with, *15*:573
Dyanap, *15*:726

Dyeing
    oxalic acid in, *16*:631
    paper, *16*:779
Dye inks, *13*:391
Dye laser, *14*:51
Dye lasers, *14*:44
Dyes
    fluorescent, *14*:548
    nickel containing, *15*:814
    from nuts, *16*:273
    for olefin fibers, *16*:361
    oxidation of, *13*:28
    polymeric, *15*:212
Dymelor, *13*:615
Dynac, *16*:78
Dynamic programming
    in operations planning, *16*:505
Dynamometers
    in failure testing, *15*:20
Dypingite [*12544-02-4*], *14*:619
Dysentery, amoebic, *15*:553
Dysentery, bacillary, *15*:553
Dyskinesia
    diagnosis, *15*:79
Dyson notation, *13*:280
Dysprosium [*7429-91-6*]
    as magnetic material, *14*:647
    from the ocean, *16*:280
Dysprosium-156 [*15720-39-5*], *13*:841
Dysprosium-158 [*14913-25-8*], *13*:841
Dysprosium-160 [*14683-25-1*], *13*:841
Dysprosium-161 [*13967-68-5*], *13*:841
Dysprosium-162 [*14834-85-6*], *13*:842
Dysprosium-163 [*14391-36-7*], *13*:842
Dysprosium-164 [*13967-69-6*], *13*:842
Dysprosium, compd with cobalt [*12052-72-1*] (2:
    17), *14*:676
Dysprosium nitride [*12019-88-4*], *15*:872

E

EADC.   See *Ethylaluminum dichloride.*
EAN.   See *Effective atomic number.*
Earth
    as insecticide carrier, *13*:416
Earth nut, *16*:250
EASC.   See *Ethylaluminum sesquichloride.*
Eastern spruce budworm
    control of with pheromones, *13*:480
EBAM.   See *Electron-beam-addressable
    memories.*
EBR-I, *16*:185

EBR-II, *16*:185
EBR-II reactor, *16*:201
Ecaria spicata, *16*:952
ECCS.   See *Emergency core-cooling system.*
Ecdysis
    insect growth regulator effect on, *13*:459
Economic indicators, *13*:309
Economic lot-size model
    of inventory, *16*:513
ECONOMICS ABSTRACTS
    INTERNATIONAL information system, *13*:
    328
E-CTFE.   See *Poly(ethylene–co-
    chlorotrifluoroethylene).*
Ectopic pregnancy
    diagnosis, *15*:79
Ecuelling, *16*:308
EDB.   See *1,2-Dibromoethane.*
EDC.   See *1,2-Dichloroethane.*
Eddy currents
    in nondestructive testing, *16*:58
Eddy-current separators, *14*:710
Edestin, *16*:249
    amino acid composition, *16*:249
Edge emitters, *14*:279
Edison battery
    lithium in, *14*:451
EDTA.   See *Ethylenediaminetetraacetic acid.*
EEA.   See *Ethylene–ethyl acrylate copolymer.*
EEC.   See *European Economic Community.*
Effective atomic number (EAN), *16*:595
Effector cells
    in immune system, *13*:169
Efflux cup, *16*:757
Eggshell formation
    DDT effect on, *13*:431
E-glass
    fiber properties, *13*:970
EHEC resin, *14*:567
1-Eicosene [*3452-07-1*]
    properties, *16*:481
Eicosoborane [*12008-84-3, 59458-35-4*], *14*:73
Einstein model
    latex rheology, *14*:89
EIS INDUSTRIAL PLANTS information
    system, *13*:329
EIS NONMANUFACTURING
    ESTABLISHMENTS information system,
    13:329
Elaeis guineensis, *16*:250
Elastic fibers
    from polyolefins, *16*:377
Elastin
    in cattlehides, *14*:203
    cross-linking in, *15*:588
Elastin impairment
    nutrients associated with, *15*:573

Elastohydrodynamic lubrication, *14*:483
Elastomers
    effect of radiation on, *16*:239
    microwave processing, *15*:515
Electrical devices
    from mica, *15*:436
Electrical Patents Index
    information system, *16*:905
Electric current conduction tests, *16*:58
Electricity
    nitric acid production, *15*:865
Electric motors
    noise control, *16*:19
Electrified-particle tests, *16*:56
Electrochemical cells
    in pH determination, *13*:2
Electrochemical processing
    metal anodes for, *15*:172
Electrocleaning
    metals, *15*:300
Electrode potentials
    mercury, *15*:157
Electrodes
    gas-sensing, *13*:727
    liquid membrane, *13*:723
    molybdenum, *15*:677
    in pH measurement, *13*:5
    of poly(sulfur nitride), *13*:407
Electrodes, ion-selective, *13*:720. (See also *Ion-selective electrodes.*)
Electrodialysis
    membranes, *15*:119
    seawater, *16*:281
Electroendosmosis
    of PVC-insulated wire, *13*:571
Electroless plating
    thin magnetic films, *14*:693
Electroluminescence
    in LEDs, *14*:273
    of phosphors, *14*:527
Electrolysis
    acrylonitrile, *15*:899
    coated anodes for, *15*:172
    of H_2O in isotope separation, *13*:852
    in nitride isolation, *15*:877
    seawater magnesium processes, *14*:577
    in TEL mfg, *14*:188
    tetraalkylleads by, *16*:581
Electrolysis, molten-salt
    lithium chloride in, *14*:461
Electrolytic manganese dioxide, *14*:864
Electrolytic processes
    for manganese, *14*:835
Electrolytic refining
    of lead, *14*:124
Electromagnet, *14*:681
Electromagnetic energy, *15*:494

Electromagnetic spectrum, *15*:494
Electromagnet iron, *14*:651
Electromagnets
    in magnetic separators, *14*:713
Electromotive force (EMF)
    and pH, *13*:1
Electron-beam-addressable memories (EBAM), *14*:686
Electron-beam-cured inks, *13*:385
Electron-beam curing
    of inks, *13*:375
Electron-beam dissociation
    for coatings, *15*:242
Electron-beam imaging tests, *16*:65
Electron-beam melting
    niobium, *15*:827
Electronic equipment
    failure model, *15*:18
Electronic states
    in dyes, *14*:554
Electron laser, *14*:61
Electrooptics
    liquid crystals used in, *14*:421
Electroosmosis, *15*:101
Electrophoretic coating, *15*:257
Electroplating
    hydroxyacetic acid in, *13*:93
    ISES used in, *13*:732
    of lead–tin alloys, *14*:156
    on magnesium, *14*:610
    nickel in, *15*:813
    salicylaldehyde uses, *13*:76
Electropolishing, *15*:303
Electroslag cladding, *15*:262
Electroslag remelted ingot process
    ingot process, *15*:262
Electroslag remelt process, *15*:799
Electrostatic inks, *13*:395
Electrowinning
    lead anodes, *14*:149
Eleocharis dulcis, *16*:251
Elettaria cardamomum M., *16*:322
ELHILL III information system, *13*:291
Elisa. See *Enzyme-linked immunosorbent analysis.*
Elk hides
    for leather, *14*:224
Elliptone [*478-10-4*]
    as insecticide, *13*:425
Elmendorf test, *16*:99
Elon [*150-75-4*]
    in phosphorus measurement, *15*:84
Elongation
    of metals, *15*:326
Elphal process, *15*:256
Elymoclavine [*548-43-6*]
    microbial transformation to agroclavine, *15*:465

Embalming fluid
    2-hydroxymethyl-2-nitro-1,3-propanediol, *15:*
    915
Emblems
    fluorescent, *14*:568
EMD.   See *Electrolytic manganese dioxide.*
Emergency core-cooling system (ECCS), *16*:230
Emery Frigid Go, *14*:498
EMF.   See *Electromotive force.*
Emission flame photometry
    in lithium analysis, *14*:457
Emissivity, *13*:342
Emit.   See *Enzyme-multiplied
    immunotechnique.*
Emley plasticimeter, *14*:354
Emplex.   See *Sodium stearoyl-2-lactylate.*
Emulsifier
    stearyl lactylates as, *13*:85
Emulsifiers
    in ice cream, *15*:565
Emulsion cleaning
    metals, *15*:297
Emulsion hardening
    in encapsulation, *15*:476
Emulsion polymerization, *14*:93
Emulsions
    stability factors, *14*:85
Emulsion sizes, *16*:820
Emulsives
    in insecticide formulations, *13*:417
Enamels
    in electric insulation, *13*:579
En40b nitriding steel
    ion implantation, *13*:717
Encapsulation.   See *Microencapsulation.*
Encephalopathy
    from lead, *14*:175
Enclosures
    for noise sources, *16*:17
ENCYCLOPEDIA OF ASSOCIATIONS
    information system, *13*:329
Encyclopedias and treatises, *13*:286
Endamoeba histolytica
    disinfection by ozone, *16*:705
Endocrine control
    enkephalins involved in, *15*:765
Endoperoxide isomerase [*59977-51-4*]
    in prostaglandin syntheses, *15*:780
Endoperoxide reductase [*59977-52-5*]
    in prostaglandin syntheses, *15*:780
α-Endorphin [*59004-96-5*], *15*:765
β-Endorphin [*61214-51-5*]
    properties, *15*:765
γ-Endorphin [*61512-77-4*], *15*:765
Endorphins
    as neuroregulators, *15*:764

Endosulfan [*115-29-7*], *13*:435
Endothall [*145-73-3*], *14*:788
Endoxan, *13*:172
Endrin [*72-20-8*], *13*:435
Ene reaction
    of maleic anhydride, *14*:776
Energy
    standardization of, *15*:40
Energy conservation
    maintenance programs for, *14*:765
    using ir thermography, *13*:353
ENERGYLINE information system, *13*:329
Energy management
    maintenance programs for, *14*:765
Energy transfer
    among fluorescent dyes, *14*:558
    in phosphors, *14*:532
Energy-transmission probes, *16*:55
Enfleurage, *16*:308, 314
Engineering design
    probabilistic, *15*:27
Engines
    noise source, *16*:9
English walnut, *16*:248
Enkamat, *16*:78
Enkephalins
    as neuroregulators, *15*:764
Enolase [*9014-08-8*]
    fluorine inhibition of, *15*:593
    in insecticide reaction, *13*:422
Enols
    of 1,3-diketones, *13*:927
Enrico Fermi FBR, *16*:184
Enrico Fermi 1 nuclear reactor
    fuel melting at, *16*:232
Entamoeba histolytica, *15*:553
Enterobacter, *15*:950
Enterobacter aerogenes, *15*:447
Enterobactin [*28384-96-5*]
    iron complex, *13*:786
Enterochromaffin cells, *15*:771
ENVIROLINE information system, *13*:329
Environmental stress-crack resistance
    of polyethylene, *16*:426
Enzose
    oxalic acid from, *16*:622
Enzyme deficiency
    encapsulated products for, *15*:489
Enzymes
    clinical tests, *15*:75
    electrodes of, *13*:729
    molybdenum in, *15*:693
Eon E-11, *14*:498
Eosin [*17372-87-1*]
    as photosensitizer, *13*:370
EP.   See *Ethylene–propylene rubber.*

EPB information system, *13*:329

EPDM. See *Ethylene–propylene copolymer.*

L-Ephedrine [*321-98-2*]
  microbial preparation, *15*:459

(1*R*,2*S*)-Ephedrine [*299-42-3*]
  microbial synthesis of, *15*:464

Ephestia elutella, *13*:481

Epichlorohydrin [*106-89-8*]
  in encapsulation, *15*:484
  epoxy resins from, *13*:973
  in gel filter preparation, *15*:445
  in papermaking, *16*:804
  toxicity, *13*:267

EPICS information system, *13*:327, 329

Epilepsy
  GABA involved in, *15*:766
  taurine associated with, *15*:781
  treatment, *13*:122; *15*:768, 781

Epileptic seizures
  treatment, *13*:134

Epinephran, *15*:757

Epinephrine [*51-43-4*], *13*:62; *15*:757.   (See also
    L-Adrenaline.)
  from adrenal glands, *15*:72
  hypoglycemia relief, *13*:608

Epirenan, *15*:757

Epitaxial films
  of silicon in ICS, *13*:625

EP lubricants.   See *Extreme pressure
    lubricants.*

EPM.   See *Ethylene–propylene copolymer.*

EPN.   See *O-Ethyl O-p-nitrophenyl
    phenylphosphonothionate.*

EPO information system, *13*:292

Epoxies
  corrosion resistant paint, *16*:755

2,3-Epoxybutyl methacrylate [*68212-07-7*]
  properties, *15*:354

3,4-Epoxybutyl methacrylate [*55750-22-6*]
  properties, *15*:354

2,3-Epoxycyclohexyl methacrylate [*76392-25-1*]
  properties, *15*:354

(3α,4α,7α,7aα)-4,7-Epoxyisobenzofuran-1,3-di-
    one [*6766-44-5*], *14*:775

cis-7,*cis*-8-Epoxy-2-methyloctadecane [*29804-
    22-6*]
  as insect pheromone, *13*:480

2,3-Epoxypropyl methacrylate [*106-91-2*], *15*:356

Epoxy resins, *16*:755
  in composites, *13*:973
  in electrode mfg, *13*:727
  as magnetic tape binder, *14*:741

cis-Epoxysuccinic acid [*16533-72-5*], *14*:780
  hydrolysis, *13*:115
  microbial hydrolysis, *15*:462

trans-Epoxysuccinic acid [*22734-83-4*]
  in *meso*-tartaric acid biosynthesis, *13*:114

2,3-Epoxy-3,5,5-trimethylcyclohexanone [*10276-
    21-8*], *13*:919

10,11-Epoxyundecyl methacrylate [*23679-96-1*]
  properties, *15*:354

Epsomite [*10034-99-8*], *14*:637

Epsom salt [*7487-88-9*]
  in leather manufacture, *14*:213
  manufacture, *14*:639

Epsom salts, *16*:277.   (See also *Magnesium
    sulfate*.)

EPT.   See *Ethylene–propylene rubber.*

Equipment replacement
  program, *16*:511

Erbium [*7440-52-0*]
  from the ocean, *16*:280

Erbium-162 [*15840-05-8*], *13*:842

Erbium-164 [*14900-10-8*], *13*:842

Erbium-166 [*14900-11-9*], *13*:842

Erbium-167 [*14380-60-0*], *13*:842

Erbium-168 [*14833-43-3*], *13*:842

Erbium-170 [*15701-24-3*], *13*:842

Erbium, compd with cobalt [*12017-59-3*] (1:5),
    14:676

Erbium, compd with cobalt [*12052-73-2*] (2:17),
    14:676

Erbium nitride [*12020-21-2*], *15*:872

Erector pili
  myosin in, *14*:203

Ergogene, *14*:864

Eric information systems, *13*:291

Erionite [*12510-42-8*], *15*:639

EROICA information system, *13*:301

E-rosette, *13*:178

Erosion control
  nonwovens in, *16*:101

Erwinia herbicola, *13*:46

Erwinia tahitica, *15*:451

Erythrocuprein [*37294-21-6*]
  copper in, *15*:588

D(−) Erythronamide [*73713-13-0*], *13*:101

L(−) Erythronamide [*74421-64-0*], *13*:101

Erythronic acid [*13752-84-6*], *13*:101

DL(−) Erythronic acid, butyl ester [*74421-62-8*],
    13:101

DL(−) Erythronic acid, γ-lactone [*28350-85-8*],
    13:101

L(−) Erythronic acid, γ-lactone [*23732-40-3*], *13*:
    101

Erythrosin [*15905-32-5*], *13*:671

Escherichia coli, *13*:55; *15*:466, 553, 950
  disinfection by ozone, *16*:705

ESCR.   See *Stress cracking resistance
    (environmental)*.

Esculetin [*305-01-1*], *16*:256

Esculin [*531-75-9*]
  in nuts, *16*:256

ESD 31
  magnetic properties, *14*:671
ESD 32
  magnetic properties, *14*:671
ESD 41
  magnetic properties, *14*:671
ESD 42
  magnetic properties, *14*:671
ESD magnets, *14*:680
Esparto
  in paper, *16*:771
ESR.   See *Electroslag remelt.*
Essences
  encapsulation of, *15*:487
Essential oils
  in perfumes, *16*:949
Estane, *13*:813
Estazolam [*29975-16-4*]
  properties of, *13*:122
Esters
  as lubricants, *14*:496
Estertins, *16*:577
17β-Estradiol [*50-28-2*], *15*:766
Estragole [*140-67-0*]
  from basil oil, *16*:321
Estrogens, *15*:766
Etching, *15*:302
  fission particle tracks, *16*:832
  by sputtering, *13*:638
Etching, aluminum
  tartrate use, *13*:117
Etching solutions
  for fission particle tracks, *16*:833
Etch-pit geometry
  of particle tracks, *16*:837
Ethamsylate, *13*:60
Ethane [*74-84-0*]
  in LPG, *14*:384
  nitration, *15*:848
  properties, *14*:384
1,2-Ethanediamine dihydriodide [*5700-49-2*], *13*: 673
Ethanedioic acid [*144-62-7*], *16*:618.   (See also *Oxalic acid.*)
Ethanediol [*107-21-1*], *14*:380
2,2′-(1,2-Ethanediyl)bisbenzoic acid, dimethyl ester (6) [*60040-81-5*], *14*:412
4,4′-(1,2-Ethanediyl)bisbenzoic acid, dimethyl ester (5) [*797-21-7*], *14*:412
Ethanenitronic acid.   See *Isonitroethane.*
Ethanol [*64-17-5*], *16*:312
  in chloral hydrate synthesis, *13*:123
Ethaverine [*486-47-5*], *13*:63
Ethchlorvynol [*113-18-8*], *14*:471
  properties of, *13*:122
4-Ethenyl-1,4-dimethylcyclohexene [*1743-61-9*]
  isoprene dimer, *13*:822

4-Ethenyl-2,4-dimethylcyclohexene [*1611-21-8*]
  isoprene dimer, *13*:822
Ethers
  decomposition of *n*-butyllithium by, *14*:467
Ethinamate [*126-52-3*]
  properties of, *13*:122
Ethion [*563-12-2*], *13*:445
Ethosuximide [*77-67-8*]
  properties of, *13*:135
2-Ethoxethoxymethyl methacrylate [*76392-17-1*]
  properties, *15*:353
Ethoxyacetylene [*927-80-0*]
  ketene reaction, *13*:886
3-Ethoxy-3-buten-2-one [*65915-73-3*]
  sodium dimethyl malonate reaction, *14*:799
1-Ethoxybutyl methacrylate [*76392-24-0*]
  properties, *15*:353
3-Ethoxy-2-cyclobutenone [*4683-54-9*]
  from ketene, *13*:887
1-Ethoxyethyl methacrylate [*51920-52-6*]
  properties, *15*:353
2-Ethoxyethyl methacrylate [*2370-63-0*]
  properties, *15*:353
(Ethoxyethynyl)lithium [*31612-88-1*]
  chlorotrimethylsilane reaction, *13*:882
(Ethoxyethynyl)trimethylsilane [*1000-62-0*]
  rearrangement to ketene, *13*:882
Ethoxymethyl methacrylate [*76392-16-0*]
  properties, *15*:353
4-Ethoxy-4-methyl-2-pentanone [*27921-36-4*], *13*:895
2-Ethoxyphenol [*94-71-3*], *13*:45.   (See also *Guethol.*)
Ethoxyquin [*91-53-2*]
  antiaging effects, *15*:138
2-Ethoxysuccinic acid [*23264-21-3*], *13*:105
Ethyl acetate [*141-78-6*]
  in inks, *13*:391
  ketene condensation, *13*:931
Ethyl acetoacetate [*141-97-9*], *13*:887
  resorcinol reaction, *13*:46, 61
Ethyl acrylate [*141-78-6*]
  copolymer with ethylene, *16*:420
  kinetics of destruction, *13*:201
  methacrylate copolymerization, *15*:384
  solubility of, *14*:83
Ethyl alcohol [*64-17-5*]
  as hypnotic, *13*:122
  in inks, *13*:390
  lactic acid reaction, *13*:83
Ethylaluminum dichloride [*563-43-9*]
  properties, *16*:565
Ethylaluminum sesquichloride [*10275-68-2*]
  properties, *16*:565
Ethylamine [*75-04-7*]
  from aziridine, *13*:151
Ethyl γ-aminobutyrate [*5959-36-4*], *15*:766

Ethyl β-aminocrotonate [3222-57-9]
   in Nenitzescu synthesis, 13:219
Ethyl amyl ketone [106-68-3]
   properties, 13:897
Ethyl n-amyl ketone [106-68-3], 13:907
2-Ethylanthraquinone [84-51-5]
   autoxidation, 13:18
1-Ethylaziridine [1072-45-3], 13:148
   from aziridine-ethylene reaction, 13:151
2-Ethylaziridine [2549-67-9]
   properties, 13:143
Ethyl 3-(1-aziridinyl)acrylate [50868-08-1], 13:
   151
Ethylbenzene [100-41-4]
   isoprene reaction, 13:825
   synthesis with zeolite catalysts, 15:667
   zeolite separation, 15:667
Ethyl benzoate [93-89-0]
   1-butene polymerization with, 16:476
Ethyl 6-bromo-2,7-dihydro-4-methyl-2,7-dioxo
      3H-dibenz[f,i,j]isoquinoline-1-carboxylic
      acid ester [51418-86-1], 14:800
Ethyl butyl ketone [106-35-4]
   properties, 13:896
2-Ethyl-2-butyl-1,3-propanediol [115-84-4]
   as repellent, 13:477
Ethyl 3-butynoate [53841-07-9]
   itaconic acid from, 13:869
(2-Ethylbutyryl)urea [2274-01-3]
   in carbomal synthesis, 13:126
Ethyl cellulose [9004-57-3]
   capsule matrix, 15:476
   carbon encapsulation by, 15:478
Ethyl chloride [75-00-3]
   aluminum reaction, 16:570
   in organolead synthesis, 14:184
   sodium–lead reaction, 16:580
Ethyl chloroacetate [105-39-5]
   in aspartic acid synthesis, 15:760
Ethyl β-chloropropionate [623-71-2]
   in L-glutamic acid synthesis, 15:767
Ethyl cyanoacetate [105-56-6], 14:795; 15:889
   properties, 15:907
Ethyl diazoacetate [623-73-4]
   indole reaction, 13:216
Ethyl 3,3-di(tert-butylperoxy)butyrate [55794-
      20-2]
   half-life of, 13:362
Ethyl p,p'-dichlorobenzilate [510-15-6]
   in insect control, 13:461
Ethyldichlorosilane [1789-58-8]
   isoprene reaction, 13:826
1-Ethyldihydro-3,5,6,S-triazene-2,4(1H,3H)
      dione, 13:797
Ethyl dimethylglycolate [33229-89-9]
   in trimethadione synthesis, 13:136
Ethyl 2,4-dioxo-6,6-
      dimethylcyclohexanecarboxylate [4029-25-8],
      13:917

O-Ethyl S,s-dipropyl phosphorodithioate
      [13194-48-4]
   as insecticide, 13:443
Ethylene [74-85-1]
   acetaldehyde addition to, 13:906
   from aziridine, 13:150
   copolymer with acrylic acid, 16:420
   copolymer with 1-butene, 16:388
   copolymer with carbon monoxide, 16:420
   copolymer with ethyl acetate, 16:420
   copolymer with 1-hexene, 16:388
   copolymer with isobutylene, 16:420
   copolymer with methacrylic acid, 16:420
   copolymer with methyl acrylate, 16:420
   copolymer with 4-methyl-1-pentene, 16:388
   copolymer with 1-octene, 16:388
   copolymer with 1-pentene, 16:388
   copolymer with propylene, 16:388, 420
   copolymer with vinyl acetate, 16:419
   high pressure polymerization, 16:385, 412
   hydroformylation, 16:639
   iodination of, 13:653
   maleic anhydride copolymers, 14:780
   methacrylates from, 15:357
   methyl propionate from, 15:364
   oxidation, 16:668
   polymerization of, 13:356
   solubility of, 14:83
   Ziegler polymerization, 16:433
N,N'-Ethylenebismethacrylamide [6117-25-5]
   properties, 15:356
Ethylene chlorobromide [107-04-0]
   as fumigant, 13:466
Ethylene cyanohydrin [109-78-4], 15:889
1,2-Ethylenediamine [107-15-3]
   tetrahydrozolin from, 15:741
Ethylenediamine dihydriodide [5700-49-2]
   feed additive, 13:673
Ethylenediaminetetraacetic acid [60-00-4]
   iron complex, 13:779
   in ISE titration, 13:731
   in metal pickling, 15:301
   paint removers, 16:763
Ethylene dibromide [106-93-4]
   as fumigant, 13:467
   in leaded automotive fuels, 16:283
   toxicity, 13:267
cis-1,2-Ethylenedicarboxylic acid [110-16-7], 14:
      770.   (See also Maleic acid.)
trans-1,2-Ethylenedicarboxylic acid [110-17-8],
      14:770.   (See also Fumaric acid.)
Ethylene dichloride [107-06-2]
   as fumigant, 13:467
   toxicity, 13:267
Ethylene dimethacrylate [97-90-5]
   in polymer curing, 13:567
   properties, 15:347

N,N-Ethylene-*N',N''*-diphenylguanidine [*74411-26-0*], *13*:155

Ethylene–ethyl acrylate copolymer
  melt index, *16*:408

Ethylene glycol [*107-21-1*]
  from hydroxyacetic acid, *13*:92
  in LPG processing, *14*:384
  in metallic coating, *15*:243
  naphthalene azeotrope, *15*:699
  nitric acid reaction, *16*:624
  oxalic acid copolymer, *16*:632
  oxalic acid from, *16*:621
  in paint, *16*:751
  in polyglycol lubricants, *14*:499

Ethylene glycol methacrylate [*97-90-5*]
  properties, *15*:354

Ethylene iodide [*624-73-7*], *13*:653

Ethylene oligomerization
  linear olefin mfg, *16*:480
  α-olefins from, *16*:487

Ethylene oxide [*75-21-8*]
  aziridine reaction, *13*:151
  from ethylene, *16*:668
  as fumigant, *13*:467
  in fumigation of nuts, *16*:263
  lime in refining of, *14*:380
  in lubricant synthesis, *14*:498
  toxicity, *13*:268

2-(Ethylenephosphito)propyl methacrylate
  [*76392-32-0*]
  properties, *15*:355

η²-Ethyleneplatinum trichloride [*12275-00-2*],
  16:592. (See also *Ziese's salt.*)

Ethylene–propylene [*9010-79-1*]
  electric insulation elastomer, *13*:565

Ethylene–propylene copolymer (EPDM) [*9010-79-1*]
  as cable insulation, *13*:554
  in electric insulation, *13*:564

Ethylene–propylene–diene [*25038-36-2*]
  electric insulation elastomer, *13*:565

Ethylene–propylene–diene rubber, *16*:584

Ethylene–propylene rubber [*9010-79-1*]
  resistance to water treeing, *13*:543

Ethylene–propylene rubber [*9010-79-1*] (EP,
  EPDM, EPM, EPT)
  gasket properties, *16*:727

Ethylene–propylene terpolymer (EPDM) [*9010-79-1*]
  in electric insulation, *13*:564

Ethylene thiourea [*96-45-7*]
  health effect, *13*:268

Ethylene–vinyl acetate copolymer [*24937-78-8*]
  in encapsulation, *15*:478
  melt index, *16*:408
  in paper coatings, *16*:791

Ethylenimines, *13*:143. (See also *Aziridines.*)

Ethyl *p*-ethoxybenzal-*p'*-aminocinnamate [*2863-94-7*]
  liquid crystalline range of, *14*:399

Ethyl 4-(4'-ethoxybenzylideneamino)cinnamate
  [*2863-94-7*]
  polymorphism of, *14*:410

Ethyl formate [*109-94-4*]
  in estazolam synthesis, *13*:131
  as fumigant, *13*:467

Ethyl glycolate [*623-50-7*]
  soap-builder from, *13*:92

2-Ethylhexanal [*123-05-7*]
  MIBK reaction, *13*:908

2-Ethyl-1,3-hexanediol [*94-96-2*]
  as repellent, *13*:477

2-Ethylhexanol [*104-76-7*]
  from oxo process, *16*:637

N-(2-Ethylhexyl)-[2,2,1]-5-heptene-2,3-dicarboximide [*113-48-4*]
  as insecticide, *13*:425

2-Ethylhexyl methacrylate [*688-84-6*]
  properties, *15*:353

Ethyl hydroxyethyl cellulose [*9004-58-4*]
  in encapsulation, *15*:479

Ethyl α-hydroxyisobutyrate [*594-61-6*]
  methacrylic acid from, *15*:346

2-Ethyl-2-hydroxymethyl-1,3-propanediol [*77-99-6*]
  in lubricants, *14*:498

4-Ethylidene-3-methyl-2-oxetanone [*5659-14-3*],
  13:883

Ethyl indole-3-acetate [*778-82-5*], *13*:216

Ethyl iodide [*75-03-6*]
  lead chloride reaction, *16*:582
  properties, *13*:670
  in tetraethyllead synthesis, *14*:180
  zinc reaction, *16*:586

Ethyl 3-isobutyl-2,2-dimethyl
  cyclopropanecarboxylate [*33419-38-4*]
  as insect attractant, *13*:481

Ethyl isobutyl ketone [*565-69-5*], *13*:908

Ethyl isocyanate [*109-90-1*], *13*:801

Ethyllactamide [*18266-55-2*], *13*:90

Ethyl lactate [*97-64-3*], *13*:87, 89
  properties, *13*:89

Ethyl levulinate [*123-76-2*]
  1,3-cyclopentanedione from, *13*:929

Ethyl-D-mannonate [*27934-94-7*]
  decomposition to lactone, *13*:100

Ethylmercuric chloride [*107-27-7*], *15*:169; *16*:587
  cottonseed bacteriostat, *16*:583

Ethyl methacrylate [*97-63-2*]
  methacrylic acid from, *15*:346
  properties, *15*:347

Ethylmethyldichlorosilane [*4525-44-4*]
  copolymer with dimethyl dichlorosilane, *13*:408

Ethyl 3-methyl-2-methylidynebutanoate [*68834-46-8*], *13*:821

7-Ethyl-2-methyl-4-undecanol [*112-72-1*], *13*:908

Ethyl *p*-nitrophenyl amylphosphonate [*3015-75-6*], *13*:452

Ethyl *p*-nitrophenyl *n*-butylphosphonate [*3015-74-5*], *13*:452

Ethyl *p*-nitrophenyl *t*-butylphosphonate [*13538-14-2*], *13*:452

Ethyl *p*-nitrophenyl ethylphosphonate [*546-71-4*], *13*:452

Ethyl *p*-nitrophenyl isopropylphosphonate [*13538-10-8*], *13*:452

Ethyl *p*-nitrophenyl methylphosphonate [*3735-98-6*], *13*:452

O-Ethyl *O*-*p*-nitrophenyl phenylphosphonothioate [*2104-64-5*], *13*:443

Ethyl *p*-nitrophenyl *n*-propylphosphonate [*3015-73-4*], *13*:452

2-Ethyl-2-nitro-1,3-propanediol [*597-09-1*] properties, *15*:910

Ethyl 5-oxohexanoate [*13984-57-1*] in catechol production, *13*:52

β-Ethylphenethylamine [*34577-88-3*], *15*:976

p-Ethylphenol [*123-07-9*] naphthalene azeotrope, *15*:699

Ethyl *p*-(*p*-phenylbenzalamino)benzoate [*3782-80-7*] liquid crystalline range of, *14*:399

O-Ethyl *S*-phenyl ethylphosphonodithioate [*944-22-9*] as insecticide, *13*:448

5-Ethyl-5-phenyl-2,4-imidazolidinedione [*631-07-2*] in mephenytoin synthesis, *13*:136

Ethyl 4-phenyl-2-oxo-3-pyrrolidine carboxylate [*52450-32-5*] in amino-2-phenyl-4Δ-1 pyrroline synthesis, *14*:803

2-Ethyl-2-phenylpentanedinitrile [*74220-50-1*] in glutethimide synthesis, *13*:125

5-Ethyl-5-phenyl-2-thiobarbituric acid [*2753-74-4*] in primidone synthesis, *13*:138

Ethyl phosphorous dichloride [*1498-40-4*], *14*:183

N-Ethylpiperidine [*766-09-6*] in maleic acid determination, *14*:785

Ethyl propiolate [*623-47-2*] aziridine reaction, *13*:151

3-Ethyl-4-propylidene-2-oxetanone [*5659-15-4*], *13*:883

Ethyl propyl ketone [*589-38-8*] properties, *13*:896

m-(1-Ethylpropyl)-phenyl *N*-methylcarbamate [*672-04-8*] as insecticide, *13*:455

Ethyl red [*634-21-9*], *13*:10

Ethylsulfinylethyl methacrylate [*3007-24-7*] properties, *15*:355

Ethylsulfonylethyl methacrylate [*25289-10-5*] properties, *15*:355

2-(Ethylthio)ethylamine [*36489-03-9*], *13*:155

N-*p*-Ethyl toluenesulfonylcarbamate butylamine reaction, *13*:617

Ethyltrichlorolead [*66300-29-6*], *14*:184

Ethyltrimethyllead [*1762-26-1*], *14*:184

Ethylvanillin [*121-32-4*], *13*:45

Ethylvinylbenzene [*28106-30-1*] in divinylbenzene, *13*:688

Ethylzinc iodide [*999-75-7*], *16*:586

1-Ethynylcyclohexanol [*78-23-3*] in ethinamate synthesis, *13*:126

Ethynylferrocene [*1271-47-2*], *15*:186

ETPE. See *Poly(ethylene–co-tetrafluoroethylene)*.

Eucalyptol [*470-82-6*] from eucalyptus oil, *16*:322

Eucalyptus dives, *16*:322

Eucalyptus macarthuri, *16*:323

Eucalyptus oil [*8000-48-4*], *16*:319, 322

α-Eudesmol [*473-16-5*] from eucalyptus oil, *16*:323

Eugenia caryophyllata, *16*:950

Eugenia caryophyllus (Sprengl), *16*:322

Eugenol [*97-53-0*], *13*:45; *16*:312, 313 in allspice oil, *16*:320 from bay oil, *16*:321 in clove oil, *16*:950 economic aspects, *13*:53 in inks, *13*:379 as insect attractant, *13*:481 in perfumes, *16*:968

Euglobulin in milk, *15*:568

Eulan CN [*4430-22-2*], *13*:479

EUROLEX information system, *13*:329

EURONET data-transmission network, *13*:292

European corn borer control of with pheromones, *13*:481

European Economic Community information sources, *13*:310

Europium [*7440-53-1*] from the ocean, *16*:280

Europium-151 [*14378-48-4*], *13*:841

Europium-153 [*13982-02-0*], *13*:841

Europium aluminum oxide [*12254-20-5*] (1:12:19) in phosphors, *14*:541

Europium(III) [*7440-53-1*] in phosphors, *14*:535

Europium magnesium aluminum oxide [*59668-40-5*] in phosphors, *14*:541

Europium nitride [*12020-58-5*], *15*:872
Europium oxide [*1308-96-9*]
    CVD coating for, *15*:264
    in magnetooptics, *14*:705
    in phosphor mfg, *14*:542
Europium oxide [*12020-60-9*] (1:1)
    Faraday effect, *14*:705
Europium selenide [*12020-66-5*]
    in magnetooptics, *14*:705
Europium selenide [*12020-66-5*] (1:1)
    Faraday effect, *14*:705
Euryale ferox, *16*:251
EVA.   See *Ethylene–vinyl acetate copolymer.*
Evaporated milk, *15*:553
Evaporation coating, *15*:265
Evaporation, thermal
    in thin-film fabrication, *14*:692
Evolution/evolution II, *16*:78
Ewens-Bassett nomenclature, *16*:35
Exaltolide [*106-02-5*]
    in perfumes, *16*:968
Exaltone
    as perfume, *13*:922
Excelsin, *16*:249
    amino acid composition, *16*:249
Exchange reactions
    tetraethyllead, *14*:183
Excimer lasers, *14*:61
Exodextranase [*52227-69-7*], *15*:441
Exopolysaccharides, *15*:439
Exoskeleton formation
    insect growth regulator effect on, *13*:459
Expectorants
    guaiacolates, *13*:62
Experimental breeder reactor-II, *16*:181
Explosion cladding, *15*:275.   (See also
        Explosively clad metals.)
Explosion welding, *15*:294
Explosive bonding, *15*:260
    metal-laminate fabrication, *13*:948
Explosive forming, *15*:334
Explosively clad metals, *15*:275.   (See also
        Metallic coatings.)
Explosives
    lead azide, *14*:169
    for metal cladding, *15*:281
    nitroparaffins in, *15*:983
Explosive welding
    metal-laminate fabrication, *13*:948
Expression
    essential oils by, *16*:314
Exraled, *14*:248
EXSTAT information system, *13*:329, 334
Extended insulin zinc suspension [*8049-62-5*],
        13:613
Extenders
    ground mica as, *15*:437

Exter press, *14*:331
Extraction
    essential oils, *16*:314
    in nuclear fuel reprocessing, *16*:175
Extraction columns
    perforated plate, *14*:967
Extractive distillation
    isoprene, *13*:832
    nitric acid, *15*:865
Extreme pressure lubricants, *14*:479
Extruders, *15*:635
    for low density polyethylene, *16*:415
Extrusion
    fabrics, *16*:94
    HDPE pipe and film, *16*:446
    of magnesium alloys, *14*:607
    of metals, *15*:330
    olefin fibers, *16*:367
    of polyethylene, *16*:395
    polypropylene film and pipe, *16*:467
Exxon donor solvent process
    for fuels from lignites, *14*:340

F

F-50, *14*:496
F-111
    laminated-glass windshield, *13*:990
Faber du Faur furnace, *14*:121
Fabric bases
    nonwoven, *16*:123
Fabric binders
    acrylic latexes for, *13*:871
Fabrics
    membrane functions, *15*:125
    urethane-coated, *14*:233
Fabrics, bonded, *16*:105.   (See also *Nonwoven
        textile fabrics.*)
Fabrics, engineered, *16*:105.   (See also *Nonwoven
        textile fabrics.*)
Fabrics, formed, *16*:105.   (See also *Nonwoven
        textile fabrics.*)
Fabric softeners
    fragrances, *16*:969
Fabry-Perot filters, *16*:524
Fabry-Perot optical filters, *16*:550
Fabrys' disease
    liquid crystals in, *14*:420
Facial tissue, *16*:772
F-acid [*92-40-0*], *15*:730, 738.   (See also *7-
        Amino-2-naphthalenesulfonic acid.*)
Facing alloys.   See *Metallic coatings.*

Fade-ometer, *16*:758

Fagus grandifolia, *16*:250

Fagus sylvatica, *16*:250

Failure

mechanisms and analysis, *15*:13

Failure-censored life tests, *15*:21

Failure-mode analysis (FMA), *15*:12

Failure-mode effects and criticality analysis
(FMECA), *15*:12

Failure rate, *15*:17

Fall armyworm

control of with pheromones, *13*:480

Fallout

radiation exposure from, *16*:217

Fan blades

fiber reinforced, *13*:943

Fanconi syndrome, *14*:198

Fanning's friction factor, *14*:953

Fans and blowers

noise control, *16*:23

Fantasy bases, *16*:968

FAO. See *Food and Agricultural Organization.*

Faraday constant

and pH, *13*:2

Faraday experiments, *14*:705

Faradiser M

maganese dioxide, *14*:864

Farbers' disease

liquid crystals in, *14*:420

FARMDOC

information system, *16*:905

Farnesol [*4602-84-0*], *16*:312

Farrel continuous mixer, *16*:396

Fast breeder reactor

gas cooled, *16*:169

Fast breeder reactors (FBRS), *16*:184. (See also
Nuclear reactors.)

Fast flux test facility (FFTF), *16*:196

Fatigue

detection, *16*:60

detection of, *16*:48

Fatigue resistance

enhancement of by ion implantation, *13*:717

Fat lime, *14*:347

Fat metabolism

sulfur in, *15*:587

Fats

from meat, *15*:71

Fatty acid nitriles, *15*:888

properties, *15*:908

Fatty acids

in grease thickeners, *14*:502

in nuts, *16*:252

Faujasite [*12173-28-3*], *15*:639

FBRS. See *Fast breeder reactors.*

FC-113. See *1,1,2-Trichloro-1,2,2-
trifluoroethane.*

Feces

nitrosamines in, *15*:994

FEDREG information system, *13*:329, 334

Feedback control, *13*:485. (See also *Automatic
control.*)

Feedstocks

hydrodesulfurization with mo catalysts, *15*:694

Feed supplement

limestone, *14*:378

Fehling's solution, *13*:118. (See also *Alkaline
cupric tartrate.*)

Fel-Pro C5-A, *14*:513

Felt

gasket properties, *16*:727

in gaskets, *16*:732

Fenchol [*1632-73-1*]

from lime, *16*:324

Fenitrothion [*122-14-5*], *13*:439

Fenoprofen [*31879-05-7*]

immunologic effects, *13*:174

Fenson [*80-38-6*], *13*:463

Fenthion [*55-38-9*], *13*:439

Fenvalerate [*51630-58-1*], *13*:458

FEP. See *Fluorinated ethylene–propylene.*

FEP teflon

membrane, *15*:122

Fermentation, *14*:811. (See also *Malts and
malting.*)

dextran production, *15*:442

in lactic acid production, *13*:80

xanthan gum process, *15*:449

Fermentation, heterolactic

of lactic acid, *13*:84

Fermentation, homolactic

of lactic acid, *13*:84

Fermi's golden rule, *14*:527

Fernbach flasks, *15*:442

Ferranti-Shirley viscometer, *16*:757

Ferrates, *13*:776

Ferredoxin, *15*:951. (See also *Iron–sulfur
protein.*)

Ferric acetylacetonate [*14024-18-1*]

in isoprene reaction, *13*:822

Ferric ammonium oxalate [*2944-67-4*], *16*:632, 633

Ferric chloride [*7705-08-0*]

in metal pickling, *15*:301

Ferric chloride diethanolate [*74930-86-2*], *13*:774

Ferric chloride tris(1,10-phenanthroline) [*40273-
22-1*] (1:3:3), *13*:781

Ferric ethylenediaminetetracetic acid, ammonium
salt [*21265-50-9*], *13*:780

Ferric hexakiscyanoferrate [*14433-93-3*] (1:1)
  in Turnbull's blue synthesis, *13*:771
Ferrichrome [*15630-64-5*], *13*:784
Ferrichrome A [*15258-20-7*], *13*:784
Ferrichrome C [*56665-78-2*], *13*:784
Ferrichromes
  structures of, *13*:784
Ferrichrysin [*18972-10-6*], *13*:784
Ferric hydroxide [*1309-33-7*]
  in limestone, *14*:350
Ferric nitrate bright pickle, *14*:610
Ferric oxalate [*2944-66-3*], *16*:632
α-Ferric oxide [*1345-25-1*]
  in nuclear reactors, *16*:159
γ-Ferric oxide [*1309-37-1*], *14*:737. (See also
  Magnetite.)
γ-Ferric oxides
  in magnetic tape, *14*:737
Ferric perchlorate bis(2,2′:6′,2″-terpyridine)
  [*21536-42-5*] (1:3:2), *13*:782
Ferric perchlorate tris(1,10-phenanthroline)
  monohydrate [*20774-81-6*] (1:3:3:1), *13*:781
Ferricrocin [*23086-46-6*], *13*:784
Ferricyanic acid [*17126-46-4*] (3:1), *13*:768
Ferrimagnetism, *14*:647
Ferrimycin A₁ [*15319-30-3*], *13*:786
Ferrioxamine A₁ [*14710-31-7*], *13*:786
Ferrioxamine B [*14836-73-8*], *13*:786
Ferrioxamine D₁ [*15349-98-1*], *13*:786
Ferrioxamine D₂ [*29825-04-5*], *13*:786
Ferrioxamine E [*20008-20-2*], *13*:786
Ferrioxamine G [*29825-05-6*], *13*:786
Ferrioxamines, *13*:784
Ferrite, *15*:338
Ferrites, *13*:776; *14*:657
  as magnetic insulators, *14*:694
  properties, *14*:666
  soft, *14*:664
Ferritin [*9007-73-2*]
  iron transport, *15*:595
  xanthine oxidase reaction, *15*:593
Ferrocene [*102-54-5*], *16*:592, 593
  acylation, *16*:604
  aldehyde reaction, *15*:196
  aminomethylation, *16*:605
  peroxy radical polymerization, *15*:199
Ferrocene polymers, *15*:184.   (See also *Metal-
  containing polymers.*)
Ferrocenesulfonic acid [*32218-90-9*], *16*:606
Ferrocenesulfonic acid, heteroannular (2), *16*:606
Ferrocenoyl azide [*1273-85-4*]
  ferrocenyl isocyanate from, *13*:801
Ferrocenylacetonitrile [*1316-91-2*], *16*:606
1,2-Ferrocenylbutadiene [*1291-68-5*]
  polymerization of, *15*:193

Ferrocenylcarbinol [*1273-86-5*], *16*:606
Ferrocenylene–silylene polymers, *15*:199
Ferrocenyl isocyanate [*74930-85-1*], *13*:801
Ferrocenyl–siloxanylene polymers, *15*:199
Ferrocenyltetrafluoroborate methylium [*62341-
  75-7*], *15*:196
Ferrochromium
  nitriding, *15*:882
Ferrocyanic acid [*17126-47-5*], *13*:766. (See also
  Tetrahydrogen hexakiscyanoferrate.)
Ferroelectrics
  niobium-containing, *15*:836
Ferromagnetism, *14*:647
Ferromanganese, *14*:824, 830
  anodic oxidation, *14*:859
  chlorination, *14*:862
  electrolysis to permanganate, *14*:875
Ferromanganese nodules
  from the ocean, *16*:291
Ferromolybdenum [*11121-95-2*], *15*:670
Ferronickel [*11133-76-9*], *15*:799
Ferroniobium, *15*:823
Ferronitrososulfate
  in nitric acid analysis, *15*:867
Ferrosilicon [*8049-17-0*]
  by-product from Mg mfg, *14*:586
  in siliconizing, *15*:251
Ferrous ammonium oxalate [*70140-03-3*]
  in printing processes, *16*:633
Ferrous ammonium sulfate [*10045-89-3*]
  encapsulation, *15*:478
  in initiator systems, *13*:365
Ferrous aspartate [*75802-64-4*], *15*:760
Ferrous chloride tris(1,10-phenanthroline)
  [*14978-15-5*] (1:2:3), *13*:781
Ferrous dihydrogen phosphate [*18718-08-6*]
  in metal surface treatment, *15*:304
Ferrous ethylenediaminetetraacetic acid,
  diammonium salt [*56174-59-5*], *13*:779
Ferrous hexakiscyanoferrate [*14460-02-7*] (2:1),
  13:771
Ferrous iodide tris(1,10-phenanthroline) [*15553-
  89-6*] (1:2:3), *13*:781
Ferrous lactate [*6047-24-1*], *13*:89
Ferrous perchlorate bis(2,2′:6′,2″-terpyridine)
  [*22079-98-7*] (1:2:2), *13*:782
Ferrous perchlorate tris(1,10-phenanthroline)
  [*14586-54-0*] (1:2:3), *13*:781
Ferrous sulfamate [*14017-39-1*]
  plutonium reduction, *16*:177
Ferrous sulfate [*7720-78-7*]
  ammonium diphosphate reduction, *15*:84
  encapsulation, *15*:476
Ferrous sulfide [*1317-37-9*]
  in iron ore processing, *13*:742

Fertilizer
  radiation exposure from, *16*:217
Fertilizers
  limestone in, *14*:377
Ferula galbaniflua
  galbanum from, *16*:950
FET. See *Field-effect transistor amplifiers.*
FFTF. See *Fast flux test facility.*
Fiber
  from wood, *14*:17
Fiber aspirators
  noise control, *16*:24
Fiberboard, *14*:4
  drums, *16*:719
  properties, *14*:22
Fiberboard, cellulose
  as thermal insulator, *13*:592
Fiber drums, *16*:719
Fiber glass, *13*:516.   (See also *Glass fiber.*)
  acoustic performance of, *13*:518
  adhesive treatment, *13*:57
  as insulation, *13*:594
Fiber, mineral
  as thermal insulator, *13*:592
Fiber optics
  laser uses, *14*:71
  in level measurement, *14*:437
Fiber-reinforced metals, *13*:963
Fibers
  for metal composites, *13*:944
  novoloid, *16*:125
  phenol–formaldehyde, *16*:125
Fibers, metal, *15*:220.   (See also *Metal fibers.*)
Fibers, nonwood
  in paper, *16*:769
Fibers, synthetic
  in papermaking, *16*:823
Fibers, wood
  in paper, *16*:769
Fibertex 200, *16*:76
Fibran, *16*:779
Fibrillation
  of polyolefin films, *16*:375
Fibrinogen
  liquid crystallinity of, *14*:418
Fibroblasts
  in synthesis of collagen, *14*:204
Fibrous glass
  toxicity, *13*:268
Fick's first law of molecular diffusion, *14*:913
Fick's law
  in membrane diffusion, *15*:93
  solute diffusion, *15*:97

Fick's second law of molecular diffusion, *14*:918
Field-effect transistor (FET) amplifiers, *13*:8
Fig moth
  control of with pheromones, *13*:481
Filaments
  from polyethylene, *16*:451
  from polypropylene, *16*:468
Filament-winding processes
  for plastic composites, *13*:975
Filariasis, *13*:414
Filbert, *16*:250
File-wrapper estoppel, *16*:879
Fillers
  in cable insulation, *13*:585
  ground mica as, *15*:437
  in insulation, *13*:543
  limestone in, *14*:376
  novoloid fibers, *16*:134
  in packing materials, gaskets, *16*:732
  in paper, *16*:769, 778
  in papermaking, *16*:819
  in thermal insulation, *13*:597
Film
  from LDPE, *16*:419
  olefin fibers from, *16*:375
  from polyethylene, *16*:399
  from polypropylene, *16*:468
Film mica, *15*:421
Film-penetration theory
  in mass-transfer, *14*:934
Films
  fibrillated, *16*:375
  as insulators, *13*:556
  from latices, *14*:90
  magnetic, *14*:686
  as membranes, *15*:102
  poly(1-butene), *16*:478
  from polyethylene, *16*:451
Filtered-particle inspection, *16*:51
Filter fabrics
  fibers for, *16*:376
Filter membranes
  track-etch, *16*:841
  track etching, *16*:846
Filters
  metal-coated, *16*:531
  nonwoven, *16*:73
  for nuclear processing, *16*:236
Filters, optical.   See *Optical filters.*
Filtration
  magnetic, *14*:730
  with membranes, *15*:106
Filtration media
  nonwoven, *16*:123
FIOR process, *13*:745
Fips process, *16*:210

Firebrick
  as insulation, *13*:597
  as insulator, *13*:591
Fire departments
  portable breathing apparatus, *16*:680
Fireworks
  lithium nitrate in, *14*:464
Fischer assay
  of oil shale, *16*:348
Fischer-assay retort, *16*:338
Fischer oxazole synthesis, *15*:892
Fischer synthesis
  of indoles, *13*:218
Fischer-Tropsch process, *16*:584
Fish farming
  oxygen use, *16*:669
Fish feeds, *15*:454
Fish oil
  in packings, *16*:734
Fission, *16*:**138**. (See also *Nuclear reactors.*)
  processes, *16*:187
Fission-track dating, *16*:843
F-100 jet engine
  metallic coating for, *15*:257
Flags
  nonwoven, *16*:94
Flakeboard, *14*:17
Flake mica, *15*:420, 423
Flame hardening, *15*:321
Flame-resistant materials
  novoloid fabrics, *16*:132
Flame-retardant coatings, *16*:755
Flame retardants
  molybdenum complexes, *15*:695
  nickel-containing, *15*:814
  oxalic acid in, *16*:632
Flame-retardant treatment
  laminated timber, *14*:38
Flame spraying, *15*:253
Flares
  lithium nitrate in, *14*:464
  matches in, *15*:5
  noise control, *16*:22
Flash pyrolysis
  of wastes, *13*:194
Flash radiography
  in nondestructive tests, *16*:55
Flavor oils
  encapsulation, *15*:476
Flax
  in packings, *16*:734
  paper from, *16*:768
Flexographic inks, *13*:389
  fluorescent pigments in, *14*:567
  nitrocellulose based, *13*:390
Flexol 3 GH [*95-08-9*]
  in laminated glass, *13*:981

Flexotallic gasket, *16*:726
Float gauges, *14*:430
Float zone crystal growth
  in IC manufacture, *13*:625
Flocculants
  from aziridines, *13*:161
  in papermaking, *16*:806
Floe process, *15*:318
Floor coatings, *16*:756
Florisil, *15*:551
Flory-Huggins theory, *15*:94
Flotation
  lead ore, *14*:102
Flour
  packaging, *16*:720
Flower extract
  as odor modifier, *16*:301
Flow-molding
  metal-laminate fabrication, *13*:948
Flue-gas desulfurization
  magnesium sulfite in, *14*:640
Fluid flow
  measurement using lasers, *14*:68
Fluidics
  in level measurement, *14*:437
Fluidized-bed combustors
  for lignites, *14*:334
Fluidized-bed incinerators, *13*:192
Fluid mechanics
  in mixing, *15*:605
Flunitrazepam [*1622-62-4*]
  properties of, *13*:122
Fluorel, *16*:727
Fluorescein [*2321-07-5*], *13*:43
  fluorescence, *14*:549
Fluorescein disodium [*518-47-8*], *14*:552
Fluorescence, *14*:527. (See also *Luminescent
    materials.*)
  conditions for, *14*:549
  excited states in, *14*:556
Fluorescent lamps
  light output, *14*:536
  phosphors for, *14*:531
Fluorescent pigments, *14*:546. (See also
    Luminescent materials.)
Fluorescent yellow [*2478-20-8*]
  fluorescent dye, *14*:547
Fluoridation, *15*:593
Fluorides
  ion exchange in, *13*:678
Fluorides, inorganic
  toxicity, *13*:268
Fluorinated ethylene-propylene [*25067-11-2*]
    (FEP)
  insulation properties of, *13*:573
Fluorine [*7782-41-4*]
  metabolic functions, *15*:593

in micas, *15*:424
nutrient properties, *15*:571
from the ocean, *16*:279
Fluorine-19 [*14792-94-8*], *13*:840
Fluorine (F$_2$) [*7782-41-4*]
  toxicosis from, *15*:598
Fluorine nitride [*13967-06-1*], *15*:883
Fluoroaluminic acid [*21340-03-4*]
  insecticides from, *13*:422
Fluoroboric acid [*16872-11-0*]
  in indole reduction, *13*:217
  nickel carbonate reaction, *15*:805
Fluorocarbon rubber (FPM)
  gasket properties, *16*:727
Fluorocarbons
  health effects of decomposition products, *13*:
    266
Fluorochemical phosphates
  paper sizes, *16*:812
Fluorochemicals
  as antioxidants, *14*:501
  in papermaking, *16*:812
Fluoroelastomers
  as organolead barrier, *14*:193
Fluorolube
  in packings, *16*:734
Fluoromicas, *15*:424
Fluorophlogopite [*12003-38-2*], *15*:416
Fluorophosphate glasses
  in lasers, *14*:56
Fluorosilicic acid [*16961-83-4*]
  in electrolytic refining, *14*:124
  insecticides from, *13*:422
N-(Fluorosulfonyl)acetoacetamide [*53440-87-2*],
    13:884
5-Fluorouracil [*51-21-8*], *13*:472
Fluo-solids kiln, *14*:366
Flurazepam hydrochloride [*1172-18-5*]
  properties of, *13*:122
Flushed colors, *13*:377
Flux
  limestone as, *14*:378
Fluxes
  in phosphor mfg, *14*:542
Fluxmeters
  in magnetic measurement, *14*:683
Fluxstone, *14*:345
Flyash
  uses, *14*:341
FMA.   See *Failure-mode analysis.*
FMECA.   See *Failure-mode effects and*
    criticality analysis.
FM 1000 film, *13*:949
Foam bonding
  nonwovens, *16*:118
Foamed plastics
  as leather substitutes, *14*:233

Foam inhibitors.   See *Defoamers.*
Foams
  as acoustic insulators, *13*:517
  as insulation, *13*:594
Foam spinning, *16*:94
Focussan, *15*:735
Foils
  in packings, *16*:738
Folding carton inks, *13*:381
Folic acid [*59-90-3*], *16*:41
  antimetabolite analogues, *13*:172
Follicle-stimulating hormone [*9002-75-9*]
  LHRH effect on, *15*:771
  zinc effect on, *15*:590
Folpet.   See *N-(Trichloromethylthio)-*
    phthalimide.
Foner magnetometer, *14*:683
Fonofos [*944-22-9*], *13*:447
Food additives
  in meat products, *15*:62
The Food and Agricultural Organization
  information sources, *13*:308
Foodboard, *16*:792
Food packaging
  tin stabilizers in, *16*:578
Food processing
  lubricants in, *14*:523
Food products
  packaging of, *16*:717
Foods Adlibra
  information service, *16*:902
Food specialities, *15*:72
Foot-and-mouth disease, *15*:553
Footware
  leather substitutes for, *14*:233
Force-balance system
  in level measurement, *14*:441
FOREIGN TRADERS information system, *13*:
    329
Forest products
  structural, *14*:1
Forging, *15*:331
  of magnesium alloys, *14*:607
  of metals, *15*:330
Formaldehyde [*50-00-0*]
  acetoacetic ester reaction, *13*:888
  as antimicrobial agents, *13*:223
  cashew oil reaction, *16*:272
  in CMME formation, *13*:695
  dihydroxybenzene reactions, *13*:45
  effect on olfaction, *16*:299
  as hardening agent, *15*:477
  from insulation foams, *13*:603
  in isocyanate synthesis, *13*:800
  in isoprene synthesis, *13*:831
  in isotope separation, *13*:857
  ketene reaction, *13*:96, 877
  in leather manufacture, *14*:216
  lecithin preservative, *14*:254

Formaldehyde (*Continued*)
  in matches, *15*:4
  in methyl methacrylate mfg, *15*:364
  in methyprylon synthesis, *13*:124
  nitro alcohols from, *15*:910
  in novoloid fiber curing, *16*:131
  in novoloid fiber mfg, *16*:129
  in paint, *16*:760
  in papermaking, *16*:817
  production, *15*:398
  toxicity, *13*:268
Formaldehyde *O*-(1-butyl-2-oxohexyl)oxime
    [*35235-30-4*]
  from 5-decyne, *15*:977
Formaldoxime hydrochloride [*3473-11-8*]
  manganese reaction, *14*:884
Formalin [*50-00-0*]
  in leather tanning, *14*:218
  as preservative for latex paints, *13*:249
2-Formamidobenzaldehyde [*25559-38-0*], *13*:217
2-Formamidobenzoic acid [*88-97-1*], *13*:217
Formate dehydrogenase [*9028-85-7*]
  Mo content, *15*:693
Formation aids
  in papermaking, *16*:806
Formetanate [*22259-30-9*], *13*:455
Formica, *13*:970
Formic acid [*64-18-6*]
  lithium formate from, *14*:460
  from oxalic acid, *16*:620
  in paint stripping, *16*:763
  removal by ion exchange, *13*:702
Formycin A [*6742-12-7*]
  microbial transformation to formycin B, *15*:465
Formycin B [*6743-51-7*]
  microbial synthesis of, *15*:465
Forsterite [*1343-88-1*]
  as insulator, *13*:562
Fortran
  in operations planning, *16*:509
Fospirate [*5598-52-7*], *13*:440
Foundry type alloys, *14*:156
Fourdrinier paper machine, *16*:771
Fourdrinier papermaking machine
  in mica paper mfg, *15*:430
Fourier sine series, *14*:921
Foxboro's model 823 transmitter
  for level measurement, *14*:446
Fox nut, *16*:251
FP fiber
  in metal laminates, *13*:953
FPM. See *Fluorocarbon rubber.*
Fracture control
  in metal laminates, *13*:960
Fragrance raw materials, *16*:948
Fragrances. See *Perfumes.*
  encapsulation, *15*:476

Frank-Caro cyanamide process, *15*:943
Frank-Condon principle, *14*:527
Frankia sp, *15*:948
Franklinite, *14*:825
  zinc ferrite as, *13*:776
Frasch process
  for sulfur recovery, *16*:287
Free erythrocyte protoporphyrin, *14*:199
Free-radical initiators, *13*:356
Free-radical reactions
  initiators for, *13*:367
Free radicals
  maleic anhydride addition, *14*:777
Freeze-dried meat, *15*:70
Freezing
  of meat products, *15*:70
  with N_2, *15*:940
French cylinder test, *16*:99
Freon
  metal cleaning, *15*:299
  in thermal insulation, *13*:596
Freon 11 [*75-69-4*]
  as electric insulation, *13*:550
Freon 12 [*75-71-8*]
  as electric insulation, *13*:550
Freon 13 [*75-72-9*]
  as electric insulation, *13*:550
Freon 14 [*75-73-0*]
  as electric insulation, *13*:550
Freon 22 [*75-45-6*]
  as electric insulation, *13*:550
Freon 113 [*76-13-1*]
  as electric insulation, *13*:550
Freon 114 [*76-14-2*]
  as electric insulation, *13*:550
Fresnel lenses, *15*:393
Fresnel's equation, *16*:522
1,3,6-Freund's acid [*6521-07-6*], *15*:730. (See
    also *4-Amino-2,7-naphthalenesulfonic acid.*)
1,3,7-Freund's acid [*6362-05-6*], *15*:730. (See
    also *4-Amino-2,6-naphthalenedisulfonic
    acid.*)
Freund's adjuvant, *13*:176
Friction
  minimization of using lubricants, *14*:477
Friction coefficient, *14*:481
Friction reduction
  enhancement of by ion implantation, *13*:717
Friction tape, *13*:578
Friedel–Crafts halides
  as initiator catalysts, *13*:371
Fries rearrangement
  of dihydroxybenzene derivatives, *13*:44
Froth stabilizers, *14*:104
Frozen meat, *15*:70

Fructose [57-48-7]
  for ion-exchange process, *13*:703
D-Fructose [57-48-7]
  dextran preparation, *15*:441
  microbial preparation, *15*:459
Fructose syrup
  manufacture, *15*:467
Fruit
  malic acid in, *13*:106
Fruit extract
  as odor modifier, *16*:301
Fruit juices
  in diuretic therapy, *15*:588
  essential oils from, *16*:314
FSTA
  information service, *16*:902
Fuchsite [12198-09-3], *15*:416
Fucose [2438-80-4]
  in zanflo, *15*:451
Fucus digitata, *16*:294
Fuel-alloy process
  for liquid hydrocarbons, *14*:845
Fuel assemblies
  in nuclear reactors, *16*:165
Fuel gases
  from lignite gasification, *14*:328
Fuel-injection systems
  reliability, *15*:20
Fuel oil
  paper coating defoamer, *16*:791
Fuels
  lignite and brown coal, *14*:313
  methanol in, *15*:413
  nitroparaffins in, *15*:984
  nuclear, *16*:140
  from wastes, *13*:193
Fuels, synthetic
  from lignite, *14*:339
Fulvic acid [479-66-3]
  ozone reaction, *16*:703
Fumarase [9032-88-6]
  in malic acid biosynthesis, *13*:107
Fumaria, *14*:770
Fumaric acid [110-17-8], *14*:770. (See also
    *Maleic anhydride, maleic acid, and fumaric
    acid.*)
  in emulsion polymerization, *14*:87
  in S(−)-malic acid biosynthesis, *13*:107
  malic acid decomposition product, *13*:105
  in R,S(±)-malic acid synthesis, *13*:107
  microbial transformation to aspartate, *15*:464
  properties, *14*:770
  in rosin size, *16*:810
  in (R*,R*)-tartaric acid synthesis, *13*:113
Fumaroyl chloride [627-63-4], *14*:773
Fume abatement
  nitric acid plants, *15*:863

Fumigants, *13*:465
  in insect control, *13*:419
  for nuts, *16*:263
Fungi
  effect on cattlehides, *14*:208
  phenolics effect on, *13*:225
Fungicides
  in magnetic tape, *14*:743
  mercury compounds, *15*:155
  mercury in, *15*:158
  naphthenates as, *15*:752
  nickel containing, *15*:814
  organoleads, *14*:193
Fungicides (agricultural), *13*:428.   (See also
    Insect control technology.)
Fungitrol 11, *13*:245
Furafil, *14*:308
Furan [110-00-9]
  adiponitrile from, *15*:898
  maleic anhydride reaction, *14*:775
2,5-Furandione [108-31-6], *14*:770.   (See also
    Maleic anhydride.)
2,5-Furandione, dihydro[(4-methylphenyl
    methyl)] [4338-65-2], *14*:774
2,5-Furandione, dihydro-3-(2-propenyl) [7539-
    12-0], *14*:774
Furfural [98-01-1]
  adiponitrile from, *15*:898
Furfuraldehyde [98-01-1]
  cashew oil reaction, *16*:272
Furfuryl alcohol [98-00-0]
  in leather tanning, *14*:216
  in rocket fuel, *15*:856
Furnace, blast
  iron oxide reduction, *13*:735
Furnace resistance element
  molybdenum, *15*:677
Furnaces
  ir inspection, *13*:350
  for lignites, *14*:332
  noise control, *16*:22
Furnaces, blast
  in DR processes, *13*:754
Furnace tube emissivities, *13*:351
Furoxans
  from nitroparaffins, *15*:978
Furyl methacrylate [3454-28-2]
  properties, *15*:354
Fused salt mixtures
  in metallic coating, *15*:252
Fused salt systems
  lithium in, *14*:459
Fuse lighters, *15*:5
Fuse lines, *15*:1
Fusible alloys, *14*:157
Fusion
  laser assisted, *14*:76

Fusion plasmas
  microwave produced, *15*:516
Fusion welding, *15*:277
Fyrquel 150, *14*:496
Fyrquel 220, *14*:496

G

GABA.   See γ-*Aminobutyric acid.*
GABA shunt, *15*:767
Gabon, Africa
  site of natural nuclear reactor, *13*:861
Gabon nut, *16*:252
G-acid [*118-32-1*], *15*:739.   (See also *7-Hydroxy-*
  1,3-naphthalenedisulfonic acid.)
Gadolinium [*7440-54-2*]
  from the ocean, *16*:280
Gadolinium-152 [*14867-54-0*], *13*:841
Gadolinium-154 [*14683-24-0*], *13*:841
Gadolinium-155 [*14333-34-7*], *13*:841
Gadolinium-156 [*14392-07-5*], *13*:841
Gadolinium-157 [*14391-32-3*], *13*:841
Gadolinium-158 [*15068-71-0*], *13*:841
Gadolinium-160 [*14834-81-2*], *13*:841
Gadolinium, compd with cobalt [*12017-61-7*] (1:
  5), *14*:676
Gadolinium, compd with cobalt [*12017-77-5*] (2:
  17), *14*:676
Gadolinium gallium garnet (GGG) [*12024-36-1*]
  in thin-film fabrication, *14*:694
Gadolinium iron garnet (GDIG) [*12063-50-2*]
  in magnetooptics, *14*:705
Gadolinium nitride [*25764-15-2*], *15*:872
  properties, *15*:873
Gadolinium oxysulfide, Tb doped [*68609-43-8*]
  x-ray phosphor, *14*:544
Galactomannans
  xanthan gum interaction, *15*:450
D(−) Galactonamide [*7256-14-6*], *13*:101
Galactonic acid [*576-36-3*], *13*:101
DL(−) Galactonic acid, γ-lactone [*2426-46-2*], *13*:
  101
Galactose [*59-23-4*]
  in soybean lecithin, *14*:251
  in zanflo, *15*:451
D-Galactose [*59-23-4*]
  in soybean phosphatides, *14*:252
  in succinoglucan, *15*:453
Galacturonic acid [*14982-50-4*]
  in indica polysaccharide, *15*:451
Galaxolide [*1222-05-5*], *16*:955, 969
  in perfumes, *16*:968

Galbanum [*8023-91-4*]
  in perfumes, *16*:950
Galbanum oil [*8023-91-4*], *16*:323
  in perfumes, *16*:967
Galbanum resin
  resorcinol from, *13*:39
Galbanum resins
  resorcinol from, *13*:39
Galena [*12179-39-4*], *14*:99, 166.   (See also *Lead*
  sulfide.)
  lead from, *14*:140
Gallaldehyde [*13677-79-7*], *13*:70
Galleria melonella, *13*:459
Gallium [*7440-55-3*]
  as lubricant, *14*:508, 515
  from the ocean, *16*:280
Gallium-69 [*14391-02-7*], *13*:840
Gallium-71 [*14391-03-8*], *13*:840
Gallium alkyls, *16*:562
Gallium antimonide [*12064-03-8*]
  properties, *14*:286
Gallium arsenide [*1303-57-5*]
  in integrated circuits, *13*:621
  ion-beam doping of, *15*:268
  laser, *14*:42
  in LEDs, *14*:276
  properties, *14*:286
Gallium–arsenide laser, *14*:51
Gallium chloride [*13450-90-3*]
  in LED mfg, *14*:288
Gallium, compd with germanium and manganese
  [*37195-97-4*] (1:1), *14*:694
Gallium nitride [*25617-97-4*], *15*:872
  properties, *15*:873
Gallium phosphide [*12063-98-8*]
  ion-beam doping of, *15*:268
  properties, *14*:286
  in semiconductor lasers, *14*:273
Gallotannin, *13*:46
Gallstone formation
  liquid crystals in, *14*:420
Galoter retort, *16*:341
Galvalume, *15*:246
Galvanic anodes
  magnesium for, *14*:611
Galvanizing, *15*:244
Galvannealing, *15*:245
Gamma methyl ionone
  in perfumes, *16*:968
Gamma radiation
  effect on components, *16*:240
Garlic
  diallyl disulfide in, *16*:311
Garlic oil [*8000-78-0*], *16*:308
Garnett cards, *16*:112
Garnierite [*12198-10-6*]
  nickel in, *15*:797

Gas centrifuge
  uranium isotope separation, *13*:855
Gas-combustion retort, *16*:338
Gas-cooled fast breeder reactor (GCFBR), *16*:203
Gas-cooled fast reactor (GCFR), *16*:192
Gas discharge
  in ion implantation, *13*:708
Gases
  as lubricants, *14*:**514**
  membrane diffusion, *15*:100
  transport through membranes, *15*:93
Gas generators
  for N₂, *15*:936
Gasification
  of oil shale, *16*:345
Gasifier
  lurgi, *14*:334
Gaskets, *16*:725.   (See also *Packing materials.*)
Gas lasers, *14*:51
Gas, liquefied petroleum, *14*:383.   (See also
    Liquefied petroleum gas.)
Gas–liquid processes
  mixing in, *15*:617
Gas lubrication, *14*:514
Gas, natural
  in DR processes, *13*:754
  LPG from, *14*:390
Gasoline
  lead content, *14*:191
  from methanol, *15*:667
GASP
  in operations planning, *16*:509
Gas permeability
  various polymers, *15*:118
Gas-phase polymerization, *16*:445
Gas purification
  with zeolites, *15*:662
Gas sensing electrodes, *13*:727
Gas separations
  membrane processes, *15*:118
Gastroenteritis, *15*:553
Gastuloric, *15*:767
Gas turbines
  noise control, *16*:21
Gator-gard, *15*:255
GATT.   See *General Agreement on Tariffs and
    Trade.*
Gattermann reaction
  of indole, *13*:216
Gauchers' disease
  liquid crystals in, *14*:420
Gauge glass
  for level measurement, *14*:428
Gauge-glass lining
  from sheet mica, *15*:437

Gaultheria procumbers L., *16*:329
GCFBR.   See *Gas-cooled fast breeder reactor.*
GCFR.   See *Gas-cooled fast reactor.*
GDIG.   See *Gadolinium iron garnet.*
Gear oil
  limit of radiation tolerance, *14*:519
Gears
  noise control, *16*:21
Geiger counter
  in level measurement, *14*:435
Geiger-Mueller tubes
  in level measurement, *14*:436
Z-Gel, *15*:85
Gelatin [*9000-70-8*]
  encapsulation with, *15*:474
  in ice cream, *15*:565
Gelatin filters, *16*:535
Gel filtration
  dextrans in, *15*:445
Gels, *16*:301
  ionotropic, *15*:104
  in zeolite mfg, *15*:654
General Agreement on Tariffs and Trade
  information sources, *13*:308
General electric mica process, *15*:432
Generator, *13*:356.   (See also *Free radical.*)
Genetic engineering
  in nitrogen fixation, *15*:953
Geneva Congress, *16*:29
Genite [*97-16-5*], *13*:463
Gentamicin [*1403-66-3*]
  inactivation, *15*:466
  measurement, *15*:85
Gentisaldehyde [*1194-98-5*], *13*:70
Gentisic acid [*490-79-9*], *13*:45
Geoarchive
  information service, *16*:902
Geochronology
  standards, *15*:36
Geometric program
  in operations planning, *16*:504
Geotextiles, *16*:101
Geothermometers, *13*:861
Geranial [*141-27-5*], *16*:312
Geraniol [*106-24-1*], *16*:312, 955
  from citronella oil, *16*:322
  from palmarosa, *16*:326
Geranium, algerian oil [*8000-46-2*], *16*:320
Geranium oil [*8000-46-2*], *16*:319, 323
Geranyl acetate [*105-87-3*], *16*:312
  from eucalyptus oil, *16*:323
Gerber test
  for milk fat, *15*:552
Geriatrics, *15*:132.   (See also *Memory-enhancing
    agents and antiaging drugs.*)

Germanium [*7440-56-4*]
  from the ocean, *16*:280
  in optical filters, *16*:526
Germanium-70 [*14687-55-9*], *13*:840
Germanium-72 [*13982-21-3*], *13*:840
Germanium-73 [*15034-58-9*], *13*:840
Germanium-74 [*15034-59-0*], *13*:840
Germanium-76 [*14687-41-3*], *13*:840
Germanium nitride [*12065-36-0*] (3:4), *15*:872
Germanium oxide [*1310-53-8*]
  in nitride formation, *15*:877
Germicides
  ozone-derived, *16*:687
Gerontology, *15*:136. (See also *Memory-enhancing agents and antiaging drugs.*)
Gerovital H$_3$
  antiaging drug, *15*:140
G fiber
  in metal laminates, *13*:953
GGG. See *Gadolinium gallium garnet.*
G GY 70 fiber
  in metal laminates, *13*:953
GHB. See *γ-Hydroxybutyric acid.*
G HM fiber
  in metal laminates, *13*:953
G HM pitch fiber
  in metal laminates, *13*:953
Gibberellic acid [*77-06-5*]
  in malting, *14*:816
Gilsonite
  in paint testing, *16*:759
Ginger oil [*8007-08-7*], *16*:323
Ginkgo, *16*:248
Ginkgoin
  globulin, *16*:249
Girdler sulfide process
  in isotope separation, *13*:849
GIX. See *1,1,1-Trichloro 2,2-bis(p-fluorophenyl)ethane.*
Glass, *13*:978. (See also *Laminated materials, glass.*)
  architectural, *13*:990
  bullet-resistant, *13*:985
  emissivity, *13*:342
  evaporative coating of, *15*:265
  leaded, *14*:158
  as lubricant, *14*:520
  metallized, *13*:989
Glass-bonded mica, *15*:429
Glass, cellular
  as thermal insulator, *13*:592
Glass-E fiber
  in metal laminates, *13*:945
Glass electrode
  for pH measurement, *13*:722
Glass electrodes, *13*:5
Glasses
  in IC doping, *13*:635

particle-track etching, *16*:826
Glass fiber
  as acoustic insulation, *13*:516
Glass fibers
  in composites, *13*:970
Glass filters, *16*:527, 535
Glassine, *16*:773
Glass, laminated. See *Laminated materials, glass.*
Glass manufacture
  lime and limestone in, *14*:380
Glass-S fiber
  in metal laminates, *13*:945
Glass transition temperature (T_g)
  in electric insulation, *13*:545
Glass wool
  in papermaking, *16*:823
Glassy carbon
  novoloid based, *16*:137
Glauber's salt [*7727-73-3*], *16*:278. (See also *Sodium sulfate decahydrate.*)
  from seawater, *16*:278
Glauconite [*1317-57-3*]
  from the ocean, *16*:289
Glazing, *13*:978. (See also *Glass.*)
  acoustical, *13*:990
  methacrylates in, *15*:393
  safety, *13*:985
Glazing materials
  laminated, *13*:985
GLDH. See *Glutamic dehydrogenase.*
Gleeble machine
  for tensile tests, *15*:327
Gleothece, *15*:950
Globaline
  water disinfectant, *13*:674
Globin, *15*:595
Globin zinc insulin [*9004-21-1*], *13*:612
Globulin
  amino acid composition, *16*:249
Globulins
  in nuts, *16*:249
Globulol [*489-41-8*], *16*:313
Gloss
  of paint films, *16*:745
Glow discharge
  in case hardening, *15*:318
Glow-discharge sputtering, *15*:268
Glucagon [*16941-32-5*]
  hypoglycemia relief, *13*:608
Glucagon USP [*16941-32-5*]
  hyperglycemic agent, *13*:619
Glucan [*9037-91-6*]
  immunotherapeutic agent, *13*:177
β-Glucan [*55965-23-6*]
  in barley, *14*:816

β-D-Glucan [*9051-97-2*], *15*:453

α-D-Glucan amylose [*9005-82-7*], *15*:454

Glucanase [*9015-78-5*]
  in scleroglucan preparation, *15*:455

D-Glucans [*9012-72-0*], *15*:447

α-D-Glucans [*9074-78-6*], *15*:440

β-D-Glucans [*9041-22-9*] capsular, *15*:454

Glucoamylase [*9032-08-0*]
  effect on dextransucrase, *15*:441

D(−) Gluconamide [*3118-85-2*], *13*:101

Gluconeogenesis
  corticosteroid effect on, *15*:762

Gluconic acid [*526-95-4*]
  in caustic baths, *16*:765

D(−) Gluconic acid [*526-95-4*], *13*:101

D(−) Gluconic acid, ethyl ester [*74421-63-9*], *13*: 101

D(−) Gluconic acid, γ-lactone [*1198-69-2*], *13*:101

L(−) Gluconic acid, γ-lactone [*74464-44-1*], *13*: 101

β-D-Glucopyranoside [*497-76-7*], *13*:46. (See also *Arbutin.*)

Glucose [*26655-34-5*], *13*:46
  from amygdalin, *16*:256
  clinical measurement, *15*:84
  detection of using ISES, *13*:728
  for ion-exchange process, *13*:703
  measurement, *15*:80
  microbial transformation, *15*:459
  nitric acid reaction, *16*:623
  oxalic acid from, *16*:621
  in zanflo, *15*:451

D-Glucose [*50-99-7*]
  polysaccharides from, *15*:440

Glucose 1-phosphate [*59-56-3*]
  in sucrose synthesis, *15*:467

Glucose-6-phosphate [*56-73-5*]
  in CPK assay, *15*:77
  in glucose assay, *15*:84

Glucose-6-phosphate dehydrogenase [*9001-40-5*]
    (G-6-PDH)
  in CPK assay, *15*:77
  in glucose assay, *15*:84

Glucose tolerance factor (GTF), *15*:591

Glucose tolerance impairment
  nutrients associated with, *15*:573

Glucose tolerance test, *15*:84

Glucose tolerance test (GTT)
  effect of chromium, *15*:592

β-D-Glucosidase [*4033-06-1*], *13*:46

Glucosylsorbose
  microbial synthesis of, *15*:467

Glucuronic acid [*6556-12-3*]
  in zanflo, *15*:451

D-Glucuronic acid [*6556-12-3*]
  in xanthan gum, *15*:448

Glucuronidases
  selenium in, *15*:591

Glue binder–vegetable fiber
  gasket properties, *16*:727

Glues
  paper sizing, *16*:778

Glues, animal
  creping adhesive, *16*:808

Glulam, *14*:34

Glutamate
  in SGOT measurement, *15*:76

L-Glutamate [*56-86-0*], *15*:767.   (See also *L-Glutamic acid.*)

Glutamic acid [*56-86-0*], *14*:802
  in diagnostic testing, *15*:76
  in nuts, *16*:249

L-Glutamic acid [*56-86-0*]
  in GABA biosynthesis, *15*:766
  as neuroregulator, *15*:767
  properties, *15*:767

L-Glutamic acid, calcium chloride complex
    [*34427-83-3*], *15*:767

Glutamic acid decarboxylase [*9024-58-2*]
  in GABA biosynthesis, *15*:766

L-Glutamic acid hydrochloride [*138-15-8*], *15*:767

Glutamic dehydrogenase [*9029-12-3*]
  zinc in, *15*:590

Glutamic dehydrogenase [*9001-46-4*] (GLDH)
  BUN reagent, *15*:81

Glutamic oxaloacetic transaminase [*9000-97-9*]
  diagnostic reagent, *15*:76

Glutamic pyruvic transaminase [*9000-86-6*]
  diagnostic reagent, *15*:79

Glutamidin, *15*:767

Glutamine synthetase
  in N$_2$ fixation, *15*:953

γ-Glutamyl-*p*-nitroanilide [*47018-63-3*]
  in enzyme assay, *15*:79

γ-Glutamyl transpeptidase [*9046-27-9*]
  in enzyme assay, *15*:79
  measurement of, *15*:78

Glutan-HCl, *15*:767

Glutaraldehyde [*111-30-8*]
  as antimicrobial agents, *13*:223
  as hardening agent, *15*:478
  in hide tanning, *14*:215
  in leather tanning, *14*:217

Glutasin, *15*:767

Glutathione [*121-24-4*]
  antiaging effects, *15*:138
  in insecticide reaction, *13*:421
  sulfur in, *15*:587

Glutathione peroxidase [*9013-66-5*]
  selenium in, *15*:590

Glutelin, *16*:249
  amino acid composition, *16*:249

Glutethimide [77-21-4]
   properties of, 13:122
Glyceroboric acid [36314-25-7]
   as insecticide, 13:422
Glycerol [56-81-5]
   in alkyd paints, 16:753
   for ion-exchange process, 13:703
   lactic acid used for, 13:87
   in lubricating greases, 14:503
   measurement, 15:83
   nitration, 15:842
   nitric acid reaction, 15:856
   ozone oxidation of, 16:710
   in packings, 16:734
Glycerolkinase [9030-66-4], 15:89
Glycerol monostearate [31566-31-1]
   in commercial lecithin, 14:261
   in lecithin manufacture, 14:260
Glycerol phosphate [12040-65-2], 15:89
α-Glycerophosphate dehydrogenase [9075-65-4]
   iron in, 15:595
Glycerophosphoric acid [12040-65-2]
   from lecithin, 14:256
   in lecithin reaction, 14:254
Glyceryl guaiacolate [93-14-1], 13:43
   economic aspects, 13:53
Glyceryl lactopalmitate [1335-49-5], 13:86
Glyceryl lactostearate [1338-10-9], 13:86
Glyceryl phosphoryl choline [563-24-6]
   from lecithin, 14:256
Glycidyl ethers
   toxicity, 13:268
Glycidyl methacrylate [106-91-2]
   aniline reaction, 15:352
   properties, 15:347
   SO_2–amine reaction, 15:356
Glycine [56-40-6]
   as neuroregulator, 15:768
   in nuts, 16:249
   ozone oxidation of, 16:710
   in papermaking, 16:817
   properties, 15:768
Glycine hydrochloride [6000-43-7], 15:769
Glycocoll.   See *Glycine.*
Glycogen [9005-79-2]
   lactic acid from, 13:80
Glycolic acid [79-14-1], 13:90.   (See also
      Hydroxyacetic acid.)
   from oxalic acid, 16:621
Glycolide [502-97-6], 13:91.   (See also
      Diglycolide.)
Glycolysis
   magnesium requirement, 15:583
Glyconitrile [107-16-4]
   hydrolysis, 13:92
Glycosides
   in nuts, 16:259

Glycosuria
   in diabetes, 13:606
   from lead exposure, 14:198
Glyoxal [107-22-2]
   oxalic acid from, 16:627
   in paper coatings, 16:790
   in papermaking, 16:818
Glyoxal bis(2-hydroxyanil) [1149-16-2]
   uranium adsorbent, 16:284
Glyoxylic acid [298-12-4], 13:91
   guaiacol, guethol reactions, 13:45
   from oxalic acid, 16:621
Glyoxylic acid cycle
   S(−)-malic acid in, 13:106
*The Gmelin Handbuch der Anorganischen
      Chemie*, 13:288
5′-GMP.   See *Guanosine-5′-monophosphate.*
Goal programming
   in operations planning, 16:504
Goatskins
   for leather, 14:224
Goethite [1310-14-1]
   in magnetic tape mfg, 14:738
   in ocean raw materials, 16:291
Goiter
   iodine deficiency in, 13:672
   nutrients associated with, 15:573
Golay cell, 13:340
Gold [7440-57-5]
   cladding of, 15:281
   in coatings, 15:241
   determination in mercury, 15:153
   emissivity, 13:342
   extraction by $HgSO_4$, 15:162
   in ICS, 13:628
   in integrated circuits, 13:641
   as lubricant, 14:508
   as magnetic material, 14:647
   from the ocean, 16:280
   from seawater, 16:284
   use in nondestructive testing, 16:52
Gold-197 [7440-57-5], 13:842
Gold compound with iron [12399-97-2] (1:1), 14:
      648
Gold cyanide [37187-64-7]
   in thiomalic acid reaction, 13:111
Gold effect
   in inks, 13:379
Gold nitride [13783-74-9], 15:872
Gold thiomalate [24145-43-5]
   in rheumatoid arthritis therapy, 13:174
Goniophotometers, 16:758
Gonorrhea
   treatment, 16:327
Gore-tex, 15:104
   fabric coating, 15:125
Goulac, 14:125

Gpc. See *Gel-permeation chromatography.*

G-6-PDH. See *Glucose-6-phosphate dehydrogenase.*

GPO MONTHLY CATALOG information system, *13*:329

GPSS
in operations planning, *16*:509

Graboxan, *14*:248

Graff's C stain, *16*:794

Grafoil, *16*:727, 731

Graft copolymers
methacrylates, *15*:389

Graft rejection, *13*:169

Grain-growth inhibitors
in magnetic materials, *14*:654

Gramine [87-52-5], *13*:216, 220
in tryptamine synthesis, *15*:782

Gram negative bacteria
antimicrobial agent's effect on, *13*:223
disinfection by *o*-phenylphenol, *13*:226
phenolics effect on, *13*:225

Gram positive bacteria
antimicrobial agent's effect on, *13*:223
disinfection by *o*-phenylphenol, *13*:226
phenolics effect on, *13*:225

Granary weevil
fumigant activity on, *13*:465

Grandjean planes
in smectogenic solids, *14*:399

Grandlure [11104-05-5], *13*:482

Grand mal
treatment, *13*:134

Granodraw SS, *14*:509

Granulators
noise control, *16*:24

Grape berry moth
control of with pheromones, *13*:480

Grapefruit oil [8016-20-4], *16*:323

Grapes
cream of tartar in, *13*:111
hydroxyacetic acid in, *13*:91

Graphite [7782-42-5]
aluminum composite, *13*:942
gasket properties, *16*:727
as grease filler, *14*:503
ion-selective electrodes, *13*:727
in magnetic tape, *14*:734
nitrogen reaction, *15*:934
as solid lubricant, *14*:508
use in lasers, *14*:53

Graphite anodes
in chlorine production, *15*:172

Graphite fibers
in composites, *13*:971

Graphite fluoride [11113-63-6]
as solid lubricant, *14*:508

Graphite foil
in packings, *16*:731

Grapholitha molesta, *13*:480

Grashof number
for mass transfer, *14*:**978**

Gravel
in iron-ore processing, *13*:740

Grave's disease
nutrients associated with, *15*:573

Gravure inks
fluorescent pigments in, *14*:567

Graybody radiation, *13*:339

Grease, *14*:477. (See also *Lubrication and lubricants.*)
in packings, *16*:734

Greases, *14*:501

Greases, synthetic, *14*:503

Greenalite
as iron source, *13*:748

Green lake
in inks, *13*:379

Green liquor process, *16*:770

Green river formation, *16*:333

Green sand. See *Glauconite.*
in ion exchange, *13*:678

Green-Yorston stain azo, *16*:794

Grignard reactions
with maleic anhydride, maleic acid, fumaric acid, *14*:777
organotins by, *16*:573

Grignard reagents, *14*:573; *16*:558. (See also *Organometallics.*)
isoprene reaction, *13*:826

Grinding
paint, *16*:744

Grisalva
in perfumes, *16*:968

Grotan, *13*:242

Grotan K [55965-84-9]
as metalworking fluid preservative, *13*:247

Ground mica, *15*:423

Growth hormone [9002-72-6]
zinc effect on, *15*:590

Growth hormone [9002-72-6] (GH)
somatostatin effect on, *15*:780

Growth regulators
in malting, *14*:816

Growth regulators, insect
in insect control, *13*:459

Growth retardation
from steroid treatment, *13*:174

GRS. See *Styrene–butadiene rubber.*

Grubs
effect on leather, *14*:208

Grunerite
as iron source, *13*:748

GS process. See *Girdler sulfide process.*

GTF. See *Glucose tolerance factor.*

G T 50 fiber
  in metal laminates, *13*:953
G T 75 fiber
  in metal laminates, *13*:953
GTT.   See *Glucose tolerance test.*
Guaiacol [*90-05-1*], *13*:43
  economic aspects, *13*:53
Guaiacol benzoate [*531-37-3*], *13*:62
Guaiacol carbonate [*533-17-3*], *13*:62
Guaiacol glyceryl ether [*93-14-1*], *13*:43.   (See
  also *Glyceryl guaiacolate.*)
Guaiacol salicylate [*87-16-1*], *13*:62
Guaiac wood oil [*8016-23-7*], *16*:309
Guaiene [*88-84-6*], *16*:326
Guanidine [*113-00-8*]
  alkyl lactate reaction, *13*:84
Guanine [*73-40-5*]
  alkylation in tumor formation, *15*:993
  xanthine oxidase reaction, *15*:592
Guano
  fertilizer, *15*:943
Guanosine-5′-monophosphate [*85-32-5*] (5′-GMP)
  microbial synthesis of, *15*:465
Guar [*9000-30-0*]
  dispersant, *16*:806
Guar gum [*9000-30-0*]
  in ice cream, *15*:565
  in paper processing, *16*:780
  xanthan gum interaction, *15*:450
Guayule, *13*:827
Guethol [*94-71-3*], *13*:43
Guiacylglycerol β-coniferyl ether [*1103-58-8*]
  in lignin, *14*:301
Guilandina bonduc, *16*:251
L-Gulopyranuronic acid [*1986-15-8*]
  in alginic acid substitute, *15*:452
L-Guluronic acid [*1986-15-8*]
  in alginate polysaccharide, *15*:452
Gum arabic [*9000-01-5*]
  in encapsulation, *15*:477
Gums
  in papermaking, *16*:815
Gun firing
  analogue control in, *13*:489
Gunnera, *15*:949
Gunpowder, *15*:851
α-Gurjunene [*489-40-7*], *16*:313
Gurvitch rule, *15*:644
Gutta percha, *13*:827
Guyton de Morveau, *16*:29
Gypsum [*13397-24-5*]
  encapsulated water reaction, *15*:491
  as insecticide carrier, *13*:416
  in lithium recovery, *14*:454
  removal with hydroxyacetic acid, *13*:94
  from seawater, *16*:280
  waterproofing with oxalic acid, *16*:633

Gypsum board, *16*:812
  sound coefficients of, *13*:513
Gypsum-gauging plaster, *14*:376
Gypsy moth
  control of with pheromones,
    13:480
Gyroscope
  laser-based, *14*:68
Gyroscopes
  ion implantation used for,
    13:717
Gyrotron, *15*:503

H

Haake viscometer, *13*:376; *16*:757
Haber-Bosch process, *15*:854, 944
H acid [*90-20-0*], *15*:740.   (See also *4-Amino-5-
  hydroxy-2,7-naphthalenedisulfonic acid.*)
Hadfield steel, *14*:841
Haertolan, *15*:735
Hafnium [*7440-58-6*]
  from the ocean, *16*:280
Hafnium-80 [*14265-78-2*], *13*:842
Hafnium-174 [*14922-49-7*], *13*:842
Hafnium-176 [*14452-48-3*], *13*:842
Hafnium-177 [*14093-09-5*], *13*:842
Hafnium-178 [*14265-77-1*], *13*:842
Hafnium-179 [*14265-76-0*], *13*:842
Hafnium carbide [*12069-85-1*]
  in nitride pseudobinary system, *15*:873
Hafnium nitride [*25817-87-2*]
  properties, *15*:874
Hafnium nitride [*25817-87-2*] (1:1), *15*:872
Hafnium nitride [*12508-69-9*] (3:2), *15*:872
Hafnium tetrachloride [*13499-05-3*]
  in nitride coatings mfg, *15*:881
Hafnocene dichloride [*12116-66-4*]
  polymers from, *15*:210
Hafnocene polymers, *15*:209
Hagen-Poiseuille equation, *15*:111
Hagen-Poiseuille relationship, *15*:99
Hair
  in packings, *16*:734
Halane, *13*:230
Hall effect
  in magnetic measurement, *14*:681
Hall-Heroult electrolyte
  lithium carbonate in, *14*:460
Hallucinations
  DMT induced, *15*:763

nutrients associated with, *15*:573
α-Haloacetophenone
  in noradrenaline synthesis, *15*:773
Halogens
  as antimicrobial agents, *13*:227
  in nitration, *15*:848
Halophosphate phosphors, *14*:535
Halowax, *15*:702
Ham
  microwave cooking, *15*:500
Hamilton Ontario incinerator, *13*:190
Hams, *15*:62. (See also *Meat products.*)
Handbooks, *13*:284
Hand lay-up/spray-up processes
  for plastic composites, *13*:975
Hansa yellow [*2512-29-0*]
  paint pigment, *16*:745
Hantzsch pyridine synthesis, *13*:888
Hardboard, *14*:4
Hardening, *15*:322
  with lasers, *14*:66
Hardening agents
  nitro alcohols, in photography, *15*:915
Hardwood
  in plywood manufacture, *14*:5
Hardwoods
  lignin content, *14*:297
Harris process
  lead refining, *14*:116
Harvest process, *16*:210
Hastelloy, *13*:184, 575; *16*:741
  cladding of, *15*:281
Hastelloy Alloy B-2 [*61608-60-4*], *15*:789
Hastelloy Alloy C-276 [*12604-59-0*], *15*:789
Hastelloy Alloy G [*39367-32-3*], *15*:789
Hastelloy alloys, *15*:792
Hastelloy B [*12605-84-4*]
  iodine resistance, *13*:656
Hastelloy C [*12605-85-5*]
  iodine resistance, *13*:656
Hastelloy C-276, *15*:792
Hastelloy N, *15*:792
Hausmannite [*1309-55-3*], *14*:825, 845
Haveg 41, *16*:629
Hawk nut, *16*:251
Hay
  lignin content, *14*:297
Hay fever
  treatment, *15*:762
Hazard function
  defined, *15*:16
Hazards
  with the use of lasers, *14*:63
Hazel nut, *16*:250
Hazel sterculia, *16*:252
Haze penetration, *16*:535

50-HB-2000, *14*:496
Hb-S. See *Hemoglobin-S.*
HCC. See *Hydroxycholecalciferol.*
HCR alloy, *14*:659
HDI. See *Hexamethylene diisocyanate.*
HDPE. See *High density polyethylene.*
Headache
  (−) adrenaline induced, *15*:758
Health effects
  of noise, *16*:7
Hearing loss
  noise-induced, *16*:1
Hearing protectors, *16*:18
Heart disease
  lactic acid elevation, *13*:80
  nutrients associated with, *15*:573
  taurine associated with, *15*:781
Hearth smelting
  lead ore, *14*:109
Heartwood
  in plywood manufacture, *14*:11
Heat exchange
  by water in nuclear reactors, *16*:150
Heat-exchange fluids
  effect of radiation on, *16*:239
Heat exchangers
  noise control, *16*:22
Heating
  dielectric, *15*:497
Heating elements
  nickel in, *15*:792
Heat sealing
  of packaging, *16*:721
Heat shields
  metallocene polymers, *15*:196
Heat sinks, *13*:598
Heat transfer
  mechanisms in insulation, *13*:592
  in mixing, *15*:628
Heat treatment
  metals, N_2 uses, *15*:939
Heavy turbine oil
  limit of radiation tolerance, *14*:519
Heavy water [*7789-20-0*]
  in nuclear reactors, *16*:162
  production, *13*:847
Heazlewoodite [*12035-71-1*], *15*:796, 805
Hectorite [*12173-47-6*]
  as grease thickener, *14*:502
  as lithium source, *14*:450
Hedione [*24851-98-7*]
  fragrance of, *16*:959
  in perfumes, *16*:968
Helianthus annuus, *16*:251
Helical ribbon mixer, *15*:633
Helicopters
  oxygen-generation system, *16*:677

Helicopters (*Continued*)
  windshields, *13*:988
Heliothis virescens, *13*:480
Heliothis zea, *13*:415, 480
Heliotropin [*120-57-0*], *16*:685
Heliotropine [*120-57-0*], *13*:63.   (See also
    Piperonal.)
  economic aspects, *13*:53
  from ocotea, *16*:325
  in perfumes, *16*:968
Heliozoa
  liquid crystallinity in, *14*:419
Helium [*7440-59-7*]
  in air, *16*:657
  in arc welding, *14*:608
  as GCFR coolant, *16*:203
  in implantation, *13*:710
  as insulation, *13*:549
  laser, *14*:42
  as lubricant, *14*:514
  from the ocean, *16*:280
  permeability in polymers, *15*:118
Helium-3 [*14762-55-1*], *13*:840
Helium-4 [*7440-59-7*], *13*:840
Helium–cadmium laser, *14*:51
Helium–neon laser, *14*:47, 51
Hematinic
  iron(II) gluconate dihydrate as, *13*:774
Hematite [*1309-37-7*]
  in ferrite mfg, *14*:675
  iron(III) oxide as, *13*:775
  in iron-ore processing, *13*:740
  as iron source, *13*:748
  in limestone, *14*:350
  in magnetic tape mfg, *14*:738
  in mica books, *15*:420
Hematology
  diagnostic reagents, *15*:75
Heme [*14875-96-8*], *15*:595
Hemicellulose [*9034-32-6*], *16*:769
  biodegradation of, *13*:225
Hemicellulose extract [*9034-32-6*], *16*:622.   (See
    also *Masonex.*)
Hemisine, *15*:757
Hemlock
  as vegetable tannin, *14*:214
Hemochromatosis, *15*:597
Hemodialysis, *15*:106
  in artificial kidney, *14*:943
Hemoglobin [*9008-02-0*], *13*:782; *15*:595
  as oxygen carrier, *16*:670
Hemoglobin-S (Hb-S)
  liquid-crystalline behavior, *14*:420
Hemopoiesis impairment
  nutrients associated with, *15*:573
Hemorrhage
  treatment, *13*:88

Hemosiderin [*9011-92-1*]
  iron transport, *15*:595
Hemostasin, *15*:757
Hemostatin, *15*:757
Hemp
  in packings, *16*:734
  paper from, *16*:768
HEMPA [*680-31-9*], *13*:473
Hempel pipet
  oxygen analysis, *16*:664
Henry's law, *13*:259
Hepalande, *15*:729
Heparin [*9005-49-6*]
  from animal intestines, *15*:72
  sulfur in, *15*:587
Hepatic fibrosis
  from iron toxicity, *15*:595
α-2 Hepatic globulin
  in angiotensin biosynthesis, *15*:759
Hepatitis
  diagnosis, *15*:80
Hepatitis, infectious
  waterborne, *16*:705
Hepatocuprein
  copper in, *15*:588
Heptachlor [*76-44-8*], *13*:434
Heptachlor epoxide [*1024-57-3*]
  as insecticide, *13*:434
1,4,5,6,7,8,8-Heptachloro-3a,4,7,7a-tetrahydro-
    4,7-methanoindene [*76-44-8*]
  as insecticide, *13*:434
2,2,5-*endo*-6-*exo*-8,9,10-Heptachlorobornane
    [*51775-36-1*]
  as insecticide, *13*:433
1-Heptadecene [*6765-39-5*]
  properties, *16*:481
4-Heptadecylidene-3-hexadecyl-2-oxetanone
    [*10126-68-8*], *13*:883
Heptamethylcyclotetrasiloxane [*15721-05-8*]
  isoprene reaction, *13*:826
Heptanal [*111-71-1*], *16*:312
2,4-Heptanedione [*7307-02-0*]
  properties, *13*:928
2,5-Heptanedione [*1703-51-1*]
  properties, *13*:931
3,4-Heptanedione [*13706-86-0*]
  properties, *13*:926
3,5-Heptanedione [*7424-54-6*]
  properties, *13*:928
2,4,6-Heptanetrione [*626-53-9*]
  from diketene, *13*:883
Heptanoic acid [*111-14-8*],
    16:497
Heptanol [*53535-33-4*], *16*:312
Heptene
  production, *16*:494
1-Heptene [*592-76-7*]
  properties, *16*:481

Heptenes
  from propylene–butylene mixtures, *16*:493
Heptyl butyrate [*5870-93-9*]
  as insect attractant, *13*:481
Herb extract
  as odor modifier, *16*:301
Herbicides, *13*:428.   (See also *Insect control technology.*)
  encapsulation, *15*:488
  from *p*-hydroxybenzaldehyde, *13*:76
Hercon, *16*:779
Hercules nitration process, *15*:847
Hercules sizing tester, *16*:814
Her Majesty's Stationery Office, *13*:313
Herpes virus
  waterborne, *16*:705
Herzberg stain, *16*:794
Hessian cloth
  in packaging, *16*:720
Heteroauxin [*87-51-4*], *13*:213
Heterogeneity index
  olefin polymers, *16*:425
Heterolactic fermentation
  of lactic acid, *13*:84
Heteropolymolybdates, *15*:688
Hevea, *13*:827
Hevea brasiliensis, *14*:82
2,2,4,4,6,6-Hexa-(1-aziridinyl)-2,4,6-triphospho-1,3,5-triazine [*52-46-0*]
  in insect control, *13*:472
Hexabutyldistannoxane [*56-35-9*]
  copolymer from, *15*:205
cis-7,*cis*-11-Hexacadienyl acetate [*52287-99-5*]
  as insect pheromone, *13*:481
1,2,3,4,10,10-Hexachloro-1,4,4a,5,8,8a-hexahydro-1,4-*endo-exo*-5,8-dimethanonaphthalene [*309-00-2*]
  as insecticide, *13*:434
6,7,8,9,10,10-Hexachloro-1,5,5a,6,9,9a-hexahydro-6,9-methano-2,4,3-benzodioxathiepin-3-oxide [*115-29-7*]
  as insecticide, *13*:435
γ-1,2,3,4,5,6-Hexachlorocyclohexane [*58-89-9*]
  as insecticide, *13*:432
Hexachlorocyclopentadiene [*77-47-4*]
  in endosulfan synthesis, *13*:435
Hexachlorocyclotriphosphazene [*940-71-6*]
  polymerization, *13*:400
1,2,3,4,10,10-Hexachloro-6,7-epoxy-1,4,4a,5,6,7,8,8a-octahydro-1,4-*endo,endo*-5,8-dimethanonaphthalene [*72-20-8*]
  as insecticide, *13*:435
Hexachloronaphthalene [*1335-87-1*]
  liver damage from, *15*:702
Hexachlorophene [*70-30-4*]
  as antimicrobial agents, *13*:226
  in bactericidal preparations, *13*:61

cis-11,*cis*-13-Hexadecadienal [*56829-23-3*]
  as insect pheromone, *13*:480
Hexadecanal [*27104-14-9*]
  as insect attractant, *13*:481
cis-7-Hexadecenal [*56797-40-1*]
  as insect attractant, *13*:481
cis-9-Hexadecenal [*56219-04-6*]
  as insect attractant, *13*:481
cis-11-Hexadecenal [*53939-28-9*]
  as insect pheromone, *13*:480
1-Hexadecene [*629-73-2*]
  properties, *16*:481
Hexadecenoic acid [*57-10-3*]
  in nut oils, *16*:252
cis-11-Hexadecenol [*53939-28-9*]
  as insect attractant, *13*:481
1,4-Hexadiene [*592-45-0*]
  from ethylene and butadiene, *16*:609
  isomerization, *15*:808
2,5-Hexadienoic acid [*6864-27-3*]
  from ketene, *13*:877
(*E,E*)-2,4-Hexadienoic acid [*110-44-1*], *13*:877
Hexa(2-ethylbutoxy) disiloxane
  as lubricant, *14*:501
Hexaethyldilead [*2388-00-3*], *14*:181
Hexaethyldiplumborane [*27761-04-2*], *14*:181
Hexafluoroacetone [*684-16-2*]
  aziridine reaction, *13*:150
Hexafluoroethane [*76-16-4*]
  in etch gas, *13*:639
Hexahydrite [*13778-97-7*], *14*:637
Hexahydroindanone [*29927-85-3*]
  microbial reduction, *15*:461
Hexahydropyrimidines
  from nitroparaffins, *15*:976
Hexahydro-1,3,5-triethyl-*s*-triazine [*7779-27-3*]
  applications of, *13*:242
Hexahydro-1,3,5-tris-(2-hydroxyethyl)-*s*-triazine [*4719-04-4*]
  applications of, *13*:242
Hexakiscyanoferrate(3−) [*13408-62-3*], *13*:766
Hexakiscyanoferrate(4−) [*13408-63-4*], *13*:766
Hexakis(β,β-dimethyl phenethyl)distannoxane [*13356-08-6*]
  as insecticide, *13*:463
Hexalon [*79-78-7*]
  in perfumes, *16*:967
Hexa(methoxymethyl)melamine [*3089-11-0*]
  in leather manufacture, *14*:216
Hexamethyldilead [*6713-83-3*], *14*:181
Hexamethyldisilazane [*999-97-3*]
  in photolithography, *13*:630
Hexamethylenediamine [*124-09-4*]
  from adiponitrile, *15*:898
Hexamethylene diisocyanate [*822-06-0*]

Hexamethylene diisocyanate (*Continued*)
  properties, *13*:789
Hexamethylene–tetramine [*100-97-0*]
  lecithin preservative, *14*:254
  nitration, *15*:844
  in noradrenaline synthesis, *15*:773
  in novoloid fiber curing, *16*:131
Hexamethylphosphoramide [*680-31-9*]
  polyamide solvent, *14*:417
Hexamethyltriphosphoramide [*680-31-9*]
  in insect control, *13*:472
n-Hexane [*110-54-3*]
  in lecithin manufacture, *14*:257
Hexanedinitrile.   See *Adiponitrile.*
2,3-Hexanedione [*3848-24-6*]
  properties, *13*:926
2,4-Hexanedione [*3002-24-2*], *13*:904
  properties, *13*:928
2,5-Hexanedione [*110-13-4*]
  mfg, *13*:931
  properties, *13*:898
3,4-Hexanedione [*4437-51-8*]
  properties, *13*:926
Hexanol [*25917-35-5*], *16*:312
Hexanoyl chloride [*142-61-0*]
  resorcinol reaction, *13*:61
4-Hexanoylresorcinol [*3144-54-5*], *13*:44
Hexaphenylethane [*17854-07-8*]
  bis(trityl)nickel from, *15*:809
3-Hexenal [*4440-65-7*], *16*:312
1-Hexene [*592-41-6*]
  copolymer with ethylene, *16*:388, 484
  hydroformylation reactivity, *16*:645
  oxidation of, *13*:934
  properties, *16*:481
2-Hexene [*592-43-8*]
  hydroformylation reactivity, *16*:645
3-Hexene [*592-47-2*]
  methyl methacrylate reaction, *15*:352
cis-3-Hexen-1-ol [*928-96-1*], *14*:471.   (See also
    Leaf alcohol.)
  in perfumes, *16*:967
cis-3-Hexenyl salicylate [*65405-77-8*]
  in perfumes, *16*:968
Hexetidine [*141-94-6*], *15*:912
Hexokinase [*9001-51-8*]
  in CPK assay, *15*:77
  in glucose assay, *15*:84
Hexone [*108-10-1*], *16*:173.   (See also *Methyl
    isobutyl ketone.*)
  in fuel processing, *16*:180
Hexone-23 process
  fuel reprocessing, *16*:181
Hexone-25 process
  fuel reprocessing, *16*:181
Hexyl acetate [*142-92-7*], *16*:312
n-Hexyl chloride [*544-10-5*]
  magnesium reaction, *16*:561

Hexylcinnamic aldehyde [*101-86-0*], *16*:955
  fragrance of, *16*:959
  in perfumes, *16*:968
p-n-Hexyl-*p′*-cyanobiphenyl [*4122-70-7*]
  liquid crystalline range of, *14*:399
Hexylene glycol [*107-41-5*]
  diacetone alcohol reaction, *13*:913
  in inks, *13*:382
n-Hexyl methacrylate [*142-09-6*]
  properties, *15*:353
Hexylresorcinol [*136-77-6*]
  economic aspects, *13*:53
  uses, *13*:60
Heyman process
  for mica paper, *15*:431
Hibernia Scholven process
  for MIBK, *13*:909
Hibiscus manihot Root, *16*:806
HIB process, *13*:745
Hickory nut, *16*:250
Hides
  curing of, *15*:73
Hides, butt-branded, *14*:225
Hides, native, *14*:225
Hiding
  by paint, *16*:745
Higgins contactor, *13*:692
High density polyethylene, *16*:421
High-impact polystyrene
  latex technology, *14*:82
High Permeability 49, *14*:659
High temperature gas-cooled reactor (HTGCR),
    16:203
Himsley contactor, *13*:692
Hiperco 27, *14*:664
  properties, *14*:662
Hiperco 35, *14*:664
Hipercos, *14*:662
HIPS.   See *High-impact polystyrene.*
Histadine decarboxylase [*9024-61-7*]
  in histamine biosynthesis, *15*:769
Histamine [*51-45-6*]
  as neuroregulator, *15*:769
  properties, *15*:769
Histamine dihydrochloride [*56-92-8*], *15*:769
Histamine dipicrate [*6026-98-8*], *15*:769
Histamine dipicrolonate [*6026-99-9*], *15*:769
Histamine monopicrate [*5990-88-5*], *15*:769
Histamine salicylate [*75802-62-9*], *15*:769
Histidine [*71-00-1*]
  in histamine biosynthesis, *15*:769
  microbial transformation to urocanic acid, *15*:
    465
  in nuts, *16*:249, 252
Hitachi magnets, *14*:680
HLB.   See *Hydrophilic–lipophilic balance.*

H-MDI. See *Dicyclohexylmethane diisocyanate.*
HMSO. See *Her Majesty's Stationery Office.*
Hoesch synthesis, 15:892
Hofmann rearrangement
  in isocyanate synthesis, 13:799
Höganäs process, 13:745
Hollander beater, 16:776
Hollandite [12008-99-0], 14:855
Hollow-fiber membranes, 15:92, 107. (See also *Membrane technology*.)
Holmium [7440-60-0]
  as magnetic material, 14:647
  from the ocean, 16:280
Holmium-165 [7440-60-0], 13:842
Holmium, compd with cobalt [12017-63-9] (1:5), 14:676
Holmium, compd with cobalt [12052-74-3] (2:17), 14:676
Holmium nitride [12029-81-1], 15:872
Holmium oxide [12281-10-6]
  in optical filters, 16:554
Holograms
  liquid crystals used in, 14:421
Holography
  lasers in, 14:69
  nondestructive testing by, 16:49
  in nondestructive testing, 16:62
Homasote, 14:4
Homeostasis, 15:574
  adrenal cortex control of, 13:173
Homocysteine [454-28-1]
  methylation reaction, 15:597
Homocystinuria
  nutrients associated with, 15:573
Homogeneous catalysts
  for oxo reaction, 16:639
Homogenization
  of milk, 15:528
Homolactic fermentation
  of lactic acid, 13:84
Homoveratric acid [93-40-3]
  from veratrole, 13:63
Homoveratronitrile [93-17-4]
  from veratrole, 13:63
Homoveratrylamine [120-20-7]
  in dopamine synthesis, 13:63
  from veratrole, 13:63
Honey
  attractant, 13:417
Honeywell DST transmitter
  for level measurement, 14:446
Hoods
  in IC manufacture, 13:624
Hooke's law, 13:682
Hopcalite
  filtration of oxygen, 16:676

nitroparaffin reaction, 15:982
Hopper cars, 16:716
Hormones
  clinical measurement, 15:84
  microbial preparation, 15:459
Horse chestnut, 16:251
  composition, 16:249
Horsehides
  for leather, 14:224
Horseradish peroxidase [9003-99-0]
  in digoxin assay, 15:87
Host lattices
  for phosphors, 14:541
Hot-transfer inks, 13:396
Hot-working
  of metals, 15:329
Houben-Hoesch reaction
  of indole, 13:216
House paints, 16:742
Howard process
  for calcium lignosulfate, 14:306
Hoya filters, 16:535
Hoya glass color-correction filters, 16:544
HQ. See *Hydroquinone.*
$HRh(CO)_2P(C_6H_5)_3$ [77520-53-7]
  oxo process catalyst, 16:642
$HRh(CO)_3P(C_6H_5)_3$ [77520-53-7]
  oxo process catalyst, 16:642
$HRh(CO)[P(C_6H_5)_3]_3$ [17185-29-4]
  oxo process catalyst, 16:648
HTGCR. See *High temperature gas-cooled reactor.*
Human ACTH [12279-41-3], 15:759
Humectant
  sodium lactate as, 13:88
Humectants
  in paper coatings, 16:791
Humic acid [1415-93-6]
  ozone reaction, 16:703
Humic acids
  in brown coal, 14:315
Humoral immune system, 13:167
Humus
  from brown coal, 14:341
  ion exchange in, 13:678
Huntington's chorea
  substance p levels, 15:781
Huntington's disease
  GABA involved in, 15:766
Hyaluronidase
  from animal thyroid glands, 15:72
Hyamine-3500, 13:234
Hycomax, 14:671
Hydeal process, 15:709
Hydrafiner, 16:777
Hydrapulper, 16:774

Hydrated iron(III) oxide [*1309-33-7*], *13*:776
Hydrated lime, *14*:346
  manufacture, *14*:367
Hydraulic fluids, *14*:477.   (See also *Lubrication and lubricants.*)
Hydraulic lime, *14*:346
Hydrazine [*302-01-2*], *13*:105
  in estazolam synthesis, *13*:131
  H_2O_2 coproduct, *13*:23
  in initiator systems, *13*:367
  iodine reaction, *13*:663
  from metal-dinitrogen complexes, *15*:955
Hydrazine hydrate [*7803-57-8*]
  malononitrile reaction, *14*:805
Hydrazine hydrochloride [*5341-61-7*]
  maleic anhydride reaction, *14*:774
Hydrazines
  toxicity, *13*:268
Hydrazobenzene [*122-66-7*], *15*:917
  autoxidation, *13*:15
  oxidation of, *13*:22
  in phenyl butazone synthesis, *14*:803
Hydrazone [*3291-00-7*]
  antimicrobial, *13*:76
Hydriodic acid [*10034-85-2*], *13*:663.   (See also *Hydrogen iodide.*)
  nickel carbonate reaction, *15*:804
Hydrionic, *15*:767
Hydroboration, *16*:564
Hydrobromic acid [*10035-10-6*]
  lead oxide, carbonate reaction, *14*:163
  lithium bromide from, *14*:462
  nickel carbonate reaction, *15*:804
Hydrocarbon distillates
  insulation properties of, *13*:551
Hydrocarbons
  in case hardening, *15*:314
  as lube oils, *14*:484
  as lubricants, *14*:499
  microbial oxidation, *15*:459
Hydrochloric acid [*7647-01-0*]
  in CMME formation, *13*:695
  in incinerator gases, *13*:185
  isoprene reaction, *13*:818
  in metal pickling, *15*:300
Hydrocobalt tetracarbonyl [*16842-03-8*]
  oxo process catalyst, *16*:638
Hydrocobalt tricarbonyl [*60105-25-1*]
  oxo process catalyst, *16*:638
Hydrocolloids
  in papermaking, *16*:822
Hydrocortisone [*50-23-7*], *13*:173; *15*:761.   (See also *Cortisol.*)
  microbial synthesis, *15*:460
Hydrocracking
  by zeolite catalyst, *15*:651
  zeolite catalysts for, *15*:665
  zeolites in, *15*:665

Hydrocyanation
  butadiene, *15*:808, 899
Hydrocyanic acid [*74-90-8*]
  isocyanate reaction, *13*:797
Hydrodealkylation
  zeolite catalysts in, *15*:652
Hydrodesulfurization
  with Mo catalysts, *15*:694
Hydrofluoric acid [*7664-39-3*]
  in IC manufacture, *13*:628
  in incinerator gases, *13*:185
  insecticides from, *13*:422
  lead hydroxide, carbonate reaction, *14*:162
  lithium fluoride from, *14*:461
  in metal pickling, *15*:301
  nickel carbonate reaction, *15*:804
  nitration reagents, *15*:841
  particle-track etching by, *16*:834
  in particle-track etching, *16*:827
Hydroformylation, *16*:637.   (See also *Oxo process.*)
  catalysts, *16*:639
  cobalt catalyst for, *16*:584
Hydrogasification
  of oil shale, *16*:345
Hydrogel processes
  for zeolite mfg, *15*:656
Hydrogen [*1333-74-0*]
  in air, *16*:657
  ammonia reaction, *15*:944
  carbon monoxide reaction, *15*:403
  in DR processes, *13*:754
  evolution by plants, *15*:948
  as insulator, *13*:549
  in iron-ore processing, *13*:741
  isotope separation processes, *13*:851
  in lignite, *14*:327
  from LiH, *14*:463
  lithium reaction, *14*:455
  as lubricant, *14*:514
  membranes for recovery, *15*:118
  in metallic coating, *15*:243
  from methane, *15*:945
  niobium pentachloride reaction, *15*:826
  from the ocean, *16*:279
  as O_2 scavenger in nuclear reactors, *16*:152
  oxygen reaction, *13*:23
  permeability in polymers, *15*:118
Hydrogen-1 [*12385-13-6*], *13*:840
Hydrogen-2 [*16873-17-9*], *13*:840
Hydrogenation
  of lignitic coals, *14*:339
  of metallocenes, *16*:602
  oxo aldehydes, *16*:639, 646
  shale oil, *16*:345
  step in H_2O_2 mfg, *13*:18
  zeolite catalysts in, *15*:652
Hydrogen bonding
  mesomorphism induced by, *14*:413

Hydrogen chloride [7647-01-0]
  butadieneiron tricarbonyl reaction, 16:602
  in IC manufacture, 13:625
  as leak tracer, 16:55
  measurement of with ISES, 13:733
Hydrogen cyanide [74-90-8], 13:266
  from acetylene, 15:934
  from amygdalin, 16:256
  as fumigant, 13:467
  fumigation of nuts, 16:263
  measurement of with ISES, 13:733
  mercuric oxide reaction, 15:158
  methacrylonitrile by-product, 15:363
  in nitrile synthesis, 15:893
Hydrogen–deuterium [12181-16-7]
  in isotope separation, 13:849
Hydrogen dichloroiodide [43413-52-1], 13:664
Hydrogen fluoride [7664-39-3]
  measurement of with ISES, 13:733
  from SF_6 breakdown, 13:551
  toxicity, 13:268
Hydrogen fluoride lasers, 14:60
Hydrogen iodide [10034-85-2]
  properties, 13:662
Hydrogen ion
  laser production, 14:75
Hydrogen-ion concentration, 13:1
Hydrogen oxide-18 [14314-42-2]
  in isotope separation, 13:849
Hydrogen peroxide [7722-84-1], 13:12; 14:356
  in brightening metals, 15:303
  in initiator systems, 13:365
  lecithin preservative, 14:254
  in missile fuel, 14:518
  as oxygen source, 16:673
  paper bleaching, 16:769
  properties, 13:12
  resorcinol reaction, 13:42
Hydrogen sulfide [7783-06-4]
  in GS process, 13:849
  health effect, 13:269
  in incinerator gases, 13:185
  in initiator reaction, 13:356
  iodine reaction, 13:663
  in isotope separation, 16:162
  as leak tracer, 16:55
  measurement of with ISES, 13:733
  mercury reaction, 15:147
  nitrile reactions, 15:889
  odor-problem chemical, 16:303
  from pulp mills, 16:795
  removal of from LPG, 14:389
  as tanning by-product, 14:225
  thiols from, 16:482
Hydrogen tetrachloroiodide tetrahydrate [12694-70-1], 13:664
Hydrogen tetrafluoroborate [16872-11-0]
  butadieneiron tricarbonyl reaction, 16:602

Hydroids
  antimicrobial agents used against, 13:238
Hydrojuglone [481-40-3]
  in walnut, 16:259
Hydrolysis lignin, 14:310
Hydromagnesite [12072-90-1], 14:619
Hydronaphthalenesulfonic acids, 15:724
Hydroperoxides, 13:21
15-Hydroperoxy E_2 [65380-05-4]
  in prostaglandin syntheses, 15:780
Hydrophilic–lipophilic balance (HLB)
  in emulsion polymerization, 14:85
Hydroprene [41096-46-2], 13:459
Hydroquinone [123-31-9], 13:39
  from aniline sulfate, 14:877
  determination, 15:368
  economic aspects, 13:53
  isoprene inhibitor, 13:833
  isoprene reaction, 13:821
  in liquid-crystal synthesis, 14:414
  from phenol, 13:29
  toxicity, 13:269
Hydroquinone monomethyl ether [150-76-5], 13:43
Hydroquinone, resorcinol, and catechol, 13:**39**
Hydroquinone sulfonic acid [88-46-0], 13:42, 60
Hydrorhodium carbonyl tris(triphenylphosphine) [17185-29-4]
  oxo process catalyst, 16:639
Hydrostannation, 16:577
Hydroxamic acids
  from nitroparaffins, 15:975
2-Hydroxpropyl methacrylate [923-26-2]
  properties, 15:347
N-Hydroxyacetamide [546-88-3], 13:91
Hydroxyacetic acid [79-14-1], 13:90. (See also *Glycolic acid.*)
Hydroxy acids, 13:95
3-Hydroxy-4-amino-1-naphthalenesulfonic acid [116-63-2], 15:740
4-Hydroxy-6-amino-2-naphthalenesulfonic acid [90-51-7], 15:740
4-Hydroxy-7-amino-2-naphthalenesulfonic acid [87-02-5], 15:740
4-Hydroxy-8-amino-2-naphthalenesulfonic acid [489-78-1], 15:740
9α-Hydroxy-4-androstene-3,17-dione [560-62-3]
  microbial synthesis of, 15:461
17β-Hydroxy-4-androsten-3-one [58-22-0], 15:462. (See also *Testosterone.*)
p-Hydroxyanisole [150-76-5], 13:43
  hydroquinone reaction, 13:43
Hydroxyapatite [1306-06-5]
  in bones, 15:584
2-Hydroxybenzaldehyde [90-02-8], 13:70. (See also *Salicylaldehyde.*)

3-Hydroxybenzaldehyde [*100-83-4*], *13*:70
4-Hydroxybenzaldehyde [*123-08-0*], *13*:70
m-Hydroxybenzaldehyde [*100-83-4*], *13*:70
p-Hydroxybenzaldehyde [*123-08-0*], *13*:70
Hydroxybenzaldehydes, *13*:**70**
p-Hydroxybenzoic acid [*99-96-7*]
   lecithin preservative, *14*:255
   preservative, *13*:247
Hydroxybenzoquinone [*2474-72-8*], *13*:41
o-Hydroxybenzyl alcohol [*90-01-7*], *13*:74.   (See
   also *Saligenin.*)
p-Hydroxybenzyl alcohol [*623-05-2*], *13*:74
4-Hydroxy-2-butanone [*590-90-9*]
   mevalonic acid from, *13*:98
2-(2-Hydroxy-3′,5′-*di-tert*-butylphenyl)-5-chloro-
   benzotriazole [*3864-99-1*]
   polypropylene light stabilization, *16*:361
α-Hydroxybutyrate [*565-70-6*]
   in myocardial infarction diagnosis, *15*:78
2-Hydroxybutyric acid [*565-70-8*], *13*:98
3-Hydroxybutyric acid [*300-85-6*], *13*:98
4-Hydroxybutyric acid [*591-81-1*], *13*:98
   as neuroregulator, *15*:770
β-Hydroxybutyric acid [*300-85-6*]
   blood ketone, *13*:605
γ-Hydroxybutyric acid [*591-81-1*]
   in ketone metabolism studies, *13*:98
α-Hydroxybutyric acid dehydrogenase (α-HBD)
   [*9028-38-0*]
   in diagnosis, *15*:75
Hydroxy carboxylic acids, *13*:**80**
Hydroxycholecalciferol [*52028-00-5*]
   in calcium homeostasis, *15*:586
25-Hydroxycholecalciferol [*36109-00-5*] (25-HCC)
   in blood-Ca^{2}, *15*:585
2-Hydroxycinnamic acid [*7400-08-0*], *13*:73
(−)-14-Hydroxycodeine [*4829-46-3*]
   microbial synthesis of, *15*:462
17-Hydroxycorticosteroids
   measurement, *15*:84
7-Hydroxycoumarin [*93-35-6*]
   fluorescence, *14*:549
2-Hydroxycyclohexanone [*533-60-8*]
   in catechol production, *13*:51
2-Hydroxycyclopentanone [*10493-98-8*], *13*:924
3-β-Hydroxydehydrogenase [*9015-81-0*]
   in corticosteroid biosynthesis, *15*:762
2-Hydroxy-4,6-diaminopyrimidine [*31458-45-4*],
   14:805
Hydroxy dicarboxylic acids, *13*:**103**
3-Hydroxy-2,4-dimethyl-2-cyclobuten-1-one
   [*20117-50-4*], *13*:883
6-Hydroxy-2,6-dimethyl-2-hepten-4-one [*5857-*
   71-6], *13*:914
5-Hydroxyelliptone [*478-07-9*]
   as insecticide, *13*:426

2-Hydroxyethyl acrylate [*818-61-1*]
   in emulsion polymerization, *14*:87
1-(2-Hydroxyethyl)aziridine [*1072-52-2*]
   from monoethanolamine, *13*:158
   properties, *13*:143
Hydroxyethyl cellulose [*9004-62-0*]
   in papermaking, *16*:822
   in paper processing, *16*:780
1-Hydroxyethylferrocene [*1277-49-2*], *15*:196
N-Hydroxyethyllactamide [*5422-34-4*], *13*:90
N-(2-Hydroxyethyl)methacrylamide [*5238-56-2*]
   properties, *15*:356
2-Hydroxyethyl methacrylate [*868-77-9*], *15*:367
   in contact lenses, *15*:126
   in emulsion polymerization, *14*:87
   properties, *15*:347
1-Hydroxyethylruthenocene [*12089-68-8*], *15*:196
Hydroxyhexamide [*3168-01-2*]
   acetohexamide metabolite, *13*:616
2-(Hydroxyimino)-*N,N*-dimethylacetoacetamide
   [*43015-36-7*]
   chlorination of, *13*:889
Hydroxyindolacetic acid [*1321-73-9*]
   measurement, *15*:84
5-Hydroxyindole [*1953-54-4*]
   in melatonin biosynthesis, *15*:772
   microbial transformation to 5-hydroxy-L-
   tryptophan, *15*:465
5-Hydroxyindole *O*-methyltransferase [*9029-77-*
   0]
   in melatonin biosynthesis, *15*:772
α-Hydroxyisobutyric acid [*594-61-6*]
   methacrylic acid from, *15*:357
2-Hydroxyisobutyronitrile [*75-86-5*]
   methacrylamide from, *15*:360
Hydroxylamine [*7803-49-8*]
   from nitric acid, *15*:856
   nitrile reactions, *15*:889
   plutonium reduction, *16*:177
Hydroxylamine hydrochloride [*5470-11-1*]
   in iron analysis, *13*:787
   as polymerization initiator, *14*:93
Hydroxylamine nitrate [*13465-08-2*]
   plutonium reduction, *16*:179
Hydroxylamines
   from nitroparaffins, *15*:976
Hydroxylammonium acid sulfate [*10039-54-0*],
   15:913
1-Hydroxylase [*9081-36-1*]
   in blood-Ca^{2}, *15*:585
24-Hydroxylase [*53112-53-1*]
   in blood-Ca^{2}, *15*:585
11β-Hydroxylase [*9029-66-7*]
   in corticosteroid biosynthesis, *15*:762
Hydroxylation
   with hydrogen peroxide, *13*:28
   of olefinic compounds, *13*:28

γ-Hydroxyleucine [*6645-45-0*], *14*:802

2-(Hydroxymethyl)aminoethanol [*34375-28-5*]
   applications of, *13*:242

2-(Hydroxymethyl)amino-2-methylpropanol
   [*52299-20-4*]
   applications of, *13*:242

5-Hydroxy-2-methylindole-3-carboxylic acid
   [*71982-15-5*], *13*:219

2-Hydroxy-6-methyl-3-(1-methylethyl)-2-cyclo-
   hexen-1-one [*54783-36-7*], *13*:925

2-Hydroxymethyl-2-nitro-1,3-propanediol [*126-
   11-4*]
   properties, *15*:910

4-Hydroxy-4-methyl-2-pentanone [*123-42-2*], *13*:
   912.   (See also *Diacetone alcohol.*)

2-Hydroxy-2-methylpropanenitrile. See
   Acetone cyanohydrin.

3-Hydroxy-2-methylquinoline-4-carboxylic acid
   [*117-57-7*], *13*:221

2-Hydroxy-1-naphthaldehyde [*708-06-5*]
   nickel pigment formation, *15*:814

Hydroxynaphthalenecarboxylic acids, *15*:744

1-Hydroxy-2-naphthalenecarboxylic acid [*86-48-
   6*], *15*:745

2-Hydroxy-1-naphthalenecarboxylic acid [*2283-
   08-1*], *15*:745

3-Hydroxy-1-naphthalenecarboxylic acid [*19700-
   42-6*], *15*:745

3-Hydroxy-2-naphthalenecarboxylic acid [*92-70-
   6*], *15*:745

4-Hydroxy-1-naphthalenecarboxylic acid [*7474-
   97-7*], *15*:745

4-Hydroxy-2-naphthalenecarboxylic acid [*1573-
   91-7*], *15*:745

5-Hydroxy-1-naphthalenecarboxylic acid [*2437-
   16-3*], *15*:745

5-Hydroxy-2-naphthalenecarboxylic acid [*2437-
   18-5*], *15*:745

6-Hydroxy-1-naphthalenecarboxylic acid [*2437-
   17-4*], *15*:745

6-Hydroxy-2-naphthalenecarboxylic acid [*16712-
   64-4*], *15*:745

7-Hydroxy-1-naphthalenecarboxylic acid [*2623-
   37-2*], *15*:745

7-Hydroxy-2-naphthalenecarboxylic acid [*613-
   17-2*], *15*:745

8-Hydroxy-1-naphthalenecarboxylic acid [*1769-
   88-6*], *15*:745

8-Hydroxy-2-naphthalenecarboxylic acid [*5776-
   28-3*], *15*:745

3-Hydroxy-2,7-naphthalenedisulfonic acid [*148-
   75-4*], *15*:739

4-Hydroxy-1,5-naphthalenedisulfonic acid [*82-
   75-7*], *15*:739

4-Hydroxy-1,6-naphthalenedisulfonic acid [*6361-
   37-1*], *15*:739

5-Hydroxy-2,7-naphthalenedisulfonic acid [*578-
   85-8*], *15*:738

7-Hydroxy-1,3-naphthalenedisulfonic acid [*118-
   32-1*], *15*:739

8-Hydroxy-1,6-naphthalenedisulfonic acid [*117-
   43-1*], *15*:738

Hydroxynaphthalenesulfonic acids, *15*:736

2-Hydroxy-1-naphthalenesulfonic acid [*567-47-5*],
   15:738

4-Hydroxy-1-naphthalenesulfonic acid [*84-87-7*],
   15:738

4-Hydroxy-2-naphthalenesulfonic acid [*3771-14-
   0*], *15*:738

5-Hydroxy-1-naphthalenesulfonic acid [*117-59-9*],
   15:738

6-Hydroxy-2-naphthalenesulfonic acid [*93-01-6*],
   15:738

7-Hydroxy-1-naphthalenesulfonic acid [*132-57-0*],
   15:738

7-Hydroxy-2-naphthalenesulfonic acid [*92-40-0*],
   15:738

8-Hydroxy-1-naphthalenesulfonic acid [*117-22-6*],
   15:738

7-Hydroxy-1,3,6-naphthalenetrisulfonic acid
   [*6259-66-1*], *15*:739

8-Hydroxy-1,3,6-naphthalenetrisulfonic acid
   [*3316-02-7*], *15*:739

1-Hydroxy-2-naphthoic acid [*84-48-6*]
   from 1-naphthalenol, *15*:734

Hydroxynaphthoic acids, *15*:745.   (See also
   Hydroxynaphthalenecarboxylic acids.)

2-Hydroxy-4-*n*-octyloxybenzophenone [*1843-05-
   6*]
   HDPE stabilizer, *16*:446
   polypropylene light stabilization, *16*:361

1-(*p*-Hydroxyphenethyl)biguanide [*17426-94-7*]
   phenformin metabolite, *13*:619

p-Hydroxyphenylacetonitrile [*14191-95-8*]
   sodium reduction, *15*:783

2-Hydroxy-3-(phenylamino)propyl methacrylate
   [*16926-84-4*], *15*:352, 356

2-Hydroxy-3-[*p*-(phenylazolanilino]propyl
   methacrylate [*16323-28-7*], *15*:355

2-Hydroxy-3-[[4-phenylazo)phenyl]amino]propyl
   methacrylate [*16323-28-7*], *15*:356

4-(*p*-Hydroxyphenyl)butanone [*5471-51-2*]
   raspberry ketone, *13*:77

p-Hydroxyphenylglycine [*938-97-6*], *13*:72

2-(*p*-Hydroxyphenyl)glycine [*22818-40-2*], *13*:77

4-Hydroxyphenyllactic acid [*39223-40-0*]
   microbial synthesis, *15*:465

17-Hydroxyprogesterone [*68-96-2*]
   in corticosteroid biosynthesis, *15*:762

11α-Hydroxyprogesterone [*80-75-1*]
   in corticosteroid synthesis, *15*:762
   microbial preparation, *15*:460

β-Hydroxypropionamide [*2043-43-8*]
  from propiolactone, *13*:96
2-Hydroxypropionic acid.   See *Lactic acid.*
3-Hydroxypropionic acid [*503-66-2*], *13*:98
  malonic acid from, *14*:795
β-Hydroxypropionic acid [*503-66-2*], *13*:95
Hydroxypropyl cellulose [*9004-64-2*]
  in encapsulation, *15*:479
N-2-Hydroxypropyllactamide [*6281-43-2*], *13*:90
2-Hydroxypropyl methacrylate [*923-26-2*], *15*:367
  properties, *15*:353
3-Hydroxypropyl methacrylate [*2761-09-3*]
  properties, *15*:353
Hydroxypropylmethyl cellulose [*9004-65-3*]
  in paint and varnish removers, *16*:763
Hydroxyquinoline metal polymers, *15*:214
5-Hydroxyrotenone [*82-10-0*]
  as insecticide, *13*:426
Hydroxysuccinic acid [*617-48-1*], *13*:103.   (See
  also *Malic acid.*)
5-Hydroxytryptamine [*50-67-9*], *15*:770.   (See
  also *Serotonin.*)
  memory in inhibiting agent, *15*:135
5-Hydroxytryptophan [*56-69-9, 4350-09-8*]
  in serotonin biosynthesis, *15*:771
5-Hydroxy-L-tryptophan [*56-69-9*]
  microbial synthesis of, *15*:465
Hygas process
  lignite gasification, *14*:338
Hylemya antiqua, *13*:422
Hylemya brassicae, *13*:422
Hylene W, *13*:807
HYL process, *13*:745, 760
Hymechrome.   See β-*Methylumbelliferone.*
HyMu80, *14*:659
HyMu800, *14*:659
Hypalon [*9008-08-6*]
  in cable insulation, *13*:585
Hyperco 27, *14*:651
Hyperfiltration, *15*:112, 115
Hyperglobulinemia
  detection, *15*:81
Hyperglycemia.   See *Diabetes mellitus.*
  neurotensin induced, *15*:772
Hyperlipidemia
  detection, *15*:82
Hyperm 52, *14*:659
Hyperm 900, *14*:659
Hyperm Co50, *14*:659
Hyperm 36M, *14*:659
Hyperm Maximum, *14*:659
Hyperm 50T, *14*:659
Hyperm 5T/7T, *14*:659
Hypernephrin, *15*:757
Hyperparathyroidism
  detection, *15*:83

diagnosis, *15*:80
Hyperphosphaturia
  from lead exposure, *14*:198
Hypersensitivity
  treatment, *13*:167
Hypersthene
  as track-etch detector, *16*:827
Hypertension
  nutrients associated with, *15*:573
  sodium role in, *15*:579
Hyperthyroidism
  nutrients associated with, *15*:573
Hypertriglyceridemia
  relation to diabetes, *13*:606
Hyphaene crinita, *16*:251
Hyphaene nut, *16*:251
Hyphomicrobium manganoxydans, *14*:856
Hypnotics, sedatives, anticonvulsants, *13*:**122**
Hypocalcemia
  nutrients associated with, *15*:573
  treatment, *15*:445
Hypochlorite cells
  metal anodes for, *15*:182
Hypochlorites
  disinfectants, *13*:229
Hypochylin, *15*:767
Hypoderma
  effect on cattlehide, *14*:208
Hypogeusia
  nutrients associated with, *15*:573
Hypoglycemia
  diagnosis, *15*:84
Hypoglycemic agents, *13*:606
Hypoiodous acid [*14332-21-9*], *13*:664
Hypokalemic alkalosis
  potassium treatment, *15*:588
Hypomanganate.   See *Individual manganate*(V)
  salts.
Hypoparathyroidism
  detection, *15*:83
Hypophosphite
  in thin-film fabrication, *14*:694
Hypophosphites
  in magnetic tape mfg, *14*:740
Hypoproteinemia
  detection, *15*:81
Hypotaurine [*300-84-5*]
  in taurine biosynthesis, *15*:781
Hypotension
  neurotensin induced, *15*:772
Hypothyroidism
  nutrients associated with, *15*:573
Hypoxanthine [*68-94-0*]
  xanthine oxidase reaction, *15*:592
HyRa49, *14*:659

HyRa80, *14*:659
Hysomer process, *15*:666
Hystazarin [*483-35-2*], *13*:43
Hysteresigraph, *14*:683
Hysteresis loop, *14*:650
Hyviz, *13*:990

I

IBIS information system, *13*:329
IBM 3850, *14*:751
Ibuprofen [*15687-27-1*]
  immunologic effects, *13*:174
Ice cream
  preparation, *15*:564
Ice-cream pudding, *15*:568
Iceland spar, *14*:345
Ice milk, *15*:566
ICS. See *Integrated circuits.*
IDC information system, *13*:329
IFI Patent Gazette Database
  information system, *16*:911
IFI–Plenum, *13*:299
Igepal, *15*:475; *16*:411
Igneous rocks
  lithium in, *14*:449
IIC. See *Impact isolation class.*
IIR. See *Isobutylene–isoprene rubber.*
Iletin [*9004-10-8*], *13*:611
Illicium verum Hook f., *16*:321
Illium alloys, *15*:792
Ilmenite [*12168-32-4*]
  from beach sand, *16*:289
  as iron source, *13*:748
ILO. See *International Labor Organization.*
Ilosvay's reagent
  acetylene analysis, *16*:665
Imac, *13*:696
IMF. See *International Monetary Fund.*
Imidazole-4-acetonitrile [*18502-05-1*]
  in histamine synthesis, *15*:769
Imidazole-4-methanol [*822-55-9*]
  in histamine synthesis, *15*:769
Imidodisulfonic acid salts
  from nitroparaffins, *15*:974
Imidosulfonates
  from nitroparaffins, *15*:974
Imines, cyclic, *13*:**142**
7,7′-Imino-bis-4-hydroxy-2-naphthalenesulfonic
  acid [*87-03-6*], *15*:742
Iminodiacetic acid [*142-73-4*]
  in ion-exchange resin mfg, *13*:687

2-Imino-5-methyl-oxazolidin-4-one [*17888-97-0*]
  from alkyl lactates, *13*:84
Immergan, *14*:217
Immunoadjuvants, *13*:176
Immunoglobulins
  as antigens, *13*:167
Immunomodulators, *13*:176
Immunopathology, *13*:167
Immunopharmacology, *13*:167
Immunostimulants, *13*:171, 176
Immunosuppressants, *13*:171
  as antiaging drugs, *15*:139
Immunotherapeutic agents, *13*:**167**
Impact isolation class (IIC), *13*:526
Impact noise rating (INR), *13*:526
Impact strength
  transparent plastics, *16*:478
Impedance-tube method
  sound absorption, *13*:513
Impellers, *15*:604. (See also *Mixing and blending.*)
  as mixers, *15*:607
Imperial smelting process
  for zinc-lead ores, *14*:125
Impingement coatings, *15*:260
Implanter, ion, *13*:707
Imuran [*446-86-6*], *15*:139. (See also *Azathioprine iv.*)
Incineration
  nuclear wastes, *16*:210
Incinerators, *13*:**182**
  catalytic, *13*:203
Incoloy, *13*:184
Incoloy 901, *15*:341
Incoloy Alloy 800 [*11121-96-3*], *15*:789
Incoloy alloy 825 [*12766-43-7*], *15*:793
Inconel, *15*:249; *16*:206, 727
  in metal-coated filters, *16*:531
Inconel 718, *15*:341
Inconel Alloy 600 [*12606-02-9*], *15*:789
Inconel alloy 601 [*12631-43-5*], *15*:793
Inconel alloy 625 [*12682-01-8*], *15*:793
Inconel Alloy 718 [*12606-10-9*], *15*:789
Inconel Alloy 754 [*62112-97-4*], *15*:789
Inconel alloy 801 [*12605-97-9*], *15*:793
Inconel alloy 802 [*51836-04-5*], *15*:793
Inconel alloy x750 [*11145-80-5*], *15*:793
Incubators
  oxygen supply for, *16*:670
Indanthrene [*81-77-6*]
  as grease thickener, *14*:502
Indene [*95-13-6*]
  from naphthalene mfg, *15*:707
Inderal, *15*:729
Indican [*16934-10-4*], *13*:220

Indicator method
    for pH determination, *13*:2
Indicators
    for pH measurement, *13*:9
Indigo [*482-89-3*]
    indole from, *13*:213
Indirubin [*479-41-4*], *13*:217
Indium [*7440-74-6*], *13*:207
    lead alloys, *14*:157
    as lubricant, *14*:482, 508
    from the ocean, *16*:280
Indium-113 [*14885-78-0*], *13*:841
Indium-115 [*14191-71-0*], *13*:841
Indium alkyls, *16*:562
Indium alloys, *13*:209
Indium and indium compounds, *13*:**207**
Indium antimonide [*1312-41-0*]
    lasers, *14*:61
    properties, *14*:286
    semiconducting compound, *13*:211
Indium–antimonide detectors, *13*:345
Indium arsenide [*1303-11-3*]
    properties, *14*:286
Indium arsenide phosphide lasers, *14*:60
Indium dibromide [*21264-43-7*], *13*:211
Indium dichloride [*13465-11-7*], *13*:210
Indium diiodide [*13779-78-7*], *13*:211
Indium hydrogen disulfate [*57344-73-7*], *13*:211
Indium hydroxide [*56108-30-6*], *13*:211
Indium methoxide [*40521-21-9*], *13*:211
Indium monobromide [*14280-53-6*], *13*:211
Indium monochloride [*13465-10-6*], *13*:211
Indium monoiodide [*13966-94-4*], *13*:211
Indium monosulfide [*12030-14-7*], *13*:211
Indium monoxide [*12136-26-4*], *13*:211
Indium nitrate hydrate [*13770-61-1*], *13*:211
Indium nitrate trihydrate [*13770-61-1*], *13*:211
Indium nitride [*25617-98-5*], *13*:211, 212; *15*:872
    properties, *15*:873
Indium oxalate [*38051-36-4*], *13*:211
Indium phosphate [*14693-82-4*], *13*:208, 211
Indium phosphide [*22398-80-7*]
    in LEDs, *14*:277
    properties, *14*:286
    semiconducting compound, *13*:211
Indium selenide [*1312-42-1*], *13*:212
Indium sesquioxide [*66525-54-0*] (In_3O_4), *13*:211
Indium suboxide [*12030-22-7*], *13*:211
Indium subsulfide [*12196-52-0*], *13*:211
Indium sulfide [*12142-00-6*] (4:5), *13*:211
Indium telluride [*1312-45-4*], *13*:212
Indium tribromide [*13465-09-3*], *13*:211
Indium trichloride [*10025-82-8*], *13*:210
Indium trifluoride [*7783-52-0*], *13*:211

Indium trifluoride tris(ammonium fluoride)
    [*15273-84-4*], *13*:211
Indium triiodide [*13510-35-5*], *13*:211
Indium trioxide [*1312-43-2*], *13*:207, 211
Indium trisulfate [*13464-82-9*], *13*:211
Indium trisulfide [*12030-24-9*], *13*:211
Indole [*120-72-9*], *13*:**213**
    microbial transformation to L-tryptophan, *15*:
        464
    in neroli oil, *16*:325
    properties, *13*:213
Indole-3-acetic acid [*87-51-4*]
    dimethylamine reaction, *15*:763
Indole 3-acetonitrile [*771-51-7*]
    in tryptamine synthesis, *15*:782
Indole-3-carboxaldehyde [*487-89-8*], *13*:216
Indole-2-carboxylic acid [*1477-50-5*], *13*:219
Indole-3-glyoxalyl chloride [*22980-09-2*], *13*:215
3-Indolemagnesium bromide
    aziridine reaction, *13*:151
Indole-3-pyruvic acid [*392-12-1*], *13*:220
Indole-1-sulfonic acid [*74220-54-5*], *13*:215
Indole-2-sulfonic acid [*74220-55-6*], *13*:215
Indoline [*496-15-1*], *13*:217
(Indol-3-ylmethyl)trimethylammonium ion
    [*19005-94-8*]
    in tryptamine synthesis, *15*:782
Indomethacin [*53-86-1*]
    controlled release of, *15*:488
    immunologic effects, *13*:174
Indophenol blue [*132-31-0*]
    color test for phenol, *15*:551
Indoxyl [*480-93-3*], *13*:217, 220
Indoxyl red [*69247-70-7*], *13*:217
1-(3'-Indoyl)-butan-3-one [*5541-89-9*], *13*:216
Indoylmagnesium bromide [*20356-50-7*]
    acylation, *13*:215
3-(3'-Indoyl)-propionic acid [*830-96-6*], *13*:220
Induction hardening, *15*:320
Inductors
    magnetic materials for, *14*:664
Indulin
    alkali lignins, *14*:307
Industrial antimicrobial agents, *13*:**223**
Industrial engineering, *16*:500.   (See also
        Operations planning.)
Industrial hygiene and toxicology, *13*:**253**
Industrial Information Sources, *13*:312
Industrialite, *14*:25
Industrial plants
    noise source, *16*:9
Inertia welding, *15*:277
Infant mortality
    as failure rate, *15*:17

INFCE. See *International Nuclear Fuel Cycle Evaluation.*
Inferential-level measurements, *14*:439
Inflammation
  treatment, *13*:174
INFOLINE information system, *13*:299
Information retrieval, *13*:**278**
Information sources
  market research, *14*:898
  materials standards, *15*:46
Information storage
  magnetic media, *14*:733
Information systems, *13*:290; *16*:889. (See also *Patents, literature.*)
Infrared-cured inks, *13*:386
Infrared drying
  of inks, *13*:376
Infrared interference filters, *16*:524
Infrared lasers, *14*:72
Infrared LEDs, *14*:289
Infrared phosphor, *14*:542
Infrared sensors
  for nondestructive testing, *16*:61
Infrared spectroscopy
  lasers in, *14*:72
Infrared technology, *13*:**337**
Infusorial earth. See *Diatomite.*
Inhalators
  oxygen supplied, *16*:670
Inhibitors
  from dihydroxybenzene, *13*:58
Initiators, *13*:**355**
  azonitriles, *15*:901
  for emulsion systems, *14*:92
Initiator systems
  for latices, *14*:91
Injection-molding
  HDPE, *16*:448
  polypropylene, *16*:467
Injection-molding processes
  for plastic composites, *13*:976
Ink-jet printing, *13*:396
Inks, *13*:**374**
  fluorescent, *14*:561
  fluorescent pigments in, *14*:567
  manufacture of, *13*:382, 389
In-line mixers, *15*:623
Inorganic high polymers, *13*:**398**
Inorganic nomenclature, *16*:30
Inorganic rubber, *15*:875
Inositol [*87-98-8*]
  soybean phosphatides associated with, *14*:252
Inositolphosphoric acid [*13004-72-3*]
  from lecithin, *14*:256
INPADOC
  information system, *16*:917

INPADOC information system, *13*:300
INPADOC International Gazette Database information system, *16*:928
INR. See *Impact noise rating.*
Insect control technology, *13*:**413**
Insect growth regulators, *13*:458
Insecticide, Fungicide, and Rodenticide Act, *13*:254
Insecticides, *13*:413, 414. (See also *Insect control technology.*)
  application of, *13*:417
  chloropicrin, *15*:984
  formulations of, *13*:416
Insecticides, microbial, *13*:**470**
Insect repellent
  lavender oil, *16*:324
Insects
  cable protection against, *13*:582
Insertion loss
  sound, *13*:516
Insertion reactions
  metallocenes, *16*:610
Insomnia
  (−) adrenaline induced, *15*:758
  treatment, *13*:122
INSPEC
  information service, *16*:902
Inspection
  nondestructive, *16*:48
"Instantized" milk products, *15*:560
Institute for Scientific Information, *13*:294
Instrumentation
  responses in, *13*:488
Instrumentation and control, *13*:**485**
Insulation
  ir inspection, *13*:348
  limestone in, *14*:376
  from polyethylene, *16*:431
Insulation, acoustic, *13*:**513**
  lead, *14*:157
  wood based, *14*:**25**
Insulation board, *14*:4
Insulation boards, *14*:22
Insulation, electric
  LDPE as, *16*:408
  properties and materials, *13*:**534**
  resonant-circuit test, *15*:421
  wire and cable coverings, *13*:**564**
Insulations, cross-linked, *13*:564
Insulations, thermoset, *13*:564
Insulation systems
  economic aspects, *13*:548
Insulation, thermal, *13*:**591**
Insulators
  metal-containing polymers, *15*:185
  from mica, *15*:436
  properties of, *13*:537

Insulin [*9004-10-8*], *13*:607
  from animal pancreas, *15*:72
  in N₂-fixing system, *15*:959
  sulfur in, *15*:587
Insulin and other antidiabetic agents, *13*:**605**
Insulin, bovine [*11070-73-8*], *13*:608
Insulin, human [*11061-68-0*], *13*:608
Insulin hydrochloride [*9004-10-8*], *13*:611
Insulin injection [*9004-10-8*], *13*:611
Insulin, porcine [*11070-73-8*], *13*:608
Insulin secretion
  adrenaline effect on, *15*:758
Insulin zinc [*8049-62-5*], *13*:612
Insulite, *14*:4
Intaglio process inks, *13*:395
Integrated circuits, *13*:**621**.   (See also *Circuits, integrated.*)
  economic aspects of, *13*:645
Integrated pest management, *13*:415
Intellectual property, *16*:851
Intelligent materials, *16*:49
Interagency Regulatory Liaison Group, *13*:260
Interchemical viscometer, *13*:376
Intercide 340A, *13*:236
Interconnection technology
  in ICS, *13*:643
Interface measurements, *14*:440
Interfacial coacervation, *15*:479
Interference filters, *16*:545
Interferences
  among patent applications, *16*:871
Interferometer
  laser-based, *14*:67
International Labor Organization
  information sources, *13*:308
International Monetary Fund
  information sources, *13*:308
International Nuclear Fuel Cycle Evaluation (INFCE), *16*:221
International patent law.   See *Patent literature.*
International Union of Pure and Applied Chemistry
  nomenclature commissions, *16*:29
Intravenous feeding
  sodium lactate in, *13*:85
  sodium lactate used in, *13*:88
Intrinsic factor (IF)
  in vitamin B₁₂ deficiency, *15*:597
Intumescent paint, *16*:756
Invar [*12683-18-0*], *14*:662; *15*:794
Invention.   See *Patents, practice and management.*
Inventor.   See *Patents, practice and management.*

Inventory management, *16*:500, 513
Inverse-transform technique
  in operations planning, *16*:508
Iodic acid [*12134-99-5*], *13*:666
  in nonlinear optics, *14*:64
Iodine [*7553-56-2*], *13*:232, 649
  as initiator catalyst, *13*:371
  metabolic functions, *15*:593
  nutrient properties, *15*:571
  from the ocean, *16*:279
  ozone reaction, *16*:703
  properties, *13*:650
  radioactive, *13*:673
Iodine-123 [*15715-08-9*]
  diagnostic use, *13*:673
Iodine-125 [*14158-31-7*]
  diagnostic use, *13*:673
  in radioimmunoassay, *15*:86
Iodine-127 [*14362-44-8*], *13*:841
Iodine-129 [*15046-84-1*]
  disposal of, *16*:208
Iodine-131 [*10043-66-0*]
  fission product, *16*:174
  thyroid therapy, *13*:673
Iodine and iodine compounds, *13*:**649**
Iodine dioxide [*13494-92-3*], *13*:665
Iodine heptafluoride [*16921-96-3*], *13*:664
Iodine(III) iodate [*66523-94-2*]
  from I₂/ozone, *16*:703
Iodine(III) oxysulfate [*11085-17-9*], *13*:665
Iodine monobromide [*7789-33-5*], *13*:664
Iodine monochloride [*7790-99-0*], *13*:663
  in organolead analysis, *14*:192
Iodine monoxide [*39319-71-6*], *13*:664
Iodine nitrate [*14696-81-2*], *13*:665
Iodine nitride [*25617-98-5*], *15*:883
Iodine oxide (4:9), *13*:664
Iodine pentafluoride [*7783-66-6*], *13*:664
Iodine pentoxide [*12029-98-0*], *13*:665
Iodine–polyethylene glycol mono(nonylphenyl) ether complex [*11096-42-7*]
  applications of, *13*:230
Iodine–poly(vinylpyrrolidinone) complex [*25655-41-8*]
  applications of, *13*:230
Iodine sesquioxide [*11085-17-9*], *13*:665
Iodine tribromide [*7789-58-4*], *13*:664
Iodine trichloride [*865-44-1*], *13*:664
Iodoacetic acid [*64-69-7*]
  potassium persulfate oxidation, *13*:669
Iodobenzene [*591-50-4*]
  in initiator systems, *13*:367
  properties, *13*:670
Iodobenzene dichloride [*932-72-9*], *13*:671
Iododimethylarsine [*676-75-5*], *13*:668
2-Iodoethyl methacrylate [*35531-61-4*]
  properties, *15*:355

2-Iodoethyltrimethylammonium iodide [*5110-69-0*], *13*:148

Iodofenfos [*18181-70-9*], *13*:440

Iodoform [*75-47-8*]
  methylene iodide from, *13*:669
  properties, *13*:670

Iodo(iodomethyl)mercury [*141-51-5*], *13*:670

Iodomethyl mercuric iodide [*141-51-5*], *13*:670

Iodometry, *13*:649

3-Iodo-1-nitropropane [*57308-87-9*], *15*:975

Iodophors
  as antimicrobial agents, *13*:232
  disinfectants, *13*:673

Iodopicrin [*39247-25-1*], *15*:975

Iodosobenzene [*536-80-1*], *13*:671

Iodosobenzene diacetate [*3240-34-4*], *13*:671

Iodox process, *16*:180

Ionac, *13*:696

Ion-beam accelerators
  in metal coating, *15*:268

Ion beams, *13*:708

Ion-beam sputtering, *14*:692

Ion chromatography, *13*:703

Ion exchange, *13*:**678**; *15*:92. (See also *Membrane technology.*)
  in electrodes, *13*:723
  molecular sieves for, *15*:668
  by zeolites, *15*:652

Ion-exchange membranes, *15*:119
  transport processes, *15*:100

Ion-exchange resins
  lignite source, *14*:341
  nitration reagents, *15*:841

Ion implantation, *13*:706
  in IC mfg, *13*:635
  in metal coating, *15*:268
  in metallic coating, *15*:265

Ionitriding, *15*:318

Ionizing radiation
  as initiator, *13*:367

Ionol [*128-37-0*]
  in inks, *13*:379

Ionomers
  salts of ethylene–methacrylic acid copolymers, *16*:420

Ionone [*8013-90-9*]
  odor modification with, *16*:304

Ionones
  in perfumes, *16*:968

Ion plating, *15*:252, 266, 878
  in thin-film fabrication, *14*:691
  in vacuum coating, *15*:265

Ion-selective electrodes
  membranes in, *15*:122

Ion-selective electrodes (ISES), *13*:**720**

Ionsiv F80, *15*:660

Ionsiv W85, *15*:660

Ion sources, *13*:707

Iosan, *13*:230

Ioxynil [*1689-83-4*], *13*:76

IPC CHEMICAL DATA BASE, *13*:327

IPC CHEMICAL DATA BASE information system, *13*:329

IPDI. See *3-Isocyanatomethyl-3,5,5-trimethylcyclohexyl isocyanate.*

IPM. See *Integrated pest management.*

Iragasan CF$_3$, *13*:240

Irati oil shale, *16*:346

Iridium [*7439-88-5*]
  in anode coatings, *15*:174

Iridium-191 [*13967-66-3*], *13*:842

Iridium-192 [*14694-69-0*]
  radiographic use, *16*:64

Iridium-193 [*13967-67-4*], *13*:842

Iridium dioxide [*12030-49-8*]
  in anode coatings, *15*:174
  electrocatalyst coating, *15*:176

Iris, concrete oil [*8002-73-1*], *16*:320

Iris diaphragm
  in cameras, *16*:533

Iris florentina, *16*:952

Iris florentina L., *16*:326

Iris germanica, *16*:952

Iris pallida, *16*:952

IRL
  information service, *16*:902

IRLG. See *Interagency Regulatory Liaison Group.*

Iron [*7439-89-6*], *13*:**735**
  allotropic modifications, *15*:338
  amalgamation with mercury, *15*:147
  emissivity, *13*:342
  ion implantation, *13*:710
  in magnesium alloys, *14*:596
  from magnesium reduction sulfide, *14*:573
  in magnetic alloys, *14*:694
  as magnetic material, *14*:647
  metabolic functions, *15*:594
  in metallic coating, *15*:242
  in mica, *15*:421
  as neutron absorber, *16*:140
  nickel alloys, *15*:792
  N$_2$ solution in, *15*:934
  nutrient properties, *15*:571
  in nuts, *16*:260
  from the ocean, *16*:292
  oxidation, *16*:656
  oxygen reaction in chlorate candle, *16*:675
  properties, *13*:736
  salts in tannages, *14*:215
  steel from, *16*:666
  surface treatment of, *15*:304
  toxicity, *15*:595

Iron-54 [*13982-24-6*], *13*:840
Iron-56 [*14903-02-8*], *13*:840; *16*:141
  neutron reaction, *16*:188
Iron-57 [*14762-69-7*], *13*:840; *16*:188
Iron-58 [*13968-47-3*], *13*:840
Iron-59 [*14596-12-4*]
  reactor corrosion product, *16*:155
Iron acetate [*2140-52-5*], *13*:764
Iron–aluminum alloys
  as magnetic materials, *14*:656
Iron–aluminum–silicon alloys
  as magnetic materials, *14*:656
Iron and steel manufacture
  limestone in, *14*:378
Iron blue [*14038-43-8*]
  in inks, *13*:378
Iron boride [*12006-84-7*] (1:1)
  in coatings, *15*:250
Iron boride [*12006-85-8*] (1:2)
  in coatings, *15*:250
Iron by direct reduction, *13*:**754**
Iron carbide [*12011-67-5*]
  in coatings, *15*:250
Iron carbonyl [*13463-40-6*]
  Ni(CO)₄ coproduct, *15*:807
Iron, cast [*11097-15-7*], *13*:735
  in packings, *16*:734
Iron chloride [*7758-94-3*]
  as antiwear agent, *14*:493
Iron citrate [*2338-05-8*], *13*:765
Iron–cobalt alloys
  as magnetic materials, *14*:662
Iron, compd with tin [*12023-01-7*] (1:2)
  in coatings, *15*:244
Iron, compd with zinc [*60383-43-9*] (1:3)
  in coatings, *15*:251
Iron, compd with zinc [*12023-07-3*] (1:7)
  in coatings, *15*:245, 251
Iron, compd with zinc [*12140-55-5*] (1:13)
  in coatings, *15*:245
Iron, compd with zinc [*12182-98-8*] (3:10)
  in coatings, *15*:245
Iron compounds, *13*:**764**
  complexes, *13*:781
Iron compound with carbon [*12169-32-3*] (3:1),
  14:651
Iron deficiency
  iron absorption during, *15*:595
Iron dextran [*9004-66-4*], *15*:445
Iron dithiophosphate
  as antiwear agent, *14*:493
Irone [*1335-94-0*]
  from orris root, *16*:326
Iron hydroxide [*1309-33-7*] (Fe(OH)₃)
  from seawater, *16*:280
Iron hydroxide oxide, hydrous [*20344-49-4*], *13*:
  775

Iron(II) acetate [*3094-87-9*], *13*:765
Iron(II) ammonium sulfate [*7783-85-9*], *13*:
  777.   (See also *Mohr's salt.*)
Iron(II) bromide [*7789-46-0*], *13*:773
Iron(II) bromide hexahydrate [*13463-12-2*], *13*:
  774
Iron(II) carbonate [*563-71-3*], *13*:765
Iron(II) chloride [*7758-94-3*], *13*:772
  as magnetic materials, *14*:647
Iron(II) chloride dihydrate [*16399-77-2*], *13*:773
Iron(II) chloride tetrahydrate [*13478-10-9*], *13*:
  773
  toxicity, *15*:574
Iron(II) citrate [*23383-11-1*], *13*:765
Iron(II) dichloride monohydrate [*23838-02-0*],
  13:773
Iron(II) ethylenediaminetetraacetic acid [*15651-
  72-6*], *13*:779
Iron(II) fluoride [*7789-28-8*], *13*:772
Iron(II) fluoride tetrahydrate [*13940-89-1*], *13*:
  772
Iron(II) formate [*3047-59-4*], *13*:771
Iron(II) formate dihydrate [*13266-73-4*], *13*:771
Iron(II) fumarate [*141-01-5, 7705-12-6*], *13*:772
Iron(II) gluconate [*299-29-6*], *13*:774
Iron(II) gluconate dihydrate [*6047-12-7*], *13*:774
Iron(II) hydrogen carbonate [*6013-77-0*], *13*:765
Iron(II) hydrogen phosphate [*31086-72-3*]
  in metal treating, *15*:305
Iron(II) hydroxide [*18624-44-7*], *13*:776
Iron(III) acetate [*1834-30-6*], *13*:765
Iron(III) acetate, basic [*10450-55-2*], *13*:765
Iron(III) ammonium alum dodecahydrate [*7783-
  83-7*], *13*:777
Iron(III) ammonium citrate [*1185-57-5*], *13*:765
Iron(III) bromide [*10031-26-2*], *13*:774
Iron(III) chloride [*7705-08-0*], *13*:773
  Wacker-process catalyst, *16*:611
Iron(III) chloride heptahydrate
  toxicity, *15*:574
Iron(III) chloride hexahydrate [*10025-77-1*], *13*:
  773
Iron(III) citrate [*28633-45-6*], *13*:765
Iron(III) citrate (1 Fe:1 citric acid) [*3522-50-7*],
  13:765
Iron(III) enterobactin [*62280-34-6*], *13*:786
Iron(III) ethanolate [*5058-42-4*], *13*:776
Iron(III) ethylenediaminetetraacetic acid [*15275-
  07-7*], *13*:779
Iron(III) fluoride [*7783-50-8*], *13*:772
Iron(III) fluoride trihydrate [*15469-38-2*], *13*:772

Iron(III) formate [*555-76-0*], *13*:771

Iron(III) fumarate [*52118-11-3*], *13*:772

Iron(III) gluconate [*38658-53-6*], *13*:774

Iron(III) hydroxide [*1309-33-7*], *13*:769, 776

Iron(III) iodide [*15600-49-4*], *13*:774

Iron(III) nitrate [*14104-77-9*]
  in iron(III) formate synthesis, *13*:772

Iron(III) nitrate hexahydrate [*13476-08-9*], *13*:775

Iron(III) nitrate nonahydrate [*7782-61-8*], *13*:774

Iron(II) iodide [*7783-86-0*], *13*:774

Iron(III) oxalate [*15321-61-6*], *13*:780

Iron(III) oxide [*1309-37-1*], *13*:775
  Prussian Blue impurity, *13*:770

Iron(III) potassium alum dodecahydrate [*13463-29-1*], *13*:777

Iron(III) sulfate [*10028-22-5*], *13*:777

Iron(III) sulfate heptahydrate [*35139-28-7*], *13*:778

Iron(III) sulfate hexahydrate [*13761-89-2*], *13*:778

Iron(III) sulfate monohydrate [*43059-01-4*], *13*:778

Iron(III) sulfate nonahydrate [*13520-56-4*], *13*:778

Iron(II) nitrate hexahydrate [*14013-86-6*], *13*:774

Iron(II) oxalate [*516-03-3*], *13*:780
  in iron(II) oxide synthesis, *13*:775

Iron(II) oxide [*1345-25-1*], *13*:775
  reduction, *14*:830

Iron(II) phthalocyanine [*132-16-1*], *13*:782

Iron(II) sulfate [*7720-78-7*]
  toxicity, *15*:574

Iron(II) sulfate heptahydrate [*7782-63-0*], *13*:777
  toxicity, *15*:574

Iron(II) sulfate tetrahydrate [*20908-72-9*], *13*:777

Iron(II)tetrachloride [*74930-87-3*], *13*:778

Ironmaking. See *Iron.*

Iron molybdate [*13718-70-2*], *15*:687

Iron–molybdenum cofactor
  in N_2 reduction, *15*:952

Iron nitride [*12023-20-0*] (2:1), *15*:872

Iron nitride [*12023-64-2*] (4:1), *15*:872

ε-Iron nitride [*37245-77-5*]
  in case hardening, *15*:320

Iron ore
  beneficiation, *13*:738
  magnetic separation, *14*:727

Iron oxide [*1345-25-1*]
  olefin fiber dye, *16*:362
  paint pigment, *16*:745

removal with hydroxyacetic acid, *13*:94
  in spent shale, *16*:351

Iron oxide [*1345-25-1*] (1:1)
  in DR processes, *13*:755

Iron oxide [*1309-37-1*] (2:3), *16*:675
  in DR processes, *13*:755
  in ferrite mfg, *14*:674
  as magnetic insulator, *14*:694
  as magnetic material, *14*:648

Iron oxide [*1317-61-9*] (3:4), *16*:675
  in DR processes, *13*:755
  magnetic impurity, *14*:730
  as magnetic insulator, *14*:694

Iron oxides
  in magnetic tape, *14*:737

Iron pentacarbonyl [*13463-40-6*]
  tetraphenylcyclobutadienepalladium
    dibromide reaction, *16*:609

Iron, pig, *13*:735

Iron porphyrins, *13*:782

Iron protein
  in nitrogenase, *15*:950

Iron sands
  from the ocean, *16*:286

Iron–silicon alloys
  as magnetic materials, *14*:653

Iron sulfate monohydrate [*17375-41-6*], *13*:777

Iron sulfide [*1317-37-9*]
  as antiwear agent, *14*:493

Iron sulfide [*11126-12-8*] (1:2)
  magnesium reaction, *14*:573

Iron sulfide [*1317-37-9*] (4:4)
  in nitrogenase, *15*:951

Iron–sulfur protein
  in nitrogenase, *15*:950

Iron tetracarbonyl (bromide) [*53173-32-3*], *16*:609

Iron, wrought, *13*:735, 754

Irradiation
  of milk, *15*:538

IRS information system, *13*:292

Isatin [*91-56-5*]
  hydrolysis, *13*:220
  indole from, *13*:213
  indoxyl reaction, *13*:217

ISES. See *Ion-selective electrodes.*

Islets of Langerhans, *13*:607
  encapsulation, *15*:490

ISM. See *Medical microwave applications.*

Isoamyl acetate [*123-92-2*]
  odor, *16*:297

Isoamyl benzoate [*94-46-2*]
  oxidation to glycolic acid, *13*:92

Isoamyl salicylate [*87-20-7*]
  as insect attractant, *13*:481

Isoapocodeine [*36507-69-4*]
  microbial synthesis of, *15*:465

Isobornyl acetate [*125-12-2*], *16*:955
  in perfume, *16*:954
Isobornyl methacrylate [*7534-94-3*]
  properties, *15*:353
Isobornyl thiocyanoacetate [*115-31-1*], *13*:429
Isobutane [*75-28-5*]
  in LPG, *14*:384
  methacrylates from, *15*:357
  properties, *14*:384
Isobutene [*115-11-7*]
  methacrylic acid from, *15*:361
4-Isobutoxy-4'-chlorobiphenyl [*41198-92-9*]
  insulation properties of, *13*:551
Isobutyl alcohol [*78-83-1*]
  oxo product, *16*:650
Isobutylaluminum sesquichloride [*12090-38-9*]
  properties, *16*:565
Isobutylene [*115-11-7*]
  catechol reaction, *13*:44
  copolymer with ethylene, *16*:420
  isoprene reaction, *13*:819
  in isoprene synthesis, *13*:831
  lube oils from, *14*:499
  methacrylic acid from, *15*:357
  polymerization of, *13*:356
  from triisobutylaluminum, *16*:569
Isobutylene–isoprene rubber (IIR) [*9010-85-9*]
  electric insulation elastomer, *13*:565
  in electric insulation, *13*:564
Isobutyl heptyl ketone [*123-18-2*]
  properties, *13*:897
Isobutyl methacrylate [*97-86-9*]
  properties, *15*:347
Isobutyraldehyde [*78-84-2*]
  acetone reaction, *13*:934
  isobutyric acid from, *15*:365
  isomerization of, *13*:906
  methacrylates from, *15*:357
  oxo intermediate, *16*:649
Isobutyramide [*563-83-7*]
  diketene reaction, *13*:884
Isobutyric acid [*79-31-2*]
  dehydrogenation, *15*:365
  methacrylates from, *15*:357
N-Isobutyroylacetoacetamide [*38367-30-5*], *13*:
  884
Isocyanates
  in leather tanning, *14*:216
  as magnetic tape binders, *14*:741
  properties, *13*:789
Isocyanates, organic, *13*:**789**
p-Isocyanatobenzenesulfonyl isocyanate [*1773-
  41-7*], *13*:801
3-Isocyanatomethyl-3,5,5-trimethylcyclohexyl
  isocyanate [*4098-71-9*] (IPDI), *13*:801
Isocyanic acid [*75-13-8*], *13*:789
Isocyanic acid, *p*-sulfamoylphenyl ester [*1773-
  42-8*], *13*:801

Isocyanurate T 1890, *13*:807
Isodecyl methacrylate [*29964-84-9*]
  properties, *15*:347
Iso E super [*54464-57-2*]
  fragrance of, *16*:959
  in perfumes, *16*:968
Isoeugenol [*97-54-1*], *13*:63
  economic aspects, *13*:53
  in perfumes, *16*:968
Isoguanine [*492-32-0*]
  microbial transformation to xanthine, *15*:465
Isoindole [*270-68-8*], *13*:213
Isolated soya protein
  in inks, *13*:392
Isolators
  acoustic, *13*:530
  microwave ferrites in, *14*:667
Isoleucine [*73-32-5*]
  in nuts, *16*:252
Isomaltose [*499-40-1*], *15*:441
Isomerization
  maleic acid, *14*:778
  zeolite catalysts in, *15*:652
Isomesityl oxide [*3744-02-3*], *13*:917
  properties, *13*:897, 917
Isonate 143-L, *13*:801
Isonate 125M, *13*:801
Isonate 136T, *13*:801
Isonitroethane [*4202-72-6*], *15*:969
1-Isonitro-1-nitroethane [*5923-46-6*], *15*:969
2-Isonitropropane [*79-46-9*], *15*:969
Isonitroso malonates, *14*:800
Isooctylmercaptoacetates
  tin derivatives, *16*:578
Isooctyl methacrylate [*28675-80-1*]
  properties, *15*:353
Isopentane [*78-78-4*]
  in isoprene synthesis, *13*:831
Isoperm, *14*:659
Isophane insulin [*9004-17-5*], *13*:612
Isophorone [*78-59-1*]
  mfg, *13*:919
  properties, *13*:898
Isophoronediamine [*2855-13-2*]
  resin from in fluorescent product, *14*:560
Isophorone diisocyanate [*4098-71-9*]
  properties, *13*:789
Isophthalic acid [*121-91-5*]
  resin from in fluorescent product, *14*:560
  in spunbondeds, *16*:79
Isophthalonitrile [*6260-60-5*], *15*:889
Isophthalonitrile [*626-17-5*] (IPN), *15*:906.   (See
  also *1,3-Dicyanobenzene.*)
Isopolymolybdates, *15*:686
Isopotential point, *13*:9
Isoprene [*78-79-5*], *13*:**818**
  citraconic acid from, *13*:870

in emulsion form, *14*:93
itaconic acid, *13*:870
properties, *13*:819
Isopropenyl acetate [*108-22-5*]
  isomerization of, *13*:930
2-Isopropenylbicyclo[2.2.1]-hept-5-ene [*13234-21-4*]
  from isoprene, *13*:821
D-*cis*-2-Isopropenyl-1-methylcyclobutaneethanol [*30820-22-5*]
  as insect attractant, *13*:482
p-Isopropenylphenol [*57829-92-2*]
  in dihydroxybenzene reaction, *13*:50
  hydroquinone from, *13*:52
2-Isopropoxyphenyl *N*-methylcarbamate [*114-26-1*]
  as insecticide, *13*:453
o-Isopropoxyphenyl *N*-methylcarbamate [*114-26-1*], *13*:452
Isopropyl alcohol [*67-63-0*]
  autoxidation, *13*:15
  autoxidation of, *13*:21
  in inks, *13*:390
  toxicity, *13*:269
Isopropylbenzene [*98-82-8*]
  isoprene reaction, *13*:825
Isopropylbiphenyl [*25460-78-2*]
  insulation properties of, *13*:551
α-Isopropyl-*p*-chlorophenylacetic acid [*2012-74-0*]
  in fenvalerate synthesis, *13*:457
4,4′-Isopropylidenediphenol. See *Bisphenol A.*
Isopropyl isocyanate [*1795-48-8*]
  properties, *13*:789
Isopropyllithium [*2417-93-8*]
  effect on *sec*-butyllithium decomposition, *14*:469
Isopropylmagnesium chloride [*1068-55-9*]
  N_2–Ti complex from, *15*:956
N-Isopropylmethacrylamide [*13749-61-6*]
  properties, *15*:356
Isopropyl methacrylate [*4655-34-9*]
  properties, *15*:353
  from propylene, *15*:366
2-Isopropyl-6-methyl-4-pyrimidinol [*2814-20-2*], *13*:884
Isopropylnaphthalene [*6158-45-8*]
  as insulator, *13*:551
2-Isopropylnaphthalene [*2027-17-0*], *15*:703
  naphthalenols from, *15*:729
2-Isopropylnaphthalene hydroperoxide [*6682-22-0*], *15*:734
Isopropylnaphthalenes, *15*:703
m-Isopropylphenyl *N*-methylcarbamate [*64-00-6*], *13*:452
o-Isopropylphenyl *N*-methylcarbamate [*2631-40-5*], *13*:452

p-Isopropylphenyl *N*-methylcarbamate [*4089-99-0*], *13*:452
Isosafrole [*120-58-1*]
  heliotropin from, *16*:685
  occurrence, *13*:63
Isosiv process, *15*:666
Isostatic pressing, *15*:344
Isotope effects
  kinetic, *13*:860
Isotopes, *13*:707, **838**. (See also *Ion implantation.*)
Isotope separation, *16*:161. (See also *Nuclear reactors.*)
  by lasers, *14*:75
Isotope shift
  use in laser separations, *14*:76
Isovaleric acid [*503-74-2*]
  diisobutyl ketone from, *13*:915
Itaconic acid [*97-65-4*], *13*:865
  copolymerization, *13*:868
  in emulsion polymerization, *14*:87
  polymerization, *13*:867
Itaconic acid and derivatives, *13*:**865**
Itaconic anhydride [*2170-03-8*]
  properties, *13*:871
ITIS information system, *13*:329
Ivadizer, *15*:267
125I-VIP, *15*:783
Ivory nut, *16*:251, 252

J

J acid [*87-02-5*], *15*:740. (See also *4-Hydroxy-7-amino-2-naphthalenesulfonic acid.*)
J-acid imide [*87-03-6*], *15*:742. (See also *7,7′-Imino-bis-4-hydroxy-2-naphthalenesulfonic acid.*)
J-acid urea [*137-47-4*], *15*:742. (See also *7,7′-Ureylene-bis-4-hydroxy-2-naphthalenesulfonic acid.*)
Jacksonian seizure
  treatment, *13*:134
Janitorial supplies
  packaging of, *16*:717
Japanese peppermint oil [*68917-18-0*] of *Mentha arvensis*, *16*:325
Jasmine [*8022-96-6*]
  in perfumes, *16*:950
Jasmine absolute [*8024-43-9*], *16*:323
Jasmine oil [*8022-96-6*], *16*:323
  indole in, *13*:213

Jasminium officinale Var. *grandiflorum*, *16*:323
Jasminum officinale, *16*:950
Jasmolin I [*4466-14-2*]
    as insecticide, *13*:424
Jasmolin II [*1172-63-0*]
    as insecticide, *13*:424
Jasmone [*488-10-8*]
    in jasmine fragrance, *13*:922
Jasper
    iron in, *13*:749
Jatropha curcas, *16*:251
Java almond, *16*:250
Jernstedt salts
    in metal surface treatments, *15*:306
Jet engine
    ester lubricants for, *14*:496
Jet engines
    film cooling in, *14*:943
    thermal insulation in, *13*:599
Jet finishing
    metallic coatings, *15*:244
Jet-fuel additive
    benzonitrile, *15*:906
Jet fuel tanks
    thermal insulation in, *13*:599
Jetting
    in metal cladding, *15*:277
JH.   See *Juvenile hormone.*
Jojoba nut, *16*:250
Jones reductor
    in iron analysis, *13*:787
Jordan refiner, *16*:777
Josephson effect, *15*:40
Journal bearings
    alloys for, *14*:156
JP-5
    in missile fuel, *14*:518
JP-4 hydrocarbon
    in propellants, *15*:984
Juglans cinerea, *16*:250
Juglans hindsii, *16*:250
Juglansin, *16*:249
Juglans nigra, *16*:250
Juglans regia, *16*:250
Juglans sieboldiana, *16*:250
Juglone [*481-39-0*]
    in walnut, *16*:259
Junction diodes, *14*:270
Juniper oil [*8012-91-7*], *16*:323
Juniperus communis L. [*8012-91-7*], *16*:308, 323
Juniperus mexicana, *16*:322
Juniperus mexicana Schiede, *16*:322
Juniperus virginiana, *16*:322
Jute
    in cable insulation, *13*:585

lignin content, *14*:297
    in packings, *16*:734
Juvenile hormone [*37246-81-4*]
    in insect control, *13*:459
Juvenile-onset diabetes, *13*:615
Juvenoids
    in insect control, *13*:459

K

Kabipor, *14*:248
K acid [*130-23-4*], *15*:740.   (See also *4-Amino-5-hydroxy-1,7-naphthalenedisulfonic acid.*)
Kainite [*67145-93-1*], *14*:637
    from the ocean, *16*:281
Kalle's acid [*486-54-4*], *15*:730.   (See also *1-Amino-2,7-naphthalenedisulfonic acid.*)
Kallikrein [*9001-01-8*], *15*:783
Kalrez, *16*:727
Kanamycin [*59-01-8*]
    inactivation, *15*:466
Kanamycin B [*4696-76-8*]
    inactivation, *15*:466
KA oil, *13*:924
Kaolin [*1314-13-2*], *15*:237; *16*:777.   (See also *China clay.*)
    micas in, *15*:418
    zeolites from, *15*:657
Kaolin clay [*1332-58-7*]
    in papermaking, *16*:819
Kaolinite [*1318-74-1*]
    as lithium source, *14*:450
Kaolins
    as insecticide carrier, *13*:416
Kapton
    tga standard, *13*:546
    thermal conductivity of, *13*:544
Karaya gum [*9000-36-6*]
    deacetylated, in paper, *16*:806
Karbutilate [*4849-32-5*], *13*:60, 62
Karl, *14*:811
Karl Fischer reagent
    for water determination, *13*:655
Karl Fisher reagent [*52365-46-5*]
    in analysis of laminated glass, *13*:986
Kastel, *13*:696
Kathon LP, *13*:245
Kawasaki process, *13*:745
KB 227, *15*:726
Keene's cement, *14*:376

Kel-F, *16*:727

Kelp, *16*:284. (See also *Seaweed*.)
  iodine source, *16*:294

Kenics in-line mixer, *15*:623

Kepone [*143-50-0*]
  health effect, *13*:269

Keratin
  in cattlehides, *14*:203

Keratosis
  treatment, *13*:118

Kerogen
  pyrolysis, *16*:337

Kerosene [*8008-20-6*]
  metal cleaning, *15*:297
  in missile fuel, *14*:518
  in organolead reactions, *14*:184
  papermaking defoamer, *16*:807

Kerr experiments, *14*:705

Ketazolam [*27223-35-4*]
  properties of, *13*:122

Ketene [*463-51-4*]
  from acetone, *13*:901
  aziridine reaction, *13*:155
  formaldehyde reaction, *13*:96
  properties, *13*:875

Ketene dimer. See *1,3-Cyclobutanedione and 4-methylene-2-oxetanone*.

Ketenes and related substances, *13*:**874**

Ketene trimer [*38425-52-4*], *13*:878

Ketenimines, *13*:882

4-Ketodioxolones
  from malic acid, *13*:106

5-Keto-D-gluconic acid [*5287-64-9*]
  in (*R*-*R**,*R**)-tartaric acid synthesis, *13*:113

α-Ketoglutaric acid [*328-50-7*], *15*:761
  diagnostic reagent, *15*:76
  in L-glutamic acid biosynthesis, *15*:767

Ketone peroxides, *13*:15

Ketones, *13*:**894**
  azeotropes of, *13*:895
  from nitriles, *15*:892
  toxicity, *13*:269

Ketosis
  in diabetes, *13*:606

17-Ketosteroids
  measurement, *15*:84

Ketoximes
  from nitroparaffins, *15*:976

Kevlar
  liquid-crystal behavior, *14*:418
  in metal composites, *13*:947

Kevlar 29
  fiber properties, *13*:970

Kevlar 49
  fiber properties, *13*:970

Keycide X-10, *13*:236

Kidney-replacement membranes, *15*:116

Kidneys
  liquid crystals in, *14*:418

Kidney stones
  and Mg deficiency, *15*:584
  nutrients associated with, *15*:573
  oxalic acid in, *16*:618

Kieserite [*14168-73-1*], *14*:637

Kilns
  in limestone mfg, *14*:362
  malting, *14*:815

Kimcloth, *16*:78

Kinchin-Pease formula, *13*:712

Kinetic isotope effects, *13*:860

Kinglor metor process, *13*:745

Kirchhoff's law, *13*:337

Kivcet process
  for lead smelting, *14*:130

Kiviter retort, *16*:340

KL5-2444, *13*:807

Klages, *14*:811

Klages barley, *14*:816

Klebsiella aerogenes, *15*:447

Klebsiellae, *15*:949

Klebsiella pneumoniae, *15*:451

Klebsiella pneumoniae nif b, *15*:952

Klystrons
  microwave tubes, *15*:504

Knapsack problem
  in operations planning, *16*:513

$K_2NbOF_5 \cdot H_2O$ [*19200-74-9*], *15*:825

Knockdown resistance, *13*:474

Knockout drops, *13*:124

Knoevenagel condensation, *15*:892

Knoevenagel reaction
  malonic acid derivatives, *14*:799

Knorr reaction, *13*:890

Knox-out 2FM, *15*:480

Knudsen transport
  in membranes, *15*:99

Koch's acid [*117-42-0*], *15*:731. (See also *8-Amino-1,3,6-naphthalenetrisulfonic acid*.)

Kodak Color Compensating Filters, *16*:541

Kodak Color Printing Filters, *16*:542

Kodak conversion filters, *16*:537

Kodak Ektacolor Paper, *16*:542

Kodak Ektamatic SC, *16*:544

Kodak Haze Cutting Filters, *16*:535

Kodak infrared interference filters, no. 480, *16*:524

Kodak Light-Balancing filters, *16*:537

Kodak Neutral Density Filter No. 5, *16*:530

Kodak Pola-Screen, *16*:534

Kodak Polycontrast filter kit, *16*:544

Kodak Polycontrast gelatin filters, *16*:544
Kodak Polycontrast papers, *16*:545
Kodak Projection Print Scale, *16*:532
Kodak Safelight Filters OA, *16*:552
Kodak Sensitometric Step Tablet No. 5, *16*:531
Kodak Wratten Filter No. 96, *16*:528
Kodak Wratten photomechanical filters, *16*:552
Kodak Wratten photometric filters, *16*:537
Kokneader compounding mixer, *15*:628
Kola, *16*:252
Kola acuminata, *16*:251
Kola nut, *16*:251
Kolbe-Schmitt carboxylation
  of dihydroxybenzenes, *13*:45
  of hydroquinone, *13*:60
Kolbe-Schmitt reaction, *15*:744
KOMPASS-FRANCE information system, *13*:
  330, 334
Koppers-Totzek gasifier, *14*:337
Koppite [*12198-49-1*], *15*:820
Kovar, *14*:681
Krabbes' disease
  liquid crystals in, *14*:420
Kraft lignins, *14*:306
Kraft-liner container inks, *13*:381
Kraft paper
  in packaging, *16*:721
Kraft process, *16*:770
Kraft pulp
  retention aids for, *16*:806
Kraft pulping
  lignin from, *14*:304
Krebs cycle
  S(−)-malic acid in, *13*:106
  rotenone effect on, *13*:427
Krebs viscometer, *16*:757
Kridee, *16*:78
Kroll process, *14*:573
  for niobium, *15*:826
Krypton [*7439-90-9*]
  in air, *16*:657
  ion implantation, *13*:715
  lasers, *14*:53
  from the ocean, *16*:280
Krypton-78 [*33580-79-9*], *13*:840
Krypton-80 [*26110-67-8*], *13*:840
Krypton-82 [*14191-81-2*], *13*:840
Krypton-83 [*13965-98-5*], *13*:840
Krypton-84 [*14993-91-0*], *13*:840
Krypton-85 [*13983-27-2*]
  disposal of, *16*:208
  in gas-penetrant inspection, *16*:52
  radiation exposure from, *16*:217
Krypton-86 [*14191-82-3*], *13*:840
Krypton-96 [*40720-79-4*], *16*:188
Krypton-97 [*53373-34-5*]
  fission product, *16*:188

Krypton fluoride laser, *14*:61
Krypton laser, *14*:51, 54
Kubelka-Munk equation, *16*:805
Kukersite, *16*:346
Kukui nut, *16*:251
Kynol, *16*:125
Kyrel, *16*:78

L

La 141 alloy, *14*:458
Labdanum [*8016-26-0*]
  in perfumes, *16*:950
Labdanum gum [*68917-77-1*], *16*:323
Labdanum oil [*8016-26-0*], *16*:323
Labdanum resin [*8016-26-0*], *16*:323
Labor
  glutethimide used during, *13*:125
LABORDOC information system, *13*:330
Laccase [*9002-10-9*]
  coniferyl alcohol reaction, *14*:301
L-acid [*117-59-9*], *15*:738.   (See also *5-Hydroxy-
    1-naphthalenesulfonic acid.*)
Lacquer
  case hardening, *15*:321
  emissivity, *13*:342
Lacquers
  in leather coatings, *14*:221
Lacrimator
  chloropicrin, *15*:984
Lactamide [*2043-43-8*], *13*:83
  properties, *13*:90
Lactation
  iron absorption during, *15*:595
Lactic acid [*50-21-5*]
  from lactose, *15*:568
  manufacture, *13*:84
  from 1-nitro-2-propanol, *15*:913
  properties, *13*:80, 81
  from propylene, *16*:625
D(−) Lactic acid [*10326-41-7*], *13*:81
Lactic acidosis
  from phenformin, *13*:617
Lactic dehydrogenase [*9001-60-9*,
    9078-32-4]
  zinc in, *15*:590
Lactic dehydrogenase (LDH) [*9001-60-9*]
  in diagnosis, *15*:75
Lactide [*95-96-5*], *13*:82
Lactobacillaceae
  in dextran production, *15*:440
Lactobacillus
  in dextran production, *15*:440

Lactobacillus bulgarcius, 13:84
Lactobacillus bulgaricus, 15:564
Lactobacillus delbrueckii, 13:80, 84
Lactobacillus leichmanii, 13:84
Lactobacillus leichmannii, 13:80
Lactoglobulin
  in casein, 15:568
Lactol [93-43-6], 15:734
  in inks, 13:393
Lactonaphthol, 15:734
Lactones
  from hydroxy acids, 13:97
Lactonitrile [78-97-7], 13:89. (See also
    Acetaldehyde cyanohydrin.)
  lactic acid from, 13:80
  in lactic acid synthesis, 13:85
Lactose [63-42-3]
  in buttermilk, 15:561
  in microbial synthesis saccharides, 15:468
  sweetening strength, 15:568
  in zanflo preparation, 15:451
Lactoyllactic acid [26811-96-1], 13:80
Laminar flow
  mass transfer, 14:**936**
Laminated and reinforced metals, 13:**941**
Laminated and reinforced plastics, 13:**968**
Laminated glass, 13:978. (See also *Laminated
    materials, glass.*)
  properties, 13:979
Laminated materials, glass, 13:978. (See also
    Glass.)
Laminated wood-based composites, 14:**1**
Laminates
  insulation properties, 13:558
Lamps
  energy-saving, 14:536
  mercury, 15:154
  phosphors for, 14:527
Landgard pyrolysis system
  for waste incineration, 13:194
Land reclamation
  nonwoven uses in, 16:94
Langbeinite [13826-56-7], 14:637
Lanolin [8006-54-0]
  as vehicle, 13:394
Lansfordite [61042-72-6], 14:619
Lanthanide oxalates, 16:632
Lanthanium nitride [25764-10-7], 15:872
Lanthanum [7439-91-0]
  from the ocean, 16:280
Lanthanum-138 [15816-87-2], 13:841
Lanthanum-139 [7439-91-0], 13:841
Lanthanum, compd with cobalt [12297-66-4] (1:
  5), 14:676
Lanthanum, compd with cobalt [53239-28-4] (2:
  17), 14:676

Lanthanum fluoride [13709-38-1]
  in fluoride electrode, 13:727
Lanthanum hydroxide [14507-19-8]
  formation of in ISES, 13:727
Lanthanum magnesium aluminum oxide [12254-
    13-6] (1:1:11:19)
  in phosphors, 14:541
Lanthanum oxide [1312-81-8]
  x-ray phosphors from, 14:544
Lanthanum oxide [12031-20-8] (1:1)
  in glass electrode, 13:5
Lanthanum oxide [56321-46-1] (1:2)
  in glass electrode, 13:5
Lanthanum oxybromide, Tb doped [68584-93-0]
  x-ray phosphor, 14:544
Lanthanum oxybromide, Tm doped [68877-14-5]
  x-ray phosphor, 14:544
Lanthanum oxysulfide, Tb doped [68611-13-2]
  x-ray phosphor, 14:544
Lanthanum ytterbium erbium fluoride (0.86:
    0.12:0.02:3)
  phosphor, 14:543
Laplace transform, 14:**932**
  in instrumentation and control, 13:498
Laray viscometer, 13:376
Lard [61789-99-9]
  from meat, 15:71
Lard oil
  in packings, 16:734
Larker, 14:811
Larvicides, 13:415
Laser
  helium–neon, 14:42
Laser alloying, 15:255
Laser cladding, 15:256
Laser coating, 15:255
Laser cutting, 14:66
  in nuclear engineering, 16:245
Laserglaze, 15:256
Laser gyroscope, 14:68
Laser light
  collimation of, 14:46
  effects on body, 14:62
Lasers, 14:**42**
  in branding cattle, 14:208
  in isotope separation, 13:856
  in nondestructive testing, 16:62
  solid state, 14:54
  stripe, 14:284
  tunable, 14:61
  use in holography, 14:69
Lasers, injection, 14:269
Lasers, semiconductor, 14:269. (See also *Lasers,
    injection.*)
Laser writing, 14:705
Laspeyresia pomonella, 13:415, 473, 480
Latex binders
  for nonwovens, 16:116

Latexes
  nonwovens from, *16*:106
  in paper, *16*:794
Latex paints, *16*:742, 753
Latex technology, *14*:**82**
Lathyrogens, *15*:138
Latices
  stability factors, *14*:85
Lattice doping
  in magnetic tape mfg, *14*:738
Laundry sour
  oxalic acid as, *16*:631
(±)-Laurene [*18452-41-0*], *13*:930
Laurent's acid [*84-89-9*], *15*:729.   (See also *5-
  Amino-1-naphthalenesulfonic acid.*)
Lauric aldehyde [*112-54-9*]
  in perfumes, *16*:967
Lauryl mercaptan [*112-55-0*]
  tin derivatives, *16*:578
Lauryl methacrylate [*142-90-5*]
  properties, *15*:347
Lautarite [*7789-80-2*], *13*:650
Lavandin oil [*8022-15-9*], *16*:324
Lavandula hybrida Reverchon, *16*:324
Lavandula latifolia Vill., *16*:328
Lavandula officinalis Chaix, *16*:324
Lavender oil [*8000-28-0*], *16*:319, 320, 324
Law of patents, *16*:851
Law of the Sea Treaty, *16*:293
Lawrencite
  iron(II) chloride as, *13*:773
LB-300-X, *14*:496
LC.   See *Lethal concentration.*
LCAO.   See *Linear combination of atomic
  orbitals.*
LCDs.   See *Liquid crystal displays.*
LDA.   See *Lithium diisopropylamide.*
LDH.   See *Lactic dehydrogenase.*
LDPE.   See *Low density polyethylene.*
LD process
  steel refinement, *14*:834
Leach electrolysis
  for lead, *14*:131
Leaching
  ores with H_2O_2, *13*:29
Lead [*7439-92-1*], *14*:**98**
  in cable insulation, *13*:583
  in coatings, *15*:243
  in diet, *14*:196
  in driers, *13*:379
  electrode, *13*:720
  environmental sources, *14*:196
  gasket properties, *16*:727
  inducement of immunodeficiency, *13*:170
  as lubricant, *14*:508
  in magnesium alloys, *14*:596

OSHA regulations, *14*:175
  in packings, *16*:737
  properties, *14*:100
  in tetraalkyllead manufacture, *14*:185
  toxicity, *13*:269
Lead-204 [*13966-26-2*], *13*:842; *14*:100
Lead-206 [*13966-27-3*], *13*:842; *14*:100
Lead-207 [*14119-29-0*], *13*:842
Lead-208 [*13966-28-4*], *13*:842; *14*:100
Lead acetate [*301-04-2*], *14*:100
  toxicity, *14*:198
Lead acetate anhydrous [*301-04-2*]
  properties, *14*:170
Lead acetate basic [*1335-32-6*]
  properties, *14*:170
Lead acetate trihydrate [*6080-56-4*]
  properties, *14*:170
Lead alkys, *16*:579
Lead alloys, *14*:**140**
Lead aluminum silicate
  in mica bonding, *15*:430
Lead anode
  in electrowinning, *15*:173
Lead antimonate [*13510-89-9*]
  properties, *14*:169
Lead–antimony alloys, *14*:143
Lead–antimony–tin alloys, *14*:156
Lead arsenate [*7784-40-9*]
  as insecticide, *13*:414
Lead arsenate, acid [*7784-40-9*]
  as insecticide, *13*:421
Lead arsenate, basic [*62648-18-4*]
  as insecticide, *13*:421
Lead azide [*13424-46-9*]
  properties, *14*:169
Lead babbitt
  coefficient of friction, *14*:481
Lead benzoate monohydrate [*6080-57-5*]
  properties, *14*:171
Lead bisilicate [*11120-22-2*], *14*:173
  properties, *14*:173
Lead borate [*14720-53-7*]
  in mica bonding, *15*:430
Lead–borate glasses, *14*:174
Lead borate monohydrate [*14720-53-7*]
  properties, *14*:173
Lead borosilicate [*56939-15-2*]
  in mica bonding, *15*:430
Lead bromide [*10031-22-8*] (1:2)
  properties, *14*:162
Lead burning
  welding process, *14*:141
Lead–calcium alloys, *14*:148
Lead–calcium compound [*12049-58-0*] (3:1), *14*:
  148
Lead–calcium–tin alloys, *14*:149

Lead carbonate [*598-63-0*]
  properties, *14*:171
  as solid lubricant, *14*:508
Lead chloride [*7758-95-4*], *14*:100
Lead chloride [*7758-95-4*] (1:2)
  properties, *14*:162
Lead chloride–lead hydroxide [*15887-88-4*] (1:1),
  14:163
Lead chromate [*7758-97-6*]
  in inks, *13*:378
Lead compounds
  lead salts, *14*:**160**
  in lubricants, *14*:479
Lead compounds, industrial toxicology, *14*:**196**
Lead compounds, organolead compounds, *14*:**180**
Lead–copper alloys, *14*:141
Lead diacetate [*546-67-8*]
  trialkylaluminum reaction, *16*:582
Lead dichloride [*7758-95-4*]
  triethylaluminum reaction, *16*:582
Lead dioxide [*1309-60-0*], *14*:102
  anode coating, *15*:173
  properties, *14*:165
Lead ferrite [*12023-90-4*]
  in bubble films, *14*:695
  as magnetic insulator, *14*:694
Lead fluoride [*7783-46-2*] (1:2)
  properties, *14*:162
Lead fluorosilicate [*25808-74-6*]
  in electrolytic refining, *14*:124
Lead glance. See *Lead sulfide.*
Lead hydroxide [*19781-14-3*]
  properties, *14*:166
Lead(II) sulfate [*7446-14-5*]
  properties, *14*:167
Lead-indium alloys, *14*:157
Lead iodide [*10101-63-0*] (1:2), *13*:661
  properties, *14*:162
Lead-lithium alloys, *14*:157
Lead-lithium-tin alloys, *14*:157
Lead molybdate [*10190-55-3*], *15*:687
Lead monosilicate [*11120-22-2*]
  properties, *14*:173
Lead monoxide [*1317-36-8*], *14*:100, 164
Lead naphthenate [*61790-14-5*], *15*:752
Lead nitrate [*10099-74-8*], *14*:100
  properties, *14*:168
Lead nitride [*58572-21-7*], *15*:872
Lead nitride [*75790-62-4*] (3:4), *15*:872
Lead orthoplumbate. See *Lead tetroxide.*
Lead oxide [*1317-36-8*]
  as solid lubricant, *14*:508
Lead oxide [*1317-36-8*] (1:1)
  properties, *14*:164
  reduction, *14*:140

Lead oxide [*1309-60-0*] (1:2)
  properties, *14*:164
Lead oxide [*1314-27-8*] (2:3)
  properties, *14*:164
Lead oxide [*1314-41-6*] (3:4)
  properties, *14*:164
Lead oxide cmpd with lead sulfate [*12765-51-4*]
  (1:1)
  properties, *14*:167
Lead perchlorate [*13637-76-8*]
  in ISE titration, *13*:731
Lead peroxide, *14*:165. (See also *Lead oxide* (*1:
  2*).)
Lead poison prevention act, *16*:760
Lead pyrosilicate. See *Lead monosilicate.*
Lead salts, *14*:160
Lead sesquioxide [*1314-27-8*], *14*:165. (See also
  Lead oxide (*2:3*).)
  properties, *14*:165
Lead silicate [*11120-22-2*]
  in lead ore processing, *14*:106
Lead silicate tribasic [*12161-62-5*]
  properties, *14*:173
Lead–silver alloys, *14*:151
Lead stannate [*1344-41-8*], *14*:162
Lead–strontium alloys, *14*:152
Lead sulfate [*7446-14-2*], *14*:100
Lead sulfate, monobasic [*12765-51-4*]
  properties, *14*:167
Lead sulfide [*1314-87-0*]
  as antiwear agent, *14*:493
  human absorption, *14*:197
  properties, *14*:166
  reduction by roasting, *14*:109
Lead telluride [*1314-91-6*],
  14:153
  properties, *14*:166
Lead–tellurium alloys, *14*:153
Lead–tellurium compound [*1314-91-6*] (1:1), *14*:
  153
Lead tetraacetate [*546-67-8*]
  properties, *14*:170
Lead tetroxide [*1314-41-6*], *14*:102, 165. (See
  also *Lead oxide* (*3:4*).)
  properties, *14*:165
Lead thiosulfate [*13478-50-7*]
  in matches, *15*:2
Lead–tin alloys, *14*:154
Lead tin telluride lasers, *14*:60
Lead titanate [*12060-00-3*]
  properties, *14*:174
Lead trioxide, *14*:165. (See also *Lead oxide* (*2:
  3*).)
Lead zirconate [*12060-01-4*]
  properties, *14*:174
Lead zirconate–titanate
  in transducers, *14*:662

Leaf alcohol [*928-96-1*], *14*:471
Leafhoppers
    insecticides, *13*:415
Leak testing
    by nondestructive methods, *16*:53
Learning, *15*:132.   (See also *Memory-enhancing*
    agents and antiaging drugs.)
Leather, *14*:**200**
    in packings, *16*:734
    properties, *14*:201
    substitutes, *14*:231
Leatherlike materials, *14*:**231**.   (See also
    Leather.)
Leather packings, *16*:739
Leather, patent, *14*:245
Leblanc process
    seaplant feeds for, *16*:294
Lecithin [*8002-43-5*], *14*:**250**; *15*:134.   (See also
    Phosphatidylcholine.)
    in encapsulation, *15*:477
    in ice cream, *15*:565
    in nuts, *16*:253
    properties, *14*:254
Lecithin hydrate [*8029-76-3*]
    processing of, *14*:264
Leclanche cells, *14*:864
Lecythis ollaria, *16*:250
Ledene [*21747-46-6*], *16*:313
Led Plate 250, *14*:513
LEDs.   See *Light-emitting diodes.*
Leghemoglobin
    in plants, *15*:947
Legumes
    in nitrogen fixation, *15*:947
Lemongras oil [*8007-02-1*], *16*:319
Lemongrass oil
    biacetyl in, *13*:925
Lemon oil [*8008-56-8*], *16*:319, 324
    encapsulation, *15*:487
Lemon oil, terpeneless [*68648-39-5*], *16*:310
Lemon oil, terpeneless, sesquiterpeneless [*68648-
    39-5*], *16*:310
Length
    standardization of, *15*:39
Lennard-Jones potential function, *15*:647
Lente iletin [*8049-62-5*], *13*:613
Lente insulin [*8049-62-5*], *13*:613
Lentinan [*9051-97-2*]
    immunotherapeutic agent, *13*:177
Leonard-Jones potential function, *14*:924
Leonite [*15226-80-9*], *14*:637
Lepidocrocite [*12022-37-6*]
    in magnetic tape mfg, *14*:738
    in ocean raw materials, *16*:291
Lepidolite [*1317-64-2*], *15*:416
    lithium in, *14*:449

Leptinotarsa decemlineata, *13*:437
Lethal concentration, *13*:258
Lethal dose, *13*:258
Letterpress inks, *13*:379; *14*:567
Leucine [*61-90-5*]
    from acetoacetamide, *13*:889
    in nuts, *16*:249, 252
Leucine amino peptidase [*9001-61-0*]
    zinc in, *15*:590
Leucine–enkephalin [*58822-25-6*]
    properties, *15*:765
Leuckart reaction
    amine preparation, *13*:901
Leucoindoxyl red [*25021-19-6*], *13*:217
Leuconostoc
    in dextran production, *15*:440
Leuconostoc citrovorum, *15*:562
Leuconostoc dextranicum, *15*:442
Leuconostoc mesenteroides
    in dextran production, *15*:440
    G-6-PDH source, *15*:87
Leukocyte acid phosphatase [*9001-77-8*]
    selenium in, *15*:591
Levamisole [*14769-73-4, 53631-68-8*]
    immunologic effects, *13*:174
    pharmacological profile, *13*:178
Levans [*9013-95-0*], *15*:445
Levarterenol, *15*:773
Leveling time
    for paint, *16*:746
Levophed, *15*:773
Levopimaric acid [*79-54-9*]
    in rosin size, *16*:810
Levopimaric acid–fumaric acid adduct [*77942-
    79-1*]
    in rosin size, *16*:810
Levoprome, *13*:122
Levorenine, *15*:757
Levulose [*57-48-7*]
    for ion-exchange process, *13*:703
Lewatit, *13*:696
Lewis acid
    zeolites as, *15*:651
Lewis base
    in isoprene polymerization, *13*:829
Lexan
    particle-track etching of, *16*:834
LFT.   See *Ligand-field theory.*
LH.   See *Luteinizing hormone.*
LHRH.   See *Luteinizing hormone releasing*
    hormone.
Licania rigida, *16*:251
Lice
    insecticide for, *13*:421
Licenses
    to patent use, *16*:886

Licensing
  patents, *16*:876
Lichee, *16*:248
Lichenic acid [*110-17-8*], *14*:770. (See also
  Fumaric acid.)
Lieberman-Storch test, *16*:793
Life expectancy
  of devices, *15*:16
Life-support
  by oxygen, *16*:670
Life testing. See *Materials reliability.*
Life tests, *15*:20
Liffarome [*67633-96-9*]
  in perfumes, *16*:967
Ligand-field theory (LFT), *16*:594
Light
  characteristics, *14*:42
Light absorption
  by dye molecules, *14*:555
Light-emitting diodes, *14*:270
Light-emitting diodes and semiconductor lasers,
  14:**269**
Lighters
  as match replacements, *15*:6
Lightguides
  fiber optic, *14*:71
Lighting
  phosphors for, *14*:527
  theatrical, color filters for, *16*:535
Lightnin in-line mixer, *15*:623
Light stabilizers
  nickel compounds as, *15*:814
  synthesis of, *14*:803
Light turbine oil
  limit of radiation tolerance, *14*:519
Light-water breeder reactor (LWBR), *16*:192, 204
Light-water reactor fuel
  reprocessing of, *16*:173
Light-water reactor (LWR)
  fuel elements, *16*:165
Light-water reactors
  water chemistry, *16*:150
Light-water reactors (LWRS), *16*:184
Lignin [*9005-53-2*], *14*:**294**, 295. (See also
  Wood.)
  biodegradation of, *13*:225
  composition, *14*:298
  humic acids from, *14*:317
  microbial oxidation, *15*:459
  removal from wood, *16*:769
Ligninsulfonic acid [*8062-15-5*], *14*:295
Lignite, *14*:313
  reserves and resources, *14*:313
Lignite and brown coal, *14*:**313**
  composition, *14*:318
Lignosulfonate [*8062-15-5*], *14*:304

Lignosulfonates
  from sulfite pulping, *14*:304
Lilial [*80-54-6*]
  in perfumes, *16*:967, 969
Lillionette 506, *16*:76
Lime [*1305-78-8*]
  in bordeaux mixture preparation, *13*:476
  carbon reaction, *15*:944
  cattlehide depilatory, *14*:209
  as insecticide carrier, *13*:416
  properties, *14*:350
  as solid lubricant, *14*:508
Lime and limestone, *14*:**343**
Lime-expressed oil, *16*:314
Lime manufacture
  operations planning for, *16*:520
Lime oil [*8008-26-2*], *16*:324
Lime putty, *14*:347
Limestone, *14*:343
  as dietary supplement, *14*:378
  in iron-ore processing, *13*:739
  properties, *14*:350
Limestones
  lithium in, *14*:449
Lime sulfur [*1344-81-6*]
  as insecticide, *13*:422
Lime-zeolite process, *13*:699
Limiting oxygen index
  novoloid fiber, *16*:126
Limonene [*138-86-3*], *16*:313
d-Limonene [*5989-27-5*]
  from grapefruit oil, *16*:323
  in orange oil, *16*:325
dl-Limonene [*7705-14-8*]
  from lemon oil, *16*:324
Limonite [*1317-63-1*]
  as iron source, *13*:748
  in limestone, *14*:350
Linalool [*78-70-6*], *16*:312, 956
  acetic anhydride reaction, *16*:310
  from bois de rose, *16*:321
  from lavender, *16*:328
  in perfumes, *16*:968
d-Linalool [*126-90-9*]
  from coriander oil, *16*:322
Linalyl acetate [*115-95-7*], *16*:956
  from bergamot oil, *16*:321
  from linalool, *16*:310
  in perfumes, *16*:968
  from sage, *16*:327
Lindane [*58-89-9*], *13*:432
Linear combination of atomic orbitals (LCAO),
  16:594
Linear congruential method
  in operations planning, *16*:508
Linear energy transfer
  fission particles, *16*:830

Linear olefins
  processes for, *16*:485
Linear-optimization models
  in operations planning, *16*:500
Linear programming
  industrial applications, *13*:508
  in operations planning, *16*:500
Line measurement, *13*:341
Linerboard, *16*:772
Linoleic acid [*60-33-3*]
  in isolation of phosphatidylcholine, *14*:251
  in milk of safflower-fed cows, *15*:491
  in nut oils, *16*:252
Linolenic acid [*60-33-3*]
  in ink driers, *13*:379
  in nut oils, *16*:252
  in soybean oil, *14*:252
Linotype alloys, *14*:156
Linseed oil [*8001-26-1*]
  in fabric for insulation, *13*:578
  in inks, *13*:381
  in packings, *16*:734
  in paint, *16*:743
  in varnishes, *13*:554
Lintrin, *15*:735
Lipase [*9004-02-8*], *15*:445
  calcium effect on, *15*:584
  in hyperlypidemia diagnosis, *15*:83
  measurement, *15*:79
Lipids
  in cattlehides, *14*:204
  in nuts, *16*:249
Lipofucin inhibitors
  antiaging effect, *15*:138
Lipofuscin
  in aging, *15*:137
Lipoic acid [*62-46-4*]
  sulfur in, *15*:587
Lipopolysaccharide (LPS)
  immunotherapeutic agent, *13*:176
Liposomes
  from lecithin, *14*:254
  in microencapsulation, *15*:476
β-Lipotropin [*9035-55-6*], *15*:764
Liquefied natural gas
  molecular-sieve dehydration, *15*:662
Liquefied petroleum gas (LPG), *14*:**383**
  properties, *14*:383
Liquid crystal displays (LCDs), *14*:423
Liquid crystals, *14*:**395**.   (See also
  Mesomorphism.)
  encapsulated, as thermometers, *15*:491
  in encapsulation, *15*:477
Liquid-level measurement, *14*:**427**
Liquid membrane electrodes, *13*:723
Liquid membranes, *15*:119
Liquid-metal fast breeder reactor (LMFBR), *16*:
  168, 184

Liquid nitrogen
  properties, hazards, *15*:938
Liquid-phase epitaxy (LPE)
  in LED manufacture, *14*:287
  in thin-film fabrication, *14*:694
Liritan process, *14*:213
Litharge, *14*:100, 164.   (See also *Lead
  monoxide.*)
Lithia [*12057-24-8*], *14*:455
N-Lithioethylenediamine
  acetylene reaction, *14*:471
Lithium [*7439-93-2*]
  in isoprene polymerization, *13*:828
  isotope separation processes, *13*:853
  lead alloys, *14*:157
  in magnesium alloys, *14*:596
  from the ocean, *16*:279
  organic derivatives, *14*:466
  properties, *14*:454
Lithium-6 [*14258-72-1*], *13*:840
Lithium-7 [*13982-05-3*], *13*:840
  in nuclear reactors, *16*:161
  separation, *16*:164
Lithium acetate [*546-89-4*], *14*:453, 459
Lithium acetate anhydrous [*546-89-4*], *14*:459
Lithium acetylide–ethylenediamine complex
  [*50475-76-8*]
  properties, *14*:471
Lithium alkyls, *16*:557
Lithium alloy x2020, *14*:457
Lithium aluminum hydride [*16853-85-3*]
  BF₃ etherate reaction, *16*:564
  initiator, *15*:194
  phenyl isocyanate reaction, *13*:798
Lithium amide [*7782-89-0*], *14*:455, 459
Lithium and lithium compounds, *14*:**448**
Lithium bromide [*7550-35-8*], *14*:461
  in batteries, *14*:458
Lithium bromide dihydrate [*13453-70-8*], *14*:462
Lithium bromide monohydrate [*23303-71-1*], *14*:
  462
Lithium bromide trihydrate [*76082-04-7*], *14*:462
Lithium bronze [*19453-81-7*]
  in organic reactions, *14*:457
Lithium n-butoxide [*1907-33-1*]
  effect on n-butyllithium decomposition, *14*:467
Lithium sec-butoxide [*4111-46-0*]
  decomposition of sec-butyllithium in, *14*:469
Lithium carbide [*1070-75-3*], *14*:451
Lithium carbonate [*554-13-2*], *14*:455
  production from brine, *14*:450
  properties, *14*:460
Lithium chlorate [*13453-71-9*]
  as oxygen source, *16*:673

Lithium chloride [7447-41-8], 14:461
  recovery of, 14:453
Lithium chloride dihydrate [16712-19-9], 14:461
Lithium chloride monohydrate [16712-20-2], 14:461
Lithium citrate [6080-58-6]
  in mania treatment, 14:460
Lithium–copper compound [76095-59-5] (1:4), 14:458
Lithium diacetylide [1070-75-3], 14:451
Lithium dialkylcuprate
  reactions, 16:585
Lithium diisopropylamide [4111-54-0] (LDA)
  in nitrosamine synthesis, 15:990
Lithium ferrite [12022-46-7], 13:776
Lithium fluoride [7789-24-4]
  in aluminum mfg, 14:460
  fission tracks in, 16:826
  in fuel reprocessing, 16:181
  properties, 14:461
Lithium formate [6108-23-2]
  properties, 14:460
Lithium hectorite
  in encapsulation, 15:482
Lithium hexafluoroarsenate [29935-35-1]
  in batteries, 14:458
Lithium hexafluorophosphate [21324-40-3]
  in batteries, 14:458
Lithium hydride [7580-67-8], 14:455, 463
  hydrogen source, 14:452
Lithium hydroxide [1310-66-3], 14:455
  atomic energy use, 14:452
  in biuret reagent, 15:81
  oxygen generation, 16:679
  pH control in nuclear reactors with, 16:152
  pH of system with boric acid, 16:152
Lithium hydroxide anhydrous [1310-65-2]
  properties, 14:463
Lithium hydroxide monohydrate [1310-66-3], 14:455
  properties, 14:464
Lithium hypochlorite [13840-33-0], 13:229
  properties, 14:464
Lithium iodide [10377-51-2], 14:462
Lithium iodide dihydrate [17023-25-5], 14:463
Lithium iodide monohydrate [17023-24-4], 14:463
Lithium iodide trihydrate [7790-22-9], 14:463
Lithium manganate(V) [12201-25-1], 14:857
Lithium metaborate [13453-69-5], 14:460
Lithium metasilicate [10102-24-6]
  properties, 14:465
Lithium methacrylate [13234-23-6], 15:367
Lithium molybdate [13568-40-6], 15:687
Lithium monoacetylide [1111-64-4], 14:456
Lithium naphthalenide [27939-69-1]

  as reductant, 15:954
Lithium niobate [12031-63-9], 15:837
  in nonlinear optics, 14:64
Lithium nitrate [7790-69-4]
  properties, 14:464
Lithium nitride [26134-62-3], 14:455; 15:872
  properties, 14:464
Lithium oxide [12057-24-8]
  in glass electrode, 13:5
  properties, 14:465
  in spodumene, 14:450
Lithium perchlorate [7791-03-9], 14:465
  in batteries, 14:458
Lithium perchlorate trihydrate [13453-78-6], 14:465
Lithium permanganate [13453-79-7], 14:845
Lithium peroxide [12031-80-0]
  lithium oxide from, 14:465
  oxygen generation, 16:679
  properties, 14:465
Lithium platinate [53989-08-5]
  anode coating, 15:173
Lithium selenide [12136-60-6], 14:451
Lithium soaps, 14:464
Lithium stearate [4485-12-5], 14:464
Lithium sulfate [10377-48-7]
  in lithium recovery, 14:452
  properties, 14:465
Lithium sulfate monohydrate [10102-25-7], 14:466
Lithium sulfide [12136-58-2], 14:451
Lithium tetraborate [1303-94-2], 14:460
Lithium tetrachloroaluminate [14024-11-4]
  in batteries, 14:458
Lithium tetrachloroferrate [15274-95-0]
  in lithium recovery, 14:454
Lithium tetrafluoroborate [14283-07-9]
  in batteries, 14:458
Lithium tetrahydridoaluminate [16853-85-3], 14:463
Lithium tetrahydridoborate [16949-15-8], 14:463
Lithium titanium sulfide [55326-82-4] (1:1:2), 14:459
  in batteries, 14:459
Lithium triisoamylborohydride [61075-97-6]
  hydroboration with, 16:565
Lithographic inks, 14:567
Litho newsprint inks, 13:379
Lithopone [1345-05-7]
  paper pigment, 16:778
Liver
  liquid crystals in, 14:418
Liver enzymes
  diagnostic tests, 15:78
Livingstonite [12532-29-5], 15:144

Living systems
  isotope substitution in, *13*:860
Ljungstrom air preheaters, *13*:203
LLDPE. See *Linear low density polyethylene.*
LMFBR. See *Liquid-metal fast breeder reactor.*
LMFBRS. See *Liquid-metal fast breeder reactors.*
LMSC
  information-retrieval system, *16*:901
LNG. See *Liquefied natural gas.*
Loco foco, *15*:5
Locus coeruleus, *15*:774
Locust bean gum [*9000-40-2*]
  dispersant, *16*:806
  in paper processing, *16*:780
  xanthan gum interaction, *15*:450
Lodex, *14*:680
Loeb-Sourirajan membranes, *15*:107
Loeweite [*16633-52-6*], *14*:637
Loft
  of nonwovens, *16*:104
Loganin [*18524-94-2*]
  in poison nut, *16*:257
Logic devices, *14*:694
Logistics time
  for spare parts, *15*:11
LOI. See *Limiting oxygen index.*
Lonchocarpus, *13*:425
Lonchocarpus arucu, *13*:426
London-van der Waal forces
  in emulsion polymerization, *14*:86
Lonza process, *14*:805
Looper
  control of with pheromones, *13*:480
Loparite [*12173-83-0*], *15*:820
Lorazepam [*846-49-1*]
  properties of, *13*:122
Lossen rearrangement
  in isocyanate synthesis, *13*:799
Loss factor
  acoustic, *13*:526
Lotes process, *16*:210
LPE. See *Liquid-phase epitaxy.*
LPG. See *Liquefied petroleum gas.*
LPG sweetening
  molecular sieves for, *15*:663
LPS. See *Lipopolysaccharide.*
LSD. See *Lysergic acid diethylamide.*
304L stainless steel, *15*:224
Lube-oil additive
  methacrylates, *15*:394
Lubricant
  phosphatides as, *14*:256
Lubricants, *14*:477. (See also *Lubrication and lubricants.*)

effect of radiation on, *16*:239
in inks, *13*:389
lead naphthenate, *15*:752
molybdenum sulfide, *15*:695
in packings, *16*:738
in paper-coating colors, *16*:822
synthetic, *14*:477
Lubricants, petroleum
  properties, *14*:485
Lubrication
  by oxalate coatings, *16*:630
Lubrication and lubricants, *14*:**477**
Lubrite SS, *14*:509
Lucite, *15*:888
Lumber, *14*:1. (See also *Laminated, wood-based composites.*)
  properties, *14*:2
Luminescence, *14*:527
  of phosphors, *14*:531
Luminescent materials, *14*:527. (See also *Phosphors.*)
  fluorescent pigments (daylight), *14*:**546**
  phosphors, *14*:**527**
Lumogen Light Yellow l [*2387-03-3*], *14*:549
Lumogen L Red Orange [*6871-91-6*], *14*:554
Lunar distance
  laser measurement, *14*:**67**
Lunar material
  particle tracks in, *16*:842
Lunar materials
  particle tracks in, *16*:826
Lunau
  protein content, *16*:249
Lunau nut, *16*:250
Lungs
  functioning of, *13*:259
Lurgi gasifier, *14*:334
Lurgi-Ruhrgas retort, *16*:340
Lurgi Spülgas process, *14*:334
Luteinizing hormone [*9034-40-6*]
  zinc effect on, *15*:590
Luteinizing hormone [*9002-67-9*] (LH)
  LHRH effect on, *15*:771
Luteinizing hormone releasing hormone [*9034-40-6*] (LHRH)
  as neuroregulator, *15*:771
  properties, *15*:771
Lutetium [*7439-94-3*]
  from the ocean, *16*:280
Lutetium-175 [*14391-25-4*], *13*:842
Lutetium-176 [*14452-47-2*], *13*:842
Lutetium, compd with cobalt [*12052-75-4*] (2:17), *14*:676
Lutetium nitride [*12125-25-6*], *15*:872
Lutrabond, *16*:78
Lutradur H7210, *16*:76
Lutrasil, *16*:78

Luzon nut, *16*:250
LVP. See *Lysine⁸-vasopressin.*
LWBR. See *Light-water breeder reactor.*
LWR. See *Light-water reactor.*
Lycra, *16*:377
Lymphogranuloma
    detection, *15*:81
Lyotropics
    liquid crystals, *14*:396
Lyral [*31906-04-4*], *16*:956, 968, 969
    fragrance of, *16*:959
Lysergic acid diethylamide (LSD) [*50-37-3*]
    serotonin interaction, *15*:771
Lysine [*56-87-1*], *15*:776
    in nuts, *16*:249, 252
L-Lysine [*56-87-1*]
    microbial synthesis of, *15*:464, 466
Lysine⁸-vasopressin [*50-57-7*] (LVP)
    avoidance behavior effect, *15*:133
Lysolecithin [*9008-30-4*], *14*:256
Lysyl oxidase [*9059-25-0*], *15*:589
D(−) Lyxonic acid [*526-92-1*], *13*:101
D(−) Lyxonic acid, γ-lactone [*15384-34-6*], *13*:101
D-Lyxose [*1114-34-7*]
    from D-galactonic acid, *13*:99

M

MAA. See *Methacrylic acid.*
MA 754 alloy, *15*:344
MABS. See *Methyl methacrylate-
    acrylonitrile–butadiene–styrene.*
Macadamia, *16*:248
Macadamia embryo
    development, *16*:260
Macadamia nut, *16*:250
Macadamia ternifolia, *16*:250
Maceration
    essential oils by, *16*:308
Machine oil
    limit of radiation tolerance, *14*:519
M acid [*489-78-1*], *15*:740. (See also *4-Hydroxy-
    8-amino-2-naphthalenesulfonic acid.*)
Macrocyclic effect, *13*:782
Macrocystis pyrifera, *16*:294
Macrolex fluorescent yellow 10GN
    fluorescent dye, *14*:547
Macromolecular nomenclature, *16*:41
Macrophages
    in immune system, *13*:168

Madelung synthesis
    of indoles, *13*:218
Maerz kiln, *14*:364
MAGAZINE INDEX information system, *13*:330, 334
Maghemite
    in magnetic tape, *14*:738
Magie 535, *14*:561
Magnefite process
    for magnesium lignosulfonate, *14*:304
Magne-rite, *15*:491
Magnesia, *14*:630
Magnesia, *14*:630
    caustic-calcined, *14*:631
    dead-burned, *14*:631
    fused, *14*:631
Magnesil, *14*:659
Magnesite [*13717-00-5*]
    in limestone, *14*:344
    magnesium ore, *14*:570
    properties, *14*:619
Magnesium [*7439-95-4*], *14*:570
    cladding of, *15*:281
    in coatings, *15*:241
    corrosion of, *14*:573
    deficiency, *15*:583
    emissivity, *13*:342
    in enzyme measurement, *15*:79
    in initiator systems, *13*:367
    in limestone, *14*:344
    lithium alloys, *14*:458
    in magnesium alloys, *14*:596
    niobium pentachloride reaction, *15*:827
    nutrient properties, *15*:571
    from the ocean, *16*:279, 292
    properties, *14*:571
    from seawater, *16*:281
    similarity to lithium, *14*:448
    surface treatment of, *15*:304
    in walnuts, *16*:261
Magnesium-24 [*14280-39-8*], *13*:840; *16*:188
Magnesium-25 [*14304-84-8*], *13*:840
Magnesium-26 [*13981-68-5*], *13*:840
Magnesium acetate [*142-72-3*], *14*:616
Magnesium acetate hydrate [*60582-92-5*], *14*:616
Magnesium acetate tetrahydrate [*16674-78-5*], *14*:616
Magnesium alkyls, *16*:558
Magnesium alloys, *14*:592
Magnesium–aluminum compound [*12254-22-7*]
    (17:12) (beta Mg–Al), *14*:603
Magnesium ammonium phosphate [*7785-21-9*], *14*:615
Magnesium bicarbonate [*2090-64-4*], *14*:619
Magnesium bisulfite [*13774-25-9*]
    in sulfite pulping, *14*:304
Magnesium bromide [*7789-48-2*]
    properties, *14*:625

Magnesium bromide, compound with diethyl ether [29906-89-6] (1:1), *14*:626

Magnesium bromide, compound with diethyl ether [17950-53-7] (1:2), *14*:626

Magnesium bromide, compound with dimethyl ketone [75198-44-6] (1:3), *14*:626

Magnesium bromide decahydrate [75198-45-7], *14*:626

Magnesium bromide diammoniate, *14*:625

Magnesium bromide ethanolate [75198-41-3], *14*:626

Magnesium bromide hexahydrate [13446-53-2], *14*:626

Magnesium bromide hexaisopentyl alcoholate [75198-42-4], *14*:626

Magnesium bromide hexammoniate [75198-46-8], *14*:625

Magnesium bromide hexapropanolate [75198-43-5], *14*:626

Magnesium bromide–magnesium oxide (1:3) dodecahydrate [75300-54-8], *14*:625

Magnesium bromide methanolate [75198-40-2], *14*:626

Magnesium bromide monoammoniate [75198-48-0], *14*:625

Magnesium carbonate [546-93-0]
in blast furnace flux, *13*:740
in inks, *13*:378
in limestone, *14*:344
properties, *14*:619

Magnesium carbonate–ammonium carbonate (1:1) tetrahydrate [22450-55-1], *14*:621

Magnesium carbonate–carbon dioxide (5:4) pentahydrate, *14*:632

Magnesium carbonate dihydrate [68973-26-2]
properties, *14*:619

Magnesium carbonate–magnesium hydroxide (1:1) trihydrate [12143-96-3], *14*:619, 621

Magnesium carbonate–magnesium bromide (2:1) octahydrate [75198-49-1], *14*:621

Magnesium carbonate–magnesium hydroxide (4:1) octahydrate [75300-49-1], *14*:619

Magnesium carbonate monohydrate [17968-26-2, 23389-33-5]
properties, *14*:619

Magnesium carbonate pentahydrate [61042-72-6], *14*:632

Magnesium carbonate–potassium carbonate (1:1) octahydrate [75198-51-5], *14*:621

Magnesium carbonate–rubidium carbonate (1:1) tetrahydrate [75198-52-6], *14*:621

Magnesium carbonate–sodium carbonate [19086-68-1] (1:1), *14*:621

Magnesium carbonate trihydrate [5145-46-0]
properties, *14*:619

Magnesium chloride [7786-30-3], *14*:571, 622
1 butene polymerization with, *16*:477
from lime, *14*:380

magnesium from by electrolysis, *14*:576
from magnesium oxide, *14*:576
toxicity, *15*:573
in Ziegler catalyst, *16*:443
Ziegler-Natta catalyst, *16*:461

Magnesium chloride ammoniate [68374-25-4], *14*:624

Magnesium chloride diammoniate [68374-24-3], *14*:624

Magnesium chloride dihydrate [19098-17-0]
decomposition, *14*:624
drying of, *14*:576

Magnesium chloride–ethanol complex [16693-00-8] (1:6)
in Ziegler catalyst, *16*:443

Magnesium chloride hexaethanolate [16693-00-8], *14*:624

Magnesium chloride hexahydrate [7791-18-6]
decomposition, *14*:624
toxicity, *15*:573

Magnesium chloride hexamethanolate [57467-93-0], *14*:623

Magnesium chloride hexammoniate [24349-22-2], *14*:624
magnesium chloride from, *14*:576

Magnesium chloride monohydrate [17638-61-8]
decomposition, *14*:624

Magnesium chloride tetrahydrate [22695-80-3]
decomposition, *14*:624

Magnesium compounds, *14*:**615**
from seawater, *16*:281

Magnesium ferrite [12068-86-9], *13*:776

Magnesium fluoride [7783-40-6], *14*:571
optical-filter coating, *16*:525, 550

Magnesium hexafluorosilicate hexahydrate [16949-65-8]
toxicity, *15*:573

Magnesium hydroxide [1309-42-8], *14*:571
from seawater, *16*:281
solubility, *14*:354
Ziegler-Natta catalyst, *16*:460

Magnesium hydroxide chloride [13759-24-5]
decomposition, *14*:624

Magnesium hydroxychloride [13759-24-0]
from magnesium chloride, *14*:576

Magnesium iodide [10377-58-9], *14*:573
properties, *14*:626

Magnesium iodide, compound with diethyl ether [29964-67-8] (1:2), *14*:627

Magnesium iodide, compound with aniline [75198-39-9] (1:6), *14*:627

Magnesium iodide hexahydrate, *14*:627

Magnesium iodide octahydrate [7790-31-0], *14*:627

Magnesium lignosulfonate [8061-54-9]
from the magnefite process, *14*:304

Magnesium limestone, *14*:344

Magnesium methylate [*109-88-6*], *14*:574

Magnesium methyl carbonate [*14171-36-9*]
  nitromethane reaction, *15*:977

Magnesium nitrate [*10377-60-3*]
  in HNO$_3$ separation, *15*:854, 865
  properties, *14*:627

Magnesium nitrate, compound with magnesium
    hydroxide [*76190-42-6*] (1:4), *14*:627

Magnesium nitrate hexahydrate [*13446-18-9*]
  properties, *14*:627

Magnesium nitride [*12057-71-2*], *15*:872

Magnesium oxide [*1309-48-4*], *14*:571, 629
  chlorine reaction, *14*:578
  in iron-ore processing, *13*:740
  in lignitic coal, *14*:334
  magnesium chloride from, *14*:576
  in quicklime, *14*:346
  reduction processes, *14*:583
  from seawater, *16*:281
  in silicon nitride processing, *15*:879
  in spent shale, *16*:351

Magnesium oxide–carbon dioxide (5:4)
    pentahydrate [*12306-51-3*], *14*:631

Magnesium oxide–magnesium nitrate (1:2)
    pentahydrate [*75300-53-7*], *14*:627

Magnesium oxide–magnesium nitrate (2:1)
    pentahydrate [*75300-52-6*], *14*:627

Magnesium oxide–magnesium nitrate (3:1)
    undecahydrate [*75300-51-5*], *14*:627

Magnesium oxy carbonate [*75300-50-4*], *14*:620

Magnesium permanganate hexahydrate [*10377-
    62-5*], *14*:845

Magnesium phosphide [*51913-28-1*], *14*:571

Magnesium silicide [*22831-39-6*], *14*:603

Magnesium soaps
  overbasing of, *14*:642

Magnesium stearate [*557-04-0*], *14*:640
  as repellent component, *13*:479

Magnesium sulfate [*7487-88-9*]
  properties, *14*:636
  toxicity, *15*:573

Magnesium sulfate diammoniate dihydrate
    [*75198-55-9*], *14*:639

Magnesium sulfate heptahydrate [*10034-99-8*],
    14:636

Magnesium sulfate hexahydrate [*17830-18-1*],
    14:636

Magnesium sulfate–magnesium oxide (1:3)
    undecahydrate [*65496-31-3*], *14*:639

Magnesium sulfate–magnesium oxide (1:5)
    octahydrate [*12286-12-3*], *14*:639

Magnesium sulfate monoammoniate trihydrate
    [*75198-54-8*], *14*:639

Magnesium sulfate monohydrate [*14168-73-1*],
    14:636

Magnesium sulfate–sulfuric acid [*10028-26-9*] (1:
    1), *14*:639

Magnesium sulfate–sulfuric acid [*39994-66-6*] (1:
    3), *14*:639

Magnesium sulfide [*12032-36-9*], *14*:573

Magnesium sulfite [*7757-88-2*], *14*:641

Magnesium sulfite hexahydrate [*13446-29-2*], *14*:
    641

Magnesium sulfite trihydrate [*19086-20-5*], *14*:641

Magnesium sulfonates
  overbased, *14*:642

Magnesium vanadate, MgV$_3$O$_8$ [*12181-49-4*], *14*:
    642

Magnesium vanadate, Mg$_2$V$_2$O$_7$ [*13568-63-3*], *14*:
    642

Magnesium vanadate, Mg$_3$V$_2$O$_8$ [*13568-68-8*], *14*:
    642

Magnesium vanadate, Mg$_{1.9}$V$_3$O$_3$, *14*:642

Magnetherm process
  for magnesium, *14*:585

Magnetic-bubble memories (MBM), *14*:686

Magnetic coating
  for recording tape, *14*:733

Magnetic disks
  materials for, *14*:701

Magnetic fluids, *14*:705

Magnetic materials, bulk, *14*:**646**

Magnetic materials, thin film, *14*:**686**

Magnetic resonance
  liquid crystals in, *14*:423

Magnetic separation, *14*:**708**

Magnetic tape, *14*:**732**.   (See also *Tape,
    magnetic.*)
  materials, *14*:701

Magnetic-tape services
  for patent information, *16*:922

Magnetic thin films
  properties, *14*:687

Magnetism
  theory and types, *14*:646

Magnetite [*1309-38-2*]
  from beach sand, *16*:289
  encapsulation, *15*:490
  in iron-ore processing, *13*:740
  as iron source, *13*:748
  in limestone, *14*:350
  in magnetic tape, *14*:737
  in mica books, *15*:420
  from the ocean, *16*:286
  triiron tetroxide as, *13*:775

Magnetization
  measurement, *14*:683

Magnetographic tests, *16*:57

Magnetohydrodynamic generators
  in nitric acid mfg, *15*:864

Magnetooptics, *14*:705

Magnetophons, *14*:733

Magnetostatic separators, *14*:709

Magnetostriction, *14*:648
Magnetrons
    for microwave ovens, *15*:504
Magnetron sputtering, *14*:693; *15*:267
Magnet steels, *14*:680
Magnicol process, *14*:673
Magnifite process, *14*:641
Maintainability, *15*:10
Maintenance, *14*:**753**
    remote, in nuclear plants, *16*:245
Maintenance coatings, *16*:755
Majac mill
    dry-grinder, *15*:423
Malabate OD-200, *16*:304
Malaccol [*478-07-9*]
    insecticide, *13*:425
Malaria, *13*:414
Malathion [*121-75-5*], *13*:111, 441
    toxicity, *13*:270
Maleic acid [*110-16-7*], *14*:770.   (See also *Maleic
    anhydride, maleic acid, and fumaric acid.*)
    in *R,S*(±)-malic acid synthesis, *13*:107
    properties, *14*:770
    from resorcinol, *13*:42
    in (*R*,R**)-tartaric acid synthesis, *13*:113
Maleic anhydride [*108-31-6*]
    copolymer, *15*:192
    Diels-Alder reaction, *16*:484
    in emulsion polymerization, *14*:87
    glycolic acid reaction, *13*:91
    in isoprene reaction, *13*:820
    malic acid decomposition product, *13*:105
    microbial transformation of, *15*:462
    properties, *14*:770
Maleic anhydride, maleic acid, and fumaric acid,
    14:**770**
Maleic hydrazide [*123-33-1*], *14*:774
Maleimide [*541-59-3*], *14*:774
Maleoyl chloride [*22542-53-6*], *14*:773
Maleyl chloride [*22542-53-6*], *14*:773
Malic acid [*6915-15-7*], *13*:103; *14*:778.   (See also
    Hydroxysuccinic acid.)
    isolation of, *14*:770
    oxidation, *14*:794
    properties, *13*:103
    resorcinol reaction, *13*:46, 62
L-Malic acid [*97-67-6*]
    microbial synthesis, *15*:464
R(+)-Malic acid [*636-61-3*], *13*:103
R,S(±)-Malic acid, *13*:103
R,S-Malic acid [*617-48-1*], *13*:103
S(−)-Malic acid [*97-67-6*], *13*:103
Malic dehydrogenase [*9001-64-3*]
    diagnostic reagent, *15*:76
    zinc in, *15*:590
Malignancy
    liquid crystals associated with, *14*:419

Malimo, *16*:120
Malipol, *16*:120
Malivlies, *16*:120
Maliwatt, *16*:120
Malnutrition
    detection, *15*:81
Malodors, *16*:303
Malonato(1,2-diamino cyclohexane)platinum II
    [*52351-07-2*], *14*:796
Malonic acid [*141-82-2*]
    in initiator systems, *13*:367
    properties, *14*:794
Malonic acid and derivatives, *14*:**794**
Malonic acid, monosodium salt [*23549-97-5*]
    toxicity, *14*:796
Malonic anhydride [*1519-48-5*], *14*:794
Malonic dinitrile [*109-77-3*], *14*:794.   (See also
    Malononitrile.)
Malonic esters
    from cyanoacetic acid, *15*:907
Malonodinitrile [*109-77-3*], *14*:794.   (See also
    Malononitrile.)
Malononitrile [*109-77-3*], *14*:794; *15*:888
Malt
    microbial preparation, *15*:459
Maltose [*69-79-4*]
    in microbial synthesis saccharides, *15*:468
Malts and malting, *14*:**810**
Malt sprouts, *14*:822
MANAGEMENT CONTENTS information
    system, *13*:330
Management science, *16*:500.   (See also
    Operations planning.)
Mandelonitrile [*532-28-5*]
    from amygdalin, *16*:257
Maneb [*12427-38-2*], *14*:**878**
Manganese [*7439-96-5*], *14*:824
    deep-sea nodules, *14*:845
    in driers, *13*:379
    in DR processes, *13*:761
    in magnesium alloys, *14*:596
    as magnetic material, *14*:647
    metabolic functions, *15*:594
    in metallic coatings, *15*:252
    nutrient properties, *15*:571
    in nuts, *16*:260
    from the ocean, *16*:280, 292
Manganese-54 [*13966-31-9*]
    reactor corrosion product, *16*:155
Manganese-55 [*7439-95-5*], *13*:840
Manganese acetate [*638-38-0*]
    toxicity, *15*:574
Manganese acetate tetrahydrate [*15243-27-3*],
    14:845
    toxicity, *15*:574
Manganese–aluminum–carbon alloys
    as magnetic materials, *14*:681
Manganese and manganese alloys, *14*:**824**

Manganese borate [*12228-91-0*], *14*:845
Manganese carbide [*75718-05-7*] (1:3), *14*:845
Manganese carbide [*75718-05-7*] (2:7), *14*:845
Manganese carbide [*12121-90-3*] (3:1), *14*:845
Manganese carbide [*12076-37-8*] (7:3), *14*:845
Manganese carbide [*12364-85-1*] (15:4), *14*:845
Manganese carbide [*12266-65-8*] (23:6), *14*:845
Manganese carbonate [*598-62-9*], *14*:845
  in phosphor mfg, *14*:536
Manganese chloride [*7773-01-5*], *14*:845
Manganese chloride dihydrate [*20603-88-7*], *14*:
  852
Manganese chloride tetrahydrate [*13446-34-9*],
  14:862
Manganese compounds, *14*:**844**
Manganese deficiency, *15*:594
Manganese dichloride dihydrate [*20603-88-7*],
  13:773
Manganese dihydrogen phosphate dihydrate
  [*18718-07-5*], *14*:845
Manganese dioxide [*1313-13-9*]
  anode coating, *15*:173
  electrolytic, *14*:865
Manganese dioxide, chemical, *14*:863
Manganese ferrite [*12063-10-4*]
  as magnetic insulator, *14*:694
Manganese–gold compound [*12006-65-4*] (1:2)
  as magnetic materials, *14*:647
Manganese heptoxide [*12057-92-0*], *14*:845
  explosive reactions, *14*:859
Manganese hexakiscyanoferrate [*14402-63-2*] (2:
  1), *13*:771
Manganese hydroxide [*18933-05-6*], *14*:845
Manganese(II) acetylacetonate [*14024-58-9*], *14*:
  852
Manganese(II) dihydrogen phosphate [*18718-07-*
  5]
  in metal surface treatment, *15*:304
Manganese(II) dithionate [*13568-72-4*], *14*:855
Manganese(III) acetate dihydrate [*19513-05-4*],
  14:845
Manganese(III) acetylacetonate [*14284-89-0*], *14*:
  845
Manganese(III) fluoride [*7783-53-1*], *14*:845
Manganese(III) fluoride dihydrate, *14*:853
Manganese(III) oxide [*1317-34-6*], *14*:845
Manganese(III) oxide hydrated [*1332-63-4*], *14*:
  845
Manganese(II) nitrate [*10377-66-9*], *14*:855, 863
Manganese(II) nitrate monohydrate [*54166-61-1*],
  14:863
Manganese(II) nitrate tetrahydrate [*20694-39-7*],
  14:863
Manganese(II) nitrate trihydrate [*55802-19-2*],
  14:863

Manganese(II) oxide [*1344-43-0*], *14*:845
  HCl reaction, *14*:862
  reduction, *14*:830
Manganese(II) phthalocyanine [*14325-24-7*]
  reduction by Li, *14*:851
Manganese(II) sulfate monohydrate [*31746-59-5*],
  14:878
Manganese(II) sulfate tetrahydrate [*10101-68-5*],
  14:878
Manganese(I) nitrate pentacarbonyl [*14488-62-1*],
  14:852
Manganese iodopentacarbonyl [*14896-42-6*]
  no reaction, *14*:851
Manganese–iron nodules. See *Ferromanganese*
  nodules.
Manganese(IV) oxide [*1313-13-9*], *14*:845
  reduction, *14*:830
  toxicosis from, *15*:598
Manganese molybdate [*14013-15-1*], *15*:687
Manganese naphthenate [*1336-93-2*], *15*:752
Manganese nitrate hexahydrate [*10377-66-9*], *14*:
  845
Manganese nitride [*36678-21-4*] (1:1), *14*:845
Manganese nitride [*12163-53-0*] (2:1), *14*:845; *15*:
  872
Manganese nitride [*12033-03-3*] (3:2), *14*:845; *15*:
  872
Manganese nitride [*12033-07-7*] (4:1), *14*:845; *15*:
  872
Manganese nitride [*64886-63-1*] (6:5), *14*:845
Manganese oxide [*1313-13-9*]
  in iron-ore processing, *13*:742
  as magnetic material, *14*:647
Manganese oxide [*1317-35-7*] (3:4), *14*:845
  reduction, *14*:830
Manganese oxide [*12163-64-3*] (5:8), *14*:845
Manganese phosphate monohydrate [*14986-93-7*],
  14:853
Manganese pneumonia, *14*:885
Manganese selenide [*1313-22-0*]
  as grain-growth inhibitor, *14*:655
Manganese sesquioxide [*1317-34-6*]
  reduction, *14*:830
Manganese silicate [*7759-00-4*], *14*:844
Manganese silicide [*12032-85-8*] (1:1), *14*:849
Manganese silicide [*12163-59-6*] (3:1), *14*:849
Manganese silicide [*12033-10-2*] (5:3), *14*:849
Manganese–silicon spring steels, *14*:842
Manganese sulfate [*7785-87-7*], *14*:845, 853
  electrolysis, *14*:865
Manganese sulfide [*18820-29-6*], *14*:852
  as grain-growth inhibitor, *14*:655
Manganese tetrafluoride [*15195-58-1*], *14*:853
Manganese trinitrosylcarbonyl [*14951-98-5*], *14*:
  851
Manganese zinc ferrites, *14*:664

Manganic acid [*54065-28-0*], *14*:858
Manganic hydroxide [*18933-05-6*]
  in initiator systems, *13*:367
Manganism, *14*:841, 885
Manganit, *14*:863
  manganese dioxide, *14*:878
Manganite [*52019-58-6*], *14*:825
Manganoan calcite, *14*:825
Manganosite [*1313-12-8*], *14*:845
Mange
  effect on cattlehide, *14*:208
Mania
  lithium salt treatments, *14*:460
Manila hemp
  nonwoven fabrics from, *16*:107
  in paper, *16*:771
Manipulators
  for nuclear plants, *16*:246
Manketti nut, *16*:251
D-Mannans [*51395-96-1*], *15*:447
Mannich bases
  from nitroparaffins, *15*:976
Mannich reaction
  of indoles, *13*:216
Mannitol [*67-56-1*]
  on ion-exchange process, *13*:703
  in nuts, *16*:252
D(−) Mannonamide [*27022-42-0*], *13*:101
L(−) Mannonamide [*67008-28-0*], *13*:101
Mannonic acid [*20248-27-5*], *13*:101
D(−) Mannonic acid, ethyl ester [*27934-94-7*], *13*:101
D(−) Mannonic acid, δ-lactone [*32746-79-5*], *13*:101
L(−) Mannonic acid, δ-lactone (10366-75-3], *13*:101
DL(−) Mannonic acid, γ-lactone [*10366-82-2*], *13*:101
L(−) Mannonic acid, γ-lactone [*22430-23-5*], *13*:101
D-Mannopyranuronic acid [*1986-14-7*]
  in alginic acid substitute, *15*:452
β-D-Mannopyranuronic acid homopolymer [*27638-01-3*], *15*:452
Mannose [*3458-28-4*]
  in ivory nut meal, *16*:272
D-Mannuronic acid [*1986-14-7*]
  in alginate polysaccharide, *15*:452
Manometers
  mercury in, *15*:155
Maps
  nonwoven, *16*:94
Maquat MC-1416, *13*:234
Marangoni effect, *14*:956
Marble, *14*:345
Marcali isocyanate test, *13*:809

Marcasite [*1317-66-4*], *16*:63
  as iron source, *13*:748
  in limestone, *14*:350
Margarine
  analysis of with ISES, *13*:732
Marine apatites.   See *Phosphorites.*
Marine engine oil
  limit of radiation tolerance, *14*:519
Marine gravels, *16*:286
Marine sands, *16*:286
Marine shells
  from the ocean, *16*:285
Marix, *16*:78
Marjoram [*8015-01-8*], *16*:307
Market and marketing research, *14*:**895**
Market information, *13*:306
Market Information Sources, *13*:316
Marketing research
  methodology, *14*:902
Market research
  methodology, *14*:897
Marking nut, *16*:251
Marking oyster, *16*:252
Markov decision process
  in operations planning, *16*:507
Markovnikov, *13*:356
Markush representations
  in information systems, *16*:911
Marl, *14*:345
Mar-M200 [*12604-85-2*], *15*:789
Marquette range
  iron ore from, *13*:751
Marrow
  liquid crystals in, *14*:418
Martempering process, *15*:323
Martensite [*12173-93-2*], *15*:340
MASC.   See *Methylaluminum sesquichloride.*
Maser, *14*:42
Masking
  in IC manufacture, *13*:628
Masks, chrome
  in IC manufacture, *13*:632
Masks, emulsion
  in IC manufacture, *13*:632
Masonex, *16*:622.   (See also *Hemicellulose extract.*)
Masonite, *14*:4
Masonry
  cleaner, *13*:93
Mason's lime, *14*:347
Mass
  standardization of, *15*:39
Mass analyzers
  in ion implantation, *13*:708
Massicot [*1317-36-8*], *14*:164
Masson and Race reagent, *13*:665

Mass transfer, *14*:**911**
  in nitration, *15*:845
Mastering tape, *14*:750.   (See also *Tape, mastering*.)
Mastitis, *15*:553
Matches, *15*:1
Maté [*68916-96-1*]
  in perfumes, *16*:950
Material requirements planning (MRP), *16*:515
Materials reliability, *15*:**9**
Materials standards and specifications, *15*:**32**
Mathematical modeling
  in operations planning, *16*:500
Mathemoglobin, *15*:595
Matrix devices, *15*:123
Matrix polymerization
  in encapsulation, *15*:483
Maturity-onset diabetes, *13*:615
Mauritia flexuosa, *16*:250
Maxilon Brilliant Flavine 10GFF [*12221-86-2*]
  fluorescent dye, *14*:547
Maxwell-Wagner interfacial polarization, *13*:537
Mayers reagent, *15*:160
  alkaloid detection, *13*:661
MBBA. See *p-Methoxybenzylidene-p′-n-butylaniline*.
MBE. See *Molecular-beam epitaxy*.
MBM. See *Magnetic-bubble memories*.
MBS. See *Methyl methacrylate–butadiene–styrene copolymer*.
MBT. See *Methylene bisthiocyanate*.
McBain-Bakr microbalance, *15*:659
MDA. See *Bis(4-aminophenyl)methane*.
MDC cell, *15*:180
MDH. See *Malic dehydrogenase*.
MDI. See *Diphenylmethane diisocyanate*.
MDPE. See *Medium density polyethylene*.
2-MEA. See *2-Mercaptoethylamine*.
Meat
  comminuted, *15*:67
Meat curing, *15*:62.   (See also *Meat products*.)
Meat-curing agents, *15*:62
Meat meal
  meat by-product, *15*:72
Meat pigments
  myoglobins, *15*:62
Meat processing. See *Meat products*.
Meat products, *15*:**62**
Meats, processed
  nitrosamines in, *15*:993
Meat tempering
  in microwave ovens, *15*:513
Meclofenoxate [*51-68-3*]
  senility treatment, *15*:138

Medical applications
  lasers, *14*:70
Medical diagnostic reagents, *15*:74
Medical equipment
  from poly(4-methyl-1-pentene), *16*:479
Medical microwave applications, *15*:496
Mediterranean flour moth
  control of with pheromones, *13*:481
MEDLARS information system, *13*:291
MEDLINE information systems, *13*:291
MEHQ.   See *p-Methoxyphenol, hydroquinone monomethyl ether*.
Meissner process, *13*:804
MEK. See *Methyl ethyl ketone*.
Melamine [*108-78-1*]
  formaldehyde reaction, *16*:817
  in leather manufacture, *14*:216
  in paper coatings, *16*:790
Melamine adhesives
  for plywood, *14*:14
Melamine–formaldehyde adhesives
  for plywood, *14*:14
Melamine–formaldehyde resin [*9003-08-1*]
  in fluorescent product, *14*:560
  as magnetic tape binders, *14*:741
  in paper, *16*:793
  in papermaking, *16*:817
Melamine–urea–formaldehyde adhesives
  for plywood, *14*:14
Melanin
  ACTH effect on, *15*:759
Melanocyte-stimulating hormone (MSH) [*9002-79-3*], *15*:759
  effect on learning and memory, *15*:132
Melatonin [*73-31-4*]
  properties, *15*:771
Melilot [*8023-73-2*]
  in perfumes, *16*:951
Melilotus officinalis, *16*:951
Mellorine, *15*:566
Melophagus ovinus
  effect on sheepskin, *14*:208
Melt-blown textiles, *16*:93
Melt-flow index
  of polyethylene, *16*:440
Melt index
  of HDPE, *16*:426
Melt-refining process
  for fuel reprocessing, *16*:182
Melt spinning
  metals for fibers, *15*:229
Membrane processes
  electrochemical, *15*:120
Membranes
  comparison of configurations, *15*:111
  controlled-release applications, *15*:123
  hydraulic permeation of, *15*:95
  liquid, *15*:475

Membranes (*Continued*)
    module configurations, *15*:108
    permeability, *15*:92
    structure of, *15*:101
Membrane separation
    for milk products, *15*:562
Membrane technology, *15*:**92**
Membrane transport
    electrochemical, *15*:100
    solution-diffusion mechanisms, *15*:93
Memories
    magnetic, *14*:687
Memories, thin film, *14*:705
Memory-enhancing agents and antiaging drugs,
    15:**132**
Memory ferrites, *14*:667
Menadione [*58-27-5*], *15*:716.   (See also *2-*
    Methyl-1,4-naphthoquinone.)
Menkes' kinky-hair syndrome
    nutrients associated with, *15*:573
Menominee range
    iron ore from, *13*:751
Mental disease
    nutrients associated with, *15*:573
Mental illness
    neuroregulators related to, *15*:755
Mental retardation
    nutrients associated with, *15*:573
Mentha arvensis [*68917-18-0*], *16*:309
Mentha arvensis L., *16*:324
p-Menthane hydroperoxide [*80-47-7*], *13*:363
Mentha piperita, *16*:326
Mentha spicata Huds., *16*:328
1-Menthol [*89-78-1*]
    from mint oil, *16*:324
l-Menthone [*14073-97-3*]
    in peppermint oils, *13*:922
Mephenytoin [*50-12-4*]
    properties of, *13*:135
Meprobamate [*57-53-4*]
    properties of, *13*:122
Meracoptobenzimidazole, zinc salt [*3030-80-6*]
    in electrical insulation, *13*:565
Meralluride [*140-20-5*], *15*:169
Merbac 35, *13*:231
Merbromin [*129-16-8*], *15*:169; *16*:588
Mercaptans
    odor-problem chemicals, *16*:303
    removal of from LPG, *14*:389
Mercaptobenzothiazole [*149-30-4*]
    in electrical insulation, *13*:565
2-Mercaptoethylamine [*60-23-1*] (2-MEA)
    antiaging effects, *15*:138
Mercaptomerin [*20223-84-1*], *15*:169
Mercaptosuccinic acid [*70-49-5*], *13*:110.   (See
    also *Thiomalic acid.*)

Mercarbolide [*90-03-9*], *15*:169
Mercerization
    nonwovens, *16*:120
Merchant oxygen, *16*:659
Mercocresol [*8063-33-0*], *15*:169
Mercuration, *15*:163
Mercurials
    as antimicrobial agents, *13*:235
    as slimicides, *16*:808
Mercuric acetate [*1600-27-7*], *15*:169
    benzene reaction, *16*:587
Mercuric amide chloride [*10124-48-8*]
    antiinfective, topical, *15*:155
Mercuric bromide [*7789-47-1*], *15*:169
Mercuric carbonate, basic [*76963-38-7*], *15*:169
Mercuric chloride [*7487-94-7*], *15*:169
    antiseptic, *15*:155
    as insecticide, *13*:422
    lecithin reaction, *14*:256
    tetraethyllead reaction, *16*:583
Mercuric cyanide [*592-04-1*], *15*:169
    antiseptic, *15*:155
Mercuric dithizonate [*14783-59-6*], *15*:153
Mercuric fulminate [*20820-45-5*], *15*:169
Mercuric iodide [*1344-45-2*], *13*:661; *15*:169
    antiseptic, *15*:155
Mercuric nitrate [*10045-94-0*], *15*:169
    radioactive iodine scrubber, *16*:180
Mercuric oxide [*21908-53-2*], *15*:169
    acetic acid reaction, *15*:158
    antiseptic, *15*:155
Mercuric oxycyanide [*1335-31-5*], *15*:169
Mercuric salicylate [*5970-32-1*], *15*:169
Mercuric succinimide [*584-43-0*], *15*:169
Mercuric sulfate [*7783-35-9*], *15*:169
Mercuric sulfate, basic [*7783-35-9*], *15*:169
Mercuric sulfide [*1344-48-5*], *15*:169
Mercurochrom, *13*:60
Mercurophen [*52486-78-9*], *15*:169
Mercurous bromide [*15385-58-7*], *15*:169
Mercurous chloride [*10112-91-1*], *13*:422; *15*:159,
    169.   (See also *Calomel.*)
    antiseptic, *15*:155
    in calomel electrode, *13*:7
Mercurous iodide [*15385-57-6*], *13*:661; *15*:169
Mercurous nitrate [*10415-75-5*], *15*:169
    sodium chloride reaction, *15*:159
Mercurous oxide [*15829-53-5*]
    explosive hydrogen peroxide reaction, *13*:27
Mercurous sulfate [*7783-36-0*], *15*:169
Mercury [*7439-97-6*], *15*:**143**
    in calomel electrode, *13*:7
    exposure limits, *15*:167
    ion-exchange removal, *13*:688
    as lubricant, *14*:515
    nitric acid reaction, *15*:161, 868
    from the ocean, *16*:280

peracetic acid reaction, *15*:158
properties, *15*:144
toxicity, *13*:270
Mercury-196 [*14917-67-0*], *13*:842; *15*:146
Mercury-198 [*13981-21-0*], *13*:842; *15*:146
Mercury-199 [*14191-87-8*], *13*:842; *15*:146
Mercury-200 [*15756-10-2*], *13*:842; *15*:146
Mercury-201 [*15185-19-0*], *13*:842; *15*:146
Mercury-202 [*14191-86-7*], *13*:842; *15*:146
Mercury-204 [*15756-14-6*], *13*:842; *15*:146
Mercury alkyls, *16*:586
in dialkylmagnesium preparation, *16*:560
Mercury bichloride. See *Mercuric chloride.*
Mercury cadmium telluride [*29870-72-2*] (0.77:
0.23:1)
lasers, *14*:61
Mercury–cadmium–telluride detectors, *13*:345
Mercury compounds, *15*:**157**
Mercury dichloride [*7487-94-7*]
tetraethyllead reaction, *16*:587
Mercury fulminate [*628-86-4*], *15*:143
Mercury lamps, *15*:154
Mercury nitride [*12136-15-1*], *15*:872
Mersalyl [*486-67-9*], *15*:169
Merthiolate [*54-64-8*], *15*:169; *16*:588
Mesabi range
iron ore from, *13*:751
Mesaconic acid [*498-24-8*]
properties, *13*:871
Mesantoin, *13*:135
Mesitylene [*108-67-8*]
from acetone, *13*:900
Mesityl oxide [*141-79-7*], *13*:908
from itaconic acid, *13*:870
mfg, *13*:916
properties, *13*:898
Mesogenic molecules, *14*:396
Mesomorphism, *14*:395
Mesoridazine besylate [*5588-33-0*]
properties of, *13*:122
Metabolic function
mineral nutrients, *15*:574
Metabolism
of benzodiazepines, *13*:126
dinitrophenols' effect on, *13*:429
Metabolism, human
lactic acid function, *13*:80
Metacinnabar [*2333-45-1*], *15*:144
Metacresol purple [*2303-01-7*], *13*:10
Metal and Nonmetallic Mine Safety Act, *13*:254
Metal anodes, *15*:**172**
Metalation, *16*:556
Metal cleaning
oxalic acid for, *16*:631
Metal coloring, *15*:310
Metal π complexes, *16*:592. (See also
Organometallics.)

Metal composites
gasket properties, *16*:727
Metal-container inks, *13*:382
Metal-containing polymers, *15*:**184**
Metaldehyde [*108-62-3*]
as insect attractant, *13*:481
in the repeat-ignition match, *15*:4
Metal deposition
in microencapsulation, *15*:474
Metal fibers, *15*:**220**
properties, *15*:220
Metal films
as lubricants, *14*:511
Metal gaskets, *16*:731
Metal ions
in zeolites, *15*:651
Metalkamate [*8065-36-9*], *13*:455
Metallic coatings, *15*:**241**
explosively clad metals, *15*:**275**
Metallic fiber
in metal laminates, *13*:945
Metallic-matrix laminates, *13*:941
Metallizing. See *Metallic coatings.*
Metallocene polymers, *15*:184
Metallocenes, *16*:592
synthesis, *16*:611
Metallography
magnesium alloys, *14*:599
Metallo-organic CVD
in LED manufacture, *14*:287
Metallurgy
limestone uses, *14*:378
powder, of superalloys, *15*:343
Metal matrices, *13*:941
Metal matrix laminates
design–analysis procedures, *13*:950
Metal–metal exchange, *16*:556
Metal-nitrogen complexes
in N_2 fixation, *15*:955
Metal oxide semiconductor (MOS), *14*:686
in ICS, *13*:621
Metals
in antimicrobial agents, *13*:235
cryoforming, *15*:940
explosive cladding, *15*:275
fiber reinforced, *13*:964
laminated and reinforced, *13*:941
in marine organisms, *16*:284
Metals, liquid
as lubricants, *14*:515
Metalsol TK-100, *13*:245
Metal sorting, *16*:58
Metals separation
MIBK for, *13*:912
Metal surface treatment
with oxalic acid, *16*:630

Metal surface treatments
  case hardening, *15*:**313**
  chemical and electrochemical conversion
    treatments, *15*:**304**
  cleaning, pickling, and related processes, *15*:**296**
  lubrication effects, *14*:509
METALS WEEK information system, *13*:327, 330
Metal treatments, *15*:**325**
  hydroxyacetic acid in, *13*:93
Metalworking
  lubrication for, *14*:513
Metamagnetism, *14*:647
Metamerism, *16*:757
Metaperiodic acid [*13444-71-8*], *13*:666
Metarsenic acid [*13768-07-5*]
  as insecticide, *13*:420
Metasol D3T, *13*:244
Metasol J-26, *13*:242
Metastatic carcinoma
  detection, *15*:83
Metco 15F
  flame spraying, *15*:256
Meteorites
  iron in, *13*:735
  particle tracks in, *16*:842
Meteoritic materials
  particle tracks in, *16*:826
Methacrolein [*75-85-3*]
  methacrylates from, *15*:357
  methacrylic acid from, *15*:361
  oxidation, *15*:365
  properties, *15*:349
Methacrylamide [*79-39-0*]
  properties, *15*:349
Methacrylamide sulfate [*3351-73-3*] (1:1), *15*:358
Methacrylamide sulfate [*29194-31-8*] (1:*n*), *15*:
    346, 893
Methacrylates
  in ion-exchange manufacture, *13*:686
  properties, *15*:346
Methacrylic acetic anhydride [*76392-13-7*]
  properties, *15*:349
Methacrylic acid [*79-41-4*]
  in cation-exchange resins, *13*:686
  copolymer with ethylene, *16*:420
  in emulsion polymerization, *14*:87
  properties, *15*:347
Methacrylic acid and derivatives, *15*:**346**
Methacrylic acid, 2,3-epoxypropyl ester [*106-91-
    2*]
  copolymers from, *15*:205
Methacrylic acid methyl ester, polymer with
    (methacryloyloxy)tripropylstannane [*30444-
    62-3*], *15*:205
Methacrylic anhydride [*760-93-0*]
  properties, *15*:349

Methacrylic polymers, *15*:**377**
Methacrylonitrile [*126-98-7*]
  from isobutylene, *15*:363
  methacrylates from, *15*:357
  properties, *15*:349
Methacryloxytributylstannane [*2155-70-6*]
  copolymers from, *15*:205
Methacryloxytripropylstannane [*4154-35-2*]
  MMA copolymer, *15*:205
Methacryloylamidoacetonitrile [*65993-30-8*]
  properties, *15*:356
4-Methacryloylamido-4-methyl-2-pentanol
    [*23878-87-1*]
  properties, *15*:356
1-Methacryloylamido-2-methyl-2-propanol
    [*74987-95-4*]
  properties, *15*:356
Methacryloyl bromide [*6997-65-5*]
  properties, *15*:349
Methacryloyl chloride [*920-46-7*]
  properties, *15*:349
[2-[Methacryloyl]ethyl]ferrocene [*34802-12-5*],
    15:187, 189
N-(2-Methacryloylethyl)-2-pyrrolidinone [*946-
    25-8*]
  properties, *15*:354
N-Methacryloylmaleamic acid [*76392-01-3*]
  properties, *15*:356
1-[[Methacryloyl]methyl]ferrocene, polymer with
    styrene [*3262-77-4*], *15*:189
N-Methacryloylmorpholine methacrylate [*5117-
    13-5*]
  properties, *15*:354
N-(Methacryloyloxyethyl)diisobutylketimine
  properties, *15*:354
2-Methacryloyloxyethylmethylcyanamide
    [*76392-28-4*]
  properties, *15*:354
(2-Methacryloyloxyethyl)trimethylammonium
    salts
  in papermaking, *16*:804
N-(Methacryloyloxy)formamide [*76392-27-3*]
  properties, *15*:354
N-(3-Methacryloyloxypropyl)-2-pyrrolidinone
    [*76747-97-4*]
  properties, *15*:354
N-Methacryloylpyrrolidinone [*23935-37-7*]
  properties, *15*:354
1-Methacryloylurea [*20602-83-9*]
  properties, *15*:356
[[Methacryoyl]methyl]ferrocene [*31566-61-7*],
    15:187
Methacryoyloxyethyltrimethylammonium
    chloride [*5039-78-1*]
  properties, *15*:354
Methadone hydrochloride [*1095-90-5*]
  measurement, *15*:88

β-Methallyl chloride [563-47-3]
  as fumigant, 13:467
Methamidophos [10265-92-6], 13:443
Methanation
  catalyst coating, 15:248
  in methanol mfg, 15:408
  nickel catalysts for, 15:813
  zeolite catalysts in, 15:652
Methane [74-82-8]
  in air, 16:657
  hydrogen from, 15:945
  from lignite, 14:338
  as lubricant, 14:514
  nitration, 15:848
  permeability in polymers, 15:118
  steam reforming, 15:401
  from waste incineration, 13:194
Methanedicarboxylic acid [141-82-2], 14:
  794.   (See also *Malonic acid.*)
Methane hydroxamic acid [546-88-3]
  from nitroethane, 15:975
Methanesulfonyl chloride [124-63-0]
  nitro alcohol reaction, 15:913
Methanethiol [74-93-1]
  from black liquor, 14:311
Methanol [67-56-1], 15:**398**
  in CMME formation, 13:695
  in emulsion polymerization, 14:83
  from fossil fuels, 16:656
  iodination, 13:668
  in LPG purification, 14:389
  oxidation with iron molybdate catalyst, 15:694
  in paint stripping, 16:762
  properties, 15:398
  from waste incineration, 13:194
Methanolferrocene [1273-86-5], 15:196
Methanolruthenocene [33270-45-0], 15:196
Methaqualone [72-21-4]
  properties of, 13:122
Methazonic acid.   See *2-Nitroacetaldehyde
  oxime.*
Methemoglobin [9008-37-1]
  from nitrobenzene poisoning, 15:922
  from nitrochlorobenzenes, 15:924
Methemoglobinemia
  from nitroparaffins, 15:981
Methiocarb [2032-65-7], 13:455
Methionine [63-68-3]
  from homocysteine, 15:597
  microbial synthesis of, 15:463
  in nuts, 16:249, 252
Methionine–enkephalin [58569-55-4]
  properties, 15:765
Methomyl [16752-77-5], 13:454
Methone [126-81-8], 13:917.   (See also
  Dimedone.)

Methoprene [4059-69-8], 13:459
Methotrexate [59-05-2]
  in neoplasm therapy, 13:172
1-Methotrimetrazine maleate [60-99-1]
  properties of, 13:122
p-Methoxybenzylidene-p'-n-butylaniline
  [26227-73-6]
  liquid crystalline range of, 14:399
Methoxychlor [72-43-5], 13:432
Methoxyethoxyethyl methacrylate [45103-58-0]
  properties, 15:353
5-Methoxyindole [1006-94-6]
  in melatonin synthesis, 15:772
p-Methoxylphenyl diethyl phosphate [5076-68-6],
  13:451
Methoxymethoxyethyl methacrylate [76392-21-7]
  properties, 15:353
Methoxymethyl isocyanate [6427-21-0], 13:799
Methoxymethylmelamine resin
  in encapsulation, 15:480
N-(Methoxymethyl)methacrylamide [3644-12-0]
  properties, 15:356
Methoxymethyl methacrylate [20363-82-0]
  properties, 15:353
4-Methoxy-4-methyl-2-pentanone [107-70-0], 13:
  916
3-(4-Methoxy-1-naphthoyl)propionic acid [3562-
  99-0], 15:729
2-Methoxyphenol [90-05-1], 13:44.   (See also
  Guaiacol.)
4-Methoxyphenol [150-76-5], 13:43.   (See also
  *Hydroquinone monomethyl ether, p-
  hydroxyanisole.*)
p-Methoxyphenol [150-76-5]
  determination, 15:368
2-Methoxyphenothiazine [1771-18-2]
  in methotrimetrazine synthesis, 13:134
N-(2-Methoxyphenyl)acetoacetamide [92-15-9],
  13:887
m-Methoxyphenyl N-methylcarbamate [3938-
  28-1], 13:452
o-Methoxyphenyl N-methylcarbamate [3938-24-
  7], 13:452
1-Methoxy-2-propanol [107-98-2]
  in inks, 13:390
6-Methoxy-1-tetralone [1078-19-9]
  in cortisone synthesis, 15:761
Methsuximide [77-41-8]
  properties of, 13:135
2-Methyacryloyloxyethyl diethyl phosphite [817-
  44-7]
  properties, 15:355
N-Methylacetoacetamide [20306-75-6], 13:890
Methyl acetoacetate [105-45-3], 13:887
Methyl α-acetoxypropionate [6284-75-9], 13:89
Methylacetylene [74-99-7]
  carbonylation, 15:366

Methylacrylate [*96-33-3*]
  aziridine reaction, *13*:151
  copolymer, *15*:192
  copolymer with ethylene, *16*:420
  isoprene reaction, *13*:821
  solubility of, *14*:83
Methylal [*109-87-5*]
  ketene reaction, *13*:876
  in methacrylate mfg, *15*:364
Methyl alcohol [*67-56-1*], *15*:398.   (See also
    Methanol.)
  toxicity, *13*:270
  use in lasers, *14*:56
Methylaluminum sesquichloride [*12542-85-7*]
  properties, *16*:565
Methylamine [*74-89-5*]
  in hydroquinone reaction, *13*:45
  in microbial ephedrine synthesis, *15*:464
  in phensuximide synthesis, *13*:137
  phosgene reaction, *13*:806
4-Methyl-2-amino-5-cyanopyrimidine [*17321-97-
    0*], *14*:806
(2-Methylaminoethyl)ferrocene [*71615-85-5*], *16*:
    606
p-Methylaminophenol [*150-75-4*], *13*:45, 57
  ammonium diphosphate reduction, *15*:84
Methyl *n*-amyl ketone [*110-43-0*]
  properties, *13*:896
Methylaniline [*100-61-8*], *13*:798
Methyl anthranilate [*134-20-3*]
  in neroli oil, *16*:325
Methylation
  by methylcobalamin, *16*:585
2-Methylaziridine [*75-55-8*]
  properties, *13*:143
Methyl (1-aziridinyl) acetate [*14745-52-9*], *13*:148
α-Methyl-1-aziridinylmethanol [*13148-34-0*], *13*:
    150
Methyl 3-(1-aziridinyl) propanoate [*1073-77-4*],
    13:151
Methyl benzoate [*93-58-3*]
  from cananga oil, *16*:322
α-Methylbenzylamine [*98-84-0*], *13*:901
  in lactic acid resolution, *13*:81
Methyl bromide [*74-83-9*]
  as fumigant, *13*:467
  fumigation of nuts, *16*:263
  lithium reaction, *14*:470
Methyl 3-bromo-4-oxo-4-[4-(pentyloxy)phenyl-
    2-butenoate [*63913-15-5*]
  sodium diethyl malonate reaction, *14*:799
2-Methyl-1,3-butadiene. See *Isoprene.*
2-Methylbutanol [*137-32-6*]
  oxo product, *16*:650
2-Methyl-1-butene [*563-46-2*]
  in isoprene synthesis, *13*:830

2-Methyl-2-butene [*513-35-9*]
  diborane reaction, *16*:564
  in isoprene synthesis, *13*:830
2-Methyl-2-butenenitrile [*4403-61-6*], *15*:889
2-Methyl-3-butenenitrile [*16529-56-9*], *15*:889,
    899
2-(Methyl-3-butenenitrile)tris[tris(2-methylphe-
    nyl)phosphite]nickel [*41686-95-7*], *15*:808
Methyl-2-butenoate [*623-43-8*]
  methyl methacrylate from, *15*:365
Methylbutenol [*115-18-4*]
  in isoprene synthesis, *13*:832
2-Methyl-3-buten-2-ol [*115-18-4*], *13*:888
3-Methyl-3-buten-2-one [*814-78-8*], *13*:904
2-(3-Methyl-2-butenyl)thiophene [*52008-16-9*]
  isoprene reaction product, *13*:826
2-Methylbutyl acetate [*624-41-9*], *16*:313
O-Methyl *O*-(4-*tert*-butyl-2-chlorophenyl)
    methylphosphoramidate [*299-86-5*]
  as insecticide, *13*:448
Methyl *tert*-butyl ether [*1634-04-4*]
  production, *15*:398
Methyl *n*-butyl ketone [*591-78-6*]
  properties, *13*:896
(−)-2-Methylbutyl *p*-(*p'*-
    methoxybenzylideneamino)cinnamate
    [*24140-30-5*]
  liquid crystalline range of, *14*:399
m-(1-Methylbutyl)-phenyl *N*-methylcarbamate
    [*2282-34-0*]
  as insecticide, *13*:455
Methylbutynol [*115-19-5*]
  in isoprene synthesis, *13*:832
Methyl butyrate [*623-42-7*]
  methyl isobutyrate from, *15*:365
Methyl 2-carbomethoxy-4-ethoxy-5-oxo
    hexanoate [*65915-74-4*], *14*:799
Methyl cellulose [*9004-67-5*]
  in encapsulation, *15*:480
  precipitation of, *14*:88
Methyl chavicol [*140-67-0*], *16*:313
Methyl chloride [*74-87-3*]
  aluminum reaction, *16*:570
  2-dimethylaminoethyl methacrylate reaction,
    15:356
  lithium reaction, *14*:470
  in organolead synthesis, *14*:184
  sodium–lead reaction, *16*:580
  tin reaction, *16*:576
Methyl 4-chloro-1,3-dimethyl-3-cyclohexenyl
    carboxylate [*76392-10-4*],
    15:349
Methylchloroform. See *1,1,1-Trichloroethane.*
Methyl chloromethyl ether [*107-30-2*]
  carcinogen, *15*:703
Methylcobalamin [*13422-55-4*]
  methylating agent, *16*:585

Methyl cyanoacetate [*105-34-0*], *15*:889
  properties, *15*:907
π-Methylcycloheptatrienylium-π-cyclopentadien-
  ylchromium [*77430-31-0*], *16*:608
1-Methylcyclohexyl methacrylate [*76392-14-8*]
  properties, *15*:353
Methylcyclopentadienyl manganese tricarbonyl
  [*12108-13-3*], *14*:845, 878; *16*:558
Methylcyclopentadienyl sodium [*1270-16-2*], *14*:
  878
Methyl 2,3-dibromo-2-methyl propanoate
  [*22262-60-8*], *15*:349
Methyldichlorosilane [*75-45-7*]
  isoprene reaction, *13*:826
Methyldiethoxymethacryloyloxyethoxysilane
  [*76392-37-5*]
  properties, *15*:355
4-Methyl-7-diethylaminocoumarin [*91-44-1*]
  photolysis, *14*:557
Methyl 2-diethylaminomethacrylate [*27316-07-0*]
  phenyl azide reaction, *15*:349
Methyl 5-diethylamino-4-methyl-1-phenyl-1,2,3-
  triazoline-4-carboxylate, *15*:352
Methyl dihydrojasmonate [*24851-98-7*]
  in perfumes, *16*:953
Methyl 2-(dimethylamino)-*N*-hydroxy-2-
  oxoethanimidothioate [*30558-43-1*]
  methyl isocyanate reaction, *13*:890
S-Methyl 1-(dimethylcarbamoyl)-*N*-
  [(methylcarbamoyl)oxy]-thioformimidate
  [*23135-22-0*]
  as insecticide, *13*:454
Methyl 1,3-dimethyl-4-cyclohexenyl carboxylate
  [*76392-11-5*], *15*:349
Methyldinitrobenzene [*25321-14-6*]
  in TDI manufacture, *13*:808
Methyldiphenylphosphine [*1486-28-8*]
  in N$_2$-containing complex, *15*:956
Methyldopa [*555-30-6*], *13*:63
(1-Methyleneallyl)ferrocene [*12504-78-8*], *15*:186
Methylene aminoacetonitrile [*109-82-0*]
  in glycine synthesis, *15*:768
N-Methyleneaniline [*100-62-9*]
  phenyl isocyanate reaction, *13*:796
4,4′-Methylene-bis-2-chloroaniline (MOCA)
  [*101-14-4*]
  in insulation, *13*:548
  toxicity, *13*:270
N,*N*″-Methylene bis[*N*′-(1-hydroxymethyl)-2,5-
  dioxo-4-imidazolinyl]-urea [*39236-46-9*]
  preservative, cosmetic, *13*:243
N,*N*′-Methylenebismethacrylamide
  [*2359-15-1*]
  properties, *15*:356

Methylenebis(4-methyl-2,1-phenylene)
  methacrylate [*18507-45-4*], *15*:350
10,11-[Methylenebis(oxy)]aporphine [*34174-86-2*]
  microbial demethylation, *15*:465
4,4′-Methylenebis(phenyl isocyanate) [*101-68-8*],
  13:808; *15*:922. (See also *4,4′-
  Diphenylmethane diisocyanate.*)
Methylene bisthiocyanate [*6317-18-6*]
  applications of, *13*:243
Methylene blue [*61-73-4*]
  encapsulation, *15*:478
  as photosensitizer, *13*:370
Methylene blue test
  for milk, *15*:550
Methylene chloride [*75-09-2*]
  catechol reaction, *13*:43
  metal cleaning, *15*:298
  in nitration, *15*:847
  paint and varnish stripping, *16*:762
  potassium iodide reaction, *13*:670
  toxicity, *13*:270; *16*:767
4,4′-Methylenedianiline [*101-77-9*]
  metallic copolymers from, *15*:205
Methylenedioxybenzene [*274-09-9*], *13*:43
1,2-Methylenedioxy-4-[2-(octylsulfinyl)propyl]-
  benzene [*120-62-7*]
  as insecticide, *13*:425
2-(3,4-Methylenedioxyphenoxy)-3,6,9-trioxoun-
  decane [*51-14-9*]
  as insecticide, *13*:425
3,4-Methylenedioxy-6-propylbenzyl
  butyldiethylene glycol ether [*51-03-6*]
  as insecticide, *13*:425
4,4′-Methylenediphenyl diisocyanate [*101-68-8*]
  from formaldehyde, *15*:412
Methylene iodide [*75-11-6*]
  properties, *13*:669
Methylene methacrylate [*4245-38-9*]
  properties, *15*:354
4-Methylene-2-oxetanone [*674-82-8*], *13*:883
  from ketene, *13*:878
Methylene succinic acid, *13*:865. (See also
  Itaconic acid.)
Methylene tetrahydrofolic acid [*18137-88-7*]
  in glycine biosynthesis, *15*:769
(1-Methylethyl)-diethyl ester butanedioic acid
  [*59322-58-6*], *14*:777
Methyl ethyl ketone [*78-93-3*]
  in paint and varnish removers, *16*:764
  properties, *13*:898, 903
Methyl ethyl ketone peroxide [*1338-23-4*], *13*:29,
  364, 905
Methyleugenol [*93-15-2*], *16*:312, 313
  economic aspects, *13*:53
  as insect attractant, *13*:481

N-Methylformamidoethyl methacrylate [*25264-39-5*]
  properties, *15*:354
N-Methylformanilide [*93-61-8*]
  ferrocene formylation, *16*:606
Methyl formate [*107-31-3*]
  as fumigant, *13*:467
  methyl methacrylate by-product, *15*:359
Methylfumaric acid [*498-24-8*], *13*:865, 866.   (See also *Mesaconic acid.*)
N-Methyl glucamine [*6284-40-8*]
  in ion-exchange resins, *13*:688
6-*O*-Methyl-D-glucose [*2461-70-3*], *16*:41
2-Methylglutaronitrile [*4553-62-2*], *15*:889
3-Methyl-2,4-heptanedione [*13152-54-0*]
  in perfume formulations, *13*:927
5-Methyl-3,4-heptanedione [*13678-56-3*]
  properties, *13*:926
6-Methyl-2,4-heptanedione [*3002-23-1*]
  properties, *13*:928
6-Methyl-2,5-heptanedione [*13901-85-4*]
  properties, *13*:931
6-Methyl-3,4-heptanedione [*3131-90-6*]
  properties, *13*:926
4-Methyl-3-heptanol [*14979-39-6*]
  as insect attractant, *13*:482
Methylheptenone [*409-02-9*], *16*:312
6-Methyl-5-hepten-2-one [*110-93-0*], *13*:888
2-(6-Methylheptyl)-4,6-dinitrophenyl crotonate [*37300-45-3*]
  as insecticide, *13*:429
Methyl heptyl ketone [*821-55-6*]
  properties, *13*:896
2-Methylhexanal [*925-54-2*]
  in perfumes, *16*:967
5-Methyl-2,3-hexanedione [*13706-89-3*]
  properties, *13*:926
5-Methyl-2,4-hexanedione [*7307-03-1*]
  properties, *13*:928
Methyl hexyl ketone [*111-13-7*]
  properties, *13*:897
Methyl hydrogen sulfate [*75-92-3*], *15*:399
Methyl *p*-hydroxybenzoate [*99-76-3*]
  as antimicrobial agents, *13*:226
1-Methyl-2-hydroxyethyl methacrylate [*4664-49-7*], *15*:367
Methyl *α*-hydroxyisobutyrate [*2110-78-3*]
  methyl methacrylate by-product, *15*:359
2-Methyl-4-hydroxyquinoline [*607-67-0*], *13*:890
4-Methyl-2-hydroxyquinoline [*607-66-9*], *13*:890
(2-Methylimidazole)tetraphenylporphinatoiron ethanol solvate [*41754-42-1*], *13*:782
2-Methylindole [*95-20-5*], *13*:218
3-Methylindole [*83-34-1*], *13*:213, 215.   (See also *Skatole.*)
1-Methylindole-3-acetic acid [*1912-48-7*], *13*:219

2-Methylindole-3-acetic acid [*1912-43-7*], *13*:219
5-Methylindole-3-acetic acid [*1912-47-6*], *13*:220
Methyl iodide [*74-88-4*]
  hydrogen iodide reaction, *13*:663
  lithium reaction, *13*:669
  magnesium reaction, *13*:668
  mercury reaction, *16*:587
  properties, *13*:668
o-Methyliodobenzene [*615-37-2*], *13*:671
o-Methyliodosobenzene [*64297-59-2*], *13*:671
o-Methyliodoxybenzene [*16825-70-0*], *13*:671
α-6-Methylionone, *16*:956
β-6-Methylionone, *16*:956
γ-6-Methylionone, *16*:956
Methyl isoamyl ketone [*110-12-3*]
  properties, *13*:896
Methyl isobutyl carbinol [*108-11-2*], *13*:895, 916
Methyl isobutyl ketone [*108-10-1*], *13*:364
  manufacture, *13*:908
  in organolead analysis, *14*:192
  properties, *13*:898
  in uranium recovery, *16*:173
Methyl isobutyl ketone [*108-10-1*] (MIBK)
  in niobium processing, *15*:825
Methylisobutylmagnesium [*17520-45-5*], *16*:559
Methyl isobutyrate [*547-63-7*]
  dehydrogenation, *15*:365
  methacrylates from, *15*:357
Methyl isocyanate [*624-83-9*]
  CO$_2$ reaction, *13*:798
  production, uses, *13*:806
  properties, *13*:789
N-Methylisoindole [*33804-84-1*], *13*:213
Methyl isopropenyl ketone [*814-78-8*], *13*:904, 933
Methylisopropylglyoxal [*7493-58-5*], *13*:908
Methyl isopropyl ketone [*563-80-4*]
  properties, *13*:896
3-Methyl-5-isopropylphenyl *N*-methylcarbamate [*2631-37-0*]
  as insecticide, *13*:455
2-Methyl-4-isothiazoline-3-one [*26172-54-3*]
  in antimicrobial agent, *13*:247
Methyl isothiocyanate [*556-61-6*]
  antimicrobial activity of, *13*:246
Methylisothiourea picrate [*2780-75-8*], *13*:669
Methyl jasmonate [*1211-29-6*]
  in perfumes, *16*:953
(*Z*)-Methyl jasmonate [*1211-29-6*], *14*:799
Methylketene [*6004-44-0*]
  properties, *13*:875
Methyl 2-keto-L-gulonate [*3031-98-9*]
  microbial oxidation, *15*:459
Methyllactamide [*51676-15-4*], *13*:90

Methyl lactate [*547-64-8*], *13*:87, 89
  acetylation, *13*:89
  hydrolysis, *13*:85
  in lactic ester preparation, *13*:83
2-Methyllactonitrile.   See *Acetone cyanohydrin.*
Methyllithium [*917-54-4*], *13*:669
  properties, *14*:470
Methylmagnesium chloride [*676-58-4*]
  in organolead synthesis, *14*:185
Methylmagnesium iodide [*917-64-6*], *13*:668
Methylmaleic acid [*498-23-7*], *13*:865, 866.   (See
  also *Citraconic acid.*)
Methylmaleic anhydride [*616-02-4*]
  radical addition, *14*:777
Methylmalonic acid [*516-05-2*]
  determination of by ion exchange, *13*:704
Methylmalonyl CoA [*1264-44-5*], *15*:597
Methyl mercaptan [*74-93-1*]
  from pulp mills, *16*:795
Methyl mercuric iodide [*143-36-2*], *13*:668; *16*:587
Methylmercury [*22967-92-6*], *15*:169
  inducement of immunodeficiency, *13*:170
N-Methylmethacrylamide [*3887-02-3*]
  properties, *15*:356
Methyl methacrylate [*80-62-6*], *13*:821; *15*:888
  copolymer, *15*:192
  in encapsulation, *15*:480
  from methanol, *15*:412
  properties, *15*:347
  solubility of, *14*:83
Methyl methacrylate–acrylonitrile–butadiene–
  styrene copolymer [*9010-94-0*] (MABS), *15*:
  390
Methyl methacrylate–butadiene–styrene
  copolymer [*25053-09-2*] (MBS), *15*:390
Methyl β-methoxyisobutyrate [*3852-11-7*]
  methyl methacrylate by-product, *15*:360
Methyl 3-methoxypropionate [*3852-09-3*], *13*:876
Methyl[(methylamino)carbonyl]carbamic
  chloride [*13188-08-4*]
  methyl mercaptan reaction, *13*:889
S-Methyl N-(methylcarbamoyloxy)-
  thioacetimidate [*16752-77-5*]
  as insecticide, *13*:454
Methyl 1-methyl-3-cyclohexenyl carboxylate
  [*6493-80-7*], *15*:349
Methyl 1-methyl-3,4-divinylcyclohexyl
  carboxylate [*56629-46-0*], *15*:352
4-Methyl-2-methylenebutanedioate [*7338-27-4*]
  properties, *13*:871
1-Methyl-3-(1-methylethenyl)cyclohexene
  [*38738-60-2*]
  isoprene dimer, *13*:822
1-Methyl-4-(1-methylethenyl)cyclohexene [*138-
  86-3*]
  isoprene dimer, *13*:822

3-Methyl-6-(1-methylethyl)-1,2-cyclohexanedione
  [*34315-76-9*], *13*:925
(*E*)-Methyl 2-methyl-2-pentenoate [*1567-14-2*],
  15:352
(*Z*)-Methyl 2-methyl-2-pentenoate [*1567-13-1*],
  15:352
2-Methyl-2-(methylthio)-propionaldehyde *O*-
  methylcarbamoyl oxime [*116-06-3*]
  as insecticide, *13*:454
1-Methylnaphthalene [*90-12-0*], *15*:716
2-Methylnaphthalene [*91-57-6*], *15*:716
2-Methyl-1,4-naphthoquinone [*58-27-5*], *15*:716
N-Methyl-N-1-naphthylfluoroacetamide [*5903-
  13-9*], *15*:727
Methylnickelocene [*1292-95-4*], *15*:807
Methyl nitrate [*598-58-3*], *15*:399
Methyl nitrite [*624-91-9*], *15*:399, 970
2-Methyl-5-nitrobenzenesulfonic acid [*32784-87-
  5*], *15*:928
1-Methyl-1-nitro 2,2-bis(*p*-chlorophenyl)ethane
  [*117-27-1*]
  as insecticide, *13*:430
5-Methyl-5-nitro-1,3-dioxane [*1194-36-1*], *15*:912
Methyl 2-nitromethylacrylate [*51914-94-4*], *15*:
  352
2-Methyl-2-nitro-1-(*p*-nitrophenyl)propane
  [*5440-67-5*]
  from sodium 2-propanenitronate, *15*:974
5-Methyl-5-nitro-2-oxo-1,3,2-dioxathiolane
  [*76530-16-0*], *15*:913
2-Methyl-2-nitro-1,3-propanediol [*77-49-6*]
  properties, *15*:910
2-Methyl-2-nitro-1-propanol [*76-39-1*]
  properties, *15*:910
2-Methyl-2-nitropropyl chloromethyl ether
  [*57039-05-1*], *15*:912
Methyl 2-nitrosomethylacrylate [*51915-02-7*], *15*:
  352
4-Methyl-4-nitrovaleronitrile [*16507-00-9*], *15*:977
2-Methylnonanal [*24424-67-7*]
  in perfumes, *16*:967
Methyl 2-nonynoate [*111-80-8*]
  in perfumes, *16*:967
2-Methyloctanal [*7786-29-0*]
  in perfumes, *16*:967
n-Methylolacrylamide [*924-42-5*]
  in bonding nonwovens, *16*:115
  in emulsion polymerization, *14*:87
4-Methyl-2-one-pentene-3 [*141-79-7*]
  ethanol reaction, *13*:900
Methyl orange [*547-58-0*], *13*:10
6-Methyl-1,2,3-oxathiazin-4(3*H*)-one, 2,2-dioxide
  [*33665-90-6*], *13*:884
4-Methyl 1,2-oxathiol-4-ene-2,2-dioxide [*64931-
  17-5*]

4-Methyl 1,2-oxathiol-4-ene-2,2-dioxide
  (*Continued*)
  isoprene reaction product, *13*:822
4-Methyl-5-oxazolecarboxamide [*4866-00-6*], *13*: 889
5-Methyl-oxazolidine-2,4-dione [*27770-23-6*]
  from alkyl lactates, *13*:84
N-Methyl-3-oxobutanamide [*20306-75-6*], *13*:887
Methylparaben [*99-76-3*]
  in insulin suspension, *13*:612
Methyl parathion [*298-00-0*], *13*:439
  encapsulation, *15*:480, 488
  toxicity, *13*:270
4-Methyl-2,3-pentadione [*7493-58-5*]
  properties, *13*:926
2-Methylpentaldehyde [*123-15-9*]
  in meprobamate synthesis, *13*:126
2-Methylpentanol [*589-35-5*]
  from oxo process, *16*:640
4-Methyl-2-pentanol [*108-11-2*], *13*:895
4-Methyl-2-pentanone [*108-10-1*], *16*:181.   (See
  also *Methyl isobutyl ketone*.)
2-Methyl-1-pentene [*763-29-1*]
  hydroformylation reactivity, *16*:645
  in isoprene synthesis, *13*:830
2-Methyl-2-pentene [*625-27-4*]
  hydroformylation reactivity, *16*:645
  in isoprene synthesis, *13*:830
4-Methyl-1-pentene [*763-29-1*]
  copolymer with ethylene, *16*:388
  hydroformylation reactivity, *16*:645
  oxidation, *13*:911
  polymerization, *16*:479
  properties, *16*:478
4-Methyl-2-pentene [*4461-48-7*]
  hydroformylation reactivity, *16*:645
4-Methyl-4-penten-2-one [*141-79-7*], *13*:
  916.   (See also *Mesityl oxide*.)
N-(2-Methylphenyl)acetoacetamide [*93-68-5*],
  13:887
p-Methylphenyl diethyl phosphate [*4877-08-1*],
  13:451
3-Methyl-1,2-phenylenediamine [*2687-25-4*]
  in isocyanate manufacture, *13*:805
3-Methyl-*o*-phenylene diisocyanate [*7373-19-5*],
  13:805
Methyl phenyl ether [*100-66-3*]
  iodine reaction, *13*:668
1-Methyl-6-phenyl-2,5-hexadiene-4-one [*55901-
  61-6*], *13*:917
N-Methyl-*N*-phenylmethacrylamide [*2918-73-2*]
  properties, *15*:356
m-Methylphenyl *N*-methylcarbamate [*1129-41-
  5*], *13*:452
o-Methylphenyl *N*-methylcarbamate [*1128-78-5*],
  13:452
3-Methyl-1-phenyl-5-pyrazolone [*41927-50-8*]
  from diketene, *13*:884

Methyl phenyl silicone oil [*9005-12-3*]
  in grease, *14*:504
  limit of radiation tolerance, *14*:519
2-(2-Methyl-1-propenyl)-2,4,6-trimethyl-4,5-di-
  hydropyran [*3682-91-5*], *13*:913
Methyl propionate [*554-12-1*]
  methacrylates from, *15*:357
  in methyl methacrylate mfg, *15*:364
Methylpropyldichlorosilane [*4518-94-9*]
  copolymer with dimethyl dichlorosilane, *13*:409
Methyl propyl ketone [*107-87-9*]
  properties, *13*:897
2-Methyl-2-propyl-1,3-propanediol [*78-26-2*]
  in meprobamate synthesis, *13*:126
L-1-Methyl-2-(3′-pyridyl)pyrrolidine [*54-11-5*]
  as insecticide, *13*:424
N-Methyl-2-pyrrolidinone [*872-50-6*]
  in paint and varnish removers, *16*:764
6-Methyl-2,3-quinoxalinedithiol cyclic carbonate
  [*2439-01-2*]
  as insecticide, *13*:463
Methyl red [*493-52-7*], *13*:10
5-Methylresorcinol [*504-15-4*], *13*:39.   (See also
  Orcinol.)
Methyl salicylate [*119-36-8*]
  from birch bark oil, *16*:321
Methyl sulfide [*75-18-3*]
  odor-problem chemical, *16*:303
Methylsulfinylmethyl methacrylate [*76392-30-8*]
  properties, *15*:355
2-(Methylsulfinyl)phenothiazine [*27612-10-8*]
  in mesoridazine synthesis, *13*:133
p-Methylsulfinylphenyl diethyl phosphate
  [*6552-21-2*], *13*:451
p-Methylsulfonylphenyl diethyl phosphate
  [*6132-17-8*], *13*:451
Methyl thiolmethacrylate [*52496-39-6*]
  properties, *15*:355
p-Methylthiophenyl diethyl phosphate [*3070-13-
  1*], *13*:451
m-Methylthiophenyl *N*-methylcarbamate [*3938-
  33-8*], *13*:452
o-Methylthiophenyl *N*-methylcarbamate [*3938-
  32-7*], *13*:452
4-Methylthio-3,5-xylyl *N*-methylcarbamate
  [*2032-65-7*]
  as insecticide, *13*:455
Methyl *p*-toluenesulfonate [*80-48-8*]
  potassium iodide reaction, *13*:668
Methyl *p*-tolylsulfonylcarbamate
  tolazamide preparation, *13*:617
Methyl transferase [*9033-25-4*], *15*:597
N-Methyltransferase [*37205-55-3*]
  in (−) adrenaline biosynthesis, *15*:758
  in DMT biosynthesis, *15*:763

Methyltricaprylammonium chloride [*5137-55-3*]
  as ion exchanger, *13*:725
2,2'-(1-Methyltrimethylenedioxy)-bis(4-methyl-
    1,3,2-dioxaborinane) [*2665-13-6*]
  as fuel preservative, *13*:247
β-Methylumbelliferone [*62252-23-7*], *13*:46
6-Methyluracil [*626-48-2*], *13*:884
Methyl vinyl ether [*107-25-5*]
  maleic anhydride copolymers, *14*:780
Methyl vinyl ketone [*78-94-4*], *13*:933
  in indole Michael reaction, *13*:216
  properties, *13*:897
1-Methyl-(2-vinyloxy)ethyl methacrylate [*76392-
    20-6*]
  properties, *15*:353
Methyl violet [*8004-87-3*], *15*:694
  in inks, *13*:379
N-Methyl, N'-2,4-xylyl-N-(N'-2,4-
    xylylformimidoyl)formamidine [*33089-61-1*]
  as insecticide, *13*:463
2-Methyoxyethylamine-methoxytetrafluoroboric
    acid complex [*75431-52-6*], *13*:151
Methyprylon [*125-64-4*]
  properties of, *13*:122
Metmyoglobin
  protein meat pigment, *15*:62
Metol [*150-75-4*], *13*:45, 57. (See also *p-
    Methylaminophenol.*)
Metrication, *15*:45
Metrol, *13*:617
Mevalonic acid [*150-97-0*], *13*:98
Mevinphos [*7786-34-7*], *13*:438, 446
Mexacarbate [*315-18-4*], *13*:455
MGK 264 [*113-48-4*]
  as insecticide, *13*:425
MI. See *Melt index.*
MIBK. See *Methyl isobutyl ketone.*
Mica [*12001-26-2*], *15*:416
  dielectric strength, *13*:541
  in electrical insulation, *13*:556
  in packings, *16*:734, 738
  particle-track etching, *16*:826
  as track-etch detectors, *16*:827
Mica paper
  reconstituted, *15*:430
Mica sheet
  reconstituted, *15*:430
Micas, natural and synthetic, *15*:**416**
Mica splittings, *15*:421
Michael reaction, *13*:895
  of indole, *13*:216
  of maleic anhydride, *14*:779
Michler's Hydrol Blue [*14844-71-4*], *14*:551
Microbial polysaccharides, *15*:**439**
Microbial transformations, *15*:**459**
Microbiology
  diagnostic reagents, *15*:75

Microcap, *15*:488
Microcapsules. See *Microencapsulation.*
Microcasing processes, *15*:318
Micrococcus MS 102, *15*:538
Microcrystalline waxes [*63231-60-7*]
  paper size, *16*:812
Microencapsulation, *15*:92. (See also *Membrane
    technology.*)
  liquid walls in, *15*:475
Microfilaments
  nematiclike organization of, *14*:419
Microfiltration, *15*:106, 115
Microfiltration membranes
  track etching, *16*:846
Micromonaspora, *13*:786
Micronizer
  dry-grinder, *15*:423
Microorganisms, *15*:459. (See also *Microbial
    transformations.*)
  sterilization by ozone, *16*:683
Microporite, *14*:376
Microprocessor
  failure rate testing, *15*:19
  in instrumentation and control, *13*:509
Microsil, *14*:659
Microspheres. See *Microencapsulation.*
  magnetic, *15*:490
Microtox system, *13*:275
Microtubules
  nematiclike organization of, *14*:419
Microwave drying
  of inks, *13*:376
Microwave ferrites
  properties, *14*:667
Microwave heating, *15*:513
Microwave ovens, *15*:494. (See also *Microwave
    technology.*)
  for food processing, *15*:512
Microwave plasmas, *15*:516
Microwave power. See *Microwave technology.*
Microwaves
  hazards, *15*:510
Microwave spectrum, *15*:495
Microwave technology, *15*:**494**
Microwave tests
  nondestructive, *16*:61
Microwave tubes, *15*:503
Mictomagnetism, *14*:647. (See also *Spin glass.*)
Midrex process, *13*:745, 760
MIF. See *Migration inhibitory factor.*
Migraine, *15*:776
Migration inhibitory factor (MIF)
  in inflammatory response, *13*:174
MIL-A-8625A, *14*:509
MIL-C-1623A, *14*:509

Mildewcide
  antimicrobial agent as, *13*:223
Mildewcides
  in paint, *16*:748
Mildew prevention
  by mercury fungicides, *15*:155
Mildew-proofing agents, *16*:587
Mildew resistance
  in paint, *16*:747
Milk
  polyunsaturated fat in, *15*:491
Milk and milk products, *15*:**522**
Milk carton, *16*:799
Milk fat, *15*:561
Milk-of-lime, *14*:347
Milk powder. See *Dry milk.*
Milk sickness, *15*:553
Milk sugar. See *Lactose.*
MIL-L-6085, *14*:496
MIL-L-7808, *14*:496
MIL-L-8937, *14*:510
MIL-L-23398, *14*:510
MIL-L-23699, *14*:496
MIL-L-46010, *14*:510
MIL-L-81329, *14*:510
Milled-wood lignin [*8068-00-6*], *14*:296
Millerite [*1314-04-1*], *15*:796, 805
Milletia, *13*:425
Millipore, *15*:104
Millon's base [*12529-66-7*], *15*:161, 169
Mills, sand
  in magnetic tape mfg, *14*:743
MIL-M-3171A, *14*:509
Milontin, *13*:135
Milori blue [*14038-43-8*]
  in inks, *13*:378
Mimeograph inks, *13*:394
Mimosa [*8031-03-6*]
  in perfumes, *16*:951
  in vegetable tanning of hides, *14*:213
Mineral dressing
  magnetic, *14*:729
Mineral nutrient
  manganese, *14*:884
Mineral nutrients, *15*:**570**
Mineral oil [*8012-95-1*]
  insulation properties of, *13*:551
  in paper insulation, *13*:576
  papermaking defoamer, *16*:807
Minerals
  in nuts, *16*:252
Mineral spirits
  in paint, *16*:746
Mineral wool
  as insulation, *13*:595
Mineral zeolites, *15*:640

Minigrain processing, *15*:341
Minimata disease, *16*:587
Mining
  hydrogen peroxide uses, *13*:29
  offshore, *16*:294
Minium [*1314-41-6*], *14*:165
Minnesotaite
  as iron source, *13*:748
Mint oil [*68917-18-0*], *16*:324
Mirafi, *16*:78
Mirex [*2385-85-5*], *13*:436
Mirrors
  in filters, *16*:549
  in lasers, *14*:45
Missiles
  fluorinated plastic cable insulation for, *13*:575
  lubricants for, *14*:517
Mitin FF [*3567-25-7*], *13*:479
Mitochondria
  manganese in, *15*:594
Mitosis
  chemotherapeutic agents' effect on, *13*:172
Mixers, *15*:604, 610. (See also *Mixing and blending.*)
  for LLDPE, *16*:396
Mixer–settlers
  in liquid–liquid extraction, *14*:973
Mixing and blending, *15*:**604**
MMA. See *Methyl methacrylate.*
3M microfragrance scents, *15*:479
MML. See *Metallic-matrix laminates.*
MMT. See *Methylcyclopentadienylmanganese tricarbonyl.*
Mobil 1, *14*:496
Mobil-Witco-Shell process
  for poly(1-butene), *16*:477
MOCA. See *4,4'-Methylene-bis-2-chloroaniline.*
Model
  inventory, *16*:513
Modeling
  in operations planning, *16*:500
Models
  for plant control systems, *13*:495
Mohr's salt [*7783-85-9*], *13*:777
Mojonnier method
  for milk fat, *15*:552
Molding
  LDPE, *16*:415
  rotational, polyethylene, *16*:400
Molding powders
  poly(methyl methacrylate), *15*:386
Molding, rotational
  polyethylene, *16*:400
Molds
  in polysaccharide preparation, *15*:439

Molecular-beam epitaxy (MBE)
in LED manufacture, *14*:287
Molecular composites, *13*:977
Molecular diffusion
transfer by, *14*:912
Molecular-orbital theory (MOT), *16*:594
Molecular sieves, *15*:**638**
in LPG drying, *14*:389
in N_2 production, *15*:936
Molecular weight
osmotic determination, *15*:121
Molecular weight averages
of polymers, determination, *16*:436
Molluscicides
organotin, *16*:579
Molozonide, *16*:684
Molten-salt cleaning
of metals, *15*:302
Molting
insect growth regulator effect on, *13*:459
Molybdate(4−) aquaheptacyanotetrapotassium
monohydrate [*76900-85-1*], *15*:691
Molybdate blue [*68409-66-5*]
in phosphorus measurement, *15*:84
Molybdate(3−) hexakis(isothiocyanato)-
tripotassium tetrahydrate [*14649-96-8*], *15*:
691
Molybdate(3−) octacyanotripotassium [*19442-
23-0*], *15*:691
Molybdate(4−) octakis(cyano-c)tetrapotassium
dihydrate [*17457-89-5*], *15*:691
Molybdate orange [*12656-85-8*]
in inks, *13*:378
Molybdate(4−)
pentacyanotrihydroxytetrapotassium
trihydrate [*76550-42-2*], *15*:691
Molybdates, *15*:687
Molybdate(4−)
tetracyanotetrahydroxytetrapotassium
[*21978-12-1*], *15*:691
Molybdenite [*1309-56-4*], *15*:670, 689
Molybdenosis, *15*:692
Molybdenum [*7439-98-7*], *15*:670
amalgamation with mercury, *15*:147
in integrated circuits, *13*:641
ion implantation, *13*:717
metabolic functions, *15*:592
as metallic coating, *15*:247
nickel alloys, *15*:792
nutrient properties, *15*:571
from the ocean, *16*:279, 292
toxicity, *15*:679
TZM alloy, *15*:679
Molybdenum-92 [*14191-67-4*], *13*:841
Molybdenum-94 [*14683-00-2*], *13*:841
Molybdenum-95 [*14392-17-7*], *13*:841
Molybdenum-96 [*14274-76-1*], *13*:841
Molybdenum-97 [*14392-19-9*], *13*:841
Molybdenum-98 [*14392-20-2*], *13*:841
Molybdenum-100 [*14392-21-3*], *13*:841

Molybdenum and molybdenum alloys, *15*:**670**
Molybdenum blue [*1313-29-7*], *15*:686
Molybdenum bromide oxide [*13595-98-7*], *15*:683
Molybdenum carbide [*12011-97-1*]
nitride form, *15*:877
Molybdenum carbonyl [*13939-06-5*]
in initiator systems, *13*:367
Molybdenum–cobalt–iron alloys
as magnetic materials, *14*:680
Molybdenum cofactor
from nitrate reductase, *15*:952
Molybdenum compounds, *15*:**683**
Molybdenum dibromide [*13446-56-5*], *15*:683
Molybdenum dicarbonyl [*12110-96-2*], *16*:592
Molybdenum dichloride [*13478-17-6*], *15*:683
Molybdenum difluoride [*20205-60-1*], *15*:683
Molybdenum diiodide [*14055-74-4*], *15*:683
Molybdenum dioxide [*18868-43-4*], *15*:685
Molybdenum dioxychloride [*20770-33-6*], *15*:683
Molybdenum dioxydichloride [*13637-68-8*], *15*:
683
Molybdenum dioxydifluoride [*13824-57-2*], *15*:683
Molybdenum diselenide [*12078-18-3*], *15*:689
Molybdenum disulfide [*1317-33-5*], *15*:689
filler in gaskets, *16*:732
as grease filler, *14*:503
lubricant, *15*:670
in metal treating, *15*:307
in packings, *16*:738
in radiation shielding, *16*:240
as solid lubricant, *14*:506
Molybdenum ditelluride [*12058-20-7*], *15*:689
Molybdenum fiber
in metal laminates, *13*:945
Molybdenum hexacarbonyl [*13939-06-5*], *15*:690
Molybdenum hexafluoride [*7783-77-9*], *15*:683
Molybdenum hydroxide [*27845-91-6*]
($MoO(OH)_3$), *15*:686
Molybdenum hydroxide [*55004-54-1*]
$Mo_2O_4(OH)_2$, *15*:686
Molybdenum hydroxide [*66589-56-8*]
$Mo_5O_7(OH)_8$, *15*:686
Molybdenum hydroxide [*66588-05-4*]
$Mo_5O_8(OH)_8$, *15*:686
Molybdenum hydroxide [*12507-76-5*]
$Mo_4O_{10}(OH)_2$, *15*:686
Molybdenum hydroxide $Mo_5O_5(OH)_{10}$, *15*:686
Molybdenum–iron protein
in nitrogenase, *15*:950
Molybdenum–iron–sulfur cluster compounds
in N_2 reduction, *15*:960
Molybdenum nitride [*12033-19-1*]
nitrogenase intermediate, *15*:693
properties, *15*:874
Molybdenum nitride [*12033-19-1*] (1:1), *15*:872

Molybdenum nitride [*12033-31-7*] (2:1), *15*:872

Molybdenum orange [*12656-85-8*], *15*:694

Molybdenum oxide [*12136-80-0*] (Mo$_3$O), *15*:685

Molybdenum oxide [*12033-38-4*] (Mo$_4$O$_{11}$), *15*:685

Molybdenum oxide [*12438-84-5*] (Mo$_5$O$_{14}$), *15*:685

Molybdenum oxide [*12058-34-3*] (Mo$_8$O$_{23}$), *15*:685

Molybdenum oxide [*12163-89-2*] (Mo$_9$O$_{26}$), *15*:685

Molybdenum oxide [*12777-01-4*] (Mo$_{17}$O$_{47}$), *15*:685

Molybdenum oxide [*12136-82-2*] (Mo$_{18}$O$_{52}$), *15*:685

Molybdenum oxychloride [*41004-72-2*], *15*:683

Molybdenum oxydichloride [*24989-40-0*], *15*:683

Molybdenum oxydisulfide [*58797-84-5*], *15*:690

Molybdenum oxyfluoride tetrahydrate [*77727-63-0*], *15*:683

Molybdenum oxytetrachloride [*13814-75-0*], *15*:683

Molybdenum oxytetrafluoride [*14459-59-7*], *15*:683

Molybdenum oxytribromide [*13596-04-8*], *15*:683

Molybdenum oxytrichloride [*13814-74-9*], *15*:683

Molybdenum pentachloride [*10241-05-1*], *15*:683

Molybdenum pentafluoride [*13819-84-6*], *15*:683

Molybdenum sesquioxide [*1313-29-7*], *15*:685

Molybdenum sesquiselenide [*12325-96-1*], *15*:690

Molybdenum sesquisulfide [*12033-33-9*], *15*:690

Molybdenum sesquitelluride [*12325-97-2*], *15*:690

Molybdenum sulfide [*1317-33-5*]
   in coatings, *15*:251
   oxo catalyst use, *16*:646

Molybdenum tetrabromide [*13520-59-7*], *15*:683

Molybdenum tetrachloride [*13320-71-3*], *15*:683

Molybdenum tetrafluoride [*23412-45-5*], *15*:683

Molybdenum tribromide [*13446-57-6*], *15*:683

Molybdenum trichloride [*13478-18-7*], *15*:683

Molybdenum trifluoride [*20193-58-2*], *15*:683

Molybdenum triiodide [*14055-75-5*], *15*:683

Molybdenum trioxide [*1313-27-5*], *14*:508; *15*:685
   manufacture, *15*:674

Molybdenum trisulfide [*12033-29-3*], *15*:690

Molybdic oxide [*1313-27-5*]
   technical, *15*:670

Molybdic(VI) acid [*7782-91-4*], *15*:686

Molybdic(VI) acid monohydrate [*13462-95-8*], *15*:686

Molybdoenzymes
   in N$_2$ reduction, *15*:960

Molybdophosphoric acid [*12026-57-2*], *15*:686

Molybdothiol systems
   for nitrogen reduction, *15*:959

Molydenum trioxide [*1313-27-5*]
   toxicity, *15*:679

Molykote, *14*:513

Monazite [*1306-41-8*]
   from beach sand, *16*:289
   oxalic acid in separations, *16*:632
   thorium source, *16*:148

Mondur CB, *13*:815

Mondur CD, *13*:801

Mondur HX, *13*:801

Mondur M, *13*:801

Mondur MR, *13*:801, 805

Mondur MRS, *13*:801

Mondur MT-40, *13*:801

Mondur O, *13*:801

Mondur P, *13*:801

Mondur PCPI, *13*:801

Mondur S, *13*:815

Mondur SH, *13*:815

Mondur TD, *13*:801

Mondur TD-80, *13*:801

Mondur TDS, *13*:801

Monel, *13*:583; *16*:727
   explosion clad, *15*:289

Monel Alloy 400 [*11105-19-4*], *15*:789

Monel Alloy K-500 [*11105-28-5*], *15*:789

Monel drums, *16*:717

Mongolism
   taurine associated with, *15*:781

Monoacetylferrocene [*1271-55-2*], *16*:604

Monoamine oxidase [*9001-66-5*], *15*:135
   inhibition of octopamine, *15*:775
   iron in, *15*:595

Monobromo malonic acid [*600-31-7*], *14*:795

Monobutyl phosphate [*1623-15-0*], *16*:180

Monochloroacetic acid [*79-11-8*]
   in dialkyl malonate synthesis, *14*:800
   diglycolic acid from, *13*:94

Monochlorobenzene [*108-90-7*]
   in isocyanate manufacture, *13*:803

Monocrotophos [*6923-22-4*], *13*:447, 890

Monoethanolamine [*141-43-5*]
   dehydration to aziridine, *13*:158

Monoethylamine [*75-04-7*]
   paint removers, *16*:763

Monoglycerides
   in ice cream, *15*:565

8-Monohydromirex [*39801-14-4*]
   from mirex, *13*:436

Mononitrochlorobenzenes, *15*:923

Mononitrotoluenes, *15*:925

Monoolefins
   metal π-complex formation, *16*:592

Monoperoxysulfuric acid [*7722-86-3*], *13*:15

Monopropellant
   nitromethane, *15*:982

Monosodium gluconate [*14906-97-9*], *14*:640

Monosodium glutamate [*142-47-2*]
   use on nuts, *16*:271

Monosodium vanadate [*13718-26-8*]
toxicity, *15*:574
Monoterpenes
microbial oxidation, *15*:459
Monothioethylene phosphorochloridite [*32847-69-1*]
isoprene reaction, *13*:826
Mono(trichloro)tetra(potassium dichloroisocyanurate) [*34651-95-1*]
applications of, *13*:230
Monotype alloys, *14*:156
Monsanto process
for acetic acid, *16*:584
Montan wax, *14*:317
from lignites, *14*:340
production, *14*:340
Monte Carlo trial and error, *13*:508
Montedison-Mitsui process
polypropylene, *16*:463
Montroydite [*21908-53-2*], *15*:144
Mood
enkephalins involved in, *15*:765
4-79 Mo Permalloy, *14*:651, 659
Moravian III, *14*:811
Mordenite [*12173-98-7*], *15*:639
Morex, *14*:811
Moringa oleifera, *16*:251
Morphine [*57-27-2*]
in lactic acid resolution, *13*:81
measurement, *15*:88
Morpholine [*110-91-8*]
in lactamide preparation, *13*:90
in maleic anhydride determination, *14*:786
Morphothion [*144-41-2*], *13*:447
Mortars, *14*:376
MOS. See *Metal oxide semiconductor.*
Mosaic membranes, *15*:121
Moss-Burstein shift, *14*:276
MOS transistor. See *Metal-oxide semiconductor transistor.*
MOT. See *Molecular-orbital theory.*
Mothballs, *15*:707
Mothproofing, *13*:479
Motion sickness
treatment, *13*:134
Motor activity
enkephalins involved in, *15*:765
Motorcycles
noise source, *16*:12
Motors
noise control, *16*:19
Mousse
frozen dessert, *15*:568
Mousse de chene [*9000-50-4*]
in perfumes, *16*:951
Mo-30W
molybdenum-base alloy, *15*:679

MSH. See *Melanocyte-stimulating hormone.*
MSO. See *Mesityl oxide.*
cis,cis-Muconic acid [*1119-72-8*], *13*:41; *16*:710
Mucopolysaccharide impairment
nutrients associated with, *15*:573
Mufflers
acoustic, *13*:520
chromized steel in, *15*:249
glass fiber insulation in, *13*:516
noise control use, *16*:15
Muguet
in perfumes, *16*:967
Mullen burst, *16*:814
Mullen test, *16*:99
Mullite [*1302-93-8*], *15*:658
Multi-hearth furnace
incinerators, *13*:191
Multiple-client studies, *14*:900
Multiple myeloma
detection, *15*:83
Multiwall bags, *16*:721
Multrathane E-516, *13*:815
Multrathane F-242, *13*:815
Mumetal plus, *14*:659
Mumps
diagnosis, *15*:79
Mundulea, *13*:425
Muriamic, *15*:767
Murumuru nut, *16*:251
Musca domestica, *13*:**456**, 458, 459
Muscle
liquid crystals in, *14*:418
Muscle relaxants
benzodiazepines, *13*:126
Muscle tremors
nutrients associated with, *15*:573
Muscle weakness
nutrients associated with, *15*:573
Muscone [*541-91-3*]
in musk oil, *13*:922
in perfumes, *16*:951
Muscovite [*1318-94-1*], *15*:416
in insulation, *13*:556
Mushrooms
bufotenine in, *13*:220
Musk ambrette, artificial [*83-66-9*], *16*:956
Musk ambrette, natural [*123-69-3*], *16*:956
Musk tonquin [*8001-04-5*]
in perfumes, *16*:951
Mustard oil [*8007-40-7*], *16*:308
Mutagenesis
in malting, *14*:816
α-Muurolene [*17627-24-6*], *16*:313
MWD. See *Molecular weight distribution.*
Mycobacterium bovis, *15*:553

Mycobacterium tuberculosis, *15*:538, 553

Mycobacterium tuberculosum
    disinfection by ozone, *16*:705

Mycorrhizae, *15*:949

Mycorrhizas
    catechol in, *13*:46

Mycotoxins
    in nuts, *16*:263

Mylar [*25038-59-9*]
    in metal-coated filters, *16*:531
    mica substitute, *15*:432
    in patent leather mfg, *14*:245
    thermal conductivity of, *13*:544

Myochrisin, *13*:111.   (See also *Sodium
    aurothiomalate.*)

Myoglobin [*9008-45-1*], *13*:782
    iron in, *15*:594
    protein meat pigment, *15*:62

Myopic policy, *16*:506

Myosin [*9000-83-3*], *14*:419
    calcium effect on, *15*:584
    in cattlehide, *14*:203
    in meat processing, *15*:67

Myrcene [*123-35-3*], *16*:312, 313
    in orange oil, *16*:325

β-Myrcene [*123-35-3*]
    in perfumes, *16*:954

Myristic acid [*544-63-8*]
    from orris root, *16*:326

Myristica fragrans Houtt., *16*:325

Myristicin [*1607-91-0*]
    in nutmeg, *16*:325

Myrobalan
    as vegetable tannin, *14*:214

Myrrh [*8016-37-3*]
    in perfumes, *16*:951

Myrrh oil [*8016-37-3*], *16*:308

Myrtle [*8008-46-6*], *16*:307

Myrustica fragrans, *16*:251

Mysoline, *13*:135

Myxedema
    nutrients associated with, *15*:573

Myzus persica, *13*:459

N

NA.   See *Naphthenic acid.*
NA 170, *15*:750

NA 210, *15*:750

Nabam [*142-59-6*], *13*:246, 475.   (See also
    Disodium ethylenebis(dithiocarbamate).)

Nacconate, *13*:801

Nacconate 5050, *13*:801

NAD.   See *Nicotinamide adenine dinucleotide.*

NADH.   See *Reduced nicotinamide adenine
    dinucleotide.*

NADH-dehydrogenase [*9079-67-8*]
    iron in, *15*:595

Nafion membranes, *15*:120

Nahcolite [*15752-47-3*], *16*:335
    thermal decomposition, *16*:337

NaK 77, *14*:515

Naled [*300-76-5*], *13*:442

Naloxone [*465-65-6*], *15*:765

Nanoencapsulation, *15*:481

Naphtachimie gas-phase process
    for polyethylene, *16*:393

Naphthalene [*91-20-3*], *15*:**698**
    in divinylbenzene, *13*:688
    as fumigant, *13*:467
    microbial preparation, *15*:460
    in mineral oils, *13*:551
    properties, *15*:698

Naphthaleneacetonitrile [*132-75-2*]
    hydrolysis, *15*:716

Naphthaleneamine, *15*:726.   (See also
    Naphthylamine.)

1-Naphthaleneamine [*134-32-7*], *15*:724, 726

2-Naphthaleneamine [*91-59-8*], *15*:726

1,2-Naphthaleneamine [*938-25-0*], *15*:726

Naphthalene azeotropes, *15*:699

Naphthalenecarboxylic acids, *15*:743

1-Naphthalenecarboxylic acid [*86-55-5*], *15*:743

2-Naphthalenecarboxylic acid [*93-09-4*], *15*:743

Naphthalene derivatives, *15*:**719**

1,4-Naphthalenediamine [*2243-61-0*], *15*:726

1,5-Naphthalenediamine [*2243-62-1*], *15*:726

1,6-Naphthalenediamine [*2243-63-2*], *15*:726

1,7-Naphthalenediamine [*2243-64-3*], *15*:726

1,8-Naphthalenediamine [*479-27-6*], *15*:726, 735

2,3-Naphthalenediamine [*771-97-1*], *15*:726

2,6-Naphthalenediamine [*2243-67-6*], *15*:726

2,7-Naphthalenediamine [*613-76-1*], *15*:726

1,2-Naphthalenedicarboxylic acid [*2088-87-1*],
    15:743

1,3-Naphthalenedicarboxylic acid [*2089-93-2*],
    15:743

1,4-Naphthalenedicarboxylic acid [*605-70-9*], *15*:
    743

1,5-Naphthalenedicarboxylic acid [*7315-96-0*],
    15:743

1,6-Naphthalenedicarboxylic acid [*2089-47-4*],
    15:743

1,7-Naphthalenedicarboxylic acid [*2089-91-0*], *15*:743

1,8-Naphthalenedicarboxylic acid anhydride [*81-84-5*], *15*:717, 737, 741, 743

2,3-Naphthalenedicarboxylic acid [*2169-87-1*], *15*:743

2,6-Naphthalenedicarboxylic acid [*1141-38-4*], *15*:716, 743

2,7-Naphthalenedicarboxylic acid [*2089-89-6*], *15*:743

1,1-Naphthalene-8,8'-dicarboxylic acid [*29878-91-9*], *15*:747

1,5-Naphthalenediisocyanate [*3173-72-6*], *15*:727
preparation, *15*:724

1,2-Naphthalenediol [*574-00-5*], *15*:732

1,3-Naphthalenediol [*132-86-5*], *15*:732

1,4-Naphthalenediol [*571-60-8*], *15*:732

1,5-Naphthalenediol [*83-56-7*], *15*:732

1,6-Naphthalenediol [*575-44-0*], *15*:732

1,7-Naphthalenediol [*575-38-2*], *15*:732

1,8-Naphthalenediol [*569-42-6*], *15*:732

2,3-Naphthalenediol [*92-44-4*], *15*:732

2,6-Naphthalenediol [*581-43-1*], *15*:732

2,7-Naphthalenediol [*582-17-2*], *15*:732

Naphthalenediols, *15*:729

2,3-Naphthalenediol-6-sulfonic acid [*92-27-3*]
2,3-naphthalenediol from, *15*:735

1,2-Naphthalenedisulfonic acid [*25167-78-6*], *15*:722

1,3-Naphthalenedisulfonic acid [*6094-26-4*], *15*:722

1,4-Naphthalenedisulfonic acid [*46859-22-7*], *15*:722

1,5-Naphthalenedisulfonic acid [*81-04-9*], *15*:722

1,6-Naphthalenedisulfonic acid [*525-37-1*], *15*:722

1,7-Naphthalenedisulfonic acid [*5724-16-3*], *15*:722

2,6-Naphthalenedisulfonic acid [*581-75-9*], *15*:722

2,7-Naphthalenedisulfonic acid [*92-41-1*], *15*:722

1-Naphthalenesulfonic acid [*85-47-2*], *15*:722
reduction, *15*:722

2-Naphthalenesulfonic acid [*120-18-3*], *15*:722

1-Naphthalenesulfonic acid dihydrate [*6036-48-2*], *15*:722

2-Naphthalenesulfonic acid hydrate [*76530-12-6*], *15*:722

Naphthalenesulfonic acids, *15*:720

2-Naphthalenesulfonic acid trihydrate [*17558-84-8*], *15*:722

Naphthalene syntan
in leather manufacture, *14*:213

1,2,4,5-Naphthalenetetracarboxylic acid [*22246-61-3*], *15*:743

1,4,5,8-Naphthalenetetracarboxylic acid [*128-97-3*], *15*:743, 744

1,3,5,7-Naphthalenetetrasulfonic acid [*6654-67-7*], *15*:722

1-Naphthalenethiol [*529-36-2*], *15*:721

1-Naphthalenethiourea [*86-88-4*], *15*:727

1,2,5-Naphthalenetricarboxylic acid [*36439-99-3*], *15*:743

1,3,8-Naphthalenetricarboxylic acid [*36440-24-1*], *15*:743

1,4,5-Naphthalenetricarboxylic acid [*28445-09-2*], *15*:743

1,3,5-Naphthalenetrisulfonic acid [*6654-64-4*], *15*:722

1,3,6-Naphthalenetrisulfonic acid [*86-66-8*], *15*:722

1,3,7-Naphthalenetrisulfonic acid [*85-49-4*], *15*:722

1,4,5-Naphthalenetrisulfonic acid [*60913-37-3*], *15*:722

1-Naphthalenol [*90-15-3*], *15*:705, 723, 729, 732. (See also *1-Naphthol.*)

2-Naphthalenol [*135-19-3*], *15*:732
naphthalene reaction, *15*:702

Naphthalenols, *15*:729

Naphthalimide [*81-83-4*], *15*:741

Naphthalophos [*1491-41-7*], *15*:741

Naphthenic. See *Alicyclic, cycloparaffinic.*

Naphthenic acid [*1338-24-5*]
in ink driers, *13*:379

Naphthenic acids [*1338-24-5*], *15*:**749**
properties, *15*:750

α-Naphthoflavone [*604-59-1*]
iodine analysis reagent, *13*:659

1-Naphthol [*90-15-3*], *15*:705, 723. (See also *1-Naphthalenol.*)

2-Naphthol [*135-19-3*], *15*:702. (See also *2-Naphthalenol.*)
lubricant antioxidant, *14*:492

2-Naphthol aniline
2-naphthol reaction, *15*:727

Naphtholsulfonic acids, *15*:733

Naphthoquinone [*130-15-4*], *15*:715

1,2-Naphthoquinone [*524-42-5*]
properties, *13*:926

1,4-Naphthoquinone [*130-15-4*], *15*:705
1,4-naphthalenediol manufacture, *15*:735

Naphthostyril [*130-00-7*], *15*:746

1-Naphthoxyacetic acid [*120-23-0*], *15*:729

1-Naphthylacetic acid [*86-87-3*], *15*:716

S-Naphthyl-*N*-alkylthiocarbamates, *15*:722

Naphthylamine, *15*:726. (See also *Naphthaleneamine.*)

1-Naphthylamine [*134-32-7*], *15*:715
2-Naphthylamine [*91-59-8*], *15*:702
  carcinogen, *15*:724
2-Naphthyl benzoate [*93-44-7*], *15*:735
1-Naphthylcarbamate [*63-25-2*], *13*:806
2-Naphthyl ethyl ether [*93-18-5*], *15*:735
2-Naphthyl lactate [*93-43-6*], *15*:735
1-Naphthyl methylcarbamate [*63-25-2*], *15*:729
1-Naphthyl *N*-methylcarbamate [*63-25-2*], *15*:714
  as insecticide, *13*:453
2-Naphthyl methyl ether [*93-04-9*], *15*:734
1-Naphthyl(phenyl)amine [*90-30-2*]
  as grease additive, *14*:502
  lubricant antioxidant, *14*:492
N-1-Naphthylphthalamic acid [*132-66-1*], *15*:727
2-Naphthyl salicate [*613-78-5*], *15*:735
1-Naphthyl salicylate [*550-97-0*], *15*:733
Naples yellow.   See *Lead antimonate.*
Naproxen [*22704-53-1*]
  immunologic effects, *13*:174
Naptalam, *15*:727
Naptopen, *15*:735
NASA/RECON
  information systems, *16*:901
National Lead Industries Brine Process
  for magnesium, *14*:580
NATIONAL NEWSPAPER INDEX, *13*:334
NATIONAL NEWSPAPER INDEX information
  system, *13*:330
National Sources Information sources, *13*:312
National Technical Information Service
  patent information, *16*:924
Natsyn, *13*:833
Natural gas
  molecular-sieve purification, *15*:663
Natural gasoline
  from LPG, *14*:385
Nausea
  nutrients associated with, *15*:573
Nauta mixer, *15*:633
Navel orangeworm
  control of with pheromones, *13*:480
NB.   See *2-Nitro-1-butanol.*
NBR.   See *Butyl rubber.*
NDEA.   See *N-Nitrosodiethylamine.*
NDELA.   See *N-Nitrosodiethanolamine.*
NDMA.   See *N-Nitrosodimethylamine.*
Necking
  in metals, *15*:326
Needle-punching
  nonwovens, *16*:120
Needling
  spunbondeds, *16*:86
Néel temperature, *14*:650

Néel walls, *14*:689
Nef reaction, *15*:974
Neimann-Pick disease
  liquid crystals in, *14*:420
Nelumbo lutea, *16*:251
Nematic liquid crystals, *14*:407
Neodiprion, *13*:471
Neodymium [*7440-00-8*]
  lasers, *14*:54
  from the ocean, *16*:280
  in optical filters, *16*:554
Neodymium-142 [*14336-82-4*], *13*:841
Neodymium-143 [*14336-83-5*], *13*:841
Neodymium-144 [*14834-76-5*], *13*:841
Neodymium-145 [*14336-84-6*], *13*:841
Neodymium-146 [*15411-67-3*], *13*:841
Neodymium-148 [*14280-26-3*], *13*:841
Neodymium-150 [*14683-22-8*], *13*:841
Neodymium, compd with cobalt [*12017-65-1*] (1:
  5), *14*:676
Neodymium, compd with cobalt [*12052-76-5*] (2:
  17), *14*:676
Neodymium–glass lasers, *14*:51, 56
  in fusion research, *14*:77
Neodymium ion lasers, *14*:56
Neodymium nitride [*25764-11-8*], *15*:872
Neodymium trifluoride [*13709-42-7*]
  in fluoride electrode, *13*:727
Neodymium–yttrium–aluminum–garnet laser,
  14:51
Neolube, *14*:513
Neomycin [*1404-04-2*]
  inactivation, *15*:466
  purification of by ion exchange, *13*:704
Neon [*7440-01-9*]
  in air, *16*:657
  ion implantation, *13*:715
  laser, *14*:42
  from the ocean, *16*:280
Neon-20 [*13981-34-5*], *13*:840
Neon-21 [*13981-35-6*], *13*:840
Neon-22 [*13886-72-1*], *13*:840
  in isotope separation, *13*:849
Neoplasia
  treatment, *13*:167
Neoplastic diseases
  chemotherapeutic agents for, *13*:172
  detection, *15*:81
Neoprene [*31727-55-6*]
  in acoustic insulation, *13*:522
  in cable coverings, *13*:585
  gasket properties, *16*:727
  as leather substitute, *14*:231
  weather resistance of, *13*:547
Neoprene–wood fiber
  gasket properties, *16*:727
Neotenin [*32766-80-6*], *13*:459

NEPD. See *2-Ethyl-2-nitro-1,3-propanediol.*
Nephridine, *15*:757
Nephritis
    detection, *15*:83
    diagnosis, *15*:80
    from lead exposure, *14*:198
Neptunium-239 [*13968-59-7*], *16*:188
Neptunium nitride [*12058-90-1*], *15*:872
Neral [*106-26-3*], *16*:312
Nernst equation, *13*:722
Nernstian response
    of pH electrode, *13*:2
Nernst-Planck equation, *15*:100
Nerol [*106-25-2*], *16*:312
Nerolin, *15*:734
Neroli oil [*8016-38-4*], *16*:324
Nerve
    liquid crystals in, *14*:418
Neryl acetate [*141-12-8*], *16*:312
Nesquehonite [*14457-83-1*], *14*:619
Nessler's reagent, *15*:160
    ammonia detection, *13*:661
Netlon, *16*:78
Netlon process, *16*:95
Net 909 P520, *16*:76
Net 909 process, *16*:94
Nets
    fibers for, *16*:376
    spunbonded, *16*:95
Network problems, *16*:506
Neurochemistry, *15*:756
Neuroleptics. See *Neuroregulators.*
Neuromodulators, *15*:754. (See also
    Neuroregulators.)
Neuropeptides
    memory effects, *15*:133
Neuropharmacology, *15*:756
Neuroregulators, *15*:**754**. (See also
    Neurotransmitters.)
Neurotensin [*39379-15-2*]
    as neuroregulator, *15*:772
    properties, *15*:772
Neurotoxins, *13*:459. (See also *Insecticides.*)
Neurotransmitters, *15*:754. (See also
    Neuroregulators.)
Neutral filters, *16*:526
Neutral protease
    zinc in, *15*:590
Neutroleum, *16*:304
Neutron barrier
    polymethacrylate, *15*:380
Neutron beams
    nondestructive testing by, *16*:49
Neutron imaging tests
    nondestructive, *16*:65

Neutron poison
    in nuclear reactors, *16*:235
Neutrons
    delayed, *16*:199
Never Seez PM5543, *14*:513
Never Seez PM5547, *14*:513
Neville-Winther acid [*84-87-7*], *15*:738. (See also
    4-Hydroxy-1-naphthalenesulfonic acid.)
Newman's plot, *14*:922
NEWSEARCH, *13*:334
NEWSEARCH information system, *13*:330
Newsprint, *16*:772, 782
NEW YORK TIMES INFORMATION BANK
    information system, *13*:330, 334
NGB
    patent information, *16*:925
Niacin [*59-67-6*]
    in dextran production, *15*:443
    in nuts, *16*:260
Nibcolloy, *14*:681
Ni-bis[*O*-monoethyl(3,5-*di-tert*-butyl-4-hydroxy-
    benzyl)]phosphonate [*30947-30-9*]
    polypropylene light stabilization, *16*:361
Nichrome [*12605-70-8*], *15*:789
Nickel [*7440-02-0*], *15*:787
    affect on iron magnetism, *13*:738
    in coatings, *15*:241
    emissivity, *13*:342
    in explosive cladding, *15*:278
    explosively clad, *15*:280
    gasket properties, *16*:727
    in magnesium alloys, *14*:596
    in magnetic alloys, *14*:694
    as magnetic material, *14*:647
    in metal treating, *15*:305
    as nutrient, *15*:598
    from the ocean, *16*:279, 292
    oxidation, *16*:656
    recovery by ion exchange, *13*:702
    from sulfide ores, *16*:668
    toxicity, *13*:270
Nickel-58 [*13981-79-8*], *13*:840
Nickel-59 [*14336-70-0*]
    neutron reaction gives He, *16*:141
Nickel-60 [*13981-80-1*], *13*:840
Nickel-61 [*14928-10-0*], *13*:840
Nickel-62 [*13981-81-2*], *13*:840
Nickel-64 [*14378-31-5*], *13*:840
Nickel acetate [*373-02-4*]
    as dye fastener, *15*:309
    toxicity, *15*:574
Nickel acetate tetrahydrate [*6018-89-9*], *15*:809
Nickel acetylacetonate [*3264-82-2*], *15*:810
    1,5,9-cyclododecatriene reaction, *15*:808
Nickel 200 alloy [*12671-92-0*], *15*:787, 789

Nickel 201 alloy, *15*:787

Nickel 205 alloy, *15*:787

Nickel 270 alloy, *15*:787

Nickel alloys
cladding of, *15*:281

Nickel amminocyanide, *15*:809

Nickel ammonium chloride hexahydrate [*16122-03-5*], *15*:806

Nickel ammonium nitrate [*22026-79-5*], *15*:803

Nickel ammonium sulfate hexahydrate [*15699-18-0*], *15*:806

Nickel and nickel alloys, *15*:**787**

Nickel (antimony) titanate [*11118-07-3*], *15*:813

Nickel arsenate octahydrate [*7784-48-7*], *15*:806

Nickel azo gold, *15*:814

Nickel azo red, *15*:814

Nickel azo yellow [*51931-46-5*], *15*:814

Nickel-base alloy
nondestructive testing of, *16*:51

Nickel boride [*12007-00-0*], NiB
hydrogenation catalyst, *13*:19

Nickel–boron alloy [*12007-00-0*], *15*:813

Nickel bromide dimethoxyethane [*29823-39-9*]
sodium cyclopentadienide reaction, *15*:807

Nickel bromide hexahydrate [*18721-96-5*], *15*:804

Nickel carbonate [*3333-67-3*], *15*:804
nickel oxide from, *15*:802

Nickel carbonate basic [*29863-10-3*], *15*:804

Nickel carbonyl [*13463-39-3*], *15*:801
in chemical deposition coating, *15*:264
properties, *15*:806
SO$_2$/H$_2$SO$_4$ reactions, *15*:803
tetraphenylcyclobutadienepalladium
dibromide reaction, *16*:609
toxicity, *13*:270

Nickel carbonyl [*13463-39-3*] (1:4)
in Ni refining, *15*:798

Nickel chloride [*7718-54-9*], *15*:808

Nickel chloride hexahydrate [*7791-20-0*], *15*:804

Nickel chromite [*12018-18-7*], *15*:813

Nickel chromium [*12433-21-9*] (1:1)
in metallic coating, *15*:254

Nickel cobalt aluminate, *15*:813

Nickel compounds, *15*:**801**

Nickel, compound with aluminum [*12003-81-5*]
(3:1), *15*:787

Nickel compound with iron [*12022-66-1*] (3:1)
magnetic behavior, *14*:657

Nickel, compound with Ti, and Al, Ni$_3$(Ti,Al)
by precipitation hardening, *15*:338

Nickel cyanide [*557-19-7*], *15*:805

Nickel cyanide tetrahydrate [*20427-77-4*], *15*:805

Nickel dibromide [*13462-88-9*], *15*:804

Nickel dibutyldithiocarbamate [*13927-77-0*], *15*:814
antioxidant, *15*:814

Nickel dichloride [*7718-54-9*], *15*:804

Nickel difluoride [*10028-18-9*], *15*:804

Nickel dihydrohexacarbonyl [*12549-35-8*], *15*:801

Nickel diiodide [*13462-90-3*], *15*:804

Nickel dimethyldithiocarbamate [*15521-65-0*], *15*:810

Nickel dimethylglyoxime [*13478-93-8*], *15*:810

Nickel ethylenediaminebisacetylacetonate [*42948-35-6*], *15*:810

Nickel ethyl xanthate [*52139-56-7*], *15*:810

Nickel ferrite [*12168-54-6*]
as magnetic insulator, *14*:694

Nickel fluoride tetrahydrate [*13940-83-5*], *15*:804

Nickel fluoroborate hexahydrate [*14708-14-6*], *15*:805

Nickel formate dihydrate [*15694-70-9*], *15*:809

Nickel halides, *15*:804

Nickel hexaammine dichloride [*10534-88-0*], *15*:806

Nickel hexaammine diiodide [*13859-68-2*], *15*:806

Nickel hexammine carbonate [*67806-76-2*], *15*:802

Nickel hexammine dibromide [*13601-55-3*], *15*:806

Nickel hydroxide [*12054-48-7*], *15*:804
in batteries, *15*:804

Nickel iodide [*13462-90-3*], *13*:661

Nickel iodide hexahydrate [*7790-34-3*], *15*:804

Nickel–iron alloys
as magnetic materials, *14*:657

Nickel laterites, *15*:796

Nickel molybdate [*12673-58-4*], *15*:687

Nickel–niobium alloy, *15*:824

Nickel nitrate [*13138-45-9*]
nickel oxide from, *15*:802

Nickel nitrate hexahydrate [*13478-00-7*], *15*:803

Nickel nitride [*12033-45-3*], *15*:872

Nickelocene [*1271-28-9*], *15*:806, 807
reduction, *16*:604
tetrafluoroethylene reaction, *16*:604

Nickel oleate [*68538-38-5*], *15*:810

Nickel oxalate [*20543-06-0*], *15*:810

Nickel oxide [*1319-99-1*], *15*:801
as magnetic material, *14*:647

Nickel oxide [*12035-36-8*] (1:2), *15*:802

Nickel oxide [*37194-86-8*] (1:4), *15*:802

Nickel oxide [*1314-06-3*] (2:3), *15*:802

Nickel oxide [*12137-09-6*] (3:4), *15*:802

Nickel oxides
  properties, *15*:801
Nickel phthalocyanine [*14045-02-8*], *15*:810
Nickel poisoning
  treatment, *15*:812
Nickel potassium sulfate hexahydrate [*10294-65-2*], *15*:806
Nickel selenite [*15060-62-5*], *15*:813
Nickel silicate [*37321-15-6*], *15*:797
Nickel silicide [*12035-57-3*] (1:1), *15*:813
Nickel silicide [*12201-89-7*] (1:2), *15*:813
Nickel silicide [*12059-14-2*] (2:1), *15*:814
Nickel silicide [*12059-22-2*] (3:1), *15*:814
Nickel silicide [*12059-27-7*] (5:2), *15*:814
Nickel silvers, *15*:792
Nickel stearate [*2223-95-2*], *15*:810
Nickel sulfamate tetrahydrate [*13770-89-3*], *15*:805
Nickel sulfate [*17786-81-4*]
  nickel oxide from, *15*:802
  oxygen-generation catalyst, *16*:679
Nickel sulfate hexahydrate [*10101-97-0*], *15*:802
Nickel sulfide [*1314-04-1*], *15*:805
  as catalyst, *15*:813
Nickel sulfide [*12137-08-5*] (2:1), *15*:806
Nickel sulfide [*12035-72-2*] (3:2), *15*:806
Nickel sulfide [*12137-12-1*] (3:4), *15*:806
Nickel sulfides, *15*:796
Nickel tetrakispyridine dichloride [*14076-99-4*]
  thermal decomposition, *15*:806
Nickel–titanium [*12035-74-4*] (3:1)
  by thermomechanical processing, *15*:342
Nickel trisethylenediamine dichloride [*13408-70-3*], *15*:806
Nickel trithiocarbonate [*39282-88-7*], *15*:810
Nickel tungstate [*14177-51-6*], *15*:813
Nickel tungsten sulfide
  as catalyst, *15*:813
Nickel zinc ferrites, *14*:664
Nicol prism, *16*:533
Nicotiana glauca, *13*:423
Nicotiana tobacum, *13*:422
Nicotinamide adenine dinucleotide (NAD) [*53-84-9*]
  diagnostic reagent, *15*:76
  in insecticide reaction, *13*:427
Nicotine [*54-11-5*]
  as insecticide, *13*:424
Nicotine sulfate [*65-30-5*]
  as insecticide, *13*:424
Nicotinic acid [*59-67-6*]
  in nuts, *16*:253
Nicotinoids
  as insecticides, *13*:422

Nicrocoat, *15*:257
NIDSH Registry of Toxic Effects of Chemical Substances, *13*:262
Nieraline, *15*:757
Nif Genes
  in nitrogen fixation, *15*:953
NIH-EPA information system, *13*:301
Nilodor, *16*:304
Nimonic 90
  hot torsion data, *15*:327
Nimonic 105, *15*:263
Nimonic 115
  hot torsion data, *15*:327
Nimonic Alloy 115 [*51204-31-8*], *15*:789
Nimonic Alloy 80A [*11068-71-6*], *15*:789
Niobates, *15*:836
Niobic acid [*12266-03-4*]
  properties, *15*:836
  salts, *15*:823
Niobium [*7440-03-1*]
  amalgamation with mercury, *15*:147
  cladding of, *15*:281
  in nuclear reactors, *16*:140
  from the ocean, *16*:280
  platinum cladding of, *15*:172
Niobium-92 [*13982-37-1*], *15*:820
Niobium-93 [*7440-03-1*], *13*:841
Niobium-94 [*14681-63-1*], *15*:820
Niobium-95 [*13967-76-5*], *15*:820
Niobium and niobium compounds, *15*:**820**
Niobium boride [*12045-19-1*], *15*:834
Niobium boride [*12344-74-0*] (2:1), *15*:834
Niobium boride [*56450-58-9*] (3:1), *15*:834
Niobium boride [*12045-89-5*] (3:4), *15*:834
Niobium carbide [*12069-94-2*], *15*:834
  in metallic coating, *15*:252
  niobium pentoxide reaction, *15*:826
Niobium, compd with tin [*12035-04-0*] (3:1)
  as magnetic material, *14*:683
Niobium, compound with aluminum [*12003-75-7*] (3:1), *15*:831
Niobium, compound with gallium [*12024-05-4*] (3:1), *15*:831
Niobium, compound with germanium [*12025-22-8*] (3:1), *15*:831
Niobium, compound with indium [*12030-07-8*] (3:1), *15*:831
Niobium, compound with tin [*12035-04-0*] (3:1), *15*:831
Niobium, compound with titanium [*12384-42-8*] (1:1), *15*:831
Niobium diboride [*12007-29-3*], *15*:834
Niobium dihydride [*13981-96-9*], *15*:835
Niobium dioxide [*12034-59-2*], *15*:834
Niobium diselenide [*12034-77-4*]
  as solid lubricant, *14*:508

Niobium fluorodioxide [*15195-33-2*], *15*:834
Niobium hydride [*13981-86-7*], *15*:834
Niobium nitride [*24621-21-4*], *15*:834, 877
  properties, *15*:874
Niobium nitride [*11092-17-4*] (1:1), *15*:872
Niobium nitride [*12033-43-1*] (2:1), *15*:872
Niobium nitride [*12163-98-3*] (4:3), *15*:872
Niobium oxide [*12034-57-0*], *15*:834
Niobium pentabromide [*13478-45-0*], *15*:834
Niobium pentachloride [*10026-12-7*], *15*:834
Niobium pentafluoride [*7783-68-8*], *15*:834; *16*:
  181
Niobium pentaiodide [*13779-92-5*], *15*:834
α-Niobium pentoxide [*1313-96-8*], *15*:834
Niobium tetraiodide [*13870-21-8*], *15*:835
Niobium tribromomonoxide [*14459-75-7*], *15*:834
Niobium trichloromonoxide [*13597-20-1*], *15*:834
NIOSH.   See *National Institute for*
  Occupational Safety and Health.
NIOSH NOHS information system, *13*:301
NIOSH RTECS information system, *13*:301
Nippon steel process, *13*:745
Nissol, *15*:726, 727
Nitramines
  production, *15*:841
Nitrate
  electrode for, *13*:725
Nitrate reductase [*9013-03-0*]
  Mo content, *15*:693
  molybdenum in, *15*:592
Nitrates
  formation, *15*:841
  ion exchange removal, *13*:700
Nitrating acid
  nitrobenzene synthesis, *15*:918
Nitration, *15*:**841**
  gas-phase, of toluene, *15*:929
  of naphthalene, *15*:702
  propane, vapor-phase, *15*:978
  of toluene for nitrotoluene mfg, *15*:928
α-Nitratolactic acid [*1860-18-0*]
  from propylene in oxalic acid mfg, *16*:625
Nitrazepam [*145-22-5*]
  properties of, *13*:122
Nitric acid [*7697-37-2*], *15*:**853**
  affect on ion-exchange matrix, *13*:684
  by air oxidation, *15*:943
  benzene reaction, *15*:916
  glucose reaction, *16*:623
  lithium nitrate from, *14*:464
  mercury reaction, *15*:161
  in metal pickling, *15*:301
  methacrylic acid from, *15*:357
  nickel carbonyl reaction, *15*:803
  nitration with, *15*:841
  paint stripper, *16*:765
  properties, *15*:854
  in purex process, *16*:176

toxicity, *13*:270
  use of Mg(NO$_3$)$_2$ in mfg, *14*:627
Nitric acid monohydrate [*13444-82-1*], *15*:854
Nitric acid trihydrate [*13444-83-2*], *15*:854
Nitric oxide [*10102-43-9*], *16*:699
  in air, *16*:657
  by air oxidation, *15*:943
  heme protein color reactions, *15*:62
  isotope separation of, *13*:847
  ozone reaction, *16*:700
Nitridation systems
  in N$_2$ fixation, *15*:955
Nitrides, *15*:**871**
Nitriding, *15*:317, 875
Nitrile–cellulose fiber
  gasket properties, *16*:727
Nitrile rubber [*9003-18-3*]
  gasket properties, *16*:727
Nitriles, *15*:**888**, 908.   (See also *Fatty acid*
  nitriles.)
  toxicity, *13*:270
α-Nitroacetaldehyde [*5007-21-6*], *15*:977
2-Nitroacetaldehyde oxime [*5653-21-4*], *15*:969
Nitroacetanilide sodium salt, *15*:974
Nitroacetic acid [*625-75-2*], *15*:977
Nitro alcohols, *15*:**910**
Nitroalkanes.   See *Nitroparaffins.*
Nitroalloy steels, *15*:877
Nitroamines, *15*:912
o-Nitroaniline [*88-74-4*], *15*:923, 924
p-Nitroaniline [*100-01-6*], *15*:924
  in enzyme assay, *15*:79
o-Nitroanisole [*91-23-6*], *15*:923
m-Nitrobenzaldehyde [*99-61-6*], *15*:927
o-Nitrobenzaldehyde [*552-89-6*], *15*:926
p-Nitrobenzaldehyde [*555-16-8*], *15*:928
Nitrobenzene [*98-95-3*]
  in hydroquinone production, *13*:52
  in isocyanate mfg, *13*:808
  production, *15*:846
  properties, *15*:916
Nitrobenzene and nitrotoluenes, *15*:**916**
4-Nitrobenzenediazonium chloride [*22719-23-9*]
  aziridine reaction, *13*:148
  indole coupling, *13*:215
m-Nitrobenzoic acid [*121-92-6*], *15*:927
o-Nitrobenzoic acid [*552-16-9*], *15*:926
p-Nitrobenzoic acid [*62-23-7*], *15*:927
Nitrobenzol.   See *Nitrobenzene.*
o-Nitrobenzyl chloride [*612-23-7*], *15*:926
p-Nitrobenzyl chloride [*100-14-1*], *15*:928
  sodium 2-propanenitronate reaction, *15*:974
4-(p-Nitrobenzyl)pyridine [*1083-48-3*]
  aziridine color reaction, *13*:159
p-Nitrobenzyltrimethylammonium iodide
  [*27953-21-5*]
  sodium 2-propanenitronate reaction, *15*:974

1-Nitrobutane [*627-05-4*]
  from butane, *15*:848
  properties, *15*:970
2-Nitrobutane [*600-24-8*]
  from butane, *15*:848
  properties, *15*:970
2-Nitro-1-butanol [*609-31-4*]
  properties, *15*:910
4-(2-Nitrobutyl)morpholine [*37304-88-4*]
  applications of, *13*:242
Nitrocellulose [*9004-70-0*], *15*:856
  in inks, *13*:389
  in leather coatings, *14*:221
  in waterproof matches, *15*:5
Nitrocellulose inks, *13*:390
m-Nitrochlorobenzene [*121-73-3*]
  properties, *15*:923
o-Nitrochlorobenzene [*88-73-3*]
  properties, *15*:923
p-Nitrochlorobenzene [*100-00-5*]
  properties, *15*:923
3-Nitro-4-chlorobenzenesulfonic acid [*121-18-6*],
  15:923
2-Nitro-4-chlorotoluene [*89-59-8*], *15*:926
2-Nitro-6-chlorotoluene [*83-42-1*], *15*:926
4-Nitro-2-chlorotoluene [*121-86-8*], *15*:928
4-Nitro-*o*-cresol [*99-53-6*]
  mercuration, *15*:164
Nitrocyclohexane [*1122-60-7*], *15*:979
  production, *15*:849
  properties, *15*:970
Nitrocyclopropane [*13021-02-8*]
  from 3-iodo-1-nitropropane, *15*:975
O-Nitro-β-dimethyl amino styrene [*58645-51-5*]
  indole preparation, *13*:218
2-Nitrodiphenylamine [*119-75-5*], *15*:923
Nitroethane [*79-24-3*]
  anhydrous acid, *15*:975
  from propane and butane, *15*:848
  properties, *15*:970
aci-Nitroethane. See *Isonitroethane.*
1-Nitro-1-ethanenitronic acid. See *2-Isonitro-*
  1-nitroethane.
N-[α-(1-Nitroethyl)benzyl]-ethylenediamine
  [*14762-38-0*]
  applications of, *13*:242
Nitroethylene [*3638-64-0*]
  isoprene reaction, *13*:821
Nitroform. See *Trinitromethane.*
Nitrogen [*7727-37-9*], *15*:883, **932**, 941
  in air, *16*:657
  chemical reduction, *15*:953
  as insulator, *13*:549
  ion implantation, *13*:717
  lithium reaction, *14*:455
  as lubricant, *14*:514
  manufacture, *15*:935

  from the ocean, *16*:279
  permeability in polymers, *15*:118
  properties, *15*:933
Nitrogen-14 [*17778-88-0*], *13*:840
Nitrogen-15 [*14390-96-6*], *13*:840
Nitrogenase [*9013-04-1*], *15*:942
  Mo content, *15*:693
  molybdenum in, *15*:592
  in N_2 fixation, *15*:950
  x-ray absorption spectroscopy, *15*:961
Nitrogen chloride [*10025-85-1*], *15*:872
Nitrogen compounds
  ozone reactions, *16*:687
Nitrogen cycle, *15*:941
Nitrogen dioxide [*10102-44-0*], *16*:699
  in air, *16*:657
  by air oxidation, *15*:943
  isoprene reaction, *13*:826
  nitrating agent, *15*:978
  from ozone, N_2 reaction, *15*:934
  paraffin reactions, *15*:848
  propylene oxidation to oxalic acid, *16*:625
Nitrogen fixation, *15*:**941**
  biological, *15*:946
  chemical, *15*:953
  industrial processes, *15*:943
  for nitric acid mfg, *15*:864
Nitrogen fluoride [*10036-47-2*] (2:4)
  S_2F_{10} reaction, *14*:74
Nitrogen iodide [*13444-85-4*], *13*:659, 660; *15*:
  872. (See also *Iodine azide.*)
Nitrogen iodide triammoniate [*15823-38-8*], *15*:
  875
Nitrogen oxide [*10102-43-9*], *15*:872
Nitrogen-15 oxide [*15917-77-8*]
  in isotope separation, *13*:849
Nitrogen oxide-18 [*15917-78-9*]
  in isotope separation, *13*:849
Nitrogen oxides
  exhaust abatement, *15*:863
  in incinerator gases, *13*:185
  ozone reactions, *16*:693
  toxicity, *13*:271
Nitrogen pentoxide [*10102-03-1*]
  for *n*-alkane nitration, *15*:978
Nitrogen reduction. See *Nitrogen fixation.*
Nitrogen sulfide [*12033-56-6*], *15*:872
Nitrogen sulfide [*25474-92-4*] (2:2)
  poly(sulfur nitride) from, *13*:407
Nitrogen sulfide [*28950-34-7*] (4:4)
  poly(sulfur nitride) from, *13*:407
Nitrogen tetroxide [*10544-72-6*]
  methacrylic acid from, *15*:357
Nitrogen triiodide [*13444-85-4*], *13*:661
Nitrogen triiodide ammoniate [*14014-86-9*], *13*:
  661
Nitroglycerin [*55-63-0*], *15*:856
  production, *15*:846

Nitroglycerin (*Continued*)
  toxicity, *13*:271
Nitrohexahydropyrimidines, *15*:912
Nitrohydrocarbons
  explosives, *15*:851
α-Nitroketones, *15*:974
γ-Nitro ketones
  from nitro paraffins, *15*:977
Nitrolic acids
  from nitroparaffins, *15*:974
Nitrolignin [*8068-08-4*]
  decay resistance, *14*:309
Nitromalondianilide sodium salt, *15*:974
Nitromersol [*133-58-4*], *15*:169
Nitrometer, *15*:868
Nitromethane [*75-52-5*]
  in explosives, *15*:983
  from propane and butane, *15*:848
  properties, *15*:970
aci-Nitromethane acid chloride [*76794-26-8*], *15*:975
1-Nitro-2-methylpropane [*625-74-1*]
  properties, *15*:970
2-Nitro-2-methylpropane [*594-70-7*], *16*:687
  properties, *15*:970
Nitron [*2218-94-2*]
  in nitric acid analysis, *15*:867
1-Nitronaphthalene [*86-57-7*], *15*:700, 702, 725
2-Nitronaphthalene [*581-89-5*], *15*:702, 725
Nitronaphthalene amines
  deamination, *15*:724
Nitronaphthalenecarboxylic acids
  decarboxylation, *15*:724
Nitronaphthalenes, *15*:724
Nitronaphthalenesulfonic acids, *15*:724
5-Nitro-1-naphthalenesulfonic acid [*17521-00-5*], *15*:726
5-Nitronaphthalene-2-sulfonic acid [*89-69-1*], *15*:723, 726
8-Nitronaphthalene-2-sulfonic acid [*117-41-9*], *15*:723
  manufacture, *15*:726
Nitronic acids, *15*:969
Nitronic anhydrides, *15*:974
Nitronium ion, *15*:842
Nitroolefins, *15*:913
2-Nitrooxyoctanoic acid [*14312-94-8*]
  hydrolysis, *13*:95
Nitroparaffins, *15*:**969**
  manufacture, *15*:848
  nitro alcohols from, *15*:910
o-Nitrophenol [*88-75-5*], *15*:923
  mercuration, *15*:164
p-Nitrophenol [*100-02-7*], *15*:924
  from parathion, *13*:439
p-Nitrophenolate [*88-75-5*]
  in diagnostic testing, *15*:79

1[(*p*-Nitrophenyl)azo]aziridine [*26439-16-7*], *13*:148
3-(4'-Nitrophenylazo)indole [*53330-79-3*], *13*:215
2-Nitro-1-phenylbutanol [*7252-51-9*], *15*:976
2-Nitro-1-phenyl-1-butene [*1202-32-0*]
  reduction, *15*:976
p-Nitrophenyl diethyl phosphate [*311-45-5*], *13*:451
2-Nitrophenyl *n*-octyl ether [*37682-29-4*]
  as ion-exchanger solvent, *13*:725
p-Nitrophenylphosphate [*330-13-2*]
  in diagnostic testing, *15*:79
2-Nitro-1-phenyl-1-propene [*705-60-2*], *15*:976
o-Nitrophenylpyruvic acid [*5461-32-5*]
  cyclization, *13*:219
1-Nitropropane [*108-03-2*]
  benzaldehyde reaction, *15*:976
  formaldehyde–sulfite reaction, *15*:976
  from propane and butane, *15*:848
  properties, *15*:970
  sulfuric acid reaction, *15*:975
2-Nitropropane [*79-46-9*]
  electrolytic oxidation, *15*:977
  in paint and varnish removers, *16*:764
  from propane and butane, *15*:848
  properties, *15*:970
aci-Nitropropane. See *2-Isonitropropane.*
2-Nitro-1,3-propanediamines, *15*:976
2-Nitro-1,3-propanediol [*1794-90-7*], *15*:910
2-Nitroresorcinol [*601-89-8*]
  in explosive detonators, *13*:60
5-Nitrosalicylic acid [*96-97-9*]
  in niobium analysis, *15*:829
N-Nitrosamines, *15*:**988**. (See also *N-Nitrosodialkylamines.*)
N-Nitrosoacetanilide [*938-81-8*]
  in initiator systems, *13*:367
Nitrosoacyloxy compounds, *15*:974
Nitrosobenzene [*586-96-9*], *15*:917
  isoprene reaction, *13*:826
  phenyl isocyanate reaction, *13*:795
N-Nitrosodialkylamines, *15*:988. (See also *N-Nitrosamines.*)
N-Nitrosodicyclohexylamine [*947-92-2*]
  toxicity, *15*:991
N-Nitrosodiethanolamine [*1116-54-7*]
  toxicity, *15*:991
N-Nitrosodiethylamine [*55-18-5*]
  toxicity, *15*:991
N-Nitrosodiisopropylamine [*601-77-4*]
  toxicity, *15*:991
N-Nitrosodimethylamine [*62-75-9*]
  toxicity, *15*:991
N-Nitrosodipropylamine [*621-64-7*]
  toxicity, *15*:991
Nitrosomethane [*865-40-7*], *15*:975

N-Nitrosomethyl(benzyl)amine [937-40-6]
  toxicity, 15:991
N-Nitrosomorpholine [59-89-2], 15:988
  toxicity, 15:991
1-Nitroso-2-naphthol [131-91-9], 15:735
2-Nitroso-2-nitropropane [5275-46-7]
  oxidation, 15:977
N-Nitrosonornicotine [16543-55-8]
  toxicity, 15:991
p-Nitrosophenol [104-91-6]
  in hydroquinone production, 13:52
N-Nitrosopiperidine [100-75-4]
  toxicity, 15:991
N-Nitrosoproline [7519-36-0]
  toxicity, 15:991
N-Nitrosopyrrolidine [930-55-3], 15:988
  toxicity, 15:991
β-Nitrostyrene [102-96-5]
  in amino-2-phenyl-4Δ-1 pyrroline synthesis,
    14:803
Nitrosyl chloride [2696-92-6]
  aziridine reaction, 13:150
  ketene reaction, 13:876
  from N_2, 15:933
m-Nitrotoluene [1321-12-6]
  properties, 15:926
  suppression in TNT mfg, 15:843
o-Nitrotoluene [88-72-2]
  properties, 15:926
p-Nitrotoluene [99-99-0]
  properties, 15:927
Nitrotoluenes, 15:916. (See also *Nitrobenzene
    and nitrotoluenes.*)
Nitrous acid [7782-77-6]
  by air oxidation, 15:943
  in hydroxyacetic acid preparation, 13:90
Nitrous oxide [10024-97-2], 15:883
  in air, 16:657
  in latex discoloration, 14:90
  from ozone, N_2 reaction, 15:934
Nitrous oxides
  from pulp mills, 16:795
Nitroxyl [14332-28-6], 15:857
Nittetu process
  for incineration, 13:200
NMFS information system, 13:301
NMP. See *2-Methyl-2-nitro-1-propanol.*
NMPD. See *2-Methyl-2-nitro-1,3-propanediol.*
N,N'-Bis(2-hydroxy benzyl)ethylenediamine-
    N,N'-diacetic acid [35998-29-9], 13:786
N,N'-Bis(4-sulfonamidophenyl)urea [1773-43-9],
    13:801
NOBIN information system, 13:313
Noble-gas formalism, 16:595
Noble-metal-coated titanium, 15:173

No-carbon-required
  carbonless copy paper, 15:477
Nocardia, 13:786
Nocardicin C [59511-12-5]
  microbial hydrolysis of, 15:463
Nodules
  ocean minerals, 16:291
Nodulizing
  iron ore, 13:739
Noise
  effects of exposure, 16:1
  insulation against, 13:513
  measurement, 16:4
Noise attenuation
  by laminated glass, 13:990
Noise control, 16:14
Noise controls
  industrial equipment, 16:19
Noise exposure
  of individuals, 16:7
Noise pollution, 16:1
Noise reduction coefficient (NRC), 13:515
Noise regulations, 16:18
Noludar, 13:122
Nomenclature, 16:28
Nomex
  in electrical insulation, 13:556
1-Nonadecene [18435-45-5]
  properties, 16:481
2,4-Nonadienoic acid [21643-39-0]
  mesomorphism of, 14:412
Nonakis(benzenethiolato)octathioxohexaferrate-
    dimolybdate(3−) [68197-68-2], 15:963
γ-Nonalactone [104-61-0]
  in perfumes, 16:967
Nonanol [28473-21-4], 16:312
Nondairy products, 15:568
Nondestructive test
  stimulated signals, 16:49
Nondestructive testing, 16:47
  standards, 15:37
Nondestructive tests
  for clad metals, 15:285
1-Nonene [174-11-8]
  properties, 16:481
Nonenes
  production, 16:494
Nonlinear optics
  in lasers, 14:64
Nonlinear program
  in operations planning, 16:504
Nonlinear programming
  industrial applications, 13:508
Nonmeasurement imaging, 13:341
Nonwoven fabrics
  in reinforced plastics, 13:970

Nonwovens
  fibers for, *16*:376
Nonwoven textile fabrics
  spunbonded, *16*:**72**
  staple fibers, *16*:**104**
Nonwoven textile fabrics, spunbonded, *16*:
  72. (See also *Textiles, nonwoven.*)
Nonylphenol [*104-40-5*], *16*:497
Nonylphenyl methacrylate [*76391-98-5*]
  properties, *15*:353
Nootkatone [*4674-50-4*]
  from grapefruit oil, *16*:323
Nootropyl [*7491-74-9*], *15*:135. (See also
  Piracetam.)
Nopcocide N-96, *16*:755
Noradrenaline [*51-41-2*], *15*:773. (See also
  Norepinephrine.)
  in (−) adrenaline biosynthesis, *15*:758
  in memory enhancement, *15*:133
Noradrenaline *d*-bitartrate [*51-40-1*]
  properties, *15*:773
Noradrenaline hydrochloride [*55-27-6*]
  properties, *15*:773
Norbornadiene [*121-46-0*]
  copper(II) halide reaction, *16*:611
Nordefrine [*61-96-1*], *13*:62
Norden, *15*:774
Norepinephrine [*586-17-4*], *13*:62; *15*:758
  properties, *15*:773
Norma-Hoffmann bomb test, *14*:503
Nornicotine [*494-97-3*]
  as insecticide, *13*:424
Norpatchoulenol [*41429-52-1*]
  in perfumes, *16*:953
Norphen, *15*:774
Norrish Type I cleavage, *13*:369
Norsk Hydro process
  for magnesium, *14*:578
Noryl, *15*:266
Nostoc, *15*:949
Novasural, *15*:166
Novaweb AB-17, *16*:76
Novelty. See *Patents, practice and
  management.*
Novolac resin, *16*:125
Novoloid fibers, *16*:**125**
  acetylated, *16*:131
Nozzle process
  in isotope separation, *13*:858
NPH iletin [*9004-17-5*], *13*:612
NPH insulin [*9004-17-5*], *13*:612
NPV. See *Nuclear polyhedrosis virus.*
NR. See *Rubber, natural.*
NRC. See *Noise reduction coefficient.*
N. rustica, *13*:422
NTIS
  information service, *16*:902

NTU retort process, *16*:338
Nuclear facilities
  safety of, *16*:216
Nuclear fuel
  reserves, *16*:143
Nuclear fuels
  reprocessing, *16*:173
Nuclear polyhedrosis virus
  (NPV)
  in insect control, *13*:471
Nuclear radiation
  in level measurement, *14*:434
Nuclear reactors
  chemical reprocessing, *16*:**173**
  fuel-element fabrication, *16*:**165**
  introduction, *16*:**138**
  natural, *13*:861
  nitrogen oxides produced in, *15*:864
  nuclear fuel reserves, *16*:**143**
  radiation exposure from, *16*:217
  safety in nuclear facilities, *16*:**216**
  special engineering for radiochemical plants,
    16:**239**
  water chemistry of light-water reactors, *16*:**150**
Nuclear reactors, fast breeder reactors, *16*:**184**
Nuclear reactors, isotope separation, *16*:**161**
Nuclear reactors, waste management, *16*:**206**
Nuclear-track detectors
  particle identification, *16*:839
Nucleation
  of emulsion polymers, *14*:83
Nucleic acids
  as antigens, *13*:167
  phosphorus in, *15*:587
Nuclepore, *15*:104
Numerical data bases, *13*:304
Nuodex, *13*:237
Nusselt number, *14*:978
Nutmeg, *16*:251
  poisoning by, *16*:319
Nutmeg oil [*8008-45-5*], *16*:319, 325
Nutrient, *14*:885
Nutrients, mineral, *15*:570
Nuts, *16*:**248**
  fatty acid values, *16*:252
  protein in, *16*:249
Nylidrin hydrochloride [*849-5-8*]
  treatment for senility, *15*:138
Nylon [*32131-17-2*]
  in insulation enamels, *13*:580
  in insulation, *13*:556
  in leatherlike materials, *14*:242
  membrane formation, *15*:104
  metallic coating of, *15*:266
  nonwovens from, *16*:108
  in nonwovens, *16*:74
  in packings, *16*:734

Nylon-1 [*32010-01-8*]
  isocyanate reaction product, *13*:794
Nylon-6 [*25038-54-4*]
  adhesive treatment, *13*:57
  nonwovens from, *16*:108
  in nonwovens, *16*:74
  properties, *16*:370
Nylon-9 [*25035-03-4*], *16*:688
Nylon-6,6 [*9011-57-6*]
  adhesive treatment, *13*:57
  from adiponitrile, *15*:897
  gas permeability, *15*:118
  moisture regain, *16*:358
  nonwovens from, *16*:108
  in nonwovens, *16*:75
  properties, *16*:358, 370
  thermal transitions of fiber, *16*:358
Nylon fibers
  in papermaking, *16*:823
Nylons
  in composites, *13*:973
Nyquist method, *13*:501

O

Oak
  catechol in, *13*:46
Oakes process, *14*:239
Oakmoss [*9004-50-4*]
  in perfumes, *16*:951
OBOG. See *On-board oxygen generation.*
Occupational Safety and Health Act, *13*:254
Ocean-mining methods, *16*:294
Ocean raw materials, *16*:**277**
trans-α-Ocimene [*6874-10-8*], *16*:313
Ocimum basilicum L., *16*:321
Oconee 1, *16*:155
Ocotea cymbarum oil [*68917-09-9*], *16*:325
Ocotea pretiosa Benth., *16*:325
2,3,4,5,6,7,8,8-Octachloro-2,3,3a,4,7,7a-hexahy-
  dro-4,7-methanoindene [*309-00-2*]
  as insecticide, *13*:434
1,3,4,5,6,7,8,8-Octachloro-3a,4,7,7a-tetrahydro-
  4,7-methanophthalan [*297-78-9*]
  as insecticide, *13*:434
2,2,5-*endo*-6-*exo*-8,9,9,10-Octachlorobornane
  [*58002-18-9*]
  as insecticide, *13*:433
Octachlorocyclopentane
  MoO$_2$ reaction, *15*:684
1-Octadecene [*112-88-9*]
  properties, *16*:481

9-Octadecene [*5557-31-3*], *16*:484
Octadecyl 3-(3,5-di-*tert*-butyl-4-hydroxyphenyl)-
  propionate [*2082-79-3*]
  LDPE antioxidant, *16*:414
  in LLDPE, *16*:395
Octadecyl isocyanate [*112-96-9*]
  properties, *13*:789
p'-n-Octadecyloxy-3'-nitrodiphenyl-*p*-carboxylic
  acid [*21351-71-3*]
  liquid crystalline range of, *14*:399
cis-3,*cis*-13-Octadienyl acetate [*53120-27-7*]
  as insect pheromone, *13*:481
trans-3,*cis*-13-Octadienyl acetate [*53120-26-6*]
  as insect pheromone, *13*:481
Octahydroindole [*4375-14-8*], *13*:217
1,1'-(1,1,3,3,5,5,7,7-Octamethyl-7-phenyltetrasil-
  oxanyl)-ferrocene [*12321-18-5*], *15*:200
Octamethyl pyrophosphoramide [*152-16-9*]
  as insecticide, *13*:444
Octanal [*124-13-0*], *16*:312
  in orange oil, *16*:325
2,3-Octanedione [*585-25-1*]
  properties, *13*:926
2,4-Octanedione [*14090-87-0*]
  properties, *13*:928
3,5-Octanedione [*6320-18-9*]
  properties, *13*:928
3,6-Octanedione [*2955-65-9*]
  properties, *13*:931
4,5-Octanedione [*5455-24-3*]
  properties, *13*:926
Octanoic acid [*124-07-2*]
  in ink driers, *13*:379
Octanol [*111-87-5*], *16*:312
2-Octanol [*123-96-6*]
  dehydrogenation of, *13*:933
1-Octene [*111-66-0*]
  copolymer with ethylene, *16*:388, 484
  disproportionation, *16*:484
  properties, *16*:481
Octopamine [*104-14-3*]
  as neuroregulator, *15*:774
Octopamine hydrochloride [*770-05-8*], *15*:774
2-*t*-Octylaminoethyl methacrylate [*14206-24-1*]
  properties, *15*:354
n-Octyl chloride [*111-85-3*]
  magnesium reaction, *16*:561
2-*n*-Octyl-4-isothiazolin-3-one [*26530-20-1*]
  applications of, *13*:245
n-Octyl methacrylate [*2157-01-9*]
  properties, *15*:353
p-n-Octyloxybenzoic acid [*2493-84-7*]
  liquid crystalline range of, *14*:399

Octylphenol poly(oxyethylene) [9036-19-5]
    latex stabilization, 14:88
4-n-Octylphenyl salicylate [2512-56-3]
    HDPE stabilizer, 16:446
Odor
    from pulp mills, 16:795
Odor control, 16:297.   (See also Odor
        modification.)
    by ozonation, 16:711
Odor counteraction.   See Odor modification.
Odor modification, 16:297
OECD.   See Organization for Economic
        Cooperation and Development.
Offenlegungsschrift, 16:891
Offset bond, 16:772
Offshore petroleum.   See Ocean raw materials.
Ohmart cell, 14:434
Ohm's law, 13:534
    in materials standards, 15:40
Oil additives
    polymethacrylates, 15:393
Oil-field equipment
    nondestructive testing of, 16:57
Oil industry
    operations planning, 16:518
Oil inks, 13:379
Oil, lubricating
    emissivity, 13:342
Oil of mirbane.   See Nitrobenzene.
Oil of wintergreen [119-36-8]
    odor, 16:301
Oil recovery
    xanthan gum in, 15:450
Oils
    nut, 16:271
Oil seals, 16:740
Oils, essential, 16:307
Oil shale, 16:333.   (See also Shale oil.)
    microwave heating, 15:516
Oils, mineral
    as insulating liquids, 13:551
Oils, synthetic, 14:496
Oil tannage, 14:217
Oiticica nut, 16:251
Oiticica oil [8016-35-1]
    in varnishes, 13:554
Olduvai gorge
    dating of, 16:844
Olefin feedstocks
    oxo process, 16:643
Olefin fibers, 16:357.   (See also Polyolefin
        fibers.)
Olefin polymers
    high pressure (low and intermediate density)
        polyethylene, 16:402
    linear (high density) polyethylene, 16:421

low pressure linear (low density) polyethylene,
    16:385
polymers of higher olefins, 16:471
polypropylene, 16:454
Ziegler process polyethylene, 16:433
Olefins
    detergent range, 16:493
    oxo process, 16:637
    separation by molecular sieves, 15:664
α-Olefins, 16:572
Olefins, higher, 16:480
    polymers of, 16:471
Olefinsiv process, 15:664
Oleic acid [112-80-1]
    in grease thickeners, 14:502
    as ink vehicle, 13:395
    in nut oils, 16:252
    ozonation, 16:686
    paint removers, 16:763
    solvent purification, 15:983
    in soybean oil, 14:251
Olex process, 15:664
Oleyl alcohol [143-28-9]
    solvent purification, 15:983
Oleyl methacrylate [13533-08-9]
    properties, 15:353
Oleyl nitrile [112-91-4]
    properties, 15:908
Olfaction, 16:297
Olibanum [8016-36-2]
    in perfumes, 16:952
Oligomerization
    higher olefins by, 16:485
Oligomerization catalysts, 16:488
Oligomerization of
    ethylene, 16:487
Oligosaccharides
    from dextran sucrose, 15:441
Olive oil emulsion
    in lipase measurement, 15:80
Olivin [6680-06-4], 14:797
Olivine [1317-71-1], 14:615
    as track-etch detectors, 16:827
OMCVD.   See Metallo-organic CVD.
Omniseal, 16:726
On-board oxygen-generation (OBOG), 16:681
Onchocerca volvulus, 13:413
Onchorhynchus suckley, 13:612
Oncology
    chemotherapeutics in, 13:172
One-step ethylene oligomerization process, 16:488
Onion
    dipropyl disulfide in, 16:311
Onion oil [8002-72-0], 16:308
Onions
    catechol in, 13:46

On-line searching, *13*:279
Onyxicide 500, *13*:231
Onyxide 200, *13*:242
Oölitic aragonite, *16*:286
Operations management, *16*:500.   (See also
    Operations planning.)
Operations planning, *16*:**500**
Operations research, *16*:500.   (See also
    Operations planning.)
  in plant control, *13*:508
Ophiocaryon paradoxum, *16*:251
Opiates
  measurement, *15*:88
Opium
  alkaloids from, *13*:63
Opopanax [*8021-36-1*]
  in perfumes, *16*:952
Optical activity
  hydroxy acids, *13*:113
Optical activity of liquid crystals, *14*:409
Optical communications
  lasers in, *14*:71
Optical comparators
  as nondestructive testing devices, *16*:49
Optical filters, *16*:**522**
Optical memory systems, *14*:705
Optical pumping, *14*:44
Optics
  nonlinear, *14*:64
Optimality principle, *16*:506
Optimization
  operations planning for, *16*:519
Orange-blossom oil [*8016-38-4*]
  indole in, *13*:213
Orange G [*1936-15-8*]
  in dye-binding test of milk, *15*:552
Orange oil [*8008-57-9*], *16*:319
Orange oil, bitter [*68916-04-1*], *16*:325
Orange oil sweet [*8008-57-9*], *16*:325
  in orange oil, *16*:325
Orbignya martiana, *16*:251
ORBIT
  information system, *16*:901
ORBIT information system, *13*:291
Orcinol [*504-15-4*], *13*:39
Ore
  leaching with hydrogen peroxide, *13*:29
OREIA-DIG kit, *15*:86
Ore leaching
  with H_2O_2, *13*:29
Org 2766, *15*:133
Organic molecules
  microbial oxidation, *15*:459
Organic nomenclature, *16*:36
Organization for Economic Cooperation and
    Development
  information sources, *13*:308

Organoalkali compounds, *16*:556
Organoaluminum compounds, *16*:565
Organoberyllium compounds, *16*:558
Organoboron compounds, *16*:562
Organo-dilithium compounds, *14*:471
Organogallium compounds, *16*:562
Organogermanium compounds, *16*:572
Organoindium compounds, *16*:562
Organolead compounds, *16*:579
  properties, *14*:181
Organolithium compounds, *14*:466; *16*:556
Organomagnesium compounds, *16*:558
Organomercurials, *16*:586
Organomercury compounds, *15*:162
Organometallics—σ-bonded alkyls and aryls, *16*:
    555
Organometallics—metal π complexes, *16*:**592**
Organomolybdenum compounds, *15*:691
Organonickel compounds, *15*:806
Organophosphates
  toxicosis from, *15*:598
Organophosphorus isocyanates, *13*:801
Organosilicon compounds, *16*:572
Organosodium compounds, *16*:557
Organosol
  in leatherlike material, *14*:236
Organosolv lignin [*8068-03-9*], *14*:309
Organothallium compounds, *16*:562
Organothiocyanates
  as insecticides, *13*:429
Organotin compounds, *16*:573
  toxicity, *13*:271
Organotins
  as antimicrobial agents, *13*:238
Organotitanium compounds, *16*:584
Organozinc compounds, *16*:585
Oriental fruitmoth
  control of with pheromones, *13*:480
Origanum oil [*8007-11-2*], *16*:325
Orinase, *13*:617
Orinase diagnostic, *13*:617
O-ring packing, *16*:733
Ornithine [*70-26-8*]
  in spermine biosynthesis, *15*:777
Ornithine decarboxylase [*9024-60-6*]
  in spermine biosynthesis, *15*:778
ORNL EMIC information system, *13*:301
ORNL ETIC information system, *13*:301
Oros pump
  for drug release, *15*:488
Orotan, *14*:216
Orotic acid [*65-86-1*]
  effect on memory, *15*:136
Orris [*8002-73-1*]
  in perfumes, *16*:952

Orris root oil [*8002-73-1*], *16*:326

Orthene.   See *o-Dichlorobenzene.*

Orthoarsenic acid [*7778-39-4*]
   as insecticide, *13*:420

Orthocide, *13*:245

Orthodontics
   methacrylates, *15*:393

Orthoesters
   from nitriles, *15*:891

Orthomumetal, *14*:659

Orthonol, *14*:659

Oryctes rhinoceros, *13*:481

Oryzaephilus surinamensis, *13*:465

Orzelith, *13*:696

OS-124, *14*:496

Oscillators
   backward-wave, *15*:503

OSHA.   See *Occupational Health and Safety Administration.*

Oskarshamn 1, *16*:160

Osmium-184 [*14922-68-0*], *13*:842

Osmium-186 [*13982-09-7*], *13*:842

Osmium-187 [*15766-52-6*], *13*:842

Osmium-188 [*14274-81-8*], *13*:842

Osmium-189 [*15761-06-5*], *13*:842

Osmium-190 [*14274-79-4*], *13*:842

Osmium-192 [*15062-08-5*], *13*:842

Osmocene [*1273-81-0*], *16*:598

Osmometry, *15*:122

Osmosis
   pressure-retarded, *15*:125

Osmotic pressure
   in ion exchange, *13*:682
   thermodynamic definition, *15*:113

Osmotic pumps, *15*:124

Osmotic shock
   of ion-exchange materials, *13*:679

Ospolot, *13*:140

Osteomalacia
   nutrients associated with, *15*:573

Osteoporosis
   nutrients associated with, *15*:573

Ostion, *13*:696

Ostrinia nubilalis, *13*:481

Oticica
   uses, *16*:270

Otophora fruticosa, *16*:250

Otto cycle engines
   nitroparaffin fuel for, *15*:984

Outokumpu flash-smelting process
   for lead, *14*:130

Ovary
   liquid crystals in, *14*:418

Overwork
   effect of, *13*:122

Ovex [*80-33-1*], *13*:463

Ovine ACTH [*39319-42-1*], *15*:758

OWA-20, *13*:275

3-Oxabicyclo[3.1.1]heptane-2,4-dione [*4462-97-9*], *14*:775

Oxalacetic acid [*328-42-7*], *13*:106

Oxalate coatings
   on metals, *16*:630

Oxalic acid [*144-62-7*], *14*:795; *16*:**618**
   in anodizing, *15*:309
   from hydroxyacetic acid, *13*:91
   in mothproofing, *13*:479
   properties, *16*:618

Oxalic acid dihydrate [*6153-56-6*], *16*:618

Oxalic acid poisoning
   treatment, *16*:629

Oxaloacetate
   in SGOT measurement, *15*:76

Oxaloacetic acid [*328-42-7*]
   in diagnostic reagent, *15*:76
   glutamic acid reaction, *15*:761

Oxalyl chloride [*79-37-8*], *16*:619

Oxamide [*471-46-5*], *16*:621

Oxamyl [*23135-22-0*], *13*:454, 889

Oxazepam [*604-75-1*]
   in temazepam synthesis, *13*:130

Oxazines
   from nitroparaffins, *15*:977

Oxazoles
   from cyanohydrins, *15*:892

Oxazolidin-2-ones
   from isocyanates, *13*:797

Oxazolidynylethyl methacrylate [*46235-93-2*]
   properties, *15*:354

Oxidant
   hydrogen peroxide as, *13*:15

Oxidation
   ammonia, *15*:856
   coals, *14*:331
   initiators used in, *13*:356
   in insulation materials, *13*:544
   metallocenes, *16*:603
   nitric oxide, *15*:857
   nitroparaffins, *15*:976
   oxygen role in, *16*:656
   by ozone, *16*:683
   by permanganates, *14*:861

Oxidative phosphorylation
   magnesium requirement, *15*:583

Oxidizing agent.   See *Oxidants.*
   iodine, *13*:654
   iodoform, *13*:670
   manganese dioxide, *14*:889
   potassium permanganate, *14*:889

Oxidizing agents
   in metal treating, *15*:305

Oximinonitriles, *15*:892

Oxindole [*59-48-3*]
   indole from, *13*:213

Oxine
  in niobium analysis, *15*:829
Oxinol, *15*:412
Oxirane process
  t-butyl alcohol from, *15*:357
Oxiranylmethyl methacrylate [*106-91-2*], *15*:352
5-Oxodigluconic acid
  oxidation to glycolic acid, *13*:92
3-Oxopentanedioc acid dichloride [*75379-06-5*]
  from ketene, *13*:876
4-Oxo-4-phenyl-2-butenoic acid [*583-06-2*], *14*:777
2-Oxo-3-piperidinecarboxylic acid, ethyl ester
  [*3731-16-6*], *14*:799
Oxo process, *16*:483, **637**
Oxo reaction, *16*:592
  catalysts, *16*:611
2-Oxo-2-[(tributylstannyl)oxy]ethanamine
  [*45213-29-4*]
  copolymers from, *15*:205
Oxotrichloromolybdenum(v) [*13814-74-9*]
  nitrogen reduction catalyst, *15*:958
1,3-Oxy-acid [*3771-14-0*], *15*:738. (See also *4-Hydroxy-2-naphthalenesulfonic acid.*)
1,4-Oxy-acid [*84-87-7*], *15*:738. (See also *4-Hydroxy-1-naphthalenesulfonic acid.*)
10,10'-Oxybisphenoxyarsine [*58-36-6*]
  applications of, *13*:236
2,2'-Oxybis(4,4,6-trimethyl-1,3,2-dioxaborinane)
  [*14697-50-8*]
  as fuel preservative, *13*:247
Oxycarbonylation
  propylene to methacrylate, *15*:365
Oxychlordane [*26880-48-8*], *13*:436
Oxychloride cements, *14*:624
Oxydemeton methyl [*301-12-2*], *13*:446
Oxydiacetic acid [*110-99-6*], *13*:94. (See also *Diglycolic acid.*)
Oxygen [*7782-44-7*], *16*:**653**
  adsorption on zeolites, *15*:659
  from air by molecular sieves, *15*:664
  in air, *16*:657
  in blood, *16*:670
  corona dissociation, *16*:691
  effect on nuclear reactor materials, *16*:152
  effect on particle tracks, *16*:836
  hydrogen reaction, *13*:23
  lithium reaction, *14*:455
  as lubricant, *14*:514
  methacrylate copolymer, *15*:384
  in missile fuel, *14*:518
  in nitration, *15*:848
  from the ocean, *16*:279
  permeability in polymers, *15*:118
  properties, *16*:653
  in steelmaking, *16*:666

Oxygen-16 [*17778-80-2*], *13*:840; *16*:654
Oxygen-17 [*13968-48-4*], *13*:840; *16*:654
Oxygen-18 [*14797-71-8*], *13*:840; *16*:654
Oxygen–ammonia process
  for copper oxidation, *16*:668
Oxygenation
  hyperbaric, *16*:670
Oxygen difluoride [*7783-41-7*], *16*:732
Oxygen equipment
  from mica, *15*:436
Oxygen-generation systems, *16*:**673**
Oxygen inhalators, *16*:670
Oxygen masks, *16*:671
Oxygen regeneration
  lithium peroxide in, *14*:465
Oxygen, singlet [*17778-80-2*], *16*:669
Oxygen systems
  in hospitals, *16*:670
Oxyhemoglobin [*9062-91-3*], *16*:670
Oxy-Koch's acid [*3316-02-7*], *15*:739. (See also *8-Hydroxy-1,3,6-naphthalenetrisulfonic acid.*)
Oxymyoglobin
  protein meat pigment, *15*:62
β-Oxynaphthoic acid [*2283-08-1*]
  nickel pigment formation, *15*:814
Oxysulfate cements, *14*:638
Oxythioquinox [*2439-01-2*], *13*:463
Oxy-Tobias acid [*567-47-5*], *15*:738. (See also *2-Hydroxy-1-naphthalenesulfonic acid.*)
Oxytocin [*50-56-6*]
  in avoidance responses, *15*:133
  sulfur in, *15*:587
Oyster nut, *16*:251
Oyster shells
  CaCO₃ source, *14*:345
Oyster-shucking
  microwaves in, *15*:514
Ozonation. See *Ozone.*
Ozone [*10028-15-6*], *16*:**683**
  in air, *16*:657
  from ink manufacture, *13*:397
  nitrogen reaction, *15*:933
Ozone monitor
  in aircraft, *16*:699
Ozonides, *16*:684
Ozonolysis
  of fumaric acid, *14*:779

P

PA. See *Pressinamide.*
PAA. See *p-Azoxyanisole.*

Pace, *13*:230
Pacemakers, cardiac
    radiation exposure from, *16*:217
Packaging
    paint, *16*:756
    paper as, *16*:768
Packaging materials, industrial, *16*:**714**
Pack carburizing, *15*:315
Packing materials, *16*:**725**
Packings, plastic, *16*:733
Pacol-Olex process, *16*:492
Pacol-Olex processes, *16*:485
Paddle mixers, *15*:629
PAG.   See *Vasopressin.*
Paget's disease
    diagnosis, *15*:80
    nutrients associated with, *15*:573
    treatment, *15*:586
Pails
    as packaging, *16*:716
Paint, *16*:**742**
    fluorescent pigments in, *14*:566
    mercury preservatives, *15*:163
Paint and varnish removers, *16*:**762**
Paint, oil
    emissivity, *13*:342
Paint preservatives, *16*:587
Pain transmission
    enkephalins involved in, *15*:765
Paints
    mercury preservatives, *15*:158
    packaging of, *16*:717
PAIS INTERNATIONAL information system,
    13:330
Palladium [*7440-05-3*]
    in coatings, *15*:241
Palladium-102 [*14833-50-2*], *13*:841
Palladium-104 [*15128-18-4*], *13*:841
Palladium-105 [*15749-57-2*], *13*:841
Palladium-106 [*14914-59-1*], *13*:841
Palladium-108 [*15749-58-3*], *13*:841
Palladium-110 [*15749-60-7*], *13*:841
Palladium black [*7740-05-3*]
    catalyst for anthraquinone autoxidation,
    13:19
Palladium(II) [*7647-10-1*]
    Wacker-process catalyst, *16*:611
Palm
    oil, *16*:252
Palmarosa oil [*8014-19-5*], *16*:326
Palmitic acid [*57-10-3*]
    in grease thickeners, *14*:502
    in nut oils, *16*:252
    ozone oxidation of, *16*:710
    in soybean oil, *14*:251
Palm nut, *16*:250

Palm oil [*8023-79-8*]
    in metallic coating, *15*:244
    in packings, *16*:734
Palpitation
    (−) adrenaline induced, *15*:758
Pamela process, *16*:210
Panax trifolium, *16*:251
Pan coating process
    for tablets, *15*:472
Pancreatic enzymes
    diagnostic tests, *15*:79
Pancreatis
    treatment, *15*:780
PANDEX information system, *13*:290
Panels
    wood derived, *14*:4
PAN HM fiber
    in metal laminates, *13*:945
PAN HTS fiber
    in metal laminates, *13*:945
Pan mixers, *15*:630
Pan muller, *15*:633
Pantothenic acid [*79-83-4*]
    in nuts, *16*:253
    sulfur in, *15*:587
Papain [*9001-73-4*]
    effect on LHRH, *15*:771
Papaveraldine [*522-57-6*], *13*:63
Papaverine [*58-74-2*], *13*:63
Papaverine hydrochloride [*61-25-6*]
    economic aspects, *13*:53
Paper, *16*:**768**
    additives, *16*:803
    aziridine-based additives, *13*:161
    in electrical insulation, *13*:555
    emissivity, *13*:342
    gasket properties, *16*:727
    mica, *15*:430
    from novoloid fibers, *16*:129
    properties, *16*:771
    as thermal insulator, *13*:592
Paper bags, *16*:800
Paper blotters
    odor modification with, *16*:304
Paperboard, *14*:4; *16*:768
    consumption, *16*:792
    properties, *14*:27
Paperchem
    information service, *16*:902
Paper coatings, *16*:774
Paper, impregnated
    as cable insulation, *13*:575
Papermakers alum [*10043-01-3*],
    16:804
Papermaking.   See *Paper.*
Papermaking additives, *16*:**803**
Papers, synthetic, *16*:823

PAPI, *13*:801, 805
PAPI 20, *13*:801
PAPI 135, *13*:801
Paprika [*8006-77-7*], *16*:314
Papyrus, *16*:768
Paradione, *13*:135
Paradise.   See *Nuts.*
Paraffin [*8002-74-2*]
   in fiberboard manufacture, *14*:18
   in neutron shielding, *16*:241
   in packings, *16*:734
   paper size, *16*:812
Paraffin oil [*8012-95-1*]
   paper abherent, *16*:808
Paraffins
   chlorination–dehydrochlorination, *16*:492
   isomerization by zeolites, *15*:666
   linear olefins from, *16*:492
   nitration, *15*:848
   separation by molecular sieves, *15*:664
Paraffin wax [*9002-74-2*]
   in matches, *15*:3
   as paper coating, *16*:791
   in varnish and paint removers, *16*:763
Paraformaldehyde [*30525-89-4*]
   nitro alcohols from, *15*:914
Paragonite [*12026-53-8*], *15*:416
Paraho retort, *16*:340
Paralactic acid, [*79-33-4*], *13*:81
Paralobesia viteana, *13*:480
Paramagnetism, *14*:647
Paramavert
   in perfumes, *16*:967
Paramethadione [*115-67-3*]
   properties of, *13*:135
Paramethasone [*53-33-8*], *13*:173
Parametric amplification, *14*:64
Paranephrin, *15*:757
Paraoxon [*311-45-5*], *13*:450
Paraperiodic acid [*10450-60-9*], *13*:666
Parasitic ferromagnetism, *14*:647
Parasitology
   diagnostic reagents, *15*:75
Parathion [*56-38-2*], *13*:439
   toxicity, *13*:271
Parathyroid hormone [*9002-64-6*]
   in calcium homeostasis, *15*:586
   magnesium regulation, *15*:583
Paratyphoid fever, *15*:553
Parazo orange [*547-57-9*], *13*:10
Parex process, *15*:664
Parfait
   frozen dessert, *15*:568
Paris green [*12770-31-9*]
   as insecticide, *13*:414
Paris union, *16*:860

Parkes desilverizing process, *14*:117
Parkinsonism
   GABA involved in, *15*:766
   treatment, *15*:764
   treatment for senility, *15*:138
Particle board, *13*:59; *14*:4
   properties, *14*:28
Particle identification
   by particle-track etching, *16*:839
   by track etching, *16*:827
   by tracks, *16*:840
Particles, coating.   See *Microencapsulation.*
Particle track
   structure of, *16*:831
Particle-track etching, *16*:**826**
Particle tracks
   formation of, *16*:830, 832
Partition
   solutes by membranes, *15*:96
Parylene [*25722-33-2*]
   in encapsulation, *15*:482
Parylene process, *15*:482
Pasania cornea, *16*:250
Paschen's law, *13*:541
Paspalum, *15*:949
Passivation
   chromating for, *15*:308
   of metals, *15*:304
Paste inks, *13*:379
Pastes
   mixing, *15*:630
Pasteurization
   in lactic acid fermentation, *13*:84
   in microbial polysaccharide preparation, *15*:446
   of milk, *15*:530
Pasteur, Louis, *15*:530
Patchouli [*8014-09-3*]
   in perfumes, *16*:952
Patchouli alcohol [*5986-55-0*], *16*:326
   in perfumes, *16*:953
Patchouli oil [*8014-09-3*], *16*:326
Patch test
   for skin sensitivity, *13*:261
PATCLASS
   information service, *16*:902
Patent and trademark office, *16*:867
Patent and Trademark Office, U.S.,
      16:898
Patent information, *16*:889
Patent infringement, *16*:874
Patent litigation, *16*:874
Patents, *16*:889.   (See also *Patents, literature.*)
   bulletins, *13*:298
   searching, *13*:298

Patentschrift, *16*:891
Patent searching.　See *Patent literature.*
Patents, literature, *16*:**889**
Patents, practice and management, *16*:**851**
Patio process
　for silver recovery, *15*:143
Patora, *14*:248
PATSEARCH
　information service, *16*:902
Pattinson process
　for magnesium compounds, *14*:620
Pattinson's lead white, *14*:162
Pauli paramagnetism, *14*:647
PBLG.　See *Poly-γ-benzyl-L-glutamate.*
PBNA.　See *N-Phenyl-2-naphthaleneamine.*
PCP.　See *Pentachlorophenol.*
PDME.　See *Pendant-drop metal extraction.*
Peach tree borer
　control of with pheromones, *13*:481
Peacock Blue [*2650-18-2*]
　in inks, *13*:378
Peanut, *16*:251
　products from, *16*:273
Peanut butter
　analysis of with ISES, *13*:732
　attractant, *13*:417
　composition of, *16*:265
　nutritive value, *16*:260
　storage, *16*:263
Peanut oil [*8002-03-7*]
　lecithin from, *14*:250
　in lecithin manufacture, *14*:257
　repellent, *13*:479
Peanuts
　oil, *16*:270
Pearlite [*12427-27-9*], *15*:339
Peat moss
　lignin content, *14*:297
Peat scours
　in cattle, *15*:593
Pebble lime, *14*:347
Pecan, *16*:250
Pechmann condensation
　of resorcinol, *13*:61
Pectin [*9000-69-5*]
　in ice cream, *15*:565
Pectinophora gossypiella, *13*:481
Pediculus corporis, *13*:414
Pedomicrobium manganicum, *14*:856
Pegmatites
　muscovite source, *15*:418
Pelargonic acid [*112-05-0*], *16*:497, 686
Pelargonium graveolens Ait, *16*:323
Pellethane, *13*:813
Pelleting
　iron ore, *13*:739

Pelletizers
　noise control, *16*:24
Pelouze reaction, *15*:892
Pemoline [*2152-34-3*]
　memory effects, *15*:135
Pendant-drop metal extraction, *15*:229
Penetrants
　in nondestructive tests, *16*:51
Penetration tests
　nondestructive, *16*:51
Penetration theory
　in mass transfer, *14*:931
Penetrometer
　in grease testing, *14*:504
P/E NEWS information system, *13*:330
Penicillamine [*52-67-5*]
　antiaging effect, *15*:139
D-Penicillamine [*52-67-5*]
　in rheumatoid arthritis therapy, *13*:174
Penicillin G [*61-33-6*], *15*:463.　(See also
　　Benzylpenicillin.)
Penicillins
　microbial hydrolysis, *15*:462
Penicillin V [*87-08-1*], *15*:463.　(See also
　　Phenylmethoxypenicillin.)
Penicillium brevi-compactum, *13*:107
Penncap-M, *15*:480, 488
Penokee–Gogebic range
　iron ore from, *13*:751
Pentac [*2227-17-0*], *13*:463
Pentacarbonyl(triethyllead)manganese [*25434-
　　98-4*], *14*:181
Pentachloromolybdenum(V) [*10241-05-1*], *15*:958
Pentachloronaphthalene [*1321-64-8*]
　liver damage from, *15*:702
Pentachlorophenol [*87-86-5*] (PCP)
　applications of, *13*:226
　clinical assay, *15*:88
　in plywood manufacture, *14*:15
　wood preservatives, *13*:480
Pentacupric sulfate
　in bordeaux mixture, *13*:476
1-Pentadecene [*13360-61-7*]
　properties, *16*:481
2,4-Pentadienonitrile [*1615-70-9*]
　aziridine reaction, *13*:151
2,3-Pentadione [*600-14-6*]
　properties, *13*:926
2,4-Pentadione, copper chelate [*13395-16-9*], *13*:929
Pentaerythritol [*115-77-5*]
　aircraft lubricant formulation, *16*:497
　from formaldehyde, *15*:412
　in lubricants, *14*:498
Pentaerythrityl tetrakis[3-(3,5-di-*tert*-butyl-4-
　　hydroxyphenyl)propionate] [*6683-19-8*]
　LDPE antioxidant, *16*:414
　in LLDPE, *16*:395

Pentafluorohydroxyniobic acid [*12062-01-0*], *15*:
834
Pentafluoroisoethyl isocyanate [*356-74-1*], *13*:800
Pentahydrite [*15553-21-6*], *14*:637
Pentamethylarsine [*51043-92-6*], *13*:669
Pentamethyleneketene [*22589-13-5*]
properties, *13*:875
Pentamethylenimines. See *Piperidines.*
Pentanal [*110-62-3*], *16*:312
n-Pentane [*109-66-0*]
in LPG, *14*:384
properties, *14*:384
2,4-Pentanedione [*123-54-6*], *13*:930. (See also
Acetylation.)
iron complex, *13*:778
nickel acetylacetonate from, *15*:810
in organolead reaction, *14*:184
properties, *13*:898
2,4-Pentanedione peroxide [*37187-22-7*], *13*:364
1-Pentanol [*71-41-0*]
oxo product, *16*:650
1-Pentene [*109-67-1*]
copolymer with ethylene, *16*:388
hydroformylation, *16*:640
2-Pentenenitrile [*13284-42-9*], *15*:889, 899
3-Pentenenitrile [*4635-87-4*], *15*:889, 899
nickel complex formation, *15*:808
4-Pentenenitrile [*592-51-8*], *15*:899
η^3-Pentenyl-2 iron tricarbonyl chloride [*58257-85-5*], *16*:603
Pentlandite [*53809-86-2*], *15*:796
Pentobarbital [*76-74-4*]
carbromal combination, *13*:125
Pentosan, *16*:252
Pentoses
soybean phosphatides associated with, *14*:252
Pentylenetetrazole [*54-95-5*]
CNS stimulant, *15*:135
3-[4-(Pentyloxy)benzoyl]-1,1,2-cyclopropanetricar-
boxylic acid 1,1-diethyl-2-methyl ester
[*67947-11-9*], *14*:799
2-(*p*-Pentylphenyl)-5-(*p*-pentyloxyphenyl)pyrim-
idine [*34913-07-0*]
liquid crystalline range of, *14*:399
Pepper black oil [*8006-82-4*], *16*:326
Peppermint oil [*8006-90-4*] of *Mentha piperita*,
16:326
Pepsdol, *15*:767
Pepsin [*9001-75-6*]
dextran-inactivated, *15*:445
effect on ACTH, *15*:758
Peracetic acid [*79-21-0*]
production, *16*:687
Perchlor, *13*:552
Perchlorates
oxygen generation from, *16*:673

Perchloric acid [*7601-90-3*]
nitration reagents, *15*:841
Perchloryl fluoride [*7616-94-6*]
odor control with, *16*:303
Perferrites, *13*:776
Perfluoroelastomer
gasket properties, *16*:727
Perfluoro(methyl vinyl ether) [*1187-93-5*]
in gaskets, *16*:726
Perfluoropolyglycols
as lubricant, *14*:501
Perfluorosulfonic acid polymeric membrane
in nitrobenzene manufacture, *15*:921
Perfumes, *16*:297, **947**
encapsulation, *15*:476
industrial, *16*:968
men's, *16*:964
safety, *16*:971
Perhydrocyclopentanophenanthrene [*482-64-4*]
microbial oxidation, *15*:461
Peri acid [*82-75-7*], *15*:729. (See also *8-Amino-
1-naphthalenesulfonic acid.*)
Periclase
from lime, *14*:380
Periodic acid [*10450-60-9*], *13*:666
for ^{131}I storage, *16*:180
Periodic acid monohydrate monosodium salt, *13*:
667
Periplaneta americana, *13*:419, 430
Periscopes
in nuclear facility design, *16*:246
Peritoneal dialysis, *15*:116
Perkin reaction
with salicylaldehyde, *13*:73
Perlite
as thermal insulator, *13*:592
45 Permalloy, *14*:651
4-79 Permalloy [*39323-53-0*], *14*:657
Permalloy films
magnetic properties, *14*:691
Permalloys
in thin-film fabrication, *14*:689
Permanent Red 2B [*3564-21-4*]
in inks, *13*:378
Permanganic acid [*13465-41-3*], *14*:859
Permanganic acid dihydrate [*24653-70-1*], *14*:859
Permanganyl chloride [*15605-27-3*], *14*:859
Permanganyl fluoride [*15586-97-7*], *14*:859
Permanox
manganese dioxide, *14*:878
MnO_2, *14*:863
Permax F, *14*:659
Permax M, *14*:659
Permax Z, *14*:659
Permeability
Darcy unit, *15*:224

Permeability coefficient
  of membranes, *15*:92
Permendur, *14*:664
Permenorm 500 H$_2$, *14*:659
Permenorm 3601 K2, *14*:659
Permenorm Z, *14*:659
Permethrin [*52645-53-1*], *13*:458
Permutit, *13*:696
Pernicious anemia, *15*:597
  diagnosis, *15*:78
Peronospora viticola, *13*:475
Peroxidase [*9001-05-2*]
  cholesterol diagnostic reagent, *15*:82
  coniferyl alcohol reaction, *14*:301
  iron in, *15*:595
  in prostaglandin syntheses, *15*:780
Peroxide [*28056-59-9*], *13*:364
Peroxides, *13*:15.   (See also *Peroxygen
    compounds.*)
  as initiators, *13*:357
  oxygen generation, *16*:678
Peroxyacetic acid [*79-21-0*]
  from acetaldehyde, *13*:22
Peroxy acids, *13*:15
Peroxy compounds
  in redox initiators, *14*:92
Peroxydisulfate [*15092-81-6*]
  in emulsion polymerization, *14*:84
Peroxydisulfuric acid [*13445-49-3*]
  hydrogen peroxide process, *13*:23
Peroxydisulfuric acid process
  for hydrogen peroxide, *13*:23
Peroxyesters, *13*:359
Peroxygen compounds, *13*:15
Peroxyhydrates, *13*:15
Perrhenic acid [*13768-11-1*]
  in nitride formation, *15*:877
PERT.   See *Program evaluation and review
    technique.*
Perthane [*72-56-0*], *13*:431
Perturbation theory
  of electromagnetic fields, *15*:499
Pervaporation, *15*:116
Pesticide Control Act, *13*:254
Pesticides, *13*:413.   (See also *Insect control
    technology.*)
  encapsulation, *15*:488
  health effect, *13*:271
  microbial oxidation, *15*:459
Petalite [*1302-66-5*]
  lithium in, *14*:449
Petex, *16*:78
Petitgrain bigarade sur fleurs d'orange [*8030-28-
    2*], *16*:326
Petit mal
  treatment, *13*:134

PETN.   See *Pentaerythritol tetranitrate.*
Petrochemicals
  from shale oil, *16*:352
Petrolatum
  in packings, *16*:734
Petroleum
  naphthalene from, *15*:706
  from the ocean, *16*:287
Petroleum lubricants, *14*:484
Petroleum oils
  in electrical insulation, *13*:565
  as insecticides, *13*:463
Petroleum refining
  nitrogen uses, *15*:939
Petroleum solvent
  toxicity, *13*:272
Petroleum sulfonates
  Mg overbased, *14*:642
Petromat, *16*:78
Petrosix retort, *16*:340
PGE$_2$.   See *Prostaglandin E$_2$.*
PGG$_2$.   See *Prostaglandin G$_2$.*
PGH$_2$.   See *Prostaglandin H$_2$.*
pH.   See *Hydrogen-ion concentration.*
PHA.   See *Phytohemagglutinin.*
Phagocytosis, *13*:169
  phosphatides in, *14*:251
Phaltan, *13*:245
Pharmaceuticals
  controlled release, *15*:488
  encapsulation, *15*:476
Phase-inversion temperature (PIT)
  in emulsion polymerization, *14*:86
Phase separation, *14*:912.   (See also *Mass
    transfer.*)
pH electrodes, *13*:4
Phellandrene [*1329-99-3*]
  from pepper, *16*:326
α-Phellandrene [*99-83-2*], *16*:313
Phenacemide [*63-98-9*]
  properties of, *13*:135
9,10-Phenanthraquinone [*84-11-7*]
  as photoinitiator, *13*:368
Phenanthrene-9,10-quinone [*84-11-7*]
  in malononitrile determination, *14*:806
1,10-Phenanthroline [*66-71-7*]
  iron phenanthroline from, *13*:780
Phenazine [*92-82-0*]
  reduction of, *13*:22
Phencylidine [*77-10-1*]
  measurement, *15*:88
Phenergan, *13*:122
Phenethyl alcohol [*60-12-8*], *16*:956
  in rose oil, *16*:954
Phenethylamine [*64-04-0*]
  dicyandiamide reaction, *13*:619

Phenethyl propionate [122-70-3]
   as insect attractant, 13:481
o-Phenetidine [94-70-2], 15:924
Phenetidone [92-43-3], 13:57.   (See also 1-
   Phenyl-3-pyrazolidone.)
Phenformin [114-86-3], 13:613
Phenformin hydrochloride [834-28-6], 13:617
Phenix, 16:196
Phenmediphan [13684-63-4], 13:60, 62
Phenol [108-95-2], 13:40
   alkylation, 16:484
   aminoacetonitrile reaction, 15:774
   cholesterol diagnostic reagent, 15:82
   dialysis of, 15:119
   formaldehyde reaction (1:1), 13:74
   hydroxylation, 13:48
   mercuration, 15:164
   microbial transformation to L-tyrosine, 15:464
   in milk, 15:551
   novoloid fibers from, 16:129
   odor-problem chemical, 16:303
   ozone reaction, in water, 16:710
   toxicity, 13:271
Phenol–formaldehyde adhesives
   for plywood, 14:14
Phenol formaldehyde polymer [9003-35-4]
   in formica, 13:969
Phenol–formaldehyde resin, 16:125.   (See also
   Novolac resin.)
   adhesive for particleboard, 14:30
Phenol–formaldehyde resins
   as magnetic tape binders, 14:741
Phenolic–resin adhesives
   for plywood, 14:14
Phenolic resins
   in novoloid fibers, 16:125
Phenolics
   as antimicrobial agents, 13:225
Phenolphthalein [77-09-8], 13:10; 16:55
Phenolphthalein sodium [518-51-4], 14:552
Phenol red [143-74-8], 13:10
Phenol–resorcinol–formaldehyde resin
   plywood adhesive, 14:14
Phenols
   toxicity, 13:225
Phenosolvan process, 14:328
Phenothiazine [92-84-2]
   derivatives, 13:133
   as oxidation inhibitor, 14:497
   in promethazine synthesis, 13:134
Phenothrin [26002-80-2], 13:458
3-Phenoxybenzyl DL-cis,trans-chrysanthemate
   [26002-80-2]
   as insecticide, 13:457
2-Phenoxyethylamine-tetrafluoroboric acid
   complex [74411-25-9], 13:151

Phenoxys
   as magnetic tape binder, 14:741
Phensuximide [86-34-0]
   properties of, 13:135
Phenurone, 13:135
Phenylacetaldehyde [122-78-1]
   in perfumes, 16:967
Phenylacetic acid [103-82-2]
   microbial hydrolysis, 15:463
   microbial synthesis of, 15:465
N-Phenylacetoacetamide [102-01-2], 13:887
Phenylacetylchloride [103-80-0]
   in phenacemide synthesis, 13:139
N-Phenyl γ-acid [119-40-4], 15:742.   (See also
   6-Phenylamino-4-hydroxy-2-naphthalene-
   sulfonic acid.)
Phenylalanine [90-41-5]
   microbial synthesis of, 15:463
   in nuts, 16:249, 252
   in phenylethylamine biosynthesis, 15:775
L-Phenylalanine [63-91-2]
   microbial synthesis of, 15:465
DL-Phenylalanine-hydantoin [54832-24-5]
   microbial transformation to L-phenylalanine,
   15:465
6-Phenylamino-4-hydroxy-2-naphthalenesulfonic
   acid [119-40-4], 15:742
7-Phenylamino-4-hydroxy-2-naphthalenesulfonic
   acid [119-40-4], 15:742
8-Phenylamino-1-naphthalenesulfonic acid [82-
   76-8], 15:729
Phenylarsine dichloride [696-28-6]
   aziridine reaction, 13:150
Phenylarsonic acid [98-05-5]
   in niobium analysis, 15:829
Phenyl azide [622-37-7]
   methyl 2-diethylaminomethacrylate reaction,
   15:349
1-Phenylaziridine [696-18-4], 13:148
N-Phenyl-(1-aziridinyl)carboxamide [13279-22-
   6], 13:155
1-Phenyl-2(1-aziridinyl)ethyl 2-aminoethyl ether
   [74411-24-8], 13:151
1-Phenyl-2(1-aziridinyl)thioethanol [74411-23-7],
   13:151
Phenyl benzoates
   synthesis of, 14:414
Phenyl 4-benzoyloxybenzoates
   synthesis of, 14:414
1-Phenyl-1,3-butanedione [93-91-4]
   properties, 13:928
2-Phenylbutanonitrile [769-68-8]
   in glutethimide synthesis, 13:124
Phenylbutazone [50-33-9], 14:803
   controlled release of, 15:488
   immunologic effects, 13:174
Phenyl(chlorocarbonyl)ketene [17118-70-6], 13:
   880

2-Phenylcyclohexanol [*1444-64-0*]
  as repellent, *13*:477
Phenyl diethyl phosphate [*2510-86-3*], *13*:451
m-Phenyl diisocyanate [*123-61-5*]
  properties, *13*:789
p-Phenyl diisocyanate [*104-49-4*]
  oxalic acid copolymer, *16*:632
Phenylenebis(dimethyl)silanol [*29036-28-0*], *15*:
  202
m-Phenylenediamine [*108-45-2*]
  metallic copolymers from, *15*:205
  in resorcinol production, *13*:52
o-Phenylenediamine [*95-54-5*]
  in digoxin assay, *15*:87
  metal coordination polymer from, *15*:217
p-Phenylenediamine [*106-50-3*], *15*:924
p-Phenylene diisocyanate [*104-49-4*], *13*:794
Phenylethanolamine [*7568-93-6*]
  in (−) adrenaline biosynthesis, *15*:758
  as neuroregulator, *15*:775
  properties, *15*:775
Phenylethyl alcohol [*60-12-8*]
  in perfumes, *16*:968
α-Phenylethyl alcohol [*98-85-1*], *16*:312
Phenylethylamine [*64-04-0*]
  as neuroregulator, *15*:775
  in phenylethanolamine biosynthesis, *15*:775
  properties, *15*:775
2-Phenylethylamine-aluminum chloride complex
  [*7441-31-7*], *13*:151
Phenylethylamine hydrochloride [*156-28-5*]
  properties, *15*:775
2-Phenyl(ethylene)sulfide [*1822-73-7*]
  aziridine reaction, *13*:151
N-Phenylhydroxylamine [*100-65-2*], *15*:917
(*R*)-1-Phenyl-1-hydroxy-2-propanone [*1798-60-3*]
  in microbial ephedrine synthesis, *15*:464
Phenyl isocyanate [*103-71-9*]
  aziridine reaction, *13*:155
  properties, *13*:789
Phenyl isocyanate dimer [*1025-36-1*], *13*:795
Phenyl isocyanate trimer [*1785-02-0*], *13*:795
Phenyl isothiocyanate [*103-72-0*]
  aziridine reaction, *13*:155
N-Phenyl J-acid [*119-40-4*], *15*:742.   (See also
    *7-Phenylamino-4-hydroxy-2-naphthalenes-
    ulfonic acid.*)
Phenylketonuria, *15*:776
  encapsulated treatment, *15*:489
Phenyllithium [*591-51-5*], *14*:471
  nickel complex reaction, *15*:809
N-Phenylmalanil, *13*:105
N-Phenylmalimide [*941-69-5*], *13*:105
Phenylmalonic acid [*2613-89-0*]
  thionyl chloride reaction, *13*:880

Phenylmalonyldichloride [*37818-48-7*], *13*:880
Phenylmercuric acetate [*62-38-4*], *15*:169
  applications of, *13*:236
  fungicide, *15*:155
Phenylmercuric acetate [*62-38-4*] (PMA), *16*:586
Phenylmercuric butyrate [*2440-29-1*], *15*:169
Phenylmercuric chloride [*100-56-1*], *15*:169
Phenylmercuric *N,N*-dimethyldithiocarbamate
  [*32407-99-1*], *15*:169
Phenylmercuric dodecenyl succinate, *15*:169
Phenylmercuric hydroxide [*100-57-2*], *15*:169
Phenylmercuric nitrate [*55-68-5*], *15*:169
Phenylmercuric oleate [*140-60-9*], *15*:169
  fungicide, *15*:155
Phenylmercuric propionate [*103-27-5*], *15*:169
Phenylmercuric sulfide [*20333-30-6*], *15*:169
Phenyl metaborate [*1095-03-0*]
  formaldehyde reaction, *13*:74
N-Phenylmethacrylamide [*1611-83-2*]
  properties, *15*:356
Phenyl methacrylate [*2177-70-0*]
  properties, *15*:353
Phenylmethoxypenicillin [*21516-76-7*]
  microbial hydrolysis of, *15*:462
Phenyl *N*-methyl carbamate [*1943-79-9*], *13*:452
2*H*-2-Phenyl-4-methyl-3,6-dihydro-1,2-oxazine
  [*26332-63-8*]
  isoprene reaction product, *13*:826
N-Phenyl-1-naphthaleneamine [*90-30-2*], *15*:727
N-Phenyl-2-naphthaleneamine [*135-88-6*]
  (PBNA), *15*:727, 728
1-Phenyl-1,3-pentanedione [*5331-64-6*]
  properties, *13*:928
1-Phenyl-2,4-pentanedione [*3318-61-4*]
  properties, *13*:928
Phenyl peri acid [*82-76-8*], *15*:729.   (See also *8-
    Phenylamino-1-naphthalenesulfonic acid.*)
o-Phenylphenol [*90-43-7*]
  applications of, *13*:227
N-Phenyl-*p*-phenylenediamine [*101-54-2*]
  in phosphorus measurement, *15*:84
N-Phenyl-*N*-phenylethylmethacrylamide
  [*76392-02-4*]
  properties, *15*:356
Phenylphosphonic dichloride [*770-12-7*]
  as flame retardant, *13*:59
1-Phenyl-1,2-propanedione-2-*O*-benzoyl oxime
  [*17292-57-8*]
  as photoinitiator, *13*:369
1-Phenyl-3-pyrazolidinone [*92-43-3*], *13*:57
1-Phenyl-3-pyrazolidone [*92-43-3*], *13*:57.   (See
  also *Phenetidone.*)
Phenylpyruvic acid [*156-06-9*]
  microbial synthesis of, *15*:465
Phenylsodium [*1623-99-0*], *16*:557

2-Phenylsuccinic acid [635-51-8]
  in phensuximide synthesis, 13:137
N-Phenylsuccinimide [83-25-0], 13:105
2-Phenylthiazolidine [4569-82-8], 13:150
Phenytoin [57-41-0]
  properties of, 13:135
Pheochromocytoma
  diagnosis, 15:770
Pheromones
  as insect attractant, 13:419
  in insect control, 13:481
Phillips catalyst, 16:385, 390
  ethylene polymerization, 16:428
Phillips process
  for polyethylene, 16:389
Philodyne, 14:864
pH jump
  laser-induced, 14:75
Phlogophite [12003-38-7]
  in insulation, 13:556
Phlogopite [12251-58-0], 15:416
Phloionic acid [137-21-3]
  properties, 13:118
(R*,R*)-(±)Phloionic acid [23843-52-9], 13:118
Phloroglucinaldehyde [487-70-7], 13:70
pH meter, 13:2
Phonons, 14:274
Phorate [298-08-2], 13:415, 438, 446
Phorone [504-20-1], 13:900, 914
Phosgene [75-44-5]
  fumaric acid reaction, 14:773
  in isocyanate mfg, 13:801
  in isocyanate synthesis, 13:799
  ketene reaction, 13:876
  lactic acid reaction, 13:84
  toxicity, 13:271
Phosmet [732-11-6], 13:443
Phosphamidon [13171-4-6], 13:446
Phosphatase
  in milk, 15:530
Phosphatase test
  for milk, 15:551
Phosphate-bonded mica, 15:430
Phosphate ester
  limit of radiation tolerance, 14:519
Phosphate glasses
  in lasers, 14:56
  as track-etch detector, 16:827
Phosphates
  ion exchange in, 13:678
  in nuclear fuel processing, 16:181
  in nuclear reactor coolant, 16:153
  as synthetic lubricants, 14:499
Phosphatides, 14:251
Phosphatidylcholine [8002-43-5], 14:250. (See also *Lecithin*.)

Phosphatidylethanolamine [5681-36-7], 14:405. (See also α-*Cephalin*.)
  in commercial lecithin, 14:250
Phosphatidylinositol [2816-11-7]
  in commercial lecithin, 14:250
  in encapsulation, 15:477
Phosphating
  of metals, 15:304
Phosphinate metal polymers, 15:214
Phosphinates
  in nuclear fuel processing, 16:181
Phosphine [7803-51-2]
  as fumigant, 13:467
  IC dopant, 13:627
  isocyanate reaction, 13:793
Phosphine metal–nitrogen complexes
  in N₂ reduction, 15:957
Phosphine oxides
  isocyanate reaction catalysts, 13:797
  in nuclear fuel processing, 16:181
Phosphines
  in oxo catalysts, 16:637
Phosphites
  as antiwear agents, 14:493
  isocyanate reaction, 13:793
Phosphoenolpyruvic acid [138-08-9]
  in hyperlypidemia diagnosis, 15:83
6-Phosphogluconate
  in glucose assay, 15:84
3-Phosphoglycerate [9001-83-6]
  in glycine biosynthesis, 15:768
Phosphokinases
  magnesium requirement, 15:579
Phospholipids, 15:586
  as liquid crystals, 14:419
  mesomorphism of, 14:405
  in nuts, 16:253
Phosphomolybdate heteropolyacid
  in phosphorus measurement, 15:84
Phosphonates
  in nuclear fuel processing, 16:181
Phosphonolipids, 15:587
Phosphorescence, 14:557
Phosphoric acid [7664-38-2]
  in metal pickling, 15:300
  in metal surface treatment, 15:304
  nickel carbonate reaction, 15:806
  in nitration ratio control, 15:929
  paint stripper, 16:765
Phosphoric anhydride [1314-56-3]
  use, 13:59
Phosphoric pickle acid [7664-38-2]
  recovery in Higgins contactor, 13:693
Phosphorite [55575-17-2]
  from the ocean, 16:289
  uranium in, 16:144
Phosphorites
  from the ocean, 16:288

Phosphorous oxychloride [*10025-87-3*]
  aziridine reaction, *13*:143
Phosphors, *14*:527.   (See also *Luminescent
    materials.*)
Phosphorus [*7723-14-0*]
  in blast furnace flux, *13*:740
  diagnostic tests, *15*:84
  in magnetic alloys, *14*:694
  in matches, *15*:1
  metabolic functions, *15*:586
  nutrient properties, *15*:571
  from the ocean, *16*:279
  ozone reaction, *16*:689
  in silicon doping, *13*:635
  in walnuts, *16*:261
Phosphorus-31 [*7723-14-0*], *13*:840
Phosphorus compounds
  in lubricants, *14*:479
Phosphorus iodide [*13455-00-0*] (2:4), *13*:661
Phosphorus nitride [*17739-47-8*], *15*:872
Phosphorus oxychloride [*10025-87-3*]
  in (−)-adrenaline synthesis, *15*:757
  amide reaction, *15*:905
  lactic ester reaction, *13*:84
  nitro alcohol reaction, *15*:913
  in TEPP synthesis, *13*:437
Phosphorus pentachloride [*10026-13-8*]
  fumaric acid reaction, *14*:773
  nitro alcohol reaction, *15*:913
  nitrophenol reaction, *15*:924
Phosphorus pentasulfide [*1314-80-3*]
  in quazepam synthesis, *13*:130
Phosphorus pentoxide [*1314-56-3*]
  in malonic acid dehydration, *14*:794
  in spent shale, *16*:351
  in TEPP synthesis, *13*:437
Phosphorus sesquioxide [*1314-24-5*]
  from ozonation phosphorus, *16*:689
Phosphorus trichloride [*7719-12-2*]
  tetramethyllead reaction, *14*:183
Phosphorus triiodide [*13455-01-1*], *13*:661
Phosphorus trisulfide [*1314-85-8*] (4:3)
  in matches, *15*:2
Phosphorylation
  microbial, *15*:466
Phosphorylation, oxidative
  phosphatides in, *14*:251
3-Phosphoserine [*7331-08-0*]
  in glycine biosynthesis, *15*:768
Phosphosilicate glass
  as IC dopant, *13*:638
Phossy jaw, *15*:2
Photar filters, *16*:528
Photobacterium fischeri, *13*:275
Photochemistry
  lasers in, *14*:73
Photocoagulation
  by lasers, *14*:70

Photodiode
  in level measurement, *14*:437
Photoexcitation, *13*:369
Photography
  infrared, *13*:341
  use of PbBr$_2$, *14*:163
Photoinitiators, *13*:367
Photoionization
  in laser separation processes, *14*:76
Photolithography
  in IC manufacture, *13*:628
Photomechanical filters, *16*:552
Photomirex [*39801-14-4*]
  from mirex, *13*:436
Photon detectors, *13*:340
Photooxidation
  of olefin fibers, *16*:360
Photopolymerization catalyst
  lead bromide, *14*:163
Photoresists
  in IC manufacture, *13*:628, 629
Photosensitized processes, *13*:370
Photosensitizers, *13*:370
PHS 149 information system, *13*:301
Phthalic acid [*88-99-3*]
  in alkyd paints, *16*:753
  from naphthalene, *15*:698, 705
Phthalic anhydride [*85-44-9*]
  dihydroxybenzene reaction, *13*:43
  from naphthalene, *15*:705
  resin from in fluorescent product, *14*:561
γ-Phthalimidobutyronitrile [*75802-63-0*]
  in GABA synthesis, *15*:766
α-Phthalimidoethane-α,α,β-tricarboxylate
  in aspartic acid synthesis, *15*:760
Phthalocyaninatobis(pyridine)iron [*20219-84-5*],
    13:782
Phthalocyaninatoiron [*132-16-1*], *13*:782
Phthalocyanine [*574-93-6*]
  iron(II) phthalocyanine from, *13*:782
Phthalocyanine blue [*147-14-8*]
  in inks, *13*:390
  paint pigment, *16*:745
Phthalocyanine green [*1328-53-6*]
  in inks, *13*:379
Phthalocyanine polymers, *15*:215
Phthalocyanines
  as grease thickeners, *14*:502
Phthaloyl chloride [*88-95-9*]
  fumaric acid reaction, *14*:773
Physic nut, *16*:251
Physostigma venenosum, *13*:453
Physostigmine [*57-47-6*]
  learning effects, *15*:134
Phytelephas seemanni, *16*:251
Phytin [*3615-82-5*]
  in nuts, *16*:255

Phytohemagglutinin (PHA) [9008-97-3]
    in lymphocyte-blast transformation, *13*:175
Phytosterols
    in nuts, *16*:255
    in soybean lecithin, *14*:251
Picket-fence porphyrins, *13*:783
Pickling, *15*:300.   (See also *Metal surface*
    treatments.)
    of hides, *14*:210
Pick strength, *16*:786
Picric acid [88-89-1]
    in ISE, *13*:728
Picrotoxin [124-87-8]
    CNS stimulant, *15*:135
Picrylpyridinium chloride [6509-51-9]
    nitromethane reaction, *15*:975
Pidgeon process
    for magnesium, *14*:584
Pierce-Smith converter
    for lead smelting, *14*:130
Piezodialysis, *15*:120
Pig iron, *13*:736
Pig lead
    in cable sheathing, *13*:583
Pigments
    fluorescent, *14*:546
    inhibitive, molybdates as, *15*:694
    in inks, *13*:377
    lead carbonate, *14*:171
    in paint, *16*:745
    in paper coatings, *16*:785
    in paper, *16*:769
    in papermaking, *16*:822
Pigment volume concentration (PVC), *16*:745
Pignolia nut, *16*:250
Pigskins
    for leather, *14*:224
Pili nut
    structure, *16*:249
Pimenta
    components, *16*:311
Pimenta diaica, *16*:320
Pimenta racemos A (mill), *16*:321
Pimento oil [8006-77-7], *16*:311
Pimpinella ansisum, *16*:321
α-Pinene [80-56-8], *16*:312, 313, 328
    in perfumes, *16*:954
β-Pinene [127-91-3], *16*:312
    from neroli oil, *16*:325
    perfumes from, *16*:954
Pine nut, *16*:248
Pine oil [8006-88-0], *16*:326
    froth stabilizer, *14*:102
    paper coating defoamer, *16*:791
Pink bollworm
    control of with pheromones, *13*:481
Pinner synthesis, *15*:891

Pinon nut, *16*:250
Pinoresinol [487-36-5, 4263-88-1]
    in lignin, *14*:301
Pinus edulis, *16*:250
Pinus mugo Turra, *16*:327
Pinus palustris Mill., *16*:326
Pinus pumilio Haenke, *16*:327
Pinus pumilio oil [8000-26-8], *16*:327
Pipe
    from polyethylene, *16*:431
Pipecolic acid [535-75-1]
    in piperidine biosynthesis, *15*:776
Pipelines
    for LPG transportation, *14*:390
    for nitrogen, *15*:937
    nondestructive testing, *16*:47
    radiographic inspection, *16*:64
Piper betle, *16*:270
Pipe resins
    polyethylene, *16*:398
Piperidine [110-89-4]
    isoprene reaction, *13*:826
    in lactamide preparation, *13*:90
    as neuroregulator, *15*:776
    properties, *15*:776
Piperidine hydrochloride [6091-44-7], *15*:776
Piperidine nitrate [6091-45-8], *15*:776
Piperidine picrate [6091-49-2], *15*:776
Piperidines, *13*:142
Piperine [94-62-2], *13*:63
    from pepper, *16*:326
Piperitone [89-81-6]
    from eucalyptus oil, *16*:323
Piper nigrum L., *16*:326
Piperonal [7797-83-3], *13*:62, 63.   (See also
    Heliotropine.)
    from ocotea, *16*:325
Piperonyl butoxide [51-03-6]
    in carbamate reaction, *13*:456
    economic aspects, *13*:53
    as insecticide, *13*:425
trans-Piperylene [2004-70-8]
    methyl methacrylate reaction, *15*:349
Piperyleneiron tricarbonyl [12080-18-1]
    HCl reaction, *16*:602
Pipes
    from LDPE, *16*:419
    poly(1-butene), *16*:478
    from polyethylene, *16*:451
    from polypropylene, *16*:467
Piping
    noise control, *16*:23
Piping insulation, *13*:597
Piptadenia macrocarpa
    DMT in, *15*:763
Piptadenia peregrina
    DMT in, *15*:763

Piracetam [*7491-74-9*]
   effects on learning, *15*:135
Piria's acid [*84-86-6*], *15*:729.   (See also *4-Amino-1-naphthalenesulfonic acid*.)
Piroline, *14*:811
Pistachio, *16*:250
Pistacia vera, *16*:250
Piston rings, *16*:740
PIT.   See *Phase-inversion temperature*.
Pitch control
   in papermaking, *16*:807
Pitch (type P) fiber
   in metal laminates, *13*:945
Pitch UHM fiber
   in metal laminates, *13*:945
PL/1
   in operations planning, *16*:509
Placidyl, *13*:122
Plague, *13*:413
Planar process
   in IC manufacture, *13*:623
Planck's law, *13*:338
Planetary mixers, *15*:630
Plant fiber
   gaskets, *16*:732
Plant location, *16*:500
Plant nutrient
   manganese, *14*:884
Plant patent, *16*:891
Plants
   lignin content, *14*:297
   process controls for, *13*:505
Plant size, *16*:500
PLASDOC
   information system, *16*:905
Plasma
   microwave, *15*:516
Plasma cells
   in antibody formation, *13*:169
Plasmalogens, *15*:587
Plasma processing
   in IC manufacture, *13*:638
Plasmas
   dielectric permittivity by conductivity, *15*:502
Plasma spraying
   metal coatings, *15*:255
Plasma torch
   in metal coating, *15*:261
Plasmids
   in gene transfer, *15*:953
Plaster, *14*:376
   emissivity, *13*:342
Plastic drums, *16*:718
Plastic filters, *16*:535
Plasticity
   measurement, *14*:354

Plasticizer alcohols
   olefins for, *16*:497
Plasticizers
   in electrical insulation, *13*:565
   in inks, *13*:389
Plastics
   effect of radiation on, *16*:239
   fluorescent pigments in, *14*:566
   in gaskets, *16*:732
   lead filled, *14*:158
   microwave processing, *15*:515
Plastics additives
   nickel-containing, *15*:814
Plastisols
   in leather substitute mfg, *14*:236
Plating, *15*:251.   (See also *Metallic coatings*.)
   fasteners, *15*:251
Platings
   nondestructive testing of, *16*:52
Platinic chloride [*16941-12-1*]
   lecithin reaction, *14*:256
Platinized titanium anodes, *15*:173
Platinum [*7440-06-4*]
   cladding of, *15*:281
   in coatings, *15*:241
   in integrated circuits, *13*:641
Platinum-190 [*15735-68-9*], *13*:842
Platinum-192 [*14913-85-0*], *13*:842
Platinum-194 [*14998-96-0*], *13*:842
Platinum-195 [*14191-88-9*], *13*:842
Platinum-196 [*14867-61-9*], *13*:842
Platinum-198 [*15756-63 5*], *13*:842
Platinum anodes
   in chlorine–caustic production, *15*:172
Platinum–cobalt
   as magnetic material, *14*:680
Platinum–iridium
   anode coatings, *15*:174
Platinum silicide [*12137-83-6*]
   in bipolar metallization, *13*:643
   in integrated circuits, *13*:641
Plattnerite [*1309-60-0*], *14*:165
Plectonema, *15*:950
Plexiglas, *15*:888
   in optical filters, *16*:532
Plictran, *16*:579
Plumbism, *14*:175, 198
Plutonium-238 [*13981-16-1*]
   as fission fuel, *16*:140
Plutonium-239 [*15117-48-3*], *16*:188
   in breeder reactors, *16*:168
   as fission fuel, *16*:140
   reprocessing of, *16*:173
   from uranium-238, *16*:187
Plutonium-240 [*14119-33-6*], *16*:188
   in fuel processing, *16*:170

Plutonium-241 [*14119-32-5*]
   in breeder reactors, *16*:168
Plutonium dioxide [*12059-95-9*]
   as fission fuel, *16*:140
Plutonium nitrate [*13823-27-3*]
   tributyl phosphate complex, *16*:176
Plutonium nitride [*12033-54-4*], *15*:872
   properties, *15*:874
Plutonium oxalate [*14448-76-1*]
   in fuel reprocessing, *16*:179
Plutonium oxide [*12059-95-9*]
   in breeder fuel, *16*:169
   in fuel reprocessing, *16*:179
   in nuclear reactors, *16*:174
Plywood, *13*:59; *14*:1, 5
   for noise isolation, *13*:526
PM. See *Preventive maintenance.*
PMA. See *Phenylmercuric acetate.*
PMA-100, *13*:236
PMDS. See *Di(phenylmercury)dodecenyl*
   succinate.
PMPPI. See *Polymethylene polyphenyl*
   isocyanate.
Pneumococcal polysaccharides, *15*:446
Pneumonia
   detection, *15*:83
   diagnosis, *15*:79
   lactic acid elevation, *13*:80
Podocarpus, *15*:949
Pogostemom cablin, *16*:952
Pogostemon cablin Benth., *16*:326
Point-bonding
   spunbondeds, *16*:86
Point radiometer, *13*:341
Poisoning
   of polymerization catalysts, *13*:828
Poison nut, *16*:251, 257
Poisons, *13*:480.   (See also *Insecticides.*)
Poisson process, *16*:520
Poisson random variable, *16*:509
Pokeweed mitogen (PWM)
   in lymphocyte-blast transformation, *13*:175
Polacolor film, *16*:541
Polarizers
   in optical filters, *16*:533
Polaroid filters, *16*:533
Polar start, *14*:499
Polcorfam, *14*:248
Polio virus
   disinfection by ozone, *16*:705
Polishing
   of metals, *15*:308
Politol
   alkali lignins, *14*:308
Pollinators
   insecticide effect on, *13*:418

Pollution, *13*:253.   (See also *Air pollution.*)
   by insecticides, *13*:414
Pollution Abstracts
   information service, *16*:902
POLLUTION ABSTRACTS information system,
   13:330
Pollution control
   hydrogen peroxide for, *13*:28
   in organolead manufacture, *14*:188
Polonovski rearrangement
   in lorazepam synthesis, *13*:129
Polyacrolein [*25068-14-8*]
   in papermaking, *16*:818
Polyacrylamide [*9003-05-8*]
   dispersant, *16*:806
   latex stabilization, *14*:88
   in *S*(−)-malic acid biosynthesis, *13*:107
   in paper strengthening, *16*:816
Polyacrylamide gel [*9003-05-8*]
   in CPK measurement, *15*:77
Polyacrylamide–gloxal resins
   in papermaking, *16*:818
Polyacrylamides
   in papermaking, *16*:804
   as retention aids, *16*:806
Polyacrylates
   in bonding nonwovens, *16*:115
Poly(acrylic acid) [*9003-01-4*]
   latex stabilization, *14*:88
Poly(acrylic acid)s
   as magnetic tape binders, *14*:741
Poly(acrylic ester)s
   as magnetic tape binders, *14*:741
Polyacrylonitrile [*25014-41-9*]
   in bonding nonwovens, *16*:115
   in cable coverings, *13*:585
   ion exchange fiber, *13*:162
   as magnetic tape binder, *14*:741
   moisture regain, *16*:358
Poly[[2-[acryloyl]ethyl]ferrocene] [*35560-98-6*],
   15:188
Polyamic acid
   metal containing, *15*:217
Polyamide inks, *13*:390
Polyamide resins
   in fluorescent pigments, *14*:561
Polyamides
   as magnetic tape binder, *14*:741
   membranes, *15*:102
   in propylene fibers, *16*:377
Polyamines
   as neuroregulators, *15*:776
Polyammonium phosphate
   flame retardant in paint, *16*:755
Poly(arylaminophosphazenes), *13*:400
Poly(arylenesiloxanes), *13*:410
Poly(aryloxyphosphazenes), *13*:400

Polyaziridine [*9002-98-6*], *13*:157

Polybenzamide [*24991-08-0*]
  mesomorphism of, *14*:416

Poly-γ-benzyl-L-glutamate (PBLG) [*25513-40-0*]
  mesomorphism of, *14*:416

Poly(benzyl methacrylate) [*25085-83-0*]
  properties, *15*:379

Poly[bis(*m*-chlorophenoxy)phosphazene] [*52233-67-6*], *13*:399

Poly[bis(*p*-chlorophenoxy)phosphazene] [*52233-66-6*], *13*:399

Poly[bis(*m*-chlorophenylamino)phosphazene] [*57629-09-1*], *13*:399

Poly[bis(*p*-chlorophenylamino)phosphazene] [*57629-08-0*], *13*:399

Poly[bis(*p*-methoxyphenylamino)phosphazene] [*57629-07-9*], *13*:399

Poly[bis(*p*-methylphenoxy)phosphazene] [*52233-91-7*], *13*:399

Poly[bis(*m*-methylphenylamino)phosphazene] [*57629-03-5*], *13*:399

Poly[bis(*p*-methylphenylamino)phosphazene] [*57629-02-4*], *13*:399

Poly[bis(phenylamino)phosphazene] [*57650-68-7*], *13*:399

Poly[bis(3,3,2,2-tetrafluoropropoxy-1,1,7-*H*-dodecafluoroheptoxy)phosphazene] [*52902-19-9*], *13*:399

Poly[bis(trifluoroethoxyheptafluoro)phosphazene] [*52902-20-2*], *13*:399

Poly[bis(trifluoroethoxy)phosphazene] [*28212-50-2*], *13*:399

Polyborazenes, *13*:410

Polyborazynes, *13*:410

Poly(butadiene-*co*-acrylonitrile) [*9003-18-3*]
  in suedelike materials, *14*:245

Poly[(1,3-butadiene)ferrocene] [*32613-29-9*], *15*:194

Polybutadiene rubber [*9003-17-2*]
  MABS preparation, *15*:389

Polybutene.   See *Poly(1-butene)*.

Poly(1-butene) [*25085-05-6*]
  fibers, *16*:357

Poly(1-butene), atactic [*9003-28-5*], *16*:477

Poly(1-butene), isotactic [*25036-29-7*]
  properties, *16*:471
  thermal transitions of fiber, *16*:358

Poly(butyl acrylate) [*9003-49-0*]
  properties, *15*:380

Polybutylene.   See *Poly(1-butene)*.

Poly(*n*-butyl methacrylate) [*9003-63-8*]
  properties, *15*:379

Poly(*sec*-butyl methacrylate) [*29356-88-5*]
  properties, *15*:379

Poly(*t*-butyl methacrylate [*25213-39-2*]
  properties, *15*:379

Polycarbonate [*24936-68-3*]
  gas permeability, *15*:118
  membrane formation, *15*:104
  microfiltration membranes, *16*:846
  particle-track etching of, *16*:836
  stability, *15*:382
  as track-etch detector, *16*:827

Polycarbonate plastic
  NaOH etching of, *16*:834

Polycarbonates
  in composites, *13*:973
  as magnetic tape binders, *14*:741

Polychlorinated biphenyl (PCB)
  health hazard of, *13*:603

Polychlorinated biphenyls
  toxicity, *13*:272

Polychloroprene (PCP) [*9010-98-4*], *13*:575

Poly(chlorotrifluoroethylene) [*9002-83-9*]
  as lubricant, *14*:501

Polychlorotrifluoroethylene-*co*-vinylidene
    fluoride [*9010-75-1*]
  gasket properties, *16*:727

Polycontrast Rapid II RC Papers, *16*:544

Poly[1-(2-cyanoethyl)aziridine] [*9016-05-1*], *13*:161

Poly(cyclohexyl methacrylate) [*25768-50-7*]
  properties, *15*:379

Polycythemia
  detection, *15*:83

Poly(1-decene), isotactic [*26746-85-0*]
  properties, *16*:471

Poly(*n*-decyl methacrylate) [*29320-53-4*]
  properties, *15*:379

Poly(diaminophosphazenes), *13*:400

Poly(dibromophosphazene) [*56529-01-2*], *13*:399

Poly[1,7-dicarbadodecaborane(12)-1,7-diyl[1,1,3,5,5-pentamethyl-3-(3,3,3-trifluoropropyl)-1,5-trisiloxanediyl]] [*62725-81-9*], *13*:399

Poly[1,7-dicarbododecaborane(12)-1,7-diyl-(1,1,3,3,5,5,7,7,9,9,11,11-dodecamethyl-1-11-hexasiloxanediyl)] [*32473-92-0*], *13*:399

Poly[1,7-dicarbododecaborane(12)-trifluoropropyl-(1,7-tetrasiloxanediyl)] [*41655-25-8*], *13*:399

Poly[dicarbonyl[(1,2,3,4,5-η)-1-vinyl-2,4-cyclopenta- dien-1-yl]nitrosylchromium] [*64552-93-8*], *15*:188

Poly(dichlorophosphazene) [*26085-02-9*], *13*:399, 400

Poly(dichlorosilane), *13*:408

Poly(diethoxyphosphazene) [*60495-46-7*], *13*:399

Poly(difluorophosphazene) [*27290-47-4*], *13*:399

Poly(dihydronaphthalene)sulfonates, *15*:724

Poly(1,2-dihydro-2,2,4-trimethylquinoline [*26780-96-1*]
  LDPE antioxidant, *16*:414

Poly(dimethoxyphosphazene) [*60998-43-8*], *13*: 399

Poly(dimethyldiallylammonium chloride) [*26062-79-3*]
    in papermaking, *16*:804

Poly(1,4-dimethylenebenzene) [*25722-33-2*]
    in encapsulation, *15*:482

Poly(dimethylsilane) [*2883-63-8*]
    solubility, *13*:408

Poly(dimethylsilane-*co*-ethylmethylsilane) [*74578-04-4*], *13*:409

Poly(dimethylsilane-*co*-methyl propylsilane) [*71092-17-6*], *13*:409

Polydimethylsiloxane [*1646-73-7*]
    in encapsulation, *15*:478
    insulation properties of, *13*:551

Poly(dimethylsilylene) methylene [*25722-25-2*]
    copolymer with dimethylsilane, *13*:408

Poly(1,2-dimethyl-5-vinylpyridinium methylsulfate) [*27056-62-8*]
    in papermaking, *16*:804

Poly[di(phenoxy)phosphazene] [*28212-48-8*], *13*: 399, 400

Poly(diphenylsilane) [*51176-28-4*], *13*:408

Poly(1,1′-divinylferrocene) [*32840-26-9*], *15*:194

Polydymite [*12137-12-1*], *15*:805

Polyelectrolyte
    xanthan gum as, *15*:448

Polyenes
    metal π-complex formation, *16*:592

Polyester
    nonwoven fabrics from, *16*:107
    in nonwovens, *16*:74
    in packings, *16*:734

Polyester fibers
    in papermaking, *16*:823

Polyester resins
    in fluorescent pigments, *14*:561

Polyesters
    in composites, *13*:972
    as liquid crystals, *14*:418
    as magnetic tape binders, *14*:741

Polyethers
    as liquid crystals, *14*:418

Poly(ethyl acrylate) [*9003-32-1*]
    in bonding nonwovens, *16*:115
    properties, *15*:380

Poly(2-ethylbutyl methacrylate) [*25087-19-8*]
    properties, *15*:379

Polyethylene [*9002-88-4*]
    alkyl magnesium catalysts, *16*:561
    coefficient of friction, *14*:481
    column-packing material, *13*:20
    in composites, *13*:973
    dielectric strength, *13*:541
    electric insulation elastomer, *13*:565
    in encapsulation, *15*:478

fibers, *16*:357
gas permeability, *15*:118
as gas-permeable membrane, *13*:728
high density, *16*:422
inks for, *13*:382
insulating properties of, *13*:556
insulation properties, *13*:571
linings, *16*:717
low density, *16*:402
low pressure, low density, linear, *16*:385
from LPG intermediates, *14*:394
membrane formation, *15*:104
membranes, *15*:102
methacrylate grafting, *15*:390
in nonwovens, *16*:74
in paper coatings, *16*:791
properties, *13*:567; *16*:402, 471
resistance to water treeing, *13*:543
treeing in, *13*:543
Ziegler process for, *16*:433

Poly(ethylene-*co*-acrylic acid) [*9010-77-9*], *16*:420

Poly(ethylene adipate) [*24938-32-2*]
    in isocyanate reaction, *13*:815

Poly(ethylene-*co*-butene) [*54570-68-2*]
    linear low density, *16*:393

Poly(ethylene-*co*-1-butene) [*54570-68-2*], *16*:388
    ir spectrum, *16*:422

Poly(ethylene-*co*-carbon monoxide) [*25052-62-4*], *16*:420

Polyethylene, chlorinated [*64754-90-1*]
    composite with novoloid fiber, *16*:134

Poly(ethylene–*co*-chlorotrifluoroethylene) [*25101-45-5*] (E-CTFE)
    insulation properties of, *13*:575

Polyethylene, cross-linked
    in electric insulation, *13*:567

Poly(ethylene-*co*-ethyl acrylate) [*9010-86-0*], *16*: 419

Polyethylene glycol [*25322-68-3*], *15*:85

Poly(ethylene-*co*-1-hexene) [*25213-02-9*], *16*:388

Polyethylene (high density) [*9002-88-4*]
    thermal transitions of fiber, *16*:358

Poly(ethyleneimine) [*9002-98-6*] (pei)
    in papermaking, *16*:818

Poly(ethylene-*co*-isobutylene) [*25639-19-4*], *16*: 420

Polyethylene, linear low density, *16*:386

Polyethylene, low density [*9002-88-4*], *16*:385
    manufacture, *16*:412

Polyethylene, medium density
    properties, *16*:403

Poly(ethylene-*co*-methacrylic acid) [*25053-53-6*], *16*:420

Poly(ethylene-*co*-methyl acrylate) [*25103-74-6*], *16*:420

Poly(ethylene-*co*-4-methyl-1-pentene) [*25213-96-1*], *16*:388

Poly(ethylene-*co*-1-octene) [*26221-73-8*], *16*:388

Poly(ethylene oxalate) [*27615-22-1*], *16*:632

Poly(ethylene oxide) [*25322-68-3*]
dispersant, *16*:806
latex stabilization, *14*:88

Polyethylene (PE) [*9002-88-4*]
in electric insulation, *13*:564

Poly(ethylene-*co*-1-pentene) [*26221-69-2*], *16*:388
^{13}C nmr spectra, *16*:423

Poly(ethylene–*co*-propylene) [*9010-79-1*], *16*:388, 420
properties, *13*:567

Poly(ethylene terephthalate) [*25038-59-9*]
in bonding nonwovens, *16*:115
fibers, *16*:111
insulating properties of, *13*:556
in magnetic tape, *14*:733
microfiltration membranes, *16*:846
moisture regain, *16*:358
nonwovens from, *16*:108
properties, *16*:370
spunbondeds, *16*:75
thermal transitions of fiber, *16*:358
as track-etch detector, *16*:827

Poly(ethylene-*co*-tetrafluoroethylene) [*25038-71-5*]
insulation properties of, *13*:575

Poly[ethylene-*co*-vinyl acetate] [*24937-78-8*], *16*:419
in bonding nonwovens, *16*:115

Polyethylene wax
in inks, *13*:390

Polyethylenimine [*151-56-4*]
in composite membranes, *15*:107
in papermaking, *16*:804

Poly(2-ethylhexyl acrylate) [*9003-77-4*]
properties, *15*:380

Poly(2-ethylhexyl methacrylate) [*25719-51-1*]
properties, *15*:379

Poly(ethyl methacrylate) [*9003-42-3*]
properties, *15*:379

Poly(ethylthioethyl methacrylate) [*27273-87-6*]
properties, *15*:379

Poly(ethynylferrocene) [*33410-56-9*], *15*:188, 194

Polyfelt, *16*:78

Poly[1,1-ferrocenediyl(1,1,3,3-tetramethyl-1,3-disiloxanediyl)-1,4-phenylene(1,1,3,3-tetra-methyl-1,3-disiloxanediyl)] [*52436-23-4*], *15*:200

Poly[1,1'-ferrocenediyl(1,1,5,5-tetramethyl-3,3-diphenyl-1,5-trisiloxanediyl)] [*52436-24-5*], *15*:200

1,1'-Poly(ferrocenylene) [*55884-62-3*], *15*:196

1,2-Poly(ferrocenylene) [*76082-20-7*], *15*:196

1,3-Poly(ferrocenylene) [*76082-21-8*], *15*:196

1,1',3,3'-Poly(ferrocenylene) [*76082-25-2*], *15*:196

Poly[1,1'-ferrocenylmethoxydimethylsilyl-1,4-phenyldimethylsiloxymethylene] [*76082-30-9*], *15*:200

Polyfon
alkali lignins, *14*:307

Polyglycolic acid [*26124-68-5*], *13*:91

Polyglycols
as lubricants, *14*:498

Polyhalite [*15278-29-2*], *14*:637

Polyhalogenated biphenyls
inducement of immunodeficiency, *13*:170

Poly(1-heptene), isotactic [*26746-83-8*]
properties, *16*:471

Poly(hexadecyl methacrylate) [*25986-80-5*]
properties, *15*:379

Poly(1-hexene), isotactic [*26746-82-7*]
properties, *16*:471

Poly(*n*-hexyl methacrylate) [*25087-17-6*]
properties, *15*:379

Poly(hydroxyacetic acid) [*26124-68-5*], *13*:91

Poly D(−)-β-hydroxybutyrate [*29435-48-1*]
D(−)-δ-hydroxybutyric acid from, *13*:98

Polyimide [*25038-81-7*]
insulating properties of, *13*:556

Poly(isobornyl methacrylate) [*28854-39-9*]
properties, *15*:379

Polyisobutylene [*9003-27-4*]
in encapsulation, *15*:478
in paper insulation, *13*:576

Polyisobutylene oils
as lubricants, *14*:499

Poly(isobutyl methacrylate) [*9011-15-8*]
properties, *15*:379

1,2-Polyisoprene [*9003-31-0*], *13*:826

3,4-Polyisoprene [*9003-31-0*], *13*:826

Poly(*cis*-1,4-isoprene) [*9003-31-0*], *13*:826
from isoprene, *13*:818

trans-1,4-Polyisoprene [*9003-31-0*], *13*:826

Poly(isopropyl methacrylate) [*26655-94-7*]
properties, *15*:379

Poly(itaconic acid) [*25119-64-6*], *13*:867

Polyketones, *13*:933
from ethylene, *13*:933

Polylactic acid [*26100-51-6*], *13*:83

Poly(*d,l*-lactic acid) [*26100-51-6*], *15*:490
capsule matrix, *15*:476

Poly(lauryl methacrylate) [*25719-52-2*]
properties, *15*:379

(Poly)maltotriose [*1109-28-0*], *15*:455

Polymer cement
nuclear waste storage, *16*:211

Polymeric diphenylmethane diisocyanate [*9016-87-9*]
  properties, *13*:789
Polymeric metal phosphinates, *15*:214
Polymerization
  anionic, *13*:370
  cationic, *13*:371
  ethylene, *16*:391
  of ethylene, *16*:412
  gas-phase HDPE, *16*:445
  gas phase, polypropylene, *16*:461
  methacrylates, *15*:383
  redox initiators, *14*:92
  solution, HDPE, *16*:445
  solvent, polypropylene, *16*:461
Polymerization catalyst
  titanium tetraiodide, *13*:672
Polymerization catalysts
  magnesium alkyls, *14*:617
  peroxygen compounds, *13*:15
Polymerization, emulsion, *14*:82.   (See also *Latex technology.*)
Polymerization initiators.   See *Initiators.*
Polymerization systems
  for latices, *14*:92
Polymer liquid crystals, *14*:414
Polymers
  fluorescent, *14*:561
  gas permeabilities of, table, *15*:118
  hollow fiber, *15*:108
  inorganic, *13*:399
  particle-track etching, *16*:826
Polymers, metal-containing.   See *Metal-containing polymers.*
Polymers, thermoplastic
  insulation properties of, *13*:559
  in nonwovens, *16*:75
Polymers, thermoset
  insulation properties of, *13*:562
Polymers, transparent, *16*:479
Poly(metallo-siloxanes), *13*:409
Poly(metal phosphinates), *13*:410
Polymethacrylates, *15*:377.   (See also *Methacrylic polymers.*)
Poly[[2-[methacryloyl]ethyl]ferrocene] [*35560-99-7*], *15*:188
Poly[[[methacryloyl]methyl]ferrocene] [*35560-97-5*], *15*:188
Poly(methyl acrylate) [*9003-21-8*]
  properties, *15*:380
Poly(3-methyl-1-butene) [*24979-98-4*]
  fibers, *16*:357
Poly(3-methyl-1-butene) (isotactic) [*26703-14-0*]
  properties, *16*:471
  thermal transitions of fiber, *16*:358
Poly(methyldiallylamine) [*55553-13-4, 29566-78-8*]

  in papermaking, *16*:817
Polymethylene [*25083-57-7*]
  HDPE as, *16*:421
Poly[4,4′-methylene bis(phenyl isocyanate)] [*9016-87-9*], *15*:923
Polymethylene polyphenyl isocyanate [*9016-87-9*] (PMPPI), *13*:801
Poly(6-methyl-1-heptene), isotactic [*77136-88-0*]
  properties, *16*:471
Poly(4-methyl-1-hexene), isotactic [*25620-38-6*]
  properties, *16*:471
Poly(5-methyl-1-hexene), isotactic [*28962-82-5*]
  properties, *16*:471
Poly(methyl isopropenyl ketone) [*25988-32-3*], *13*:933
Poly(methyl methacrylate) [*9011-14-7*]
  contact lenses of, *15*:126
  dielectric strength, *13*:541
  properties, *15*:379
  thermal cracking, *15*:367
  weather resistance of, *13*:547
Poly(4-methyl-1-pentene) [*25038-59-9*]
  fibers, *16*:357
Poly(3-methyl-1-pentene), isotactic [*28575-86-2*]
  properties, *16*:471
Poly(4-methyl-1-pentene) (isotactic) [*24979-98-4*]
  properties, *16*:471
  thermal transitions of fiber, *16*:358
Poly(methyl sorbate) [*25987-01-3*]
  in olefin fibers, *16*:380
Poly(methyl vinyl ether) [*9003-09-2*]
  precipitation, *14*:88
Poly(methyl vinyl ketone) [*25033-87-3*], *13*:933
Polymorphism
  of liquid crystals, *14*:410
Poly(nitrogen sulfide) [*56422-03-8*], *13*:406
Poly(1-octadecene), isotactic [*26476-89-4*]
  properties, *16*:471
Poly(octadecyl methacrylate) [*25639-21-8*]
  properties, *15*:379
Poly(*n*-octyl methacrylate) [*25087-18-7*]
  properties, *15*:379
Polyolefin fibers, *16*:357.   (See also *Olefin fibers.*)
  in papermaking, *16*:823
Polyolefins
  C-4 and higher, *16*:471
Poly(oxydecamethylene) [*28726-55-8*], *16*:43
Poly[oxy(3,3-dimethyl-1,1-diphenyl-1,3-disiloxanediyl)-1,4-phenylene(dimethylsilylene)] [*41205-85-0*], *15*:200
Poly[oxy(dimethylsilylene)-1,4-phenylene(dimethylsilylene)] [*54811-84-6*], *15*:200

Poly[oxy(1,1,3,3-tetramethyl-1,3-disiloxanediyl)-1,4-phenylene(dimethylsilylene)] [*41205-84-9*], *15*:200

Poly(parabanic acid) [*60092-26-4*]
   insulating properties of, *13*:556

Polyparabanic acids, *13*:797

Poly(1-pentene), isotactic [*25587-79-5*]
   properties, *16*:471

Poly(perfluorosulfonic acid)
   membrane in nitrobenzene mfg, *15*:920

Polyphenylchromium compounds, *16*:592

Poly(phenylene oxide) [*9041-80-9*]
   gas permeability, *15*:118

Poly(*p*-phenyl ether) [*25718-67-6*]
   as lubricant, *14*:500

Polyphenyl ethers
   in lube oils, *14*:500

Poly(phenyl methacrylate) [*25189-01-9*]
   properties, *15*:379

Polyphosphates
   in papermaking, *16*:822

Poly(phosphazenes), *13*:399

Polyphosphazines
   as liquid crystals, *14*:418

Poly(1,4-piperazinediyltrimethylene
      iminoadipolyimino trimethylene) [*28730-14-5*], *16*:43

Polypropylene [*9003-07-0*], *13*:391; *16*:454
   column-packing material, *13*:20
   in composites, *13*:973
   insulating properties of, *13*:556
   insulation properties, *13*:571
   in leatherlike materials, *14*:242
   from LPG intermediates, *14*:394
   membrane formation, *15*:104
   membranes, *15*:102
   nonwovens from, *16*:108
   in nonwovens, *16*:74
   oxidation of, *13*:544

Poly(propylene glycol) [*25322-69-4*]
   in lubricants, *14*:498

Polypropylene (isotactic) [*25085-53-4*]
   fibers, *16*:357
   properties, *16*:370, 471
   thermal transitions of fiber, *16*:358

Poly(propylene oxide) [*29322-69-4*]
   limit of radiation tolerance, *14*:519

Poly(*n*-propyl methacrylate) [*25609-74-9*]
   properties, *15*:379

Polyresearch
   information service, *16*:925

Polyresearch Service
   patent information, *16*:925

1,1′-Poly(ruthenocenylene) [*76082-22-9*], *15*:196

1,2-Poly(ruthenocenylene) [*76082-23-0*], *15*:196

1,3-Poly(ruthenocenylene) [*76082-24-1*], *15*:196

1,1′,3,3′-Poly(ruthenocenylene) [*76028-25-2*], *15*:196

Polysaccharides, *15*:79, 439.   (See also *Microbial polysaccharides.*)
   as antigens, *13*:167
   biosynthesis, *15*:445

Polysilanes, *13*:407.
   (See also *Polysilylenes.*)

Polysilanyl ferrocenes, *15*:199

Polysiloxanes
   as liquid crystals, *14*:418

Polysiloxanyl ferrocenes, *15*:199

Polysilylene [*32028-95-8*], *13*:407

Poly(stearyl methacrylate) [*9086-85-5*]
   properties, *15*:379

Polystyrene [*9003-53-6*]
   capsule matrix, *15*:476
   in composites, *13*:973
   gas permeability, *15*:118
   inks for, *13*:382
   membrane formation, *15*:104
   methacrylate grafting, *15*:390
   in papermaking, *16*:819
   in propylene fibers, *16*:376
   specific heat, *16*:423
   thermal conductivity of, *13*:595

Poly(styrene-*co*-butadiene carboxylates)
   in bonding nonwovens, *16*:115

Polystyrene (isotactic) [*9003-53-6*]
   properties, *16*:370, 471

Poly(styrene-*co*-maleic anhydride) [*9011-13-6*]
   in leather tanning, *14*:216

Poly(styrene-*co*-methyl vinyl ketone) [*100-42-5*], *13*:933

Poly(styrene-*co*-phenyl vinyl ketone) [*27340-61-0*], *13*:933

Poly(*p*-styrene sulfonic acid) [*28210-41-5*]
   in ion exchange, *13*:680

Polysulfide rubber
   gasket properties, *16*:727

Polysulfone fibers
   in hydrogen recovery, *15*:118

Polysulfones
   in composite membranes, *15*:107
   in composites, *13*:973
   as magnetic tape binders, *14*:741

Poly(sulfur nitride) [*56422-03-8*], *13*:406

Poly(tetradecyl methacrylate) [*30525-99-6*]
   properties, *15*:379

Polytetrafluoroethylene [*9002-84-0*]
   in ball bearings, *14*:517
   in cation-exchange membranes, *15*:120
   coefficient of friction, *14*:481
   column-packing material, *13*:20
   in fabric membranes, *15*:125
   in hook-up wire, *13*:554

insulating properties of, *13*:556
in insulation enamels, *13*:580
iodine resistance, *13*:656
membrane formation, *15*:104
Poly(tetrafluoroethylene-*co*-perfluoropropylene) [*25067-11-2*]
as lubricant, *14*:501
Poly(tetrafluoroethylene) [*9002-84-0*] (PTFE)
insulation properties of, *13*:573
Poly(2,2,6,6-tetramethyl-4-piperidylamino-1,3,5-triazine) [*77136-90-4*] (CR 141 and 144)
polypropylene light stabilization, *16*:361
Polythiazyl [*56422-03-8*], *13*:406
Polytran, *15*:455
Poly[tricarbonyl[(η^6-phenylmethyl)acrylate]chromium(2+)] [*33338-60-2*], *15*:188
Poly[tricarbonyl[(1,2,3,4,5-η)-1-vinyl-2,4-cyclopenta-diene-1-yl]manganese] [*35443-49-3*], *15*:188
Poly(3,3,5-trimethycyclohexyl methacrylate) [*75673-26-6*]
properties, *15*:379
Poly(trimethyldihydroquinoline) [*26780-96-1*]
in electrical insulation, *13*:565
Polyurea [*37955-36-5*]
as grease thickener, *14*:502
Polyurethane [*27416-86-0*]
in acoustic foams, *13*:517
foam catalysts, *16*:579
as insulation, *13*:595
in packings, *16*:739
thermal conductivity of, *13*:595
Polyurethane coatings
removal from rubber, *16*:765
Polyurethanes [*68400-67-9*]
in leather coatings, *14*:221
as magnetic tape binders, *14*:741
in papermaking, *16*:820
Poly(vinyl acetal) [*27790-26-7*]
as magnetic tape binder, *14*:741
Poly(vinyl acetate) [*9003-20-7*]
in bonding nonwovens, *16*:115
in emulsion polymerization, *14*:84
in gel filter preparation, *15*:445
latexes in paint, *16*:748
as magnetic tape binder, *14*:741
nonwoven binder, *16*:106
in paint, *16*:748
in paper coatings, *16*:790
in papermaking, *16*:822
Poly[vinyl acetate-*co*-acrylic acid] [*24980-58-3*]
in bonding nonwovens, *16*:115
Poly(vinyl alcohol) [*9002-89-5*]
in adhesives with polyaziridine, *13*:161
in encapsulation, *15*:480
latex stabilization, *14*:88
methacrylate grafting, *15*:390

nonwovens from, *16*:108
in papermaking, *16*:814
Poly(vinyl alcohol) [*9002-89-5*] (PVA)
in paper coatings, *16*:790
Poly(3-vinyl-1,1″:1,1‴-bisferrocene) [*53775-36-3*], *15*:186, 188
Poly(vinyl butyral) [*63148-65-2*]
aircraft grade, *13*:989
as magnetic tape binder, *14*:741
Poly(vinyl butyral) (PVB) [*63148-65-2*], *13*:978
Poly(vinyl chloride) [*9002-86-2*]
alkyltin stabilizers, *16*:573
in bonding nonwovens, *16*:115
column-packing material, *13*:20
as insulation, *13*:595
as ion exchange membrane, *13*:725
latex technology, *14*:82
in leatherlike materials, *14*:236
from LPG intermediates, *14*:394
as magnetic tape binder, *14*:741
in magnetic tape, *14*:734
MBS in, *15*:390
membrane formation, *15*:104
as organolead barrier, *14*:193
Poly(vinyl chloride) (PVC) [*9002-86-2*]
antimicrobial agents in, *13*:238
in electric insulation, *13*:571
Poly(vinylcyclohexane), isotactic [*26951-20-2*]
properties, *16*:471
Poly(vinylferrocene) [*34801-99-5*], *15*:188, 195
Poly(vinylidene chloride) [*9002-85-1*]
in paper coatings, *16*:792
Poly(vinylidene dichloride) [*9002-85-1*]
as magnetic tape binder, *14*:741
Poly(vinylidene fluoride) [*24937-79-1*] (PVDF)
insulation properties of, *13*:575
Poly(vinylidine fluoride) [*24937-79-9*]
as insulator, *13*:554
Poly[(vinylimino) carbonyl] [*55553-09-8*], *13*:797
Poly(vinyl isocyanate) [*55553-09-8*], *13*:796
Poly(vinylpyridine) [*9003-47-8*]
dyeing of, *16*:362
Poly(2-vinylpyridine) [*25014-15-7*]
with $Co_2(CO)_8$ as oxo catalyst, *16*:641
Poly(vinylpyridine) copolymers
as magnetic tape binders, *14*:741
Poly(2-vinylpyridine-*co*-styrene-*co*-butadiene) [*9019-71-0*]
adhesive with RF, *13*:58
Poly(*N*-vinylpyrrolidinone) [*9003-39-8*]
in adhesives with polyaziridine, *13*:161
Polyvinyl pyrrolidinone–iodine complex [*25655-41-8*]
disinfectant, *13*:674
Poly(vinylpyrrolidinone) (PVP) [*25655-41-8*]
in antimicrobial agents, *13*:232

Poly(vinylruthenocene) [76082-17-2], 15:188, 195
Polyweb, 16:78
α-Poly-p-xylylene [25951-90-0]
    as liquid crystal, 14:418
Pomanders
    odor modification with, 16:304
Popillia japonica, 13:476, 481
Population inversion
    in lasers, 14:44
Porapak N, 13:86
Porcelain [1332-58-7]
    thermal conductivity of, 13:544
Pork, 15:62.   (See also *Meat products.*)
Poromerics, 14:233
    sueded, 14:245
Porphyrins, iron, 13:782
Porphyrins, picket fence, 13:783
Porthetria dispar, 13:420, 480
Portia nut, 16:251
Portland cement
    limestone in, 14:374, 376
Porvair, 14:234
Potash [17353-70-7]
    from seaweed, 16:284
Potassium [7440-09-7]
    in black walnut, 16:261
    electrode, 13:720
    as lubricant, 14:515
    metabolic functions, 15:574
    nutrient properties, 15:571
    from the ocean, 16:279, 292
    in walnuts, 16:261
Potassium-39 [14092-91-2], 13:840
Potassium-40 [13966-00-2], 13:840
    radiation exposure from, 16:217
Potassium-41 [13965-93-0], 13:840
Potassium acetate [127-08-2]
    in isocyanate trimerization, 13:794
    in lithium recovery, 14:453
    toxicity, 15:573
Potassium acid iodate [13455-24-8], 13:666
Potassium acid tartrate [868-14-4], 13:118
Potassium alum [10043-67-1]
    in mothproofing, 13:479
Potassium aluminum sulfate dodecahydrate
    [7784-24-9]
    from aluminum sulfate, 14:860
Potassium amide [7440-09-7]
    in isotope separation, 13:852
Potassium antimony tartrate sesquihydrate
    [28300-74-5]
    properties, 13:118
Potassium aspartate [14007-45-4], 15:760
Potassium bitartrate [868-14-4]
    in cream of tartar production, 13:115
    in tartar emetic production, 13:116

Potassium borate [13709-94-9], 16:33
    in metallic coating, 15:252
Potassium bromate [7758-01-2]
    in malting, 14:816
Potassium carbonate [584-08-7]
    in liquid membranes, 15:119
    malonic acid from, 14:795
    toxicity, 15:573
Potassium chlorate [3811-04-9]
    in matches, 15:1
    as oxygen source, 16:673
Potassium chloride [7447-40-7]
    controlled release of, 15:488
    in metallic coating, 15:252
    from seawater, 16:284
    toxicity, 15:573
Potassium cobalt(II) hexacyanoferrate [14874-
    73-8], 13:769
Potassium copper(II) hexacyanoferrate [53295-
    15-1], 13:769
Potassium cyanate [590-28-3]
    in case hardening, 15:318
Potassium cyanide [151-50-8]
    in case hardening, 15:318
    nickel sulfate reaction, 15:805
Potassium 2,4-cyclopentadienide [30994-24-2],
    16:598
Potassium dichloroisocyanurate [2244-21-5]
    applications of, 13:230
Potassium dichromate [7778-50-9]
    in iron analysis, 13:787
    in matches, 15:5
Potassium diethyl malonate [37892-24-3], 14:800
Potassium difluoroiodate, 13:666
Potassium dihydrogen citrate [866-83-1]
    pH standard, 13:3
Potassium dihydrogen iodate, 13:665
Potassium dihydrogen phosphate [7778-77-0]
    pH standard, 13:3
Potassium diphenylphosphine [15475-27-1]
    aziridine reaction, 13:150
Potassium eugenol [76963-40-1], 16:310
Potassium ferrate [13718-66-6], 13:776
Potassium ferricyanide [13943-58-3]
    blueprinting use, 16:633
    resorcinol reaction, 13:41
Potassium ferrous hexakiscyanoferrate [25869-
    98-1] (1:1:1), 13:771
Potassium fluoride [7789-23-3]
    in metallic coating, 15:252
    toxicity, 15:573
Potassium fluoroborate [14075-53-7], 16:33
Potassium fluorophosphate [17084-13-8], 16:33
Potassium fluorosilicate [16871-90-2]
    fluorophlogopite from, 15:425
Potassium guaiacol sulfonate [1321-14-8], 13:62

Potassium heptafluoro niobate [*36354-32-2*], *15*: 825

Potassium heptafluoro tantalate [*16924-00-0*], *15*:825

Potassium heptaiodide monohydrate [*74685-93-1*], *13*:650

Potassium hexachloromanganate(IV) [*16923-41-4*], *14*:856

Potassium hexachloroplatinate(IV) [*16921-30-5*], *16*:35

Potassium hexachlororhenate(IV) [*16940-97-9*], *16*:35

Potassium hexacyanoferrate(III) [*13746-66-2*]
  m-nitrotoluene reaction, *15*:927

Potassium hexafluoroiodide [*20916-97-6*], *13*:665

Potassium hexafluoro niobate [*16919-14-5*], *15*: 825

Potassium hexafluorosilicate [*16871-90-2*], *14*:860
  toxicity, *15*:573

Potassium hexafluorotantalate [*20137-53-5*], *15*: 825

Potassium hydrogen malonate [*34346-13-9*], *14*: 794

Potassium hydrogen oxalate [*125-95-7*], *16*:632

Potassium hydrogen phthalate [*877-24-7*]
  pH standard, *13*:3

Potassium hydrogen tartrate [*868-14-4*]
  pH standard, *13*:3
  from wine fermentation, *13*:114

Potassium hypochlorite [*7778-66-7*]
  in potassium ferrate synthesis, *13*:776

Potassium iodate [*74869-66-2*], *13*:665, 666

Potassium iodide [*7681-11-0*], *13*:661
  toxicity, *15*:574

Potassium iodide–iodine complex [*12298-68-9*]
  stain for poly(vinyl acetate) binder in nonwoven fabric, *16*:106

Potassium iodomercurate [*7783-33-7*], *13*:661; *15*:169

Potassium manganate(III) [*12142-17-5*], *14*:853

Potassium manganate(IV) [*12142-27-7*], *14*:845, 855; *16*:35

Potassium manganate(V) [*12142-41-5*], *14*:845, 857

Potassium manganate(VI) [*10294-64-1*], *14*:845

Potassium manganese(II) hexacyanide [*28299-84-5*], *14*:853

Potassium manganese(II) hexaperiodate [*55866-55-2*], *14*:853

Potassium metabisulfate [*16731-55-8*]
  in antiaging drug, *15*:140

Potassium metaperiodate [*7790-21-8*], *13*:667

Potassium methylamide [*13427-02-6*]
  in isotope separation, *13*:852

Potassium molybdate [*13446-49-6*], *15*:687

Potassium monobromomonochloroiodide [*15859-97-9*], *13*:665

Potassium monoperoxysulfate [*10058-23-8*], *13*:29

Potassium monopersulfate [*25482-78-4*], *13*:674

Potassium-2-naphthalenolate [*36294-21-0*]
  2-hydroxy-1-naphthalenecarboxylic acid, *15*: 746

Potassium nickelate [*50811-97-7*], *15*:802

Potassium nickel(II) hexacyanoferrate [*53295-14-0*], *13*:769

Potassium nickel molybdate [*59228-72-7*], *15*:813

Potassium niobate [*12502-31-7*] (4:3), *15*:837

Potassium nitrate [*7757-79-1*]
  in meat products, *15*:66
  nitric acid from, *15*:853

Potassium nitride [*29285-24-3*], *15*:872

Potassium nitrite [*7758-09-0*]
  in meat products, *15*:66

Potassium nitroborate [*76081-99-7*], *16*:33

Potassium nitroform [*14268-23-6*]
  nitryl chloride reaction, *15*:974

Potassium octacyanodinickelate(VI), *15*:801

Potassium oxalate [*127-96-8*], *16*:632
  pH standard, *13*:3

Potassium oxide [*20762-60-1*]
  in isotope separation, *13*:852
  in lignitic coal, *14*:334
  in spent shale, *16*:351

Potassium ozonide [*12030-89-6*], *16*:688

Potassium palmitate [*57-10-3*]
  styrene solubilization, *14*:85

Potassium perchlorate [*7778-74-7*]
  as oxygen source, *16*:673

Potassium permanganate [*7722-64-7*], *14*:845
  in iron analysis, *13*:787
  reduction by hydrogen peroxide, *13*:16

Potassium peroxydisulfate [*7727-21-1*]
  in initiator systems, *13*:365

Potassium persulfate [*7727-21-1*]
  iodoacetic acid reaction, *13*:670

Potassium phthalimide [*1074-82-4*]
  γ-chlorobutyronitrile reaction, *15*:766

Potassium salts
  from seaweeds, *16*:284

Potassium soap rosin [*61790-50-9*]
  paper size, *16*:809

Potassium sodium tartrate tetrahydrate [*6381-59-5*]
  in Fehling's solution, *13*:118
  properties, *13*:118

Potassium sorbate salt [*590-00-1*]
  preservative, *13*:247

Potassium superoxide [*12030-88-5*]
  oxygen generation, *16*:679

Potassium tartrate [*868-14-4*]
  determination of with ISES, *13*:729

Potassium tetrachloroiodide [*14323-44-5*], *13*:665

Potassium tetrachloroplatinate(II) [*10025-99-7*], *16*:35

Potassium tetracyanonickelate(II) [*14220-17-8*], *15*:805

Potassium tetracyanopalladate [*76082-10-5*], *16*:35

Potassium thioborate [*14355-24-9*], *16*:33

Potassium triiodide [*7790-42-3*], *13*:661

Potassium triiodide monohydrate [*74869-66-2*], *13*:665

Potato chips
  analysis of with ISES, *13*:732

Potomac yellow
  fluorescent dye, *14*:547

Pour-point depressants, *14*:494

Powder coating, *15*:256

Powder metallurgy
  of superalloys, *15*:342

Powder rolling, *15*:259

Powders
  as insulation, *13*:598
  odor modification with, *16*:304

Powders, wettable
  as insecticide formulations, *13*:417

Powellite [*14020-51-0*], *15*:670

Power generation
  from lignite, *14*:315
  from lignites, *14*:332
  by pressure-retarded osmosis, *15*:125

Power-metallurgy methods
  metal-laminate fabrication, *13*:948

Pragmoline, *15*:756.   (See also *Acetylcholine bromide.*)

Prandtl number, *14*:978

Prandtl's one seventh power law, *14*:951

Praseodymium [*7440-10-0*]
  from the ocean, *16*:280
  in optical filters, *16*:554

Praseodymium-141 [*7440-10-0*], *13*:841

Praseodymium, compd with cobalt [*12017-67-3*] (1:5), *14*:676

Praseodymium, compd with cobalt [*12052-77-6*] (2:17), *14*:676

Praseodymium nitride [*25764-09-4*], *15*:872

Precious metal anodes, *15*:173

Precipitation hardening
  of metals, *15*:335

Precipitators
  in lead refining, *14*:108

Precursor, *13*:356.   (See also *Free radical.*)

PREDICASTS
  business data bases, *13*:333

Prednisolone [*50-24-8*]
  microbial synthesis of, *15*:460

Prednisone [*53-03-2*], *13*:173
  microbial synthesis of, *15*:460

Pregnancy
  detection, *15*:83
  iron absorption during, *15*:595

5-Pregnene [*18872-17-8*]
  isomerization, *15*:467

Pregnenolone [*145-13-1*]
  in corticosteroid biosynthesis, *15*:762

Presdwood, *14*:304

Preservation
  of books, *16*:586

Preservative
  antimicrobial agent as, *13*:223

Preservative, food
  lactic acid as, *13*:87

Preservatives
  Hg compounds in paint, *15*:163
  mercury, for paint, *15*:158
  salt as, *16*:279
  for wood, *13*:480
  wood laminates, *14*:38

Pressinamide [*27559-18-8*] (PA)
  avoidance behavior effect, *15*:133

Pressing
  essential oils by, *16*:309

Pressure
  standardization of, *15*:39

Pressure-sensing devices
  mercury in, *15*:155

Pressure-swing adsorption
  air-separation process, *16*:658
  in N_2 mfg, *15*:937

Pressure vessels
  explosion-clad plates for, *15*:285

Pressurized-water reactor (PWR), *16*:150

Pressurized-water reactors
  fuel assembly, *16*:165

Preventive maintenance, *14*:753

Primers, *16*:751

Primidone [*125-33-7*]
  properties of, *13*:135

Print-bonding, *16*:90
  nonwovens, *16*:117

Printers
  laser, *14*:285

Printing
  fluorescent inks for, *14*:567
  type alloys, *14*:156

Printing inks, *13*:374

Printing plates
  photomechanical process for, *13*:881

Prior art
  for patents, *16*:880

Prism separators, *15*:118

Prisons
  laminated glass used in, *13*:991

Proangiotensin [*484-42-4*], *15*:759.   (See also *Angiotensin I.*)

Proanthocyanidin [*18206-61-6, 27567-66-4*]
  in malting, *14*:816

Probabilistic engineering design, *15*:27

Probability
  of material failures, *15*:9

Procainamide [*614-39-1*]
  learning effects, *15*:140

Procaine hydrochloride [*51-05-8*]
  as possible antiaging drug, *15*:140

Process control
  by pH measurement, *13*:10

Processing
  instrumentation and control of, *13*:505

Processing aids, *16*:804

Prodenia, *13*:471

Product Information Sources, *13*:318

Production
  computer control in, *13*:504

Production flow
  queuing theory applied to, *16*:520

Production planning, *16*:513

Production scheduling, *16*:500

Progesterone [*57-83-0*]
  in corticosteroid synthesis, *15*:762
  microbial oxidation, *15*:460

Program evaluation and review technique (PERT), *16*:516

Program languages
  in operations planning, *16*:509

Programming, linear
  industrial applications, *13*:508

Programming, nonlinear
  industrial applications, *13*:508

Programs
  in operations planning, *16*:500

Proinsulin (ox) [*11062-00-3*], *13*:608

Proinsulin (pig) [*11062-03-6*], *13*:608

Prokaryotes
  in nitrogen fixation, *15*:946

Prolactin [*9046-05-3*]
  somatostatin effect on, *15*:780

Prolamide [*7531-52-4*]
  in TRH synthesis, *15*:782

Proline [*147-85-3*]
  in nuts, *16*:249

Promecarb [*2631-37-0*], *13*:455

Promethazine, *13*:122

Promethazine hydrochloride [*58-33-3*]
  properties of, *13*:122

Prometryne [*7287-19-6*], *13*:674

Prompt insulin zinc suspension [*8049-62-5*], *13*:612

Propane [*74-98-6*]
  in LPG, *14*:384
  nitration, *15*:848

properties, *14*:384
specifications, *14*:393
storage, *14*:393
vapor-phase nitration, *15*:978

Propanedinitrile [*109-77-3*], *14*:794.   (See also *Malononitrile.*)

Propanedioic acid.   See *Malonic acid.*

1,2-Propanediol methacrylate [*7559-82-2*]
  properties, *15*:354

Propane HD-5, *14*:383

2-Propanenitronic acid.   See *2-Isonitropropane.*

1-Propanol [*71-23-8*]
  from oxo process, *16*:637

n-Propanol [*71-23-8*]
  in inks, *13*:390

Propargite [*2312-35-8*], *13*:463

Propargyl chloride [*624-65-7*]
  from itaconic acid, *13*:870

Propargyl methacrylate [*13861-22-8*]
  mannich reaction, *15*:356

Propellant
  hydrogen peroxide, *13*:29

Propellants
  nitroparaffins, *15*:983

Propellers, *15*:609.   (See also *Impellers.*)

2-Propenal [*107-02-8*]
  isoprene reaction, *13*:821

2-Propenenitrile [*107-13-1*]
  isoprene reaction, *13*:821

η^3-2-Propenyl iodonickel
  in isoprene polymerization, *13*:829

Prophos [*13194-48-4*], *13*:443

Propiolactone [*57-57-8*]
  from ketene, *13*:877

β-Propiolactone [*57-57-8*], *13*:96

Propionaldehyde [*123-38-6*]
  methacrylates from, *15*:357
  in methyl methacrylate mfg, *15*:364

Propionamide [*79-05-0*]
  salicylaldehyde reaction, *13*:72

1-Propionaphthone [*33744-50-2*], *13*:894

2-Propionaphthone [*21567-68-0*], *13*:894

Propionic acid [*79-09-4*]
  methacrylates from, *15*:357
  in methyl methacrylate mfg, *15*:364

Propionyl chloride [*79-03-8*]
  benzene reaction, *13*:934

Propiophenone [*93-55-0*]
  in mephenytoin synthesis, *13*:136
  properties, *13*:897

Propoxur [*114-26-1*], *13*:62, 453

Proprietary 425, *14*:510

Proprietary 905, *14*:511

Proprietary 2123, *14*:511

Proprietary AFSL-28, *14*:510

n-Propyl acetate [109-60-4]
  in inks, 13:390
n-Propyl alcohol [71-23-8]
  in inks, 13:390
Propyl 1,4-benzodioxan-2-carboxylate [24902-02-1]
  as insect attractant, 13:481
Propyl 2,2-dichloro-1-methylcyclopropyl carboxylate [35749-14-5], 15:349
n-Propyl N,N-diethylsuccinamate [5834-44-4]
  as repellent, 13:477
Propylene [115-07-1], 13:85
  copolymer with ethylene, 16:388, 420
  dimerization, 16:479
  epoxidation with Mo catalyst, 15:694
  hydroformylation, 16:639
  in isoprene synthesis, 13:830
  lime in refining of, 14:380
  malonic acid from, 14:795
  naphthalene reaction, 15:703
  olefins from, 16:480
  oxalic acid from, 16:621
  oxycarbonylation to methacrylate, 15:365
  in propane HD-5, 14:383
  properties, 14:384
1,2-Propylenediamine [78-90-0]
  salicylaldehyde reaction, 13:72
Propylene glycol [57-55-6]
  in paint, 16:751
  resin from in fluorescent product, 14:561
Propylene oxide [75-56-9]
  HCl scavenger, 16:763
  in lubricant synthesis, 14:498
Propylene sulfide [1072-43-1]
  odor-problem chemical, 16:303
Propylene tetramer [6842-15-5]
  H$_2$S reaction, 16:482
Propylenimine, 13:143.   (See also 2-Methylaziridine.)
Propyl isocyanate [110-78-1]
  chlorpropamide preparation, 13:616
  properties, 13:789
Propyl isome [83-59-0]
  as insecticide, 13:425
n-Propyllactamide [74421-70-8], 13:90
Propyl methacrylate [2210-28-8], 15:349
  properties, 15:353
Propylpseudonitrole.   See 2-Nitroso-2-nitropropane.
Propyl red [2641-01-2], 13:10
2-Propynyl methacrylate [13861-22-8]
  properties, 15:353
2-Propynyl-2-naphthalenecarboxylate [53548-27-9], 15:737
Prostaglandin E$_2$ [363-24-6], 15:778
Prostaglandin F$_{2\alpha}$ [551-11-1], 15:778

Prostaglandin G$_2$ [51982-36-6], 15:780
Prostaglandin H$_2$ [42935-17-1], 15:780
Prostaglandins
  as neuroregulators, 15:778
Prostanoic acid [25151-81-9]
  prostaglandins derived from, 15:778
Prostate disease
  diagnosis, 15:80
Prostatic acid phosphatase [9001-77-8]
  measurement, 15:80
Protactinium [7440-13-3]
  from the ocean, 16:280
Protactinium-233 [13981-14-1], 16:180
  in thorex process, 16:180
Protactinium nitride [75733-54-9], 15:872
Protamine sulfate [9009-65-8]
  in isophane insulin, 13:612
Protamine zinc insulin [9004-17-5], 13:613
Protect, 15:741
Protein
  measurement, 15:80
  in nuts, 16:249
Protein A
  in microspheres, 15:490
Proteins
  as antigens, 13:167
  in nuts, 16:249
Protocatechualdehyde [139-85-5], 13:70
Protocatechuic aldehyde [139-85-5], 13:45
Protonation
  of metallocenes, 16:602
Proton transfer
  with glass electrode, 13:5
Protoporphyrin, 15:595
Protosynthex
  information system, 16:901
Proxcel CRL, 13:245
Prunasin [138-53-4]
  from amygdalin, 16:256
  amygdalin splitting, 16:256
Prunus amygdalus, 16:250
Prussian blue [14038-43-8], 13:765.   (See also Turnbull's blue.)
  in inks, 13:378
Pseudoglobulin
  in milk, 15:568
Pseudomonas aeruginosa, 15:460
Pseudomonas saccharophila, 15:467
Pseudonitroles
  from sec-nitroparaffins, 15:974
Pseudoplusia includens, 13:480
Pseudorandom numbers, 16:508
Psilocin [520-53-6], 13:213
Psilocybin [520-52-5], 13:213, 220
Psilomelane [12322-95-1], 14:825, 855
Psoriasis
  treatment, 14:266

Psychophysical law, *16*:298

Psychoses, manic-depressive
treatment, *14*:460

Psychotria, *15*:949

Pteridine [*91-18-9*]
from γ-chloroacetoacetyl chloride, *13*:890

PTFE. See *Polytetrafluoroethylene*.

PTH. See *Parathyroid hormone*.

PTSA. See *p-Toluenesulfonic acid*.

PTS FEDERAL INDEX, *13*:334

PTS FEDERAL INDEX information system, *13*:330

PTS F+S INDEXES information system, *13*:330

PTS INTERNATIONAL ANNUAL TIME
SERIES information system, *13*:330

PTS INTERNATIONAL STATISTICAL
ABSTRACTS information system, *13*:330

PTS PREDALERT information system, *13*:330

PTS PROMPT information system, *13*:331

PTS US ANNUAL TIME SERIES information
system, *13*:331

PTS US STATISTICAL ABSTRACTS
information system, *13*:331

Puddling process
in wrought iron production, *13*:735

Pulley
magnetic, *14*:718

Pull match, *15*:4

Pullulan [*9057-02-7*], *15*:455

Pullularia, *15*:455

Pulp
hydroxyacetic acid in, *13*:91

Pulp and paper industry
limestone uses, *14*:380

Pulping
liquor composition, *14*:306

Pulps, chemical, *16*:770

Pulps, mechanical, *16*:769

Pulps, semichemical, *16*:770

Pulpy butt, *14*:208

Pultrusion
for plastic composites, *13*:976

Pumps
noise control, *16*:23

Pumps, oil well
cable insulation for, *13*:571

Punch
frozen dessert, *15*:568

Punch mica, *15*:420

Punking
of novoloid fibers, *16*:127

Purafil
for odor control, *14*:889

Purchasing research, *14*:908

Purex process
fuel reprocessing, *16*:173

Purine [*120-73-0*]
antimetabolite analogues, *13*:172

Purofer process, *13*:745

Puromycin [*53-79-2*]
amnesia from, *15*:133
protein synthesis inhibitor, *15*:136

Purox system
for waste incineration, *13*:194

Purpurol [*83-55-6*], *15*:740. (See also *5-Amino-
1-naphthalenol*.)

Putrescine [*110-60-1*]
properties, *15*:776

Putrescine 3-aminopropyl transferase [*37277-82-
0*]
in spermine biosynthesis, *15*:777

Putrescine dihydrochloride [*323-93-7*], *15*:776

PVA. See *Poly(vinyl alcohol)*.

PVB. See *Poly(vinyl butyral)*.

PVC. See *Poly(vinyl chloride)*.

PVDF. See *Poly(vinylidene fluoride)*.

PVP. See *Poly(vinylpyrrolidinone)*.

PVP–iodine, *13*:230

PWM. See *Pokeweed mitogen*.

PWR. See *Pressurized-water reactor*.

Pycnocoma macrophylla, *16*:251

Pyrazines
odor-problem chemicals, *16*:303

Pyrene [*129-00-0*]
1,4,5,8-naphthalenetetracarboxylic acid from,
15:744

Pyrethrin I [*121-21-1*]
as insecticide, *13*:424

Pyrethrin II [*121-29-9*]
as insecticide, *13*:424

Pyrethroids
as insecticides, *13*:424

Pyrethroids, synthetic
as insecticides, *13*:456

Pyrethrum, *13*:62

Pyridine [*110-86-1*]
electrolytic reduction, *15*:776
odor, *16*:301
odor-problem chemical, *16*:303

Pyridineiodine *13*:665

Pyridinium bromide perbromide [*39416-48-3*]
in indole halogenation, *13*:215

Pyridoxal phosphate [*54-47-7*]
in aspartic acid biosynthesis, *15*:761

L-2-(3′-Pyridyl)piperidine [*494-52-0*]
as insecticide, *13*:424

2-(3′-Pyridyl)pyrrolidine [*494-97-3*]
as insecticide, *13*:424

Pyrimicarb [*23103-98-2*], *13*:455

Pyrimidanthrones
fluorescence, *14*:553

Pyrimidine [*289-95-2*]
antimetabolite analogues, *13*:172

Pyrite [*1309-36-0*]
as iron source, *13*:748

Pyrite (*Continued*)
  in limestone, *14*:350
Pyroarsenic acid [*13453-15-1*]
  as insecticide, *13*:420
o-Pyrocatechualdehyde [*24677-78-9*], *13*:70
Pyrocatechuic aldehyde [*24677-78-9*], *13*:45
Pyrochemistry
  of matches, *15*:2
Pyrochlore [*12174-36-6*], *15*:820
Pyroelectric detectors, *13*:340
Pyroelectric vidicon tube (PEV), *13*:344
Pyrogallol [*87-66-1*]
  in niobium analysis, *15*:829
  in tannin, *14*:214
Pyrolusite [*14854-26-3*], *14*:825, 845
Pyrolysis
  hydrocarbons, *16*:486
  in insulation materials, *13*:545
  oil shale, *16*:336
  of wastes, *13*:188
Pyrolysis systems
  for fuels from waste, *13*:194
Pyrophosphatases
  magnesium requirement, *15*:579
Pyrophosphoryl chloride [*13498-14-1*]
  IC dopant, *13*:638
Pyrophyllite [*12269-78-2*]
  as insecticide carrier, *13*:416
Pyrotartaric acid [*498-21-5*], *13*:112
Pyroxenes
  lithium in, *14*:449
Pyroxenite
  phlogopite source, *15*:418
Pyrrhotite [*1310-50-5*], *15*:796
  as iron source, *13*:748
Pyrrolidine [*616-45-5*]
  from *σ*-butyrolactone, *13*:97
  isoprene reaction, *13*:826
  in lactamide preparation, *13*:90
Pyrrolidines, *13*:142
  from aziridines, *13*:155
Pyrrolidone [*123-75-1*]
  GABA reaction product, *15*:766
Pyrrolines
  from aziridines, *13*:155
Pyruvatekinase [*9001-59-6*], *15*:89
Pyruvic acid [*127-17-3*], *13*:112
  in polysaccharides, *15*:447
Pyruvic carboxylase
  manganese in, *15*:594
PZT.   See *Lead zirconate–titanate.*

Q

Q-5700, *13*:234
Q-BOP furnace, *14*:378
Q fever, *15*:553
Qiana, *13*:807
Q-meter
  for resonant-circuit test, *15*:421
Q-S process
  for lead smelting, *14*:130
Q-switching
  in lasers, *14*:50
Quaalude [*72-44-6*], *13*:124, 125
Quantum efficiency
  phosphors, *14*:531
Quartz [*14808-60-7*]
  in gaskets, *16*:731
  in insulation, *13*:598
  in mica books, *15*:420
  as track-etch detectors, *16*:827
Quats, *13*:233.   (See also *Quaternary ammonium compounds.*)
  as antimicrobial agents, *13*:233
Quazepam [*36735-22-5*]
  properties of, *13*:122
QUEBEC-ACTUALITE information system, *13*:331
Quench baths, *15*:323
Quenching
  of fluorescence, *14*:557
  in phosphors, *14*:535
Queuing theory
  operations planning, *16*:520
Quicklime, *14*:343, 346.   (See also *Lime and limestone.*)
  slaking, *14*:346
Quicksilver.   See *Mercury.*
QUIC–LAW Project, *13*:292
Quinacridone [*1047-16-1*], *13*:890
Quinaldine red [*117-92-0*], *13*:10
Quinazarin [*81-64-1*], *13*:43
Quindex, *13*:236
Quinhydrone [*106-34-3*], *13*:42
Quinic acid [*77-95-2*]
  hydroquinone from, *13*:39
Quinine [*130-95-0*]
  in lactic acid resolution, *13*:81
Quinoline [*91-22-5*]
  odor-problem chemical, *16*:303
8-Quinolinol [*148-24-3*]
  in niobium analysis, *15*:829
8-Quinolinol sulfate [*2149-36-2*]
  in calcium, *15*:84
Quinone [*106-51-4*]
  hydroquinone from, *13*:39
Quinone monochloroimide [*637-61-6*]
  lignin reaction, *14*:299

p-Quinquephenyl [*3073-05-0*]
  liquid crystalline range of, *14*:399

R

R-20458 [*37571-83-8*], *13*:459
R-acid [*148-75-4*], *15*:739.   (See also *3-Hydroxy-2,7-naphthalenedisulfonic acid.*)
2R acid [*90-40-4*], *15*:740.   (See also *3-Amino-5-hydroxy-2,7-naphthalenedisulfonic acid.*)
Radiation
  affect on ion-exchange resins, *13*:684
  determination of with liquid crystals, *14*:421
  exposure effects, *16*:217
  infrared, *13*:337
  as initiator, *13*:367
  lubricant effects, *14*:518
  nonionizing, *15*:494
  through insulators, *13*:592
Radiation-cure inks, *13*:385
Radiation curing
  in magnetic tape mfg, *14*:743
Radiation damage
  from ion implantation, *13*:712
  in particle tracks, *16*:832
  by phosphors, *14*:535
Radiation exposure
  shielding for, *16*:240
Radiation resistance
  of materials, *16*:239
Radiation shielding
  lead, *14*:158
Radiation shields
  ruthenocene polymers, *15*:196
  thermal, *13*:594
Radiation wastes
  removal of by ion exchange, *13*:700
Radiative decay, *14*:527
Radicals
  as initiators, *13*:356
Radioactivity dating, *16*:843
Radio emissions
  in nondestructive tests, *16*:53
Radio frequency
  in drying of inks, *13*:376
Radiography
  of clad metals, *15*:285
  nondestructive testing, *16*:63
Radioisotopes
  production, *13*:859
Radiometal 36, *14*:659
Radiometers, *16*:524

  for infrared, *13*:341
Radiopharmaceuticals
  radiation exposure from, *16*:217
Radium [*7440-14-4*]
  from the ocean, *16*:280
Radium-226 [*13982-63-3*], *16*:206
  disposal of, *16*:211
  in level measurement, *14*:436
Radon [*10043-92-2*]
  in air, *16*:657
  dosimetry, *16*:829
  from the ocean, *16*:280
  radiation exposure from, *16*:217
Radon-220 [*15756-71-5*]
  disposal of, *16*:211
Radon dosimetry, *16*:845
Raffinose [*512-69-6*]
  in nuts, *16*:252
  in soybean lecithin, *14*:251
Railroad ballast, *14*:375
Railroad cars
  insulation in, *13*:595
Railroad fusee, *15*:5
Rail systems
  noise source, *16*:9
RAM.   See *Random-access memories.*
Raman lasers, *14*:61
Raman spectroscopy
  lasers in, *14*:72
Ramie
  in packings, *16*:734
2R-amino acid [*118-03-6*], *15*:731.   (See also *7-Amino-1,3,6-naphthalenetrisulfonic acid.*)
Ramsdellite [*12032-73-4*], *14*:854
Ram tensile test, *15*:286
Rancidity
  of nuts, *16*:262
Random-access memories (RAM), *14*:686
Random numbers, *16*:508
Raney cobalt
  nitrile reduction catalyst, *15*:889
Raney nickel [*7740-02-0*], *15*:812
  hydrogenation catalyst, *13*:19
Raoult's law
  in isotope separation, *13*:847
Rapeseed oil [*8002-13-9*]
  lecithin from, *14*:250
Rapid patent international
  patent information, *16*:924
RAPRA Abstracts
  information service, *16*:902
RAPRA ABSTRACTS information system, *13*:331
Rare earth
  as phosphors, *14*:531
Rare earth activators
  in phosphors, *14*:540

Rare-earth cobalt
    magnetic properties, *14*:671
Rare-earth elements
    deficiences, *16*:253
    in hickory, *16*:252
Rare-earth iron garnets
    as magnetic insulator, *14*:694
Rare-earth magnets, *14*:675
Rare-earth orthoferrites
    as magnetic insulator, *14*:694
Rare earths
    in magnesium alloys, *14*:596
Rare gases
    discharge-lamp phosphors, *14*:544
Raspail test, *16*:793
Raspberry ketone [*5471-51-2*], *13*:77
Rat brain
    acetylcholine in, *15*:756
Rats
    repellant for, *16*:579
Ravensara nut, *16*:251
Rayleigh instability, *15*:473
Rayon
    in leatherlike materials, *14*:242
    moisture regain, *16*:358
    nonwoven fabrics from, *16*:107
    in packings, *16*:734
    properties, *16*:358
    spunbonded, *16*:96
Rayon (T50) fiber
    in metal laminates, *13*:945
γ-Ray testing methods
    nondestructive, *16*:49
RDA.   See *Recommended dietary allowance.*
RDF.   See *Refuse-derived fuel.*
RDX.   See *Cyclotrimethylenetrinitramine.*
Reaction injection molding (RIM)
    for plastic composites, *13*:976
Reax
    alkali lignins, *14*:308
Recommended dietary allowances, *15*:571
Recorders, helical, *14*:733
Recording
    magnetic, *14*:732
Recording heads, *14*:694, 701
    alloys for, *14*:657
Recording materials
    magnetic, *14*:701
Records
    in PM program, *14*:758
Rectisol purification
    gasified lignite, *14*:327
Recycling
    lead, *14*:131, 140
    of lubricating oils, *14*:522
Red-banded leafroller
    control of with pheromones, *13*:481

Red blood cells
    artificial, *15*:489
Red carbon
    carbon suboxide polymer, *13*:879
Redistribution
    of organometallics, *16*:556
    tetraalkyllead compounds, *14*:188
Red Lake C [*2092-56-0*]
    in inks, *13*:378
Red lead.   See *Lead tetroxide.*
Red-lye process
    for K_2O_3 from KOH, *14*:875
Red mountain formation
    iron ore from, *13*:752
Redox initiators, *13*:364
Redox process
    fuel reprocessing, *16*:173
Redox systems
    for emulsion polymerization, *14*:92
Reduced nicotinamide adenine dinucleotide
        (NADH) [*606-68-8*]
    rotenone effect on, *13*:427
Reducing agent
    hydrogen peroxide as, *13*:16
    iodides, *13*:654
Reducing sugars
    in nuts, *16*:252
Reduction
    magnesium oxide, *14*:583
    manganese ores, *14*:863
    metallocenes, *16*:604
    nitrogen, *15*:941
    nitroparaffins, *15*:976
    of oxalic acid, *16*:620
Redundancy
    nuclear reactor design, *16*:228
Reemay, *16*:78
Reference data
    standard, *15*:36
Reference materials
    standard, *15*:34
Refiner mechanical pulp (RMP), *16*:769
Refining
    paper, *16*:774
Reflectance
    of optical filters, *16*:522
Reformatsky reaction
    hydroxy acids by, *13*:95
Refractories
    magnesia in, *14*:636
Refractory lime, *14*:347
Refrigerants
    molecular-sieve drying, *15*:663
Refrigeration
    with liquid N_2, *15*:939
    lubricants in, *14*:485
Refuse
    beneficiation, *13*:191

combustion of, *13*:182
incineration of, *13*:182
Refuse-derived fuel, *13*:191
Regeneration
in electrical feedback, *13*:489
R. E. Ginna, *16*:155
Registry numbers
for information retrieval, *13*:280
Regulator
in automatic control, *13*:486
Reimer-Tiemann reaction, *13*:62
with dihydroxybenzenes, *13*:45
of indole, *13*:216
with phenol, *13*:74
Reinforced metals, *13*:941
Reinforced plastics, *13*:968
Reinforcement fibers
properties, *13*:944
Reissert synthesis
of indoles, *13*:219
Reiter's syndrome
treatment, *13*:179
Release agents
for paper, *16*:808
Reliability
of systems, *15*:9
Reliability analysis.　See *Materials reliability.*
Relite, *13*:696
Remalloy, *14*:679, 680
magnetic properties, *14*:671
Remanence
in magnetic tape, *14*:735
Remendur, *14*:681
Renaglandin, *15*:757
Renal disease
diagnosis, *15*:78
Renaleptine, *15*:757
Renalina, *15*:757
Renal rickets
detection, *15*:83
Rendering
fats, *15*:71
René 41 [*11068-84-1*], *15*:789
Renin [*9015-94-5*]
in angiotensin biosynthesis, *15*:759
Rennet
from animal stomachs, *15*:72
Rennin [*9001-98-3*]
aldosterone and, *15*:579
Renoform, *15*:757
Renostypticin, *15*:757
Renostyptin, *15*:757
Renshaw cells, *15*:756
Renuzit solid super odor killer, *16*:304
Repellents
against insects, *13*:419

in insect control, *13*:475
organotin, *16*:579
Reppe process, *15*:805; *16*:584
Resazurin test
for milk, *15*:550
Research disclosure
patent information, *16*:923
Research publications
patent information, *16*:926
Reserpine [*50-55-5*], *13*:213
Reservoir devices, *15*:124
Resinoids
in perfumes, *16*:949
Resins
in electrical insulation, *13*:565
ion exchange in, *13*:678
paper sizing, *16*:778
Resins, ion exchange, *13*:679.　(See also *Ion exchange.*)
Resins, ion-retardation, *13*:688
Resistance to insecticides, *13*:473
Resistivity
in insulation materials, *13*:540
Resmethrin [*10453-86-8*], *13*:458
Resole [*9003-35-4*]
composite with novoloid fiber, *16*:134
Resonant-circuit test
for insulation, *15*:421
Resorcinol [*108-46-3*], *13*:39.　(See also *Hydroquinone, resorcinol, and catechol.*)
economic aspects, *13*:53
properties, *13*:40
Resorcinolarsonic acid [*6269-96-7*]
uranium adsorbent, *16*:284
Resorcinol–formaldehyde resin
in adhesives, *13*:57
plywood adhesive, *14*:14
Resorcinol, monoacetate [*102-29-4*]
economic aspects, *13*:53
use, *13*:62
β-Resorcylic acid [*89-86-1*], *13*:45
γ-Resorcylic acid [*303-07-1*], *13*:45
α-Resorcylic aldehyde [*26153-38-8*], *13*:70
β-Resorcylic aldehyde [*95-01-2*], *13*:70
γ-Resorcylic aldehyde [*387-46-2*], *13*:70
Resource allocation, *16*:512
Resource Conservation and Recovery Act, *13*:184, 254
Respiration
enkephalins involved in, *15*:765
Respiratory devices
oxygen supply for, *16*:670
Retention aids
in papermaking, *16*:804
Reticulin
in cattlehides, *14*:203

Retinal defects
    laser treatment, *14*:70
Retinitis pigmentosa
    taurine associated with, *15*:781
Retinopathy
    laser treatment, *14*:70
Retorting
    oil shale, *16*:336
Retorts
    classes, *16*:338
Retrolental fibroplasia, *16*:670
Reverberation-room method
    sound absorption, *13*:514
Reverberatory process
    for lead, *14*:113
Reverse osmosis, *15*:92, 112.   (See also
    Membrane technology.)
    membranes for, *15*:107
    for milk products, *15*:563
    thermodynamics, *15*:113
Reviews
    technical, *13*:290
Reyn
    definition, *14*:485
RF.   See *Resorcinol–formaldehyde resin.*
RFI.   See *Radio-frequency interference.*
RG acid [*578-85-8*], *15*:738.   (See also 5-
    Hydroxy-2,7-naphthalenedisulfonic acid.)
RG-H process, *14*:655
Rhagoletis completa, *13*:473, 481
Rhamnose [*3615-41-6*]
    in indica polysaccharide, *15*:451
Rhenium [*7440-16-6*]
    in coatings, *15*:241
    from the ocean, *16*:280
Rhenium-185 [*14391-28-7*], *13*:842
Rhenium-187 [*14391-29-8*], *13*:842
Rhenium nitride [*12033-55-5*], *15*:872
Rheology modifiers
    in paint, *16*:753
Rheumatoid arthritis
    treatment, *13*:172; *14*:803
Rheumatoid factor (RF), *13*:169
Rhinaspray, *15*:726
Rhizobium–legume associations
    in N$_2$ fixation, *15*:947
Rhizobium leguminosarum, *15*:953
Rhizobium meliloti, *15*:453
Rhizopus arrhizus, *15*:460
Rhizopus nigricans, *15*:460
Rhodamine 101 [*41175-43-3*], *14*:554
Rhodamine B [*81-88-9*]
    fluorescent dye, *14*:547
Rhodamine F5G [*989-38-8*]
    fluorescent dye, *14*:547

Rhodamine 6G [*989-38-8*], *14*:552
    lasers, *14*:56
Rhodenocenium ion
    reduction, *16*:604
Rhodinol [*6812-78-8*], *16*:327.   (See also *1-*
    Citronellol.)
    from geranium oil, *16*:323
    in perfumes, *16*:968
Rhodium-103 [*7440-16-6*], *13*:841
Rhodium carbonyl triphenylphosphine
    acetylacetonate [*25470-96-6*]
    oxo process catalyst, *16*:639
Rhodium catalyst
    for methanol carbonylation, *16*:584
Rhodium chloride [*10049-07-7*]
    olefin insertion catalyst, *16*:609
Rhodium oxide [*12036-35-0*]
    in oxo process, *16*:639
Rhodochrosite [*598-62-9*], *14*:825, 845
Rhodocrosite
    MnO$_2$ from, *14*:865
Rhodonite [*14567-57-8*], *14*:825
Rhodospirillum, *15*:950
Rhodylene, *13*:57
Rhone-Poulenc process
    for oxalic acid, *16*:627
Rhopalosiphum padi, *13*:415
Ribbon blenders, *15*:632
Ribbon mixers, *15*:629
Ribonucleic acid
    in adenosine biosynthesis, *15*:757
    in memory enhancement, *15*:133
D-Ribose [*50-69-1*]
    in nucleic acids, *15*:587
Ribose-1-phosphate [*4300-28-1*]
    in adenosine biosynthesis, *15*:757
Rice hulls
    lignin content, *14*:297
Ricinodendron rautanenii, *16*:251
Rickets
    nutrients associated with, *15*:573
Rickettsiae burneti, *15*:553
Riedl-Pfleiderer process, *13*:16
RIM.   See *Reaction injection molding.*
Ring Index code
    in information systems, *16*:908
Risk assessment
    of power sources, *16*:220
Rivets
    for magnesium alloys, *14*:608
RMP.   See *Refiner mechanical pulp.*
Roadstone, *14*:**375**
Roccal II, *13*:234
Rochelle salt [*304-59-6*]
    manufacture of, *13*:114
Rock
    nuclear waste storage in, *16*:213

Rocket-fuel oxidizer
  lithium perchlorate, *14*:465
Rocket fuels, *15*:856
Rock wool
  as acoustic insulation, *13*:517
  insulation, *14*:376
  in papermaking, *16*:823
Rodenticide, *15*:726
Rodents
  cable protection against, *13*:582
Rohypnol, *13*:128
Roll bonding
  metal coatings, *15*:260
  metal-laminate fabrication, *13*:948
Roller dryers
  for milk, *15*:557
Rolling
  of metals, *15*:330
Romancheite, *14*:855
Ronnel [*299-84-3*], *13*:439
Roof decks, acoustic, *13*:518
Roofing
  lead alloys, *14*:144
Ropes
  fibers for, *16*:376
Rosa alba L., *16*:327
Rosa centifolia L., *16*:327
Rosa damascena Mill. forma *trigintipelala* Dieck,
  16:327
Roscoelite [*12271-44-2*], *15*:416
Roscogel Filter No. 800, *16*:530
Roscolene plastic filters, *16*:535
Rose [*8007-01-0*], *16*:307
Rose bengale [*11121-48-5*], *13*:671
Rosemary oil [*8000-25-7*], *16*:327
  in perfumes, *16*:952
Rose oil [*8007-01-0*], *16*:319, 327
Rose oxide [*16409-43-1*], *16*:315
cis-Rose oxide [*3033-23-6*], *16*:312
trans-Rose oxide [*876-18-6*], *16*:312
Rosewood oil [*8015-77-8*], *16*:320
Rosin [*8050-09-7*]
  in grease thickeners, *14*:502
  paper size, *16*:808
  paper sizing, *16*:778
Rosmarinum officinalis, *16*:952
Rosmarinus officinalis L., *16*:327
Rotary-hearth process
  for lead, *14*:109
Rotary heat-set publication inks, *13*:381
Rotary kiln
  incinerators, *13*:191
Rotenaids
  as insecticide, *13*:425
Rotenone [*83-79-4*], *13*:62
  as insecticide, *13*:425

Rotoformer, *16*:114
Rotogravure inks, *13*:392
Rotothene, *16*:400
Round Permalloy 80, *14*:659
Royalties
  patent licenses, *16*:876
RR acid [*90-40-4*], *15*:740. (See also *3-Amino-
  5-hydroxy-2,7-naphthalenedisulfonic acid.*)
RTECS. See *NIDSH Registry of Toxic Effects
  of Chemical Substances.*
Rubber [*9006-04-6*]
  effect of radiation on, *16*:239
  in gaskets, *16*:726
  gas permeability, *15*:118
  odor compensation of, *16*:303
  in vibration isolation, *13*:532
Rubber antioxidant, *15*:726
Rubber latexes
  in paper coatings, *16*:790
Rubber, natural [*9006-04-6*] (NR)
  gasket properties, *16*:727
Rubber products
  microwave treatment of, *15*:515
Rubber, synthetic
  lining for bulk containers, *16*:715
Rubeanic acid [*79-40-3*]
  metal polymers, *15*:214
Ruben-Mallory dry cell, *15*:162
Rubidium [*7440-17-7*]
  as lubricant, *14*:515
  from the ocean, *16*:279
  oxidation, *16*:656
Rubidium-85 [*13982-12-2*], *13*:840
Rubidium-87 [*13982-13-3*], *13*:840
  radiation exposure from, *16*:217
Rubidium manganate(V) [*12438-62-9*],
  14:857
Rubidium manganate(VI) [*25583-21-5*],
  14:859
Rubidium molybdate [*13718-22-4*], *15*:687
Rubidium nitride [*12316-85-5*], *15*:872
Rubidium ozonide [*12060-04-7*], *16*:688
Rubidium permanganate [*13465-49-1*], *14*:845
Rubidium trifluoronickelate [*14105-01-2*]
  in magnetooptics, *14*:705
Rubidium triiodide [*12298-69-0*], *13*:665
Rubinate M, *13*:801
Ruby laser, *14*:42, 51, 54
Ruby mica, *15*:420
Ruhröl-Veba process, *14*:782
Rule 66
  solvent emission controls, *13*:902
Rundle deposit, *16*:347
Russell-Saunders ground-state term, *13*:764
Rust inhibitors
  in lubricants, *14*:492

Rust stains
  removal from fabrics, *16*:631
Ruthenium [*7440-18-8*]
  in fuel reprocessing, *16*:179
Ruthenium-96 [*15128-32-2*], *13*:841
Ruthenium-98 [*18393-13-0*], *13*:841
Ruthenium-99 [*15411-62-8*], *13*:841
Ruthenium-100 [*14914-60-4*], *13*:841
Ruthenium-101 [*14914-61-5*], *13*:841
Ruthenium-102 [*14914-62-6*], *13*:841
Ruthenium-104 [*15766-01-5*], *13*:841
Ruthenium oxide [*12036-10-1*]
  in anodic coatings, *15*:173
Ruthenium pentafluoride [*14521-18-7*], *16*:181
Ruthenium–titanium oxides
  anode coatings, *15*:174
Ruthenocene [*1287-13-4*], *16*:598
  aldehyde reaction, *15*:196
Ruthenocene polymers, *15*:184.   (See also *Metal-containing polymers.*)
Ruthenocenylmethyl borontetrofluoride [*76082-15-0*], *15*:196
Rutile [*14913-80-5*]
  from beach sand, *16*:289
  paper pigment, *16*:778
Ryania speciosa
  insecticides in, *13*:428
Ryanodine [*15662-33-9*], *13*:428

S

Sabadilla
  insecticidal action of, *13*:427
Sabadine [*28903-30-2*], *13*:427
Sabatier reaction, *16*:681
Sabinene [*2009-00-9*]
  in nutmeg, *16*:325
Sabin unit
  sound absorption, *13*:514
Saccharin [*81-07-2*]
  from *o*-toluenesulfamide, *14*:863
2S acid [*82-47-3*], *15*:740.   (See also *4-Amino-5-hydroxy-1,3-naphthalenedisulfonic acid.*)
Sacrolactic acid [*79-33-4*], *13*:81
SAE 10 automotive oil
  limit of radiation tolerance, *14*:519
SAE 20 automotive oil
  limit of radiation tolerance, *14*:519
Safe Drinking Water Act, *13*:254
Safety
  fluorescent color uses, *14*:564
  as a maintenance function, *14*:765
  in nuclear facilities, *16*:225
Safety match, *15*:2
SAFETY SCIENCE ABSTRACTS information
  system, *13*:331
Safflower oil [*8001-23-8*]
  emulsion with na-caseinate, *15*:491
  lecithin from, *14*:250
Saflex PT, *13*:989
Safranine dyes, *15*:728
Safrole [*94-59-7*], *13*:62; *16*:320
  from camphor oil, *16*:322
  in nutmeg, *16*:325
Sage muscatel.   See *Sage oil clary.*
Sage oil clary [*8016-63-5*], *16*:327
Sage oil, Dalmatian [*8022-56-8*], *16*:327
Sales analysis.   See *Market research.*
Salicylaldehyde [*90-02-8*], *13*:70, 76
Salicylaldehyde, copper salt [*14523-25-2*]
  as chelate, *13*:73
Salicylaldehyde, lithium salt [*27761-48-4*]
  as chelate, *13*:73
Salicylaldehyde phenylhydrazone [*614-65-3*]
  antimicrobial, *13*:76
Salicylaldehyde, potassium salt [*33838-32-3*]
  as chelate, *13*:74
Salicylaldehyde, sodium salt [*3116-83-4*]
  as chelate, *13*:74
Salicylaldoxime [*94-67-1*]
  antioxidant, *13*:78
Salicylanilide [*87-17-2*]
  applications of, *13*:240
Salicylates
  measurement, *15*:88
Salicylic acid [*69-72-7*]
  microbial preparation, *15*:460
  in perfumes, *16*:954
  preservative, *13*:247
Salicylidenepropionamide, *13*:72
Saligenin [*90-01-7*], *13*:74.   (See also *o-Hydroxybenzyl alcohol.*)
Saligenin process, *13*:74
Salinaphthol, *15*:734
Saliva
  nitrite in, *15*:994
Salmo linne, *13*:612
Salmonella paratyphi, *15*:553
Salmonella typhi, *15*:553
Salmonella typhimurium, *15*:952
Salmonellosis, *15*:553
Salt [*7647-14-5*], *16*:279.   (See also *Sodium chloride.*)
  from seawater, *16*:279
Salt deposits
  nuclear waste storage in, *16*:213
Salt formations
  disposal of nuclear wastes in, *16*:209

Salting
  meat, *15*:62
Salt, molten
  in paint stripping, *16*:765
Saltpeter. See *Sodium nitrate.*
  nitric acid from, *15*:853
Salvia officinalis L., *16*:327
Salvia sclarea L., *16*:327
SAM. See *S-Adenosylmethionine.*
Samarium [*7440-19-9*]
  from the ocean, *16*:280
Samarium-144 [*14981-82-9*], *13*:841
Samarium-147 [*14392-33-7*], *13*:841
Samarium-148 [*14913-64-5*], *13*:841
Samarium-149 [*14392-34-8*], *13*:841
Samarium-150 [*14907-33-6*], *13*:841
Samarium-152 [*14280-32-1*], *13*:841
Samarium-154 [*14833-41-1*], *13*:841
Samarium, compd with cobalt [*12017-68-4*] (1:5),
    14:676
  in magnets, *14*:676
Samarium, compd with cobalt [*12052-78-7*] (2:17),
    14:676
Samarium magnets, *14*:676
Samarium nitride [*25764-14-1*], *15*:872
Samarium trifluoride [*13765-24-7*]
  in fluoride electrode, *13*:727
Samica process
  for mica paper, *15*:431
Samples
  standard, *15*:34
SAN. See *Styrene–acrylonitrile.*
Sandalwood oil [*8006-87-9*], *16*:319
  in perfumes, *16*:952
Sandalwood oil, east Indian [*8006-87-9*], *16*:327
Sand blasting
  in metal treating, *15*:305
Sand casting
  of magnesium alloys, *14*:607
Sandmeyer reaction, *15*:737
  for nitrochlorobenzenes, *15*:924
SANI information system, *13*:331
Sanitary napkins
  nonwoven, *16*:123
3A Sanitary standards
  for milk processing, *15*:539
Sanninoidea exitiosa, *13*:481
San Onofre 1, *16*:155
SANSS information system, *13*:301
Santalol [*11031-45-1*], *16*:328
Santalum album, *16*:952
Santalum album L., *16*:327
Sapindus saponaria, *16*:251
Sapium sebiferum, *16*:251
Saponin
  as surfactant, *15*:475

Saponins
  in nuts, *16*:255
Sapwood
  in plywood manufacture, *14*:11
Saran, *16*:629
  iodine resistance, *13*:657
Sasil, *15*:659
Satin white [*12004-14-7*]
  paper coating, *16*:789
  in papermaking, *16*:822
Saugus iron works, *13*:736
Sausage casings
  collagen in, *14*:227
Sausages, *15*:62. (See also *Meat products.*)
  classification, *15*:67
Save-alls, *16*:806
Saw-toothed grain weevil
  fumigant activity on, *13*:465
SBR. See *Styrene–butadiene rubber.*
SBR-bonded cotton
  gasket properties, *16*:727
Scandium [*7440-20-2*]
  from the ocean, *16*:280
Scandium-45 [*7440-20-2*], *13*:840
Scandium nitride [*25764-12-9*], *15*:872
Scarabaeid beetles
  mesomorphism in exocuticle of, *14*:420
Scarlet fever, *15*:553
Scattering coefficient
  of paper, *16*:805
Schaeffer's acid [*93-01-6*], *15*:738. (See also *6-Hydroxy-2-naphthalenesulfonic acid.*)
Scheelite [*14940-68-2*]
  from beach sand, *16*:289
Schiff bases
  in liquid-crystal synthesis, *14*:414
Schiff's base
  formation of in aldehyde tanning, *14*:218
Schistosoma japonium, *13*:118
Schizophrenia, *15*:776
  GABA involved in, *15*:766
  treatment, *13*:133; *15*:776
Schizotoxin
  DMT, *15*:763
Schlieren methods of optical imaging, *16*:63
Schlieren texture
  of smectic structures, *14*:401
Schmidt number, *14*:941, 978
Schoellkopf's acid [*82-75-7*], *15*:739. (See also *4-Hydroxy-1,5-naphthalenedisulfonic acid.*)
Schoenite [*15491-86-8*], *14*:637
  from seawater, *16*:284
Schoenocaulon drummondi, *13*:427

Schoenocaulon officinale, 13:427
Schoenocaulon texanum, 13:427
Scholler lignin, 14:309
Schotten-Baumann reaction
    in interfacial polymerization, 15:480
Schott filter glasses, 16:527
Schott filters, 16:535
Schottky diodes
    in ICS, 13:641
Schradan [152-16-9], 13:438, 445
Schumacher process
    for electrolytic MnO₂, 14:867
Schusspol, 16:120
Scirtothrips citri, 13:473
SCISEARCH
    information service, 16:902
Sclair, 16:400
Scleroderma
    treatment, 13:179
Scleroglucan [39464-87-4], 15:454
Scleron, 14:448
Sclerotinias, 13:106
Sclerotium glucanicum, 15:454
Sclerotium rolfsii, 15:455
Scolytus multistriatus, 13:482
Scotophobin (rat) [33579-45-2], 15:132
Scott bond strength, 16:816
Scrap mica, 15:422
Screen process inks, 13:393
Scrubbers
    in incinerators, 13:186
Scumming
    of ink, 13:385
Scurenaline, 15:757
SDC
    information-retrieval system, 16:901
SDI.   See *Selective Dissemination of*
    Information.
Sea-floor spreading
    track dating of, 16:844
Sealants, 16:732
    for laminated glass, 13:991
Sealing liquids
    in gauges, 14:442
Seals, 16:725, 740.   (See also *Packing materials.*)
Sea plants
    inorganic materials from, 16:293
Search check, 13:300
    patent information, 16:924
Seawater
    desalination of, 14:943
    electrodialysis of, 16:280
    elements in, 16:278
    magnesium from, 14:577
    ʻranium in, 16:147

Seawater process
    for magnesium hydroxide, 14:634
Seaweed
    potash source, 16:284
Seaweed alginates, 15:452
Sebacic acid [111-20-6]
    lubricant from, 14:496
Seborrheic keratosis
    treatment, 13:60
Sec.   See *Size-exclusion chromatography.*
Securiflex, 13:988
Security glass, 13:991
Sedatives, 13:122.   (See also *Hypnotics,*
    sedatives, anticonvulsants.)
Seed disinfectants, 16:587
Seeds
    packaging, 16:720
Seed treatment
    molybdates for, 15:695
Sefor reactor, 16:200
Seizures
    treatment, 13:122
Selective Dissemination of Information (SDI),
    13:279
Selectivity
    of organophosphorus insecticides, 13:451
    in toluene nitration, 15:929
Selectoforming process, 15:666
Selenic acid [7783-08-6]
    nitration reagents, 15:841
Selenium [7782-49-2]
    isocyanate catalyst, 13:808
    metabolic functions, 15:590
    nutrient properties, 15:571
Selenium-74 [13981-33-4], 13:840
Selenium-76 [13981-32-3], 13:840
Selenium-77 [14681-72-2], 13:840
Selenium-78 [14833-16-0], 13:840
Selenium-80 [14681-54-0], 13:840
Selenium-82 [14687-58-2], 13:840
Selenium diethyl dithiocarbamate [144-34-3]
    in electric insulation, 13:565
Selenium nitride [12033-59-9], 15:872
Selenium sulfide [7488-56-4]
    applications of, 13:237
Selenoaminoacids
    antiaging effects, 15:138
α-Selinene [473-13-2], 16:313
Selsun, 13:237
Semecarpus anacardium, 16:251
Semibenzoquinone radical, 13:40
meta-Semibenzoquinone radical, 13:40
Semibulk containers, 16:716
Semicarbazide [563-41-7]
    antiaging effect, 15:139
Semiconducting compounds, 13:212

Semiconductor lasers, *14*:58, 269
Semiconductors
 ion implantation for, *13*:707
 in LEDs, *14*:270
 metal-containing polymers, *15*:211
 properties, *14*:286
Semidine [*101-54-2*]
 in phosphorus measurement, *15*:84
Semilente iletin [*8049-62-5*], *13*:612
Semiphorone [*5857-71-6*], *13*:914
Sendust [*12606-95-0*], *14*:657
Sendust alloy, *14*:694
Sendust alloy films, *14*:703
Sendzimir line
 Al coating, *15*:242
Senesence, *15*:136. (See also *Memory-enhancing agents and antiaging drugs.*)
Senile dementia
 GABA involved in, *15*:766
Senility
 treatment, *15*:138
Sensitizers
 lead halides, *14*:163
 in phosphors, *14*:532
Separation
 magnetic, *14*:709
 zeolites in, *15*:649
Separation processes, *15*:92. (See also *Membrane technology.*)
 for isotopes, *13*:839
 membranes in, *15*:112
Separators
 for cream from milk, *15*:528
Sephadex
 immunoassay use, *15*:85
Sepiolite [*15501-74-3*], *14*:615
Serentil, *13*:122
D,L-Serin [*56-45-1*]
 L-DOPA from, *13*:46
Serine [*56-45-1*]
 in glycine biosynthesis, *15*:768
 in nuts, *16*:249
Serine transhydroxy-methylase [*9029-83-8*]
 in glycine biosynthesis, *15*:768
Serology
 diagnostic reagents, *15*:75
Serotonin [*50-67-9*], *13*:213; *15*:770. (See also *5-Hydroxytryptamine.*)
Serotonin hydrochloride [*153-98-0*], *15*:770
Serotonin picrate [*6106-56-5*], *15*:770
Serpak process
 for aluminum nitride, *15*:944
Serpentine [*12168-92-2*], *14*:615
Serum ferritin [*9007-13-2*], *15*:595
Serum oxaloacetic transaminase [*9000-97-9*] (SGOT)
 in diagnosis, *15*:75

Serum sickness
 treatment, *15*:762
Serviceability, *15*:11
Servings
 in cable and wire insulation, *13*:584
Servomanipulators, *16*:247
Servomechanisms, *13*:489
Sesamex [*51-14-9*]
 as insecticide, *13*:425
Set point, *13*:486
Setup boxes, *16*:798
Sevin, *13*:806; *15*:714, 729
Sewage treatment
 lime in, *14*:379
SF-96, *14*:496
SGB information system, *13*:331
SGOT. See *Serum oxaloacetic transaminase.*
SGPT. See *Alanine transferase.*
Shale oil, *16*:333. (See also *Oil shale.*)
Shales
 lithium in, *14*:449
 nuclear waste storage in, *16*:213
 uranium in, *16*:147
Shampoo
 fragrances, *16*:970
Shaped filters, *16*:537
Sharnet, *16*:78
Shavings board, *14*:17
SHC 629, *14*:496
SHC 824, *14*:496
Shear rate, *15*:606
Shear response
 olefin polymers, *16*:427
Shear stress, *15*:606
Shechter-Kaplan reaction, *15*:979
Sheep ked
 effect on sheepskin, *14*:208
Sheep skins
 processing of, *15*:73
Sheet-fed commercial inks, *13*:381
Sheeting
 properties of pmma, *15*:391
Sheet mica, *15*:421
Sheet-molding compound, *13*:975
Shellac [*9000-59-3*]
 stripping of, *16*:764
Shell dredging, *16*:285
Shell oligomerization process, *16*:491
Shells. See *Marine shell.*
Sheradizing, *15*:247
Sherardizing, *15*:251
Sherbets, *15*:566
Sherbine, *15*:566
Sherwood number, *14*:917, 978
Shielding
 lead, *14*:158

Shielding (*Continued*)
    nuclear, *16*:240
Shigella dysenteriae, *15*:553
Shikimic acid [*138-59-0*], *13*:46
Shimane, *16*:160
Shipboard cables, *13*:571.   (See also *Cables, shipboard.*)
Shipping machinery lubricants
    rust inhibitors for, *14*:493
Ships
    analogue control in, *13*:489
    metal cladding, *15*:258
    metal-clad fabrication, *15*:294
Shirlan, *13*:240
Shirley-Ferranti Cone-Plate viscometer, *13*:376
Shock
    treatment, *15*:444
Shock waves
    in nitric acid mfg, *15*:864
Shoe manufacturing
    costs, *14*:248
Shot alloys
    lead–antimony, *14*:146
Shulze-Hardy rule, *14*:87
Sialon, *15*:883
Siam benzoin [*9000-72-0*], *16*:949
SIBB information system, *13*:331
SiC fiber
    in metal laminates, *13*:953
Sickle cell anemia
    liquid crystals implicated in, *14*:418
SiC whisker fiber
    in metal laminates, *13*:953
Siderite [*14808-60-7*]
    as iron source, *13*:748
Siderophones, iron, *13*:783
Siderosis, *15*:597
    from iron(III) oxide poisoning, *13*:787
Sideryl, *15*:760
Sierracin 900, *13*:990
Sierracin windshields, *13*:990
Sierracote 3, *13*:990
Sight-glass gauges, *14*:428
Silane [*7803-62-5*], *14*:463
    laser purification of, *14*:75
    in plasma doping by Si_3N_4, *13*:639
    in Si epitaxy, *13*:626
    silicon nitride from, *15*:877
Silastic microspheres, *15*:483
Silectron, *14*:659
Silencers, *16*:23
    insertion loss of, *13*:516
Silheim, *16*:78
Silica [*7631-86-9*]
    electric insulation filler, *13*:565
    as filler, *13*:558
    filler in plastics, *13*:973

    in iron ore, *13*:738
    in limestone, *14*:350
    as lubricant, *14*:520
    in metal coatings, *15*:251
    in papermaking, *16*:819
    paper pigment, *16*:789
    in spent shale, *16*:351
    as thermal insulation, *13*:592
    as track-etch detector, *16*:827
    zeolites from, *15*:654
Silica aerogel
    as insulation, *13*:595
Silica, crystalline
    toxicity, *13*:272
Silica gel [*1343-98-2*]
    in LPG drying, *14*:389
Silicalite, *15*:642
Silicate concrete, *14*:376
Silicate esters
    as lubricants, *14*:500
Silicate glasses
    in lasers, *14*:56
Silicates
    glasses as IC dopants, *13*:638
    ion exchange in, *13*:678
Silicic acid [*1343-98-2*]
    in ion-exchange removal, *13*:687
    in leather tanning, *14*:216
    zeolites from, *15*:654
Silicomanganese, *14*:833
Silicon [*7440-21-3*]
    in chufa nuts, *16*:252
    in coatings, *15*:241
    dolomite reaction, *14*:584
    in DR processes, *13*:761
    in integrated circuits, *13*:621
    in magnesium alloys, *14*:596
    as nutrient, *15*:598
    from the ocean, *16*:279, 292
    optical-filter coating, *16*:526
    properties of, *13*:626
    purification for IC use, *13*:625
    as semiconductor material, *14*:269
Silicon-28 [*14276-58-5*], *13*:840
Silicon-29 [*14304-87-1*], *13*:840
Silicon-30 [*13981-69-6*], *13*:840
Silicon carbide [*409-21-2*]
    in case hardening, *15*:319
    as semiconductor, *14*:286
    in siliconizing, *15*:251
Silicon carbide–carbon fiber
    in metal laminates, *13*:945
Silicon carbide–tungsten fiber
    in metal laminates, *13*:945
Silicon chloride [*409-21-2*]
    in coatings, *15*:241
Silicon dioxide [*7631-86-9*]
    in Burrus diode,
        14:277
    in ICS, *13*:623

in iron-ore processing, *13*:740
properties of, *13*:626
Silicone [*68083-20-5*]
electric insulation elastomer, *13*:565
properties, *13*:567
weather resistance of, *13*:547
Silicone oils
as paper abherents, *16*:808
Silicone rubber [*9016-00-6*]
gasket properties, *16*:727
ion-selective electrodes, *13*:727
Silicones
gas permeability, *15*:118
in packings, *16*:734
paper coating defoamers, *16*:791
as synthetic lubricants, *14*:500
Siliconizing, *15*:319
metals, *15*:251
Silicon nitride [*12033-89-5*], *15*:872
in IC mfg, *13*:639
in ICS, *13*:628
microwave application, *15*:516
properties, *15*:873
Silicon nitride [*12033-89-5*] (3:4)
from ion implantation, *13*:717
α-Silicon nitride [*12033-89-5*]
properties, *15*:877
β-Silicon nitride [*12033-89-5*]
properties, *15*:877
Silicon oxides
as insecticide carrier, *13*:416
Silicon steel
anisotropy energy of, *14*:648
as magnetic material, *14*:653
Silicon tetrachloride [*10026-04-7*]
in case hardening, *15*:319
glycolic acid reaction, *13*:93
Silicon tetraoxide [*12359-25-0*]
in ICS, *13*:626
Silicosis, *16*:731
Silk-screen inks
fluorescent pigments in, *14*:567
Silomelane
as iron source, *13*:748
Silver [*7440-22-4*]
cladding of, *15*:281
in coatings, *15*:241
determination in mercury, *15*:153
electrode, *13*:7
electrode for, *13*:726
emissivity, *13*:342
extraction by $HgSO_4$, *15*:162
in glass filters, *16*:551
lead alloys, *14*:151
as lubricant, *14*:508
in magnesium alloys, *14*:596
as magnetic material, *14*:647
from the ocean, *16*:280
photographic, *16*:531

recovery from lead, *14*:120
Silver-107 [*14378-37-1*], *13*:841
Silver-109 [*14378-38-2*], *13*:841
Silver bromide [*7785-23-1*]
reduction by hydroquinone, *13*:42
in silver electrodes, *13*:727
Silver chloride [*7783-90-6*]
electrode, *13*:7
ion exchange in, *13*:678
in silver electrodes, *13*:726
Silver cyanate [*3315-16-0*]
in isocyanate synthesis, *13*:801
Silver cyanide [*506-64-9*]
in cyanide electrode measurements, *13*:726
Silver Goop, *14*:513
Silver iodide [*7783-96-2*], *13*:661
as electrode membrane, *13*:726
Silver iodomercurate [*7784-03-4*], *13*:661; *15*:169
Silver molybdate [*13765-74-7*], *15*:687
Silver nitrate [*7761-88-8*]
for ^{131}I disposal, *16*:180
iodine reaction, *13*:665
Victor Meyer reaction, *15*:841
Silver nitride [*20737-02-4*], *15*:872
Silver oxide [*1301-96-8*] (1:1), *16*:688
Silver permanganate [*7783-98-4*], *14*:845
Silver powder
in inks, *13*:379
Silver sulfide [*21548-73-2*]
as electrode membrane, *13*:726
in silver electrode, *13*:725
use in ISE, *13*:720
Silver zeolite
for ^{131}I scrubbing, *16*:180
Silyl chromate
in Ziegler catalyst, *16*:443
Simmondsia californica, *16*:250
SIMSCRIPT
in operations planning, *16*:509
Simulation
in instrumentation and control, *13*:503
in operations planning, *16*:507
Simulium damnosum, *13*:413
Sinapyl alcohol [*537-33-7*]
in lignin, *14*:295
Singhara nut, *16*:251
Single-screw extruders
for mixing, *15*:633
Singlet state
and fluorescence, *14*:555
Sintered Alnico 2
magnetic properties, *14*:671
Sintered Alnico 5
magnetic properties, *14*:671
Sintered Alnico 6
magnetic properties, *14*:671

Sintered Alnico 8
  magnetic properties, *14*:671
Sintered Alnico 8 HC
  magnetic properties, *14*:671
Sintering
  iron ore, *13*:739
Sintering furnace
  for nuclear fuel, *16*:167
Sirius
  color temperature of, *13*:339
Sisomicin [*32385-11-8*]
  inactivation, *15*:466
Sitophilus granarius, *13*:465
β-Sitosterol [*83-46-5*]
  microbial oxidation, *15*:461
β-Sitosterol-degrading bacteria, *15*:461
Sitotroga cerealella, *13*:481
SI units
  as international standard, *15*:45
Size precipitate, *16*:810
Size press, *16*:819
Sizing
  paper, *16*:778
Sizing agents
  ketene dimers, *13*:886
  in papermaking, *16*:808
Skane M-8, *13*:245; *16*:755
Skatole [*83-34-1*], *13*:213.   (See also *3-Methylindole*.)
Skin
  contact with toxicants, *13*:260
Skin allergy, *13*:261
Skin cancer
  uv radiation, *16*:696
Skin pigmentation
  induced by bergamot oil, *16*:320
Skive inlaying, *15*:258
Skydrol 500b, *14*:496
Skydrols, *16*:727
Slaked lime, *14*:347
Sleep
  GHB effect on, *13*:98
  induced, *13*:122
  noise interference, *16*:1
Slimacide V-10, *13*:231
Slime control, *16*:578
Slime formation
  in pulp mill machinery, *16*:797
Slimicide
  antimicrobial agent as, *13*:223
Slimicides, *16*:587
  mercury in, *15*:158
  in papermaking, *16*:797, 807
Slow release
  in insecticide formulations, *13*:417
SL-RN process, *13*:745

Slurry coatings, *15*:257
Slush castings, *14*:156
SMC.   See *Sheet molding compound*.
Smectic liquid crystals, *14*:399
Smelting
  battery scrap, *14*:132
  lead ore, *14*:106
Smith and Ewart theory, *14*:83
Smog
  from ink manufacture, *13*:397
Smoke
  from incinerators, *13*:185
Smoke detectors
  radiation exposure from, *16*:217
Smokeless fuel
  from lignite, *14*:334
Smoking
  effect of asbestos fiber exposure, *13*:603
  meat, *15*:62
  of meat products, *15*:68
Smoking (meat), *13*:63
S_N2
  for β-propiolactone, *13*:96
Snake nut, *16*:251
Sneakers, *14*:248
Sneeze powder, *13*:427
Snuff
  bufotenine in, *13*:220
Soap
  fragrances, *16*:969
  in packings, *16*:734
Soap nut, *16*:251
Soda ash [*497-14-8*]
  lime in, *14*:380
Sodalime [*8006-28-8*]
  filtration of oxygen, *16*:676
  as track-etch detector, *16*:827
Sodalite [*1302-90-5*]
  as ion-exchange material, *13*:685
Soda process
  for lead smelting, *14*:110
Soda pulping
  lignin from, *14*:304
Sodium [*7440-23-5*]
  electrode, *13*:720
  in lead purification, *13*:583
  as LMFBR coolant, *16*:168
  as lubricant, *14*:515
  in N_2 fixation, *15*:955
  nutrient properties, *15*:571
  from the ocean, *16*:279, 292
  reactor coolant, *16*:193
  as reactor coolant, *16*:184
  reduction of K_2NbF_7, *15*:826
  in tetraalkyllead manufacture, *14*:185
Sodium-23 [*7440-23-5*], *13*:840
  neutron reaction, *16*:188

Sodium-24 [*13982-04-2*], *16*:188
Sodium acetate [*127-09-3*]
  in lithium recovery, *14*:453
  toxicity, *15*:573
Sodium alginate [*9005-38-3*]
  in encapsulation, *15*:481
  in ice cream, *15*:565
  in papermaking, *16*:822
Sodium alkyls, *16*:557
Sodium aluminate [*1302-47-7*]
  zeolites from, *15*:654
Sodium aluminosilicate [*1344-00-9*]
  in epoxide isomerization, *13*:20
Sodium aluminum silicate [*1344-00-9*]
  zeolites from, *15*:655
Sodium γ-aminobutyrate [*6610-05-5*], *15*:766
Sodium antimonate heptahydrate [*13517-15-2*]
  in lead refining, *14*:117
Sodium arsenate dodecahydrate [*13510-46-8*]
  in lead refining, *14*:117
Sodium arsenite [*7784-46-5*]
  as insecticide, *13*:421
  iodoform reaction, *13*:669
Sodium ascorbate [*134-03-2*]
  meat-curing agent, *15*:62
Sodium aurothiomalate [*12224-57-4*]
  in arthritis treatment, *13*:111
Sodium A zeolite [*68989-22-0*], *15*:647
Sodium azide [*26628-22-8*]
  nitroparaffin reaction, *15*:974
Sodium benzoate [*532-32-1*]
  as food preservative, *13*:110
  in paint and varnish removers, *16*:763
Sodium bicarbonate [*144-55-8*]
  toxicity, *15*:573
Sodium bisulfite [*7681-38-1*]
  in initiator systems, *13*:365
  isocyanate reaction, *13*:793
  in papermaking, *16*:817
  in sulfite pulping, *14*:304
Sodium borohydride [*16940-66-2*]
  BF₃ etherate reaction, *16*:564
  in initiator systems, *13*:367
  oxo product refining, *16*:647
  reduction organolead salts, *14*:189
Sodium *tert*-butoxide [*465-48-5*]
  n-butyllithium reaction, *16*:558
Sodium carbonate [*497-19-8*]
  in bleaching nuts, *16*:265
  in case hardening, *15*:315
  in leak detection, *16*:55
  in lithium recovery, *14*:454
  nitrogen reaction, *15*:934
  pH standard, *13*:3
  from pulp mills, *16*:795
  toxicity, *15*:573
Sodium carbonate perhydrate [*3313-92-6*], *13*:29

Sodium carbonate peroxyhydrate [*16884-59-6*], *13*:15
Sodium carboxymethyl cellulose [*9004-32-4*] (CMC)
  in papermaking, *16*:814
Sodium caseinate [*9005-46-3*]
  safflower oil emulsifier, *15*:490
Sodium chlorate [*7775-09-9*]
  chlorine dioxide from, *13*:232
  metal anodes in mfg, *15*:174
  in niobium processing, *15*:823
  as oxygen source, *16*:673
  in wool dyeing, *16*:631
Sodium chloride [*7647-14-5*], *16*:279. (See also *Salt*.)
  dielectric strength, *13*:541
  in ion exchange, *13*:681
  as leatherlike material filler, *14*:241
  meat-curing agent, *15*:62
  in metallic coating, *15*:252
  optical-filter coating, *16*:526
  toxicity, *15*:573
Sodium chlorite [*7758-19-2*]
  chlorine dioxide from, *13*:229
Sodium citrate [*68-04-2*]
  added to milk, *15*:554
Sodium cyanate [*917-61-3*]
  in case hardening, *15*:318
Sodium cyanide [*143-33-9*]
  in case hardening, *15*:316
  dialkyl malonate from, *14*:800
Sodium cyanoacetate [*1071-36-9*]
  dialkyl malonate from, *14*:800
Sodium cyclopentadienide [*4984-82-1*]
  nickel bromide reaction, *15*:807
  nickel halide reaction, *15*:806
Sodium deoxycholate [*302-95-7*]
  emulsifier, *15*:80
Sodium dichloroisocyanurate [*2893-78-9*]
  applications of, *13*:230
Sodium dihexyl sulfosuccinate [*577-11-7*]
  in freeze–thaw stabilization, *14*:88
Sodium 2,4-dihydroxy-3,5-dihydroxymercuribenzophenone-2-sulfonate [*6060-47-5*], *15*:169
Sodium diisopropylnaphthalenesulfonate [*1322-93-6*], *15*:724
Sodium dimethyldithiocarbamate [*128-04-1*]
  applications of, *13*:243
Sodium dimethyl malonate [*18424-76-5*], *14*:798
Sodium diphenylphosphinate [*22214-04-6*]
  cobalt chloride reaction, *15*:217
Sodium disulfide
  p-nitrochlorobenzene reaction, *15*:924
Sodium dithionite [*7775-14-6*]
  nitrogenase support, *15*:951
  in tris(1,10-phenanthroline)iron(II) preparation, *13*:780

Sodium dodecyl sulfate [151-41-7]
  in emulsion polymerization, 14:84
Sodium erythorbate [7378-23-8]
  meat-curing agent, 15:62
Sodium ethanenitronate [25854-39-1]
  Nef reaction, 15:974
Sodium 6-(2-ethoxy-1-naphthamido)penicillanate
    [985-16-0], 15:735
Sodium ethylenebis(dithiocarbamate) [142-59-6],
    14:878
Sodium ethylmercurithiosalicylate [54-64-8], 15:
    164. (See also *Merthiolate.*)
Sodium ferrate [13773-03-0], 13:777
Sodium ferrite [12062-85-0], 13:776
Sodium ferrocyanide [1360-19-9]
  in lactic acid refining, 13:85
Sodium fluoride [7681-49-4]
  in fuel reprocessing, 16:181
  as insecticide, 13:421
  lecithin preservative, 14:254
  toxicity, 15:573
Sodium fluoroaluminate [1331-71-1]
  toxicity, 15:574
Sodium fluorosilicate [16893-85-9]
  as insecticide, 13:421
  mothproofing use, 13:480
Sodium formate [141-53-7]
  in iron(II) formate dihydrate synthesis, 13:771
  in lithium recovery, 14:454
  oxalic acid from, 16:621
Sodium fumarate [7704-73-6]
  in iron(II) fumarate synthesis, 13:772
  in S(−)-malic acid biosynthesis, 13:107
Sodium gluconate [527-07-1]
  in metal pickling, 15:301
Sodium ʟ-glutamate [142-47-2], 15:767
Sodium glycerophosphate [1555-56-2]
  phosphatase reagent, 15:80
Sodium glycinate [6000-44-8], 15:769
Sodium glycolate [2836-32-0], 13:94. (See also
    Sodium hydroxyacetate.)
Sodium hexacyanoferrate(II) [13601-19-9], 16:35
Sodium hexafluorosilicate [16893-85-9]
  toxicity, 15:573
Sodium hexametaphosphate [10124-56-8]
  in meat processing, 15:68
  in vegetable tanning of hides, 14:213
Sodium hydrogen carbonate [144-55-8]
  pH standard, 13:3
Sodium hydrogen sulfate [7681-38-1]
  in metal pickling, 15:301
Sodium hydrogen sulfide [16721-80-5]
  reaction with maleic acid, 13:111
Sodium hydroquinone sulfonate [88-46-0], 13:
    42. (See also *Hydroquinone sulfonic acid.*)

Sodium hydrosulfite [7775-14-6]
  in metal pickling, 15:301
  paper bleaching, 16:769
  stabilization by oxalic acid, 16:633
Sodium hydroxide [1310-73-2]
  in digoxin assay, 15:87
  etchant for plastic, 16:834
  in ion-exchange resin manufacture, 13:694
  kraft process use, 16:770
  in metal coloring, 15:311
  paint stripper, 16:765
  in pulping, 14:307
  toxicity, 13:272
Sodium hydroxyacetate [2836-32-0], 13:94. (See
    also *Sodium glycolate.*)
Sodium 4-hydroxybutyrate [502-85-2], 15:770
Sodium 3-hydroxy-2,7-naphthalenedisulfonate
    [135-59-3], 15:736
Sodium 7-hydroxy-2-naphthalenesulfonate [135-
    55-7], 15:736
Sodium hypochlorite [7681-52-9], 13:229
  in bleaching nuts, 16:265
  metal anodes in mfg, 15:174
  pulp bleaching, 16:770
  reduction by hydrogen peroxide, 13:16
Sodium hypophosphite [7681-53-0]
  nickel sulfate reaction, 15:803
Sodium iodate [7681-55-2], 13:665, 666
  sodium bisulfate reaction, 13:656
Sodium iodide [7681-82-5]
  toxicity, 15:574
Sodium iodomercurate [7784-03-4], 15:169
Sodium lactate [72-17-3], 13:85, 88
Sodium lauryl sulfate [151-21-3]
  in emulsion polymerization, 14:84
  in paint and varnish removers, 16:763
Sodium–lead alloy [12740-44-2]
  alkyl chloride reaction, 16:580
Sodium–lead alloy [12740-44-2] (1:1)
  in TEL mfg, 14:185
Sodium lignate [37203-80-8], 14:308
Sodium manganate dodecahydrate, 14:857
Sodium manganate(II) [12179-11-2], 14:852
Sodium manganate(V) [12163-41-6], 14:845, 857
Sodium manganate(V) heptahydrate, 14:857
Sodium manganate(VI) [15702-33-7], 14:845
Sodium manganese(I) hexacyanide [75535-10-3],
    14:852
Sodium metaniobate [67211-31-8] (1:1), 15:837
Sodium metaperiodate [7790-28-5], 13:667
Sodium metaperiodate trihydrate [13472-31-6],
    13:667
Sodium metarsenite [7784-46-5]
  as insecticide, 13:420
Sodium metasilicate [6834-92-0]
  paint stripper, 16:765
  toxicity, 15:574

Sodium methanenitronate [25854-38-0], 15:969
  phenyl isocyanate reaction, 15:974
Sodium molybdate [7631-95-0], 15:687
Sodium molybdate dihydrate [10102-40-6], 15:688
Sodium molybdate(VI) [7631-95-0]
  nitrogen reduction catalyst, 15:958
Sodium monohydrogen phosphate heptahydrate
    [7782-85-6]
  toxicity, 15:573
Sodium 1,6-naphthalenedisulfonate [1655-43-2],
    15:723
Sodium 2,6-naphthalenedisulfonate [1655-45-4],
    15:723
Sodium 2,7-naphthalenedisulfonate [1655-35-2],
    15:723
Sodium 1-naphthalenesulfonate [130-14-3], 15:
  720
Sodium 2-naphthalenesulfonate [532-02-5]
  2-naphthalenol from, 15:734
Sodium 1-naphthalenolate [3019-88-3]
  1-hydroxy-2-naphthalenecarboxylic acid from,
    15:745
Sodium-2-naphthalenolate [875-83-2]
  2-hydroxy-1-naphthalenecarboxylic acid from,
    15:746
Sodium niobate [12201-59-1] (7:6), 15:837
Sodium nitrate [7631-99-4]
  in meat products, 15:66
  in metal coloring, 15:311
  nitric acid from, 15:853
Sodium nitride [12136-83-3], 15:872
Sodium nitrite [7631-99-4]
  as grease additive, 14:503
  meat-curing agent, 15:62
  nitration reagent, 15:841
Sodium 2-nitrobutane-1-sulfonate [76794-27-9],
    15:976
Sodium Omadine, 13:244
Sodium orthoarsenite [7631-89-2]
  as insecticide, 13:420
Sodium oxide [1313-59-3]
  in glass electrode, 13:5
  in lignitic coal, 14:334
  in spent shale, 16:351
Sodium oxybate [502-85-2], 15:770. (See also
  Sodium 4-hydroxybutyrate.)
Sodium ozonide [12058-54-7], 16:688
Sodium paraperiodate [18122-77-0], 13:667
Sodium pentachlorophenoxide [131-52-2]
  applications of, 13:226
Sodium pentobarbital [57-33-0]
  estazolam similarity, 13:131
Sodium perborate [10486-00-7], 13:15. (See also
  Sodium peroxyborate trihydrate.)
Sodium perchlorate [7601-89-0]
  as oxygen source, 16:673

Sodium periodate [7790-28-5]
  in *cis,cis*-muconic acid synthesis, 13:41
Sodium permanganate [10101-50-5], 14:845
Sodium peroxide [25277-93-4]
  acid reaction, 13:16
  H_2O_2 from, 13:23
  oxygen generation, 16:679
Sodium peroxyborate trihydrate [28962-65-4],
    13:15. (See also *Sodium perborate.*)
Sodium phenolate [139-02-6]
  dialysis product, 15:119
Sodium *o*-phenylphenoxide [132-27-4]
  applications of, 13:227
Sodium phenylsulfamate [15790-84-8], 15:917
Sodium phosphate [7558-79-4]
  meat-curing agent, 15:62
Sodium phosphomolybdate [1313-30-0]
  toxicity, 15:573
Sodium polyacrylate [9003-04-7]
  in nonwovens, 16:109
  in papermaking, 16:822
Sodium–potassium
  as lubricant, 14:515
Sodium–potassium pump, 15:574
Sodium 2-propanenitronate [12384-98-4]
  chlorovinyl ketone reaction, 15:978
  p-nitrobenzyl chloride reaction, 15:974
  p-nitrobenzyltrimethylammonium iodide
    reaction, 15:974
Sodium propionate [137-40-6]
  preservative, 13:247
Sodium pyridinethione [3811-73-2]
  applications of, 13:244
Sodium pyroarsenite [15195-05-8]
  as insecticide, 13:420
Sodium pyrophosphate [7722-88-5]
  H_2O stabilizer, 13:14
Sodium retention
  and hypertension, 15:579
Sodium selenate [13410-01-0]
  as insecticide, 13:422
  toxicity, 15:573
Sodium selenite [10102-18-8]
  toxicity, 15:573
Sodium silicate [15593-82-5], 14:509
  insulation coating, 13:603
  zeolites from, 15:655
Sodium silicate [13870-28-5] SiO_2/Na_2O (2:1)
  toxicity, 15:574
Sodium silicate [13870-30-9] SiO_2/Na_2O (3:1)
  toxicity, 15:574
Sodium soap rosin [61790-51-0]
  paper size, 16:809
Sodium stannate [12058-66-1]
  H_2O stabilizer, 13:14
Sodium stannate trihydrate [12209-98-2]
  in lead refining, 14:117

Sodium stearoyl-2-lactylate [*25383-99-7*], *13*:86
Sodium sulfate [*7681-38-1*]
  by-product of lithium recovery, *14*:453
  in pulping, *14*:307
  from pulp mills, *16*:795
Sodium sulfate decahydrate [*7727-73-3*]
  in lithium recovery, *14*:453
  from seawater, *16*:277
Sodium sulfhydrate [*16721-80-5*]
  in hair removal from cowhide, *14*:209
Sodium sulfide [*1313-82-2*]
  in chemical conversion coating, *14*:509
  in hair removal from cowhide, *14*:209
  kraft process use, *16*:770
  in pulping, *14*:307
Sodium sulfite [*7757-83-7*]
  hydroquinone reaction, *13*:42
  TMP processing, *16*:770
Sodium 2-sulfoethyl acrylate, *14*:87
Sodium sulfonates
  as rust inhibitors, *14*:493
Sodium superoxide [*12034-12-7*]
  oxygen generation, *16*:679
Sodium tartrate [*868-18-8*]
  in biuret reagent, *15*:81
Sodium tetraborate [*12258-53-6*]
  as insecticide, *13*:422
Sodium tetraborate decahydrate [*1303-96-4*]
  pH standard, *13*:3
Sodium tetrachloraluminate [*7784-16-9*]
  in cementation coatings, *15*:248
Sodium tetrachloroaluminate [*7784-16-9*], *16*:587
Sodium tetracyanonickelate(0) [*76082-11-6*],
  16:35
Sodium tetralinsulfonate [*37837-69-7*], *15*:724
Sodium thiosulfate [*7772-98-7*]
  in chemical conversion coating, *14*:509
  as cyanide antidote, *14*:806
  in metal pickling, *15*:301
  in TEL manufacture, *14*:186
Sodium *p*-toluenesulfonchloramide [*127-65-1*]
  applications of, *13*:230
Sodium trichlorophenoxide [*1320-79-2*]
  in leather preservation, *13*:226
Sodium 2,4,5-trichlorophenoxide [*136-32-3*]
  applications of, *13*:226
Sodium triethylmercaptolead [*75214-14-1*], *14*:
  181
Sodium tripolyphosphate [*7758-29-4*]
  in meat processing, *15*:67
Sodium *p*-vinylbenzenesulfonate [*98-70-4*]
  in emulsion polymerization, *14*:87
Sodium α-vinylbenzylsulfonate, *14*:87
Sodium vinylsulfonate [*1184-84-5*]
  in emulsion polymerization, *14*:87

Sodospun, *16*:78
Soft-drink containers
  ozone sterilization of, *16*:709
Soft lumbang nut, *16*:251
Soft rot, *13*:225
Softwood
  in plywood manufacture, *14*:5
Softwoods
  lignin content, *14*:297
Soil
  emissivity, *13*:342
  nitrate measurement in, *13*:731
  nitrogen cycle, *15*:941
Soil additives
  from lignite, *14*:341
Soil conditioners
  from brown coal, *14*:341
Soil extract
  odor modification of, *16*:304
Soils
  liming of, *14*:377
  molybdenum-deficient, treatment of, *15*:693
Soil stabilization
  with lime, *14*:377
Solar-cell technology
  organogallium uses, *16*:562
Solar energy
  lithium fluoride heat-sink for, *14*:461
Solar heating
  maintenance programs and, *14*:766
Solar-power satellite
  microwave power for, *15*:517
Solders
  indium in, *13*:210
  lead–silver alloys, *14*:151
  lead–tin alloys, *14*:154
  lithium in, *14*:458
Solenoid, *14*:681
Solid bitumen
  electric insulation filler, *13*:565
Solids
  mixing, *15*:629
Solid-state devices
  microwave power output of, *15*:504
Solvent cleaning, *15*:296
Solvents
  effect on liquid crystals, *14*:397
  emission controls, *13*:902
  for inks, *13*:386
  in ion exchange, *13*:683
  ketones, *13*:902
  in metal cleaning, *15*:298
  paint and varnish stripping, *16*:762
Somastic, *13*:583
Somatostatin [*51110-01-0*], *15*:**780**
Soot
  from incinerators, *13*:185

Sopor, *13*:122
Sorbic acid [*110-44-1*]
  from ketene, *13*:877
  preservative, *13*:247
Sorbitol [*50-70-4*]
  for ion-exchange process, *13*:703
D-Sorbitol [*50-70-4*]
  microbial oxidation, *15*:459
Sorbose [*87-79-6*]
  in microbial synthesis, *15*:467
L-Sorbose [*87-79-6*]
  microbial oxidation, *15*:459
Sorrel
  oxalate in salt of, *16*:621
Souffle
  frozen dessert, *15*:568
Sound
  insulation against, *13*:513
  insulation materials for, *13*:513
  noise-related pressure of, *16*:2
Sound attenuation
  lead sheet for, *14*:157
Sound level meter, *16*:5
Sound transmission class (STC), *13*:525
Southern armyworm
  control of with pheromones, *13*:480
Soya oil [*8001-22-7*]
  in inks, *13*:381
Soybean oil [*8001-22-7*]
  antimicrobial agent in, *13*:238
  lecithin from, *14*:250
  in varnishes, *13*:554
Soybeans
  binders from, *16*:789
Soy meal [*68153-28-6*]
  protein from, in papermaking, *16*:822
  in Zanflo preparation, *15*:451
Soy protein [*68153-28-6*]
  paper binder, *16*:789
Spacecraft
  oxygen generation, *16*:673
Space flights
  atmosphere regeneration in, *14*:463
Space-radiation dosimetry, *16*:844
Space shuttle
  metallic coating for, *15*:257
Span 85 [*26266-58-0*]
  in encapsulation, *15*:479
Spartocine, *15*:760
Spathulenol [*6750-60-3*], *16*:313
SPD. See *Spectral power distribution.*
Spearmint oil [*8008-79-5*], *16*:328
Specialty chemicals, *14*:904
Specifications, *15*:32, 41. (See also *Materials standards and specifications.*)
Spectinomycin [*1695-77-8*]
  inactivation, *15*:466

Spectral power distribution
  in fluorescent lamps, *14*:537
Spectral wedges, *16*:552
Spectrophotometers
  in filter measurement, *16*:524
Spectroscopy
  lasers in, *14*:72
Speech
  noise interference, *16*:1
Speiss, *14*:107
Sperm
  mesomorphism of, *14*:419
Spermidine [*124-20-9*]
  properties, *15*:776
Spermidinephosphate [*1945-32-0*], *15*:777
Spermidine synthase [*9077-12-7*]
  in spermine biosynthesis, *15*:777
Spermidine trihydrochloride [*334-50-9*], *15*:776
Spermidinetripicrate, *15*:777
Spermine [*71-44-3*], *15*:777
  properties, *15*:776
Spermine synthase [*9077-13-8*]
  in spermine biosynthesis, *15*:777
Spermine tetrahydrochloride [*306-67-2*], *15*:777
Spermine tetrapicrate [*38693-36-6*], *15*:777
Sphalerite [*12169-28-7*]
  lead ore separation, *14*:102
Spider mites
  antifeeding agent, *16*:579
Spiegeleisen, *14*:824
Spike lavender oil [*8016-78-2*], *16*:328
Spinel [*1302-67-6*], *14*:615
Spin glass, *14*:648. (See also *Mictomagnetism.*)
Spinodal alloys, *15*:881
Spiral elevator mixers, *15*:630
Spirotallic gasket, *16*:726
Spleen
  liquid crystals in, *14*:418
Spodoptera, *13*:471
Spodoptera eridania, *13*:480
Spodoptera exigua, *13*:480
Spodoptera frugiperda, *13*:480
Spodoptera littoralis, *13*:481
Spodoptera litura, *13*:481
Spodumene [*1302-37-0*]
  lithium in, *14*:449
Sporiline, *15*:735
Spray-chilling
  in microencapsulation, *15*:476
Spray dryer
  for dry milk, *15*:559
Spray-drying
  in microencapsulation, *15*:475
Spring reliability, *15*:22
Spring water
  packaging, *16*:722

Sprue
  detection, *15*:83
Spülgas process, *14*:334
Spunbondeds, *16*:72.   (See also *Nonwoven
    textiles, spunbonded.*)
Spunlace textiles, *16*:121
Sputtering, *15*:267
  in IC manufacture, *13*:638
  in ion implantation, *13*:708
  in metal coating, *15*:265
  in thin-film fabrication, *14*:692
Sputtering yield, *13*:713
Squalene [*111-02-4*]
  from mevalonic acid, *13*:99
Square 50, *14*:659
Square 80, *14*:659
Square Permalloy, *14*:659
Square Permalloy 80, *14*:659
SRI International, *13*:323
SS acid [*82-47-3*], *15*:740.   (See also *4-Amino-5-
    hydroxy-1,3-naphthalenedisulfonic acid.*)
Stabilizers
  for HDPE, *16*:446
  for hydrogen peroxide, *13*:14
  lead monoxide for resins, *14*:164
  for LLDPE, *16*:394
  nitro alcohols, *15*:915
  nitroparaffins, *15*:983
  for polypropylene, *16*:466
  tin alkyls as, *16*:573
Stachyose [*470-55-3*]
  in soybean lecithin, *14*:251
Stainless steel
  bottom-blowing method, *16*:668
  cladding of, *15*:275
  emissivity, *13*:342
  gasket properties, *16*:727
  thermal conductivity of, *13*:544
316 Stainless steel [*11107-04-3*]
  coating of, *15*:268
409 Stainless steel, *15*:243
Stainless steels
  cladding of, *15*:281
  iodine resistance, *13*:656
  manganese alloy in, *14*:842
  nondestructive testing, *16*:51
Stairs information system, *13*:291
Stalactites, *14*:345
Stalagmites, *14*:345
Stamp-pad inks, *13*:394
Standard cell potential, *15*:39
Standard ferromanganese, *14*:830
Standardization
  of plant facilities, *14*:757
Standards, *15*:32.   (See also *Materials standards
    and specifications.*)

Stannic chloride [*7646-78-8*]
  Grignard reaction, *16*:575
  as initiator catalyst, *13*:371
Stannous chloride [*7772-99-8*]
  ammonium diphosphate reduction, *15*:84
Stannous octanoate [*301-10-0*]
  encapsulation catalyst, *15*:483
Staphylea trifolia, *16*:251
Staphylococcal infections, *15*:553
Staphylococcus aureas, *15*:490
Staphylococcus aureus, *15*:553
Staple fibers, *16*:105.   (See also *Nonwoven
    textile fabrics.*)
Starch [*9005-25-8*]
  in encapsulation, *15*:481
  in matches, *15*:5
  in nuts, *16*:252
  oxalic acid from, *16*:621, 622
  in papermaking, *16*:790
Starch, acetylated [*9045-28-7*]
  in papermaking, *16*:822
Starch, enzyme-converted [*65996-64-7*]
  in papermaking, *16*:820
Starches
  in papermaking, *16*:815
  paper sizing, *16*:778
Starch, hydroxyethylated [*9005-27-0*]
  in papermaking, *16*:820
Starch–iodine complex [*9005-83-8*]
  in iodine analysis, *13*:659
Starch, oxidized [*65996-64-7*]
  in papermaking, *16*:820
Starch xanthate [*11116-64-6*]
  in encapsulation, *15*:483
Starkeyite [*24378-31-2*], *14*:637
Static fatigue
  glass fibers, *13*:971
Static-pressure-measuring system, *14*:442
Stationary-tank mixers, *15*:631
Statistisches Bundesamt, *13*:316
Status epilepticus
  treatment, *13*:136
STC.   See *Sound transmission class.*
Steam cleaning
  metals, *15*:300
Steam cylinder oil
  limit of radiation tolerance, *14*:519
Steam distillation
  essential oils, *16*:313
Steam generators
  nuclear, corrosion control, *16*:152
Steam turbines
  noise control, *16*:22
Stearic acid [*57-11-4*]
  in electrical insulation, *13*:565
  in grease thickeners, *14*:502
  in nut oils, *16*:252
  in soybean oil, *14*:251

Stearic anhydride [638-08-4]
  sizing agent, 16:811
Stearoptenes
  in rose oil, 16:311
Stearoyl chloride [112-76-5]
  malic acid reaction, 13:105
Stearyl methacrylate [32360-05-7]
  properties, 15:347
Stearyl nitrile [638-65-3]
  properties, 15:908
Steatite [14807-96-6]
  as insulator, 13:562
Steel [12597-69-2]
  carbonitriding, 15:316
  carburizing, 15:313
  emissivity, 13:342
  heat treatment of, 15:338
  ion implantation, 13:716
  laser hardening of, 14:66
  limestone flux, 14:378
  manganese uses in, 14:841
  melt spinning, 15:229
  niobium in, 15:830
  nitriding, 15:317
  N_2 solution in, 15:934
  oxygen in mfg, 16:666
  painting of, 16:755
  surface treatment of, 15:304
  terne-metal, 14:155
  wear coefficient, 14:482
Steel, blue
  triiron tetroxide coating on, 13:776
Steel die inks, 13:395
Steel fiber
  in metal laminates, 13:945
Steels
  electrical properties, 14:656
  high temperature, 14:842
  molybdenum in, 15:670
  stainless, 14:842
Steel wool, 15:229
Steeping systems
  for malting, 14:816
Stefan-Boltzmann law, 13:339
Stegobium paniceum, 13:465
Stellite 6 [11105-35-4]
  iodine resistance, 13:656
Stellite Alloy No. 1
  laser cladding, 15:256
Stellite 6B, 15:281
Sterile filtration, 15:115
Sterility
  nutrients associated with, 15:573
Sterilization
  in insect control, 13:471
Sterol glycosides
  in soybean lecithin, 14:251

Stevens' law, 16:298
Stibine [7784-42-1]
  from calcium antimonides, 14:150
  nickel carbonyl reaction, 15:807
Stigmasterol [83-48-7]
  soybean phosphatides associated with, 14:252
Stilbenes
  from fumarates, 14:776
Stimulated emission
  in lasers, 14:43
  in semiconductors, 14:280
Stirofos [22248-79-9], 13:443
Stock nomenclature, 16:35
Stoke's law
  in fluorescence, 14:556
Stokes shift, 14:531
Stomach poisons
  as insecticides, 13:418
Stone groundwood process, 16:769
Stopping power
  fission particles, 16:830
Storage devices, 14:694
STORET information system, 13:301
Stormer viscometer, 13:376; 16:757
Strain
  on metals, 15:326
Strain hardening, 15:326
Stratosphere
  ozone effect in aircraft, 16:698
Straw
  lignin content, 14:297
Streaming potentials, 13:8
Strecker synthesis
  with p-hydroxybenzaldehyde, 13:72
Streptococcus
  in dextran production, 15:440
Streptococcus agalactiae, 15:553
Streptococcus cremoris, 15:561
Streptococcus fecalis
  disinfection by ozone, 16:705
Streptococcus lactis, 15:561
Streptococcus pneumoniae, 15:446
Streptococcus thermophilus, 15:564
Streptococcus viridans, 15:443
  dextran preparation, 15:442
Streptomyces, 13:786
Streptomyces albogriseolus, 15:465
Streptomyces galilaeus, 13:786
Streptomyces griseoflavus, 13:786
Streptomyces lavendulae, 13:786
Streptomyces pilosus, 13:786
Streptomyces spp, 15:467
Streptomycin [57-92-1]
  inactivation, 15:466
  purification of by ion exchange, 13:704
Stress
  effect of, 13:122

Stress (*Continued*)
on metals, *15*:326
Stress-cracking resistance, environmental
polyethylene, *16*:411
Stribeck curve, *14*:478
Strip-mining
lignite, *14*:326
Strippers, *16*:762.   (See also *Paint and varnish
removers.*)
Strontium [*7440-24-6*]
lead alloys, *14*:152
from the ocean, *16*:279
Strontium-84 [*15758-49-3*], *13*:840
Strontium-86 [*13982-14-4*], *13*:840
Strontium-87 [*13982-64-4*], *13*:840
Strontium-88 [*14119-10-9*], *13*:840
Strontium-90 [*10098-97-2*]
in nuclear waste, *16*:209
Strontium-100 [*15117-48-3*], *16*:188
Strontium aluminum oxide [*12254-24-9*] (1:12:19)
in phosphors, *14*:542
Strontium apatite [*64476-38-6*]
in phosphors, *14*:536
Strontium borate [*12007-66-8*]
phosphor, *14*:527, 542
in phosphor mfg, *14*:542
Strontium carbonate [*1633-05-2*]
in ferrite mfg, *14*:674
in phosphor mfg, *14*:536
Strontium chlorapatite [*12515-50-3*]
phosphor, *14*:542
Strontium chloride [*10476-85-4*]
flux, *14*:542
Strontium chlorophosphate, Eu doped [*68784-
77-0*]
phosphor, *14*:527
Strontium fluoride [*7783-48-4*]
in phosphor mfg, *14*:536
Strontium fluoride phosphate, Sb and Mn doped
[*68784-78-1*]
phosphor, *14*:527
Strontium hydrogen phosphate [*13450-99-2*]
in phosphor mfg, *14*:536
Strontium magnesium aluminum oxide [*12428-
80-7*] (1:1:10:19)
in phosphors, *14*:541
Strontium magnesium pyrophosphate
phosphor, *14*:543
Strontium manganate(VI), *14*:858
Strontium manganate(V)–strontium hydroxide,
14:857
Strontium molybdate [*13470-04-7*], *15*:687
Strontium nitride [*12033-82-8*], *15*:872
Strontium orthophosphate [*7446-28-8*]
in phosphors, *14*:539
Strontium perferrite [*12022-69-4*], *13*:776

Strontium phosphate [*7446-28-8*]
phosphor, *14*:543
Strontium pyrophosphate, Eu doped [*68784-61-2*]
phosphor, *14*:543
Strontium silicon chloride oxide [*51882-30-5*] (5:
4:6:10)
phosphor, *14*:542
Strychnic acid [*509-52-4*]
in poison nut, *16*:257
Strychnine [*57-24-9*], *13*:220
CNS stimulant, *15*:135
in lactic acid resolution, *13*:81
in poison nuts, *16*:257
source, *16*:270
Strychnos nuxvomica, *16*:251,
270
from amygdalin, *16*:257
Strychnos potatorum, *16*:251
Strychnus nuxvomica, *16*:270
Stucco, *14*:376
Stuckofen, *13*:735
Styphnic acid [*82-71-3*]
in explosive detonators, *13*:60
Styptirenal, *15*:757
Styralyl acetate [*93-92-5*]
in perfumes, *16*:968
Styrene [*100-42-5*]
copolymer, *15*:192
in encapsulation, *15*:480
in grafting to SBR, *15*:390
in ion exchange resins, *13*:686
isoprene reaction, *13*:821
maleic anydride copolymers, *14*:780
methacrylate copolymerization, *15*:384
in paint, *16*:760
polymerization of, *13*:356
in resin insulators, *13*:558
solubility of, *14*:83
Styrene–acrylonitrile [*9003-56-9*]
in composites, *13*:973
latex technology, *14*:82
Styrene–butadiene copolymer [*9003-55-8*]
in papermaking, *16*:822
Styrene–butadiene latexes
in paint, *16*:748
Styrene–butadiene rubber [*9003-55-8*], *13*:829
electric insulation elastomer, *13*:565
properties, *13*:567
Styrene–butadiene rubber [*9003-55-8*] (Buna S,
GRS, SBR)
gasket properties, *16*:727
Styrene–butadiene rubber (SBR) [*9003-55-8*]
in electric insulation, *13*:564
leather substitute, *14*:248
in paint, *16*:748
in paper coatings, *16*:790
Styrene–maleic anhydride copolymer [*9011-13-6*]

in papermaking, *16*:820

Sualen, *16*:78

Subfornical organ, *15*:760

Submarine cables
  steel sheathing of, *13*:585

Submarines
  atmosphere regeneration in, *14*:463
  oxygen generation systems, *16*:677

Substance P [*11035-08-8, 12769-48-1*], *15*:780

Substitute natural gas
  from lignites, *14*:336

Subtilisin [*1393-38-0*]
  effect on LHRH, *15*:771

Sub zero fluid, *14*:499

Success testing, *15*:27

Succinate dehydrogenase [*9002-02-2*]
  iron in, *15*:595

Succinic acid [*110-15-6*]
  in succinoglucan, *15*:453

Succinic semialdehyde [*692-29-5*]
  in GABA shunt, *15*:767

Succinoglucan [*9014-37-3*], *15*:452

Succinyl chloride [*543-20-4*]
  aziridine reaction, *13*:148

Succinyl CoA [*604-98-8*], *15*:597

Sucrose [*57-50-1*]
  in dextran production, *15*:440
  for ion-exchange process, *13*:703
  meat-curing agent, *15*:62
  in nuts, *16*:252
  oxalic acid from, *16*:621
  sweetening strength, *15*:568

Sucrose phosphorylase [*9074-06-0*], *15*:467

Suede
  synthetic, *14*:244

Sugar
  attractant, *13*:417
  dextran effect in processing, *15*:440
  effect on lime solubility, *14*:353
  lime in refining, *14*:381
  from wood, *14*:309

Sugar beets
  hydroxyacetic acid in, *13*:91

Sugar metabolism
  role of chromium, *15*:592

Sugar of lead. See *Lead acetate trihydrate.*

Sugars
  fermentation of, *14*:810
  as insect attractants, *13*:481

Sugar syrup
  analysis of with ISES, *13*:732

Sulfamic acid [*5329-14-6*]
  in anodizing, *15*:309
  nickel or nio reaction, *15*:805

Sulfamide [*7803-58-9*]
  potassium diethyl malonate reaction, *14*:800

Sulfate
  titration procedure, *13*:731

Sulfate process
  for pulping, *14*:307

Sulfatoceric acid [*17106-39-7*]
  in *n*-butyllithium analysis, *14*:468

2-Sulfatoisobutyramide sulfate [*56862-61-4*]
  methacrylamide from, *15*:359

2-Sulfatoisobutyronitrile [*10230-31-6*]
  methacrylamide from, *15*:359

Sulfide
  electrode, *13*:720

Sulfide oxidase [*9029-38-3*]
  Mo content, *15*:693

Sulfite lignins, *14*:304

Sulfite oxidase [*9029-38-3*]
  iron in, *15*:595
  molybdenum in, *15*:592

Sulfite-pulping process, *14*:304

Sulfomaleic anhydride [*6007-87-0*], *14*:781
  ethyl glycolate reaction, *13*:92

Sulfon Acid Blue G [*6406-30-0*], *15*:740

p-Sulfonamidophenyl isocyanate [*1773-42-8*], *13*: 801

Sulfonation
  metallocenes, *16*:606
  of naphthalene, *15*:702

Sulfonic acids
  in nitration ratio control, *15*:929

Sulfonyl chloride [*7791-25-5*]
  in leather tanning, *14*:217

Sulfonylureas
  as antidiabetic agents, *13*:613

Sulfosuccinic acid [*5138-18-1*], *13*:110

Sulfotepp [*3689-24-5*], *13*:439

Sulfoxide [*120-62-7*]
  as insecticide, *13*:425

Sulfur [*7704-34-9*]
  in blast furnace flux, *13*:740
  in bleaching nuts, *16*:265
  in electric insulation, *13*:565
  as insecticide carrier, *13*:416
  in lead processing, *14*:110
  in lignite, *14*:327
  in matches, *15*:2
  metabolic functions, *15*:587
  nitrogen reaction, *15*:933
  nutrient properties, *15*:571
  in nuts, *16*:260
  from the ocean, *16*:279
  offshore sources, *16*:287
  recovery from lignite, *14*:327

Sulfur-32 [*13981-57-2*], *13*:840

Sulfur-33 [*14257-58-0*], *13*:840

Sulfur-34 [*13965-97-4*], *13*:840

Sulfur colors
  in paper, *16*:779

Sulfur compounds
    in lubricants, *14*:479
    ozone reactions, *16*:686
    removal by molecular sieves, *15*:663
Sulfur decafluoride [*5714-22-7*]
    from SF$_6$ breakdown, *13*:551
Sulfur difluoride [*13814-25-0*] (1:2)
    from SF$_6$ breakdown, *13*:551
Sulfur difluoride [*13709-35-8*] (2:2)
    from SF$_6$ breakdown, *13*:551
Sulfur dioxide [*7446-09-5*]
    in air, *16*:657
    extraction from walnuts, *16*:270
    in incinerator gases, *13*:185
    isoprene reaction, *13*:822
    in leak detection, *16*:55
    in malting, *14*:822
    membrane monitoring device, *15*:122
    odor-problem chemical, *16*:303
    tetraethyllead reaction, *14*:183
    toxicity, *13*:272
    in vitamin C, *16*:266
Sulfur fluoride [*2551-62-4*] (1:6)
    in sand casting magnesium, *14*:608
Sulfur fluoride [*5714-22-7*] (2:10)
    N$_2$F$_4$ reaction, *14*:74
Sulfur hexafluoride [*2551-62-4*]
    dielectric strength, *13*:541
    as insulator, *13*:550
    laser dissociation, *14*:75
    laser isotope separation in, *14*:76
Sulfuric acid [*7664-93-9*]
    electrolysis of, *13*:15
    hydrogen peroxide reaction, *13*:15
    in ion exchange, *13*:686
    in metal pickling, *15*:300
    nickel reaction, *15*:803
    nickel recovery, *15*:799
    in nitration, *15*:847
    nitration reagents, *15*:841
    paint stripper, *16*:765
    toxicity, *13*:272
Sulfur nitride [*28950-34-7*], *15*:883
Sulfurous acid [*7782-99-2*]
    in sulfite pulping, *14*:304
Sulfur oxides
    from pulp mills, *16*:795
Sulfur tetrafluoride [*7783-60-0*]
    from SF$_6$ breakdown, *13*:551
Sulfur trioxide [*7446-11-9*]
    coal additive, *14*:332
    hydrogen peroxide reaction, *13*:15
    in ion exchange, *13*:686
    isoprene reaction, *13*:822
    maleic anhydride reaction, *14*:780
    naphthalene reaction, *15*:720
    in spent shale, *16*:351
Sulfuryl chloride [*7791-25-5*]
    in organolead decontamination, *14*:184

Sulfuryl fluoride [*2699-79-8*]
    as fumigant, *13*:467
Sulindac [*38194-50-2*]
    immunologic effects, *13*:174
Sulphenone [*80-00-2*], *13*:463
Sulthiame [*61-56-3*]
    properties of, *13*:135
Sumatra benzoin [*9000-73-1*], *16*:949
Sumatrol [*82-10-0*]
    insecticide, *13*:426
Sumitomo process, *13*:745
Sunflower, *16*:251, 252
Sunflower oil [*8001-21-6*]
    lecithin from, *14*:250
Sunscreen
    in polymethacrylates, *15*:380
Suntan age, *16*:843
Supac, *16*:78
Super-Ad-It, *13*:236
Superalloys, *15*:343, 793
Superconductivity
    effect of ion implantation, *13*:717
Superconductors
    niobium-containing, *15*:831
Superhybrid composites, *13*:947
Superhybrids
    in fan blades, *13*:944
Superior oil retort, *16*:342
Supermalloy, *14*:651, 659
Supermendur, *14*:651, 659, 663, 664
Supermumetal, *14*:659
Superoxide dismutase, *15*:588
Superoxide dismutases
    metals in, *15*:570
    zinc in, *15*:590
Superoxides
    oxygen generation, *16*:678
Superperm 49, *14*:659
Superperm 80, *14*:659
Superphenix, *16*:194
Superposition principle
    in control systems, *13*:503
Superradiometal, *14*:659
Supersonic transport
    radiation exposure from, *16*:217
Supersuede, *14*:231
Supracapsulin, *15*:757
Supranephrane, *15*:757
Suprarenaline, *15*:757
Suprarenin, *15*:757
Surface Coating Abstracts
    information service, *16*:902
Surfactants
    in emulsion polymerization, *14*:85
    packaging of, *16*:717
    zeolites in, *15*:668

Surgery
  laser uses, *14*:71
Surgical packs and gowns
  nonwoven, *16*:123
Surrenine, *15*:757
Suspension polymerization
  HDPE, *16*:443
Sutures
  from animal intestines, *15*:72
Sweco mills, *14*:743
Swimming pools
  iodine microbicide, *13*:674
Switching devices
  mercury, *15*:155
Switching elements, *14*:694
Sydrapulper, *16*:774
Synacthen, *15*:759
Synanthedon pictipes, *13*:481
Synergism
  in patents, *16*:864
Synesstic 68, *14*:496
Synovitis
  induced by immune complexes, *13*:169
Syntans, *14*:213, 216.   (See also *Tannins,
    synthetic.*)
Syntex, *16*:78
Synthesis gas, *15*:946.   (See also *Water gas.*)
  ammonia from, *15*:946
  for methanol mfg, *15*:401
  oxo process use, *16*:643
  oxygen in mfg, *16*:668
  production, *15*:945
Synthetic fuels
  from lignite, *14*:339
Synthetic hydrocarbon oil
  limit of radiation tolerance, *14*:519
Synthetic micas.   See *Micas, natural and
    synthetic.*
Synthetol process, *14*:328
Syrups
  as insect attractants, *13*:481
Systemic lupus
  treatment, *13*:179
Systemic lupus erythematosus
  treatment, *13*:172
System reliability models, *15*:13
Systems analysis, *16*:500.   (See also *Operations
    planning.*)

T

T₃.   See *3,5,3'-Triiodothyronine.*
T₄.   See *3,5,3',5'-Tetraiodothyronine.*

Tablecloths
  nonwoven, *16*:93
Tablets
  pan coating process for, *15*:472
Tack spinning
  textiles, *16*:96
Taconite
  iron in, *13*:749
Taenothrips simplex, *13*:473
Tafnel, *16*:78
Tagua nut, *16*:251
Takamina, *15*:757
Talc [*14807-96-6*], *14*:615
  electric insulation filler, *13*:565
  fragrances, *16*:970
  as grease filler, *14*:503
  as insecticide carrier, *13*:416
  mica substitute, *15*:432
  in packings, *16*:738
  in papermaking, *16*:819
  paper pigment, *16*:778
  as solid lubricant, *14*:508
Tall oil [*8002-26-4*]
  in inks, *13*:381
Tallow [*61789-97-7*]
  as lubricant, *14*:477
  from meat, *15*:71
  in packings, *16*:734
Tallow nitrile [*61790-28-1*]
  properties, *15*:908
Tallow nut, *16*:251
Tamarind gum [*39386-78-2*]
  in papermaking, *16*:815
Tampons
  nonwoven, *16*:123
Tangiers' disease
  liquid crystals in, *14*:420
Tank cars, *16*:715
Tankers
  for LPG transportation,
    14:390
Tank linings
  lead alloys, *14*:147
Tannage, aldehyde, *14*:217
Tannage, aluminum, *14*:215
Tannage, chrome, *14*:208
Tannage, iron, *14*:215
Tannage, oil, *14*:217
Tannage, resin, *14*:216
Tannage, silica, *14*:216
Tannage, sulfonyl chloride,
    14:217
Tannage, zirconium, *14*:214
Tannic acid [*1401-55-4*]
  ozone reaction, *16*:704
  from pecan shells, *16*:273

Tannin [1401-55-4]
  in leather manufacture, 14:212
  in niobium analysis, 15:829
Tanning
  lactic acid used in, 13:87
  leather, 14:208
Tanning agents, 15:724
Tannins
  in nuts, 16:252, 259
Tannins, synthetic, 14:213
Tannins, vegetable
  in leather manufacture, 14:212
Tantalic acid [76347-19-8]
  salts, 15:823
Tantalite [1310-23-2], 15:821
Tantalo-niobates, 15:821
Tantalum [7440-25-7], 15:258
  amalgamation with mercury, 15:147
  cladding of, 15:281
  in coatings, 15:241
  platinum cladding of, 15:172
  in thermal insulation, 13:599
Tantalum-180 [15759-29-2], 13:842
Tantalum-181 [7440-25-7], 13:842
Tantalum carbide [12070-06-3]
  nitride form, 15:877
Tantalum–copper clads, 15:275
Tantalum fiber
  in metal laminates, 13:945
Tantalum nitride [12033-62-4] (1:1), 15:872
Tantalum nitride [12033-63-5] (2:1), 15:872
Tantalum nitride [12033-94-2] (3:5), 15:872
δ-Tantalum nitride [12033-62-4]
  properties, 15:874
ε-Tantalum nitride [12033-62-4]
  properties, 15:874
Tantalum pentachloride [7721-01-9]
  in chemical deposition coating, 15:263
Tantalum–steel clads, 15:275
Tape
  electric insulation, 13:575
  fibers for, 16:376
Tape, friction
  in cable insulation, 13:578
Tape, magnetic, 14:732
Tape recorder, 14:733. (See also *Magnetic tape.*)
Tapyrus, 16:78
Tar
  in DR processes, 13:754
Tar sands
  microwave heating, 15:517
Tartar emetic [28300-74-5], 13:118. (See also *Potassium antimony tartrate sesquihydrate.*)
Tartaric acid [87-69-4], 13:111; 14:778
  microbial synthesis of, 15:462

in niobium analysis, 15:828
  properties, 13:104
meso-Tartaric acid [147-73-9], 13:112
(R*,R*)-Tartaric acid [133-37-9], 13:111
(R-R*,R*)-Tartaric acid [87-69-4]
  properties, 13:111
(S-R*,R*)-Tartaric acid [147-71-7], 13:112
L-Tartrate [87-69-4]
  phosphatase reagent, 15:80
Tartronic acid [80-69-3], 13:112
  properties, 13:118
Tasmanite, 16:347
Tattoos
  laser removal, 14:70
Taurine [107-35-7]
  as neuroregulator, 15:781
  sulfur in, 15:587
Taurocholic acid [81-24-3]
  from animal bile, 15:72
Taylor's series, 15:29
Tay Sachs' disease
  liquid crystals in, 14:420
TBC. See *tert-Butylcatechol.*
4-TBC. See *4-tert-Butylcyclohexanone.*
TBP. See *Tributyl phosphate.*
TBTF. See *Tributyltin fluoride.*
TBTO. See *Tributyltin oxide.*
TCC. See *Trichlorocarbanilide.*
TCDD. See *Tetrachlorodibenzo-p-dioxin.*
T-cells
  in antibody formation, 13:168
TCF, 16:75
TCL. See *Toxic concentration low.*
TCP. See *Trichlorophenol.*
TDE. See *1,1-Dichloro-2,2-bis(p-chlorophenyl)-ethane.*
TDI. See *Toluene diisocyanate.*
TDL. See *Toxic dose low.*
Tea bags
  nonwoven, 16:109
TEA laser, 14:54
Teart
  cattle disease, 15:593
Technical service, 14:905
Teeth, artificial
  radiation exposure from, 16:217
Teflon, 15:894; 16:727
Teflon, microporous, 13:728
Tegretol, 13:135
Tektamer 38, 13:231
Tektites
  dating of, 16:844
TEL. See *Tetraethyllead.*
Telangiectasia
  treatment, 14:266
TEL aviation mix
  formulation of, 14:189

Telenet
  communications system, *16*:901
Television
  in nuclear facility design, *16*:246
  radiation exposure from, *16*:217
Telfairia pedata, *16*:251
Tellurium [*13494-80-9*]
  in fuel reprocessing, *16*:182
  lead alloys, *14*:153
Tellurium-120 [*15125-47-0*], *13*:841
Tellurium-122 [*14390-71-7*], *13*:841
Tellurium-123 [*14304-80-4*], *13*:841
Tellurium-124 [*14390-72-8*], *13*:841
Tellurium-125 [*14390-73-9*], *13*:841
Tellurium-126 [*14390-74-0*], *13*:841
Tellurium-128 [*14390-75-1*], *13*:841
Tellurium-130 [*14390-76-2*], *13*:841
Tellurium nitride [*59641-84-8*], *15*:883
Telodrin [*297-78-9*], *13*:434
TEM. See *Triethylenemelamine.*
Temasept IV, *13*:240
Temazepam [*846-50-4*]
  properties of, *13*:122
Temephos [*3383-96-8*], *13*:440
Temperature
  standardization of, *15*:38
Temperature measurement
  by encapsulated chromogens, *15*:491
  in microwave heating, *15*:509
Tempered glass, *13*:990
Tempering, *15*:323
Tenax GC, *13*:86
Tendon
  liquid crystals in, *14*:418
Tenebrio molitor, *13*:459
Tenera, *14*:248
Tensile tests
  metals, *15*:326
Ten-twenty, *13*:990
TEPA [*57-39-6*], *13*:473
Tephrosia (Cracca) Virginiana, *13*:425
Tephrosin [*76-80-2*]
  insecticide, *13*:425
TEPP. See *Tetraethyl pyrophosphate.*
Terbacil [*5902-51-2*], *13*:888
Terbium [*7440-27-9*]
  as magnetic material, *14*:647
  from the ocean, *16*:280
Terbium-159 [*7440-27-9*], *13*:841
Terbium, compd with cobalt [*12017-69-5*] (1:5),
  14:676
Terbium, compd with cobalt [*12052-79-8*] (2:17),
  14:676
Terbium(III) [*22541-20-4*]
  phosphor activator, *14*:540
Terbium(III,IV) oxide [*12738-76-0*]
  phosphors from, *14*:541

Terbium magnesium aluminum oxide (1:1:11:19)
  in phosphors, *14*:541
Terbium nitride [*12033-64-6*], *15*:872
Terbufos [*13071-79-9*], *13*:447
Terbutryne [*886-50-0*], *13*:674
Teredos
  cable protection against, *13*:582
Terephthalic acid [*100-21-0*]
  as binder, *13*:60
  in liquid-crystal synthesis, *14*:414
Terlinguaite [*12394-37-5*], *15*:144
Terminalia catappa, *16*:250
Termites
  cable protection against, *13*:582
  protection against in fiberboard, *14*:18
Terne coatings, *14*:155
Terne metal [*39428-85-8, 54938-78-2*]
  in coatings, *15*:243
Terne sheets, *14*:155
Terpenes, chlorinated
  as insecticides, *13*:433
cis-Terpine hydrate [*565-48-0*]
  microbial transformation to α-terpineol, *15*:465
α-Terpinene [*99-86-5*], *16*:313
γ-Terpinene [*99-85-4*], *16*:313
1-Terpinen-4-ol [*562-79-2*], *16*:312
  from lime, *16*:324
Terpineol [*8000-41-7*], *16*:329
  from pine, *16*:327
α-Terpineol [*98-55-5*]
  from lime, *16*:324
  microbial synthesis of, *15*:465
  in perfume, *16*:954
Terpinolene [*586-62-9*], *16*:313
Terpinyl acetate [*80-26-2*], *16*:956
2,2':6',2''-Terpyridine [*1148-79-4*]
  iron complex, *13*:781
Terram 1000, *16*:76
Tesa test, *15*:552
Testing, nondestructive. See *Nondestructive
    testing.*
Testosterone [*58-22-0*]
  microbial synthesis, *15*:462
  as neuroregulator, *15*:759
Tetany
  detection, *15*:83
  nutrients associated with, *15*:573
Tetraalkyllead compounds
  properties, *16*:580
Tetraammonium hexakiscyanoferrate [*14481-29-
    9*], *13*:766
Tetraarylleads
  properties, *16*:580
Tetraaryl silicate
  limit of radiation tolerance, *14*:519

Tetra(1-aziridinyl)copper chloride [*28174-42-7*], *13*:151

Tetrabasic lead sulfate [*52732-72-6*], *14*:161, 168

Tetrabutyl ammonium permanganate [*35638-41-6*], *14*:860

Tetra-*n*-butyltin [*1461-25-2*]
properties, *16*:573

Tetrabutyl titanate [*5593-70-4*]
in isoprene reaction, *13*:823

Tetracarbonylcobaltate, sodium salt [*14878-28-5*], *16*:613

Tetra-*μ*-3-carbonyldodecacarbonylhexarhodium [*28407-51-4*]
oxo process catalyst, *16*:639

Tetracarbonylhydrocobalt [*16842-03-8*]
oxo process catalyst, *16*:611

Tetrachlorodibenzo-*p*-dioxin [*1746-01-6*] (TCDD)
inducement of immunodeficiency, *13*:170

2,3,7,8-Tetrachlorodibenzo-*para*-dioxin (TCDD) [*1746-01-6*]
in 2,4,5-TCP, *13*:226

2,4,5,4'-Tetrachlorodiphenyl sulfone [*116-29-0*]
in insect control, *13*:461

1,1,2,2-Tetrachloroethane [*79-34-5*]
toxicity, *13*:272

Tetrachloro[1,2-ethanediylbis[diphenylphosphine]-*P,P'*]molybdenum [*40219-67-8*]
in N$_2$ fixation, *15*:958

Tetrachloroethylene [*127-18-4*]
as insulator, *13*:552
toxicity, *13*:272

2,3,5,6-Tetrachlorohydroquinone [*87-87-6*]
developer, *13*:60

Tetrachloroisophthalanonitrile [*1897-45-6*]
applications of, *13*:248

Tetrachloroisophthalonitrile [*1897-45-6*], *15*:906

2,3,5,6-Tetrachloro-4-(methylsulfonyl)pyridine [*1308-52-6*]
applications of, *13*:244

Tetrachloro-*o*-quinone [*2435-53-2*]
ketene reaction, *13*:877

Tetrachlorosilane [*10026-04-7*]
in chemical deposition coating, *15*:263

Tetrachlorosilicate [*7646-78-8*]
in Si epitaxy, *13*:626

Tetracobalt dodecacarbonyl [*17786-31-1*]
oxo process catalyst, *16*:639

Tetracupric sulfate
in bordeaux mixture, *13*:476

Tetracyanoethylene [*670-54-2*]
ozonide reaction, *16*:686

Tetracyanonickelates, *15*:805

cis-9,*trans*-11-Tetradecadienyl acetate [*30562-09-5*]
as insect pheromone, *13*:481

cis-9,*trans*-12-Tetradecadienyl acetate [*31654-77-0*]
as insect pheromone, *13*:481

Tetradecanal [*124-25-4*]
as insect attractant, *13*:481

1-Tetradecanol [*112-72-1*], *13*:908

cis-9-Tetradecenal [*60671-78-5*]
as insect attractant, *13*:481
as insect pheromone, *13*:480

trans-11-Tetradecenal [*35746-21-5*]
as insect pheromone, *13*:480

1-Tetradecene [*1120-36-1*]
properties, *16*:481

7-Tetradecene [*10374-74-0*], *16*:484

cis-9-Tetradecenyl acetate [*16725-53-4*]
as insect pheromone, *13*:480

cis-11-Tetradecenyl acetate [*20711-10-8*]
as insect pheromone, *13*:481

trans-11-Tetradecenyl acetate [*39298-60-7*]
as insect pheromone, *13*:481

Tetradecyl methacrylate [*2549-53-3*]
properties, *15*:353

Tetradifon [*116-29-0*], *13*:463

Tetraethylammonium dibromoiodide, *13*:664

Tetraethylammonium tetrachloronickelate [*5964-71-6*], *15*:808

Tetraethylammonium triphenylphosphinetribromonickelate [*41828-60-8*], *15*:808

Tetra(2-ethylbutyl) silicate [*78-13-7*]
as lubricant, *14*:501

Tetraethyldiborane [*12081-54-8*], *16*:562

Tetraethyl dithionopyrophosphate [*3689-24-5*]
as insecticide, *13*:438

Tetraethylene glycol dimethacrylate [*109-17-1*]
in adhesive, *15*:491

Tetraethylenepentamine [*112-57-2*]
in papermaking, *16*:817

Tetra(2-ethylhexyl) silicate [*115-82-2*]
as lubricant, *14*:501

Tetraethyllead [*78-00-2*]
mercury dichloride reaction, *16*:587
stabilization of by lecithin, *14*:266

Tetraethyllead (TEL) [*78-00-2*], *14*:180; *16*:579

O,O,O',O'-Tetraethyl *S,S'*-methylenebis(phosphorodithioate) [*563-12-2*]
as insecticide, *13*:443

Tetraethyl pyrophosphate [*107-49-3*], *13*:439

Tetraethyl pyrophosphate (TEPP) [*107-49-3*]
as insecticide, *13*:437

Tetraethyl silicate [*78-10-4*]
as lubricant, *14*:501

Tetrafluoroethylene [*116-14-3*]
nickelocene reaction, *16*:604

Tetrafluoroethylene resin
in packings, *16*:726

Tetrafluoroethylene (TFE) (solid) [*116-14-3*]
gasket properties, *16*:727

Tetrafluoromethane [*75-73-0*]
in etch gas, *13*:639

Tetrahydroanthraquinone [*28758-94-3*]
from naphthoquinone, *15*:715

Tetrahydrocitral [*5988-91-0, 25795-46-4*]
fragrance compound, *16*:969

Tetrahydro-3,5-dimethyl-2*H*-1,3,5-thiadiazine-
2-thione [*533-74-4*]
applications of, *13*:244

Tetrahydrofolic acid [*109-99-0*]
in glycine biosynthesis, *15*:769

Tetrahydrofuran [*109-99-9*]
adiponitrile from, *15*:899
n-butyllithium reaction, *14*:467
in liquid membrane electrode mfg, *13*:725
from maleic anhydride, *14*:780
in paint and varnish removers, *16*:764

Tetrahydrofuryl methacrylate [*2455-24-5*]
properties, *15*:353

Tetrahydrogen hexakiscyanoferrate [*17126-47-5*],
13:766. (See also *Ferrocyanic acid*.)

4,5,6,7-Tetrahydroindole [*13618-91-2*], *13*:218

Tetrahydromuguol [*18479-57-7, 78-69-3*], *16*:956

1,2,3,4-Tetrahydronaphthalene [*119-64-2*], *15*:
704. (See also *Tetralin*.)

1,4,5,8-Tetrahydronaphthalene [*493-04-9*], *15*:703

2-(5,6,7,8-Tetrahydro-1-naphthaleneamine)-2-
imidazoline [*1082-57-1*], *15*:727

1,2,3,4-Tetrahydro-1-naphthoic acid [*1914-65-4*]
tetrahydrozodin from, *15*:737

Tetrahydro-1,3-oxazines
from nitroparaffins, *15*:976

1*a*,2,7,7*a*-Tetrahydro-3-(phenylmethoxy)
naphtho[*2,3-b*]oxirene [*51927-62-9*]
sodium diethyl malonate reaction, *14*:798

3*a*,4,9,9*a*-Tetrahydro-5-(phenylmethoxy)
naphtho[*2,3-b*]furan-2(3*H*)-one [*67317-47-9*],
14:798

3*a*,4,9,9*a*-Tetrahydro-β-(phenylmethoxy)
naphtho[*2,3-b*]furan-2(3*H*)-one [*67317-46-8*],
14:798

Tetrahydrophthalic anhydride [*85-43-8*], *14*:774

2,3,4,5-Tetrahydroxybenzaldehyde [*74186-02-0*],
13:70. (See also *Gallaldehyde*.)

Tetrahydrozoline [*84-22-0*], *15*:737

Tetraiodogermane [*13450-95-8*]
in chemical deposition coating, *15*:263

3,5,3′,5′-Tetraiodothyronine [*55-03-8*] (T$_4$), *15*:
593

Tetrairon(II) tris(hexakiscyanoferrate) [*14038-
43-8*], *13*:770
in Prussian Blue synthesis, *13*:769

Tetrakis(η^5-2,4-cyclopentadien-1-yl)[μ-(dinitro-
gen-*N,N′*)] dititanium [*11136-31-5*], *15*:955,
957

Tetrakis(η^5-2,4-cyclopentadien-1-yl)[μ-(dinitro-
gen-*N,N′*)]bis(1-methylethyl)dititanium
[*12769-71-0*], *15*:957

Tetrakis(η^5-2,4-cyclopentadien-1-yl)[μ-(dinitro-
gen-*N,N′*)]bis-(3-methylphenyl)dititanium
[*42613-11-6*], *15*:957

Tetrakis(η^5-2,4-cyclopentadien-1-yl)[μ-(dinitro-
gen-*N,N′*)]bis-(4-methylphenyl)dititanium
[*42613-10-5*], *15*:957

Tetrakis(isocyanobenzene)nickel(0) [*23411-45-2*],
15:801

Tetrakis(methylphosphorous dichloride)nickel(0)
[*76404-15-4*], *15*:808

Tetrakistrichlorophosphinenickel(0) [*36421-86-
0*], *15*:808

Tetrakistrifluorophosphinenickel(0) [*13859-65-9*],
15:808

Tetrakistrimethylphosphinenickel(0) [*28069-69-
4*], *15*:808

Tetrakis(trimethylphosphite-*P*)nickel(0) [*14881-
35-7*], *15*:801

Tetrakistriphenylphosphinenickel(0) [*15133-82-
1*], *15*:808

Tetrakis(tritolylphosphite-*P*)nickel [*35884-66-3*],
15:808

Tetralin [*119-64-2*], *15*:704. (See also *1,2,3,4-
Tetrahydronaphthalene*.)
1-naphthalenol from, *15*:729
properties, *15*:705

1-Tetralol [*529-33-9*]
1-naphthalenol from, *15*:729

1-Tetralone [*529-34-0*]
1-naphthalenol from, *15*:729
properties, *13*:896

1,2,4,4-Tetramethyl-5-acetylcyclohexene [*74631-
05-3*], *13*:917

Tetramethylammonium hydroxide [*75-59-2*]
zeolites from, *15*:655

Tetramethyl ammonium nonaiodide [*3345-37-7*],
13:665

Tetramethylarsonium iodide [*5814-20-0*], *13*:669

1,2,3,4-Tetramethylcyclobutadienenickel
dichloride [*76404-16-5*], *15*:808

Tetramethylcyclobutanedione [*29714-52-3*]
in perfume formulations, *13*:927

Tetramethyl-1,3-cyclobutanedione [*933-52-8*]
P$_2$S$_5$ reaction, *13*:881

2,2,4,4-Tetramethyl-1,3-cyclobutanedione [*933-
52-8*], *13*:883

Tetramethyl-1,3-cyclobutanedithione [*29718-52-
3*]
dimethylthio ketene from, *13*:881

2,2,2,4-Tetramethyl-1,3-cyclobutanedithione
[*10181-56-3*], *13*:883

Tetramethylcyclobutanone [933-52-8]
  photolysis, 13:885
Tetramethylcyclopropanone [2610-40-4], 13:886
1,1'-(1,1,3,3-Tetramethyl 1,3-disiloxanediyl)-
  ferrocene [1272-10-2], 15:205
Tetramethylenediamine [110-18-9]
  in spermine synthesis, 15:777
Tetramethylene glycol [110-63-4]
  in γ-butyrolactone production, 13:97
Tetramethylenimines. See *Pyrrolidines.*
Tetramethylethylene [563-79-1], 13:886
N,N,N',N'-Tetramethylethylenediamine
  (TMEDA) [110-18-9]
  n-butyllithium complex, 14:468
2,4,4,6-Tetramethylhexahydropyrimidine
  [74966-86-2]
  color stabilizer, 13:916
3,3,4,4-Tetramethyl-2,5-hexanedione [23328-38-
  3]
  properties, 13:931
1,2,3,3-Tetramethylindoleninium iodide [5418-
  63-3], 13:215
Tetramethyllead (TML) [75-74-1], 14:180; 16:579
2,2,6,6-Tetramethyl-γ-pyrone [1197-66-6], 13:
  914. (See also *Phorone.*)
Tetramethylsuccinonitrile [3333-52-6] (TMSN)
  VAZO 64 decomposition product, 15:904
3,3,6,8-Tetramethyl-1-tetralone [5409-55-2], 13:
  921
O,O,O',O'-Tetramethyl O,O'-thio-di-p-
  phenylene phosphorothionate [3383-96-8]
  as insecticide, 13:440
Tetramethylthiuram disulfide [137-26-8]
  as repellent, 13:475
Tetramethylthiuram monosulfide [97-74-5]
  in electrical insulation, 13:565
Tetramethyltin [594-27-4]
  properties, 16:573
Tetraneopentyltitanium [36945-13-8], 16:583
Tetranitromethane [509-14-8]
  from potassium nitroform, 15:974
  properties, 15:970
1,3,5,7-Tetranitronaphthalene [60619-96-7], 15:
  725
1,3,5,8-Tetranitronaphthalene [2217-58-5], 15:725
1,3,6,8-Tetranitronaphthalene [28995-89-3], 15:
  725
1,4,5,8-Tetranitronaphthalene [4793-98-0], 15:725
3,6,9,12-Tetraoxatetradecan-1-ol [5650-20-4],
  16:39
Tetrapak system, 15:549
Tetraphenyl arsonium permanganate [4312-28-1],
  14:860
1,1,4,4-Tetraphenyl-2-butene-1,4-diol [59856-52-
  9], 14:778

Tetraphenylcyclobutadienepalladium dibromide
  [75811-08-4], 16:609
  metal carbonyl reactions, 16:609
  metal complexes, 16:609
π-Tetraphenylcyclobutadienecobalt dicarbonyl
  bromide [12157-16-3], 16:608
2,2,4,4-Tetraphenyl-1,3-cyclobutanedione [3469-
  15-6], 13:883
Tetraphenyllead [595-89-1], 14:182
Tetraphenyl phosphonium permanganate
  [34209-26-2], 14:860
Tetraphenylporphinatoiron [16591-56-3], 13:782
Tetrapotassium cyanonickelate(I) [40810-33-1],
  15:801
Tetrapotassium hexakiscyanoferrate trihydrate
  [14459-95-1], 13:767, 771
Tetrapropylammonium hydroxide [4499-86-9]
  zeolites from, 15:655
Tetrarhodium dodecacarbonyl. See *Tri-
  µ-carbonylnonacarbonyltetrarhodium.*
Tetrasilver hexakiscyanoferrate [14038-75-6], 13:
  771
Tetrasodium hexakiscyanoferrate decahydrate
  [14434-22-1], 13:767
Tetrasodium lead [12346-61-1]
  in organolead synthesis, 14:188
Tetrasodium phthalocyaninetetrasulfonatoferrate
  [41867-66-7], 13:782
Tetrasodium pyrophosphate [7722-88-5]
  in electroplating solution, 14:610
Texaco gasifier
  lignite, 14:338
Texin, 13:813
Texizol, 16:78
Textile finishes
  polyaziridines, 13:161
Textile finishing
  oxalic acid in, 16:631
Textiles
  flash spun, 16:93
  melt blown, 16:93
  microwave treatment of, 15:515
  tack spun, 16:96
Textiles, nonwoven, 16:72. (See also *Nonwoven
  textile fabrics.*)
Text processors, 13:281. (See also *Word
  processors.*)
Tga. See *Thermogravimetric analysis.*
Thalassemia, 15:595
Thallium [7440-28-0]
  electrode, 13:7
  as lubricant, 14:508
Thallium-203 [14280-48-9], 13:842
Thallium-205 [14280-49-0], 13:842
Thallium molybdate [34128-09-1], 15:687
Thallium nitride [12033-67-9], 15:872

Thallium sulfate [7446-18-6]
  as insecticide, 13:422
Thallous chloride [13453-32-2]
  electrode, 13:7
Thermal bonding
  spunbondeds, 16:86
Thermal conversion processes, 13:193
Thermal detectors, 13:340
Thermal insulation, 13:591
Thermal measurement, 13:347
Thermistor bolometer, 13:340
Thermocouple detectors, 13:340
Thermography, 13:337
  liquid crystals in, 14:395
Thermogravimetric analysis (tga)
  of insulation materials, 13:546
Thermolysin [9073-78-3]
  effect on LHRH, 15:771
  zinc in, 15:590
Thermomagnetostrictive recording, 14:705
Thermomechanical pulp (TMP), 16:769
Thermometers
  encapsulated liquid crystals as, 15:491
  mercury in, 15:155
Thermomolding
  polypropylene, 16:467
Thermonuclear fusion
  laser assisted, 14:76
Thermopile, 13:341
Thermoplast brilliant yellow 10G [2744-50-5],
  14:553
Thermoplastic elastomers (TPE)
  in wire and cable coverings, 13:587
Thermoplastic materials
  in composites, 13:973
Thermoplastics
  in latices, 14:82
Thermotropics
  liquid crystals, 14:396
Thesaurus-controlled vocabulary
  in information systems, 16:913
Thespesia populnea, 16:251
THF. See *Tetrahydrofuran*.
2H-1,2,6-Thiadiazine-3,5(4H,6H)-dione, 1,1-
  dioxide [68556-58-1], 14:800
Thiamine [67-03-8], 16:260
  in dextran production, 15:443
  sulfur in, 15:587
Thiamine pyrophosphate [154-87-0]
  magnesium requirement, 15:583
Thiaxanthone [492-22-8]
  as photoinitiator, 13:368
2-(4-Thiazodyl)benzimidazole [148-79-8]
  applications of, 13:245
Thiazolidines
  from aziridines, 13:150

2-Thiazolidinethione [96-53-7], 13:155
Thickeners
  in greases, 14:502
Thimerosal [54-64-8], 15:169
Thin-layer chromatography (tlc)
  in diagnostic testing, 15:88
Thinsulate, 16:78
Thioamides
  from nitriles, 15:889
1,1'-Thiobis(aziridine) [2881-79-0], 13:150
4,4'-Thiobis(2-*tert*-butyl-5-methylphenol)[96-
  69-5]
  LDPE antioxidant, 16:414
2,2'-Thiobis(ethylamine) [871-76-1], 13:155
Thio-1,1-bis(2-naphthol) [17096-15-0], 15:734
4-Thiocyanatobutyl methacrylate [76392-31-9]
  properties, 15:355
Thiocyanatomethyl methacrylate [76392-29-5]
  properties, 15:355
3,3'-Thiodipropionic acid [30421-51-3]
  as stabilizer, 13:60
Thioflavine T [2390-54-7], 14:546
Thioglycerol [96-27-5]
  in N_2-reducing system, 15:959
Thioglycolate
  titration procedure, 13:731
6-Thioguanosine [85-31-4]
  microbial phosphorylation of, 15:467
Thioketenes, 13:878
Thiokinases
  magnesium requirement, 15:579
Thiokol, 16:727
Thiolacetic acid [507-09-5]
  isoprene reaction, 13:824
Thiomalic acid [70-49-5], 13:110.   (See also
  Mercaptosuccinic acid.)
Thionaphthene [95-15-8], 15:704
Thionyl chloride [7719-09-7]
  fumaric acid reaction, 14:773
  lactic ester reaction, 13:84
  nitro alcohol reaction, 15:913
Thiophene [110-02-1]
  isoprene reaction, 13:826
η^5-Thiophenechromium tricarbonyl [12078-15-8],
  16:593
Thiophenol [108-98-5]
  maleic anhydride reaction, 14:779
  malic acid reaction, 13:105
Thiophosgene [463-71-8]
  malonic ester reaction, 13:881
Thiophosphates
  as antiwear agents, 14:493
Thiostat, 13:243
Thiostop N, 13:243
Thiotepa [52-24-4], 13:472
Thiourea [62-56-6]
  aziridine reaction, 13:155

Thioxanthone [492-22-8]
  as photosensitizer, 13:370
Thiram [137-26-8], 13:475
THMs.   See *Trihalomethanes.*
Thompson, J. J., 13:838
Thompson-Stewart process
  for lead carbonate, 14:171
Thorex process
  fuel reprocessing, 16:173
Thoria [1314-20-1], 15:345.   (See also *Thorium dioxide.*)
  dispersion-strengthened materials, 15:345
Thorium [7440-29-1]
  as fission fuel, 16:147
  in magnesium alloys, 14:596
  resources, 16:148
  uranium-233 from, 16:169
Thorium-232 [7440-29-1], 13:842; 16:188
  in FBRS, 16:187
  radiation exposure from, 16:217
  reprocessing of, 16:173
Thorium-233 [15720-66-8], 16:188
Thorium, compd with cobalt [12017-70-0] (1:5), 14:676
Thorium, compd with cobalt [12187-56-3] (2:17), 14:676
Thorium, compd with nickel [12423-55-1] (2:17)
  crystal structure, 14:676
Thorium dioxide [1314-20-1]
  as fission fuel, 16:140
Thorium nitride [12033-65-7]
  properties, 15:874
  reactors of, 15:882
Thorium nitride [12033-65-7] (1:1), 15:872
Thorium nitride [12033-90-8] (3:4), 15:872
Thornel 75 (T75) fiber
  in metal laminates, 13:945
Thread sealants, 16:725, 732
Three Mile Island accident, 16:222
Three Mile Island nuclear accident, 16:138
Three Mile Island nuclear power station
  core damage at, 16:233
DL(−) Threonamide [74464-43-0], 13:101
L(−) Threonamide [74421-65-0], 13:101
Threonic acid [3909-12-4], 13:101
D(−) Threonic acid, γ-lactone [23732-41-4], 13:101
L(−) Threonic acid, γ-lactone [21730-93-8], 13:101
Threonine [72-19-5]
  in nuts, 16:252
Threshold limit value, 13:258
Threshold shift
  noise-induced, 16:2
Thuja occidentalis L., 16:328
Thuja oil [8007-20-3], 16:328

β-Thujaplicine [499-44-5], 16:328
α-Thujene [2867-05-2], 16:313
Thujone [546-80-5]
  from Dalmatian sage, 16:327
  toxicity, 16:320
Thulium [7440-30-4]
  from the ocean, 16:280
Thulium-169 [7440-30-4], 13:842
Thulium, compd with cobalt [12214-15-2] (1:5), 14:676
Thulium, compd with cobalt [12052-80-1] (2:17), 14:676
Thullium nitride [12033-68-0], 15:872
Thyme oil [8007-46-3], 16:319, 328
Thymine [65-71-4]
  in nucleic acid, 15:587
Thymol [89-83-8], 16:328
Thymol blue [76-61-9], 13:10
Thymolphthalein [125-20-2], 13:10
Thymus
  atrophy from immunological deficiency, 13:170
Thymus zygis L. var. *gracilis* Bois, 16:328
Thyroglobulin [9010-34-8], 13:673
Thyroid gland
  iodine effect on, 15:593
  iodine requirements, 13:672
Thyroid hormones, 15:593
Thyroid-stimulating hormone [9002-71-5], 15:594
  zinc effect on, 15:590
Thyrotropin [9002-71-5], 13:673
  activity, 15:594
  somatostatin effect on, 15:780
Thyrotropin-releasing hormone [24305-27-9] (TRH)
  as neuroregulator, 15:781
Thyroxin [51-48-9]
  iodine requirement, 15:593
Thyroxine [51-48-9], 13:673; 15:594.   (See also *T₄*.)
  determination of by ion exchange, 13:704
L-Thyroxine [51-48-9]
  from animal thyroid glands, 15:72
Tiber hybrids, 13:947
Ticks
  effect on hides, 14:208
Ticonal, 14:671
Tiffen Decamired filters, 16:541
Tiffen filters, 16:541
Tiffen photar filters, 16:528
Tiles
  acoustic performance of, 13:517
Timbers
  laminated, 14:35
Time
  standardization of, 15:39

Time-censored life tests, *15*:21
Timepieces
  radiation exposure from, *16*:217
Time-weighted average, *13*:258
Tin [*7440-31-5*]
  in coatings, *15*:243
  emissivity, *13*:342
  lead alloys, *14*:150, 154
  as lubricant, *14*:508
  in magnesium alloys, *14*:596
  as nutrient, *15*:598
  from the ocean, *16*:280
  offshore deposits, *16*:285
  surface treatment of, *15*:307
Tin-112 [*15125-53-8*], *13*:841
Tin-114 [*14998-72-2*], *13*:841
Tin-115 [*14191-72-1*], *13*:841
Tin-116 [*14191-70-9*], *13*:841
Tin-117 [*13981-59-4*], *13*:841
Tin-118 [*14914-65-9*], *13*:841
Tin-119 [*14314-35-3*], *13*:841
Tin-120 [*14119-17-6*], *13*:841
Tin-122 [*14119-18-7*], *13*:841
Tin-124 [*14392-29-1*], *13*:841
Tinactin, *15*:735
Tinaderm, *15*:735
Tin alkyls, *16*:573
Tin-containing polymers, *15*:205
Tinctures
  in perfumes, *16*:949
Tin dichloride [*7772-99-8*]
  acrylate reaction, *16*:577
Tin esters, *16*:577
Tin(II) chloride [*7772-99-8*]
  p-nitrochlorobenzene reaction, *15*:924
  toxicity, *15*:574
Tin nitride [*75790-61-3*] (3:2), *15*:872
Tin nitride [*12142-83-5*] (3:4), *15*:872
Tinstone. See *Cassiterite.*
Tin tetrachloride [*7464-78-8*]
  alkylation, *16*:573
Tin tetrachloride pentahydrate [*10026-06-9*]
  isoprene reaction, *13*:821
TIP. See *Total isomerization process.*
Titanium [*7440-32-6*]
  as an anode, *15*:173
  cladding of, *15*:281
  in coatings, *15*:241
  in explosive cladding, *15*:278
  in integrated circuits, *13*:641
  in metallic coatings, *15*:243
  nondestructive testing, *16*:51
  from the ocean, *16*:279, 292
  platinum cladding of, *15*:172
  from titanium chloride, *14*:573
Titanium-46 [*14304-89-3*], *13*:840

Titanium-47 [*14304-90-6*], *13*:840
Titanium-48 [*14304-91-7*], *13*:840
Titanium-49 [*14304-92-8*], *13*:840
Titanium-50 [*14304-93-9*], *13*:840
Titanium alkyls, *16*:584
Titanium–aluminum clads, *15*:275
Titanium–aluminum–tin alloy (1:5:2.5)
  stress–strain diagram, *15*:327
Titanium carbide [*12070-08-5*], *15*:264
  in metallic coating, *15*:252
Titanium chloride [*56780-56-4*]
  in coatings, *15*:241
Titanium dioxide [*13463-67-7*]
  in anode coatings, *15*:176
  as delustering agent, *16*:111
  filler in plastics, *13*:973
  in inks, *13*:377, 390
  olefin fiber dye, *16*:361
  paint pigment, *16*:744
  in papermaking, *16*:819
  paper pigment, *16*:777
Titanium(III) chloride [*7705-07-9*]
  nitrobenzene analysis, *15*:922
Titanium(III) sulfate [*10343-61-0*]
  nitrobenzene analysis, *15*:921
Titanium(IV) chloride [*10049-06-6*]
  nitrogen reaction, *15*:934
Titanium lactate [*58676-69-0*], *13*:88
Titanium nitride [*25583-20-4*], *15*:264
  in integrated circuits, *13*:641
  microwave application, *15*:516
  properties, *15*:874
Titanium nitride [*25583-20-4*] (1:1), *15*:872
Titanium nitride [*12169-08-3*] (2:1), *15*:872
Titanium oxide [*13463-67-7*]
  in metallic coating, *15*:252
Titanium oxide hydrate [*12214-43-6*]
  uranium adsorbent, *16*:284
Titanium phosphate [*13765-94-1*]
  in metal treating, *15*:306
Titanium–steel clads, *15*:275
Titanium tetrachloride [*7550-45-0*]
  encapsulation co-catalyst, *15*:482
  as initiator catalyst, *13*:371
  in isoprene polymerization, *13*:818, 827
  in LLDPE catalyst, *16*:391
  magnesium reaction, *14*:573
  in nitride deposition, *15*:878
  nitrogen reduction catalyst, *15*:954
  propylene polymerization catalyst, *16*:460
  removal from silicon tetrachloride, *15*:906
Titanium tetraiodide [*7720-83-4*]
  polymerization catalyst, *13*:672
Titanium trichloride [*7705-07-9*]
  1-butene polymerization with, *16*:477
  nitrogen reduction catalyst, *15*:958
  propylene polymerization cocatalyst, *16*:458

Titanocene [1271-29-0]
 polyethers of, 15:209
Titanocene dichloride [1271-19-8]
 polymers from, 15:209
Titanocene polymers, 15:209
Titano-niobates, 15:820
TITLEX information system, 13:331
Titration
 electrodes for, 13:731
Titration, acid-base
 for pH, 13:1
TITUS
 information service, 16:902
TLV. See *Threshold limit value.*
T-lymphocytes, 13:168
TMEDA. See *N,N,N′,N′-*
 Tetramethylethylenediamine.
TMI. See *Three Mile Island (Nuclear Power Station).*
TML. See *Tetramethyllead.*
TMM. See *Trimethylolmelamine.*
TMP. See *Thermomechanical pulp.*
TMSN. See *Tetramethylsuccinonitrile.*
TMV. See *Tobacco mosaic virus.*
TNO
 patent information, 16:925
TNT. See *Trinitrotoluene.*
Toads
 bufotenine in, 13:220
Tobacco
 binding agent, 15:454
Tobacco budworm
 control of with pheromones, 13:480
Tobacco mosaic virus
 liquid-crystal phases of, 14:420
Tobacco moth
 control of with pheromones, 13:481
Tobias acid [81-16-3], 15:729. (See also *2-Amino-1-naphthalenesulfonic acid.*)
α-Tocopherol [59-02-9]
 nitrosamine inhibition, 15:989
Toggling, 14:220
Toilet odors, 16:305
Tokuyama soda process, 13:910
Tolazamide [1156-19-0], 13:616
Tolbutamide [64-77-7], 13:617
 antidiabetic agent, 13:614
o-Tolidine [119-93-7]
 health effect, 13:273
 in methaqualone synthesis, 13:125
Tolidine diisocyanate [91-97-4]
 properties, 13:789
Tolinase, 13:617
Tolmetin [76171-23-3]
 immunologic effects, 13:174
Tolnaftate [2398-96-1], 15:735

Tolu balsam oil [8024-03-1], 16:308
Toluene [108-88-3]
 ammoxidation, 15:905
 control of nitration ratios, 15:929
 isoprene reaction, 13:825
 nitration, 15:928
 odor-problem chemical, 16:303
 in paint and varnish removers, 16:764
 in TDI manufacture, 13:804
 toxicity, 13:273
Toluenechromium tricarbonyl [12083-24-8], 16:608
 acylation, 16:607
Toluene diisocyanate [26471-62-5], 13:794; 15:931
 in encapsulation, 15:480
 mfg, 13:804
 toxicity, 13:273
2,4-Toluene diisocyanate [584-84-9]
 properties, 13:789
2,6-Toluene diisocyanate [91-08-7], 13:801
65/35-Toluene diisocyanate [26471-62-5]
 properties, 13:789
80/20-Toluene diisocyanate [26471-62-5]
 properties, 13:789
Toluene diisocyanate dimer [26747-90-0], 13:794
o-Toluenesulfonamide [88-19-7]
 resin from in fluorescent product, 14:560
 saccharin from, 14:863
p-Toluenesulfonamide [70-55-3]
 diethyl malonate from, 14:802
 ethyl chloroformate reaction, 13:617
 resin from in fluorescent product, 14:560
p-Toluenesulfondichloramide [473-34-7]
 applications of, 13:230
p-Toluenesulfonyl isocyanate [4083-64-1]
 diethyl malonate from, 14:802
p-Toluic acid [99-94-5]
 microbial preparation, 15:460
m-Toluidine [108-44-1], 15:927
o-Toluidine [95-53-4], 15:926
p-Toluidine [106-49-0], 15:928
p-Tolyldiiodomethyl sulfone [20018-09-1]
 applications of, 13:230
Tolyl red [6410-10-2], 13:10
N-(p-Tolylsulfonyl)molonimide [1888-29-5]
 diethyl malonate from, 14:802
Tomato paste
 analysis of with ISES, 13:732
Tomato soup
 analysis of with ISES, 13:730
Tonalid [21145-77-7]
 in perfumes, 16:968
Toners
 electric, 13:396
 fluorescent, 14:561
 in inks, 13:378

Toolebuc formation, *16*:347

Tools
    magnesium in, *14*:611
Tororo-aoi, *16*:806. (See also *Hibiscus manihot root*.)
Torreya nucifera, *16*:250
Torreya nut, *16*:250
Tortuosity factor
    of membranes, *15*:98
TOSCA. See *Toxic Substances Control Act*.
Tosco II retort, *16*:340
Total isomerization process, *15*:667
Totes, *14*:231
Towels
    nonwoven, *16*:123
Toxaphene [*8001-35-2*]
    toxicity, *13*:433
Toxic concentration low, *13*:258
Toxic dose low, *13*:258
Toxicity, *13*:253; *14*:885. (See also *Industrial hygiene and toxicology*.)
    acephate, *13*:443
    acetone, *13*:901
    acetone cyanohydrin, *15*:894
    acetonitrile, *14*:806; *15*:897
    acetylcholine chloride, *15*:756
    acrylamide, *13*:263
    acrylonitrile, *13*:263, 466
    adiponitrile, *15*:900
    aflatoxins, *16*:263
    aldicarb, *13*:453
    aldrin, *13*:434
    aldrin/dieldrin, *13*:263
    alkanes (C_5–C_8), *13*:263
    allethrin, *13*:458
    allethrin isomers, *13*:458
    allyl chloride, *13*:263
    allyl glycidyl ether, *13*:268
    aminocarb, *13*:455
    amitraz, *13*:463
    ammonia, *13*:263
    ammonium vanadate, *15*:574
    anise oil, *16*:319
    antimony, *13*:263
    apholate, *13*:473
    arsenic, *13*:263
    arsenic trioxide, *13*:421
    asarone, *16*:320
    asbestos, *13*:263
    azinphos methyl, *13*:441
    aziridine, *13*:160
    barium fluorosilicate, *13*:421
    bay oil, *16*:319
    bendiocarb, *13*:453
    benzene, *13*:264
    benzil, *13*:476
    benzonitrile, *15*:905
    benzoyl peroxide, *13*:264
    benzyl benzoate, *13*:476

    benzyl chloride, *13*:264
    bergamot oil, *16*:319
    beryllium, *13*:264
    biopermethrin, *13*:458
    bioresmethrin, *13*:458
    2,3,4,5-bis(butyl-2-ene) tetrahydrofurfural, *13*:476
    bitter almond oil, *16*:320
    bois de rose oil, *16*:319
    bromophos, *13*:440
    2-(2-butoxyethyl)-ethylthiocyanate, *13*:429
    butoxypolypropylene glycol, *13*:476
    N-butylacetanilide, *13*:476
    t-butylaminoethyl methacrylate, *15*:372
    n-butyl 6,6-dimethyl-5,6-dihydro-1,4-pyrone-2-carboxylate, *13*:476
    1,3-butylene dimethacrylate, *15*:372
    n-butyl glycidyl ether, *13*:268
    butyl isocyanate, *13*:812
    n-butyl methacrylate, *15*:372
    cade, *16*:320
    cadmium, *13*:264
    calamus, *16*:320
    calcium acetate monohydrate, *15*:574
    calcium arsenate, *13*:421
    calcium arsenate, basic, *13*:421
    calcium carbonate, *15*:574
    calcium chloride, *15*:574
    calcium fluoride, *15*:574
    calcium molybdate, *15*:574
    carbaryl, *13*:264, 453
    carbofuran, *13*:453
    carbon black, *13*:264
    carbon dioxide, *13*:264
    carbon disulfide, *13*:466
    carbon monoxide, *13*:265
    carbon tetrachloride, *13*:265
    carbophenothion, *13*:443
    carbophenothion methyl, *13*:443
    carbromal, *13*:125
    cassia, redistilled oil, *16*:320
    of catechol, *13*:55
    cevine, *13*:427
    chloral hydrate, *13*:124
    chlorbenside, *13*:463
    β-chlordane, *13*:434
    chlordecone, *13*:436
    chlordimeform, *13*:463
    chlorfenethol, *13*:461
    chlorfenvinphos, *13*:443
    chlorine, *13*:265
    chlorobenzilate, *13*:461
    chloronaphthalenes, *15*:703
    p-chlorophenyl diethyl phosphate, *13*:451
    m-chlorophenyl *N*-methylcarbamate, *13*:452
    o-chlorophenyl *N*-methylcarbamate, *13*:452
    chloropicrin, *13*:466
    chloroprene, *13*:265
    chlorpyrifos, *13*:440
    chlorpyrifos methyl, *13*:440

Toxicity (*Continued*)
chlorthion, *13*:439
chromic acid, *13*:265
chromium(III) chloride, *15*:574
chromium(III) chloride hexahydrate, *15*:574
chromium(VI), *13*:265
cinnamon ceylon oil, *16*:319
coal-tar products, *13*:265
coke-oven emissions, *13*:266
copper acetoarsenite, *13*:421
copper arsenate, basic, *13*:421
copper(II) chloride, *15*:574
copper sulfate, *15*:574
copper sulfate pentahydrate, *15*:574
coriander oil, *16*:320
cotton dust, *13*:266
coumaphos, *13*:448
crotoxyphos, *13*:448
crufomate, *13*:448
cryolite, *13*:421
cyanoacetic acid, *15*:907
cyanogen chloride, *14*:806
p-cyanophenyl diethyl phosphate, *13*:451
cyclodienes, *13*:437
DDT, *13*:266, 431
DDT analogues, *13*:430
decahydronaphthalene, *15*:705
decalin, *15*:705
decamethrin, *13*:458
decamethyl triphosphoramide, *13*:445
demeton, *13*:446
demeton methyl, *13*:446
dermal, *13*:260
desmodur N, *13*:812
dialifor, *13*:443
diazinon, *13*:441
dibromochloropropane, *13*:266
dibutyl adipate, *13*:476
dibutyl phthalate, *13*:476
di-*n*-butyl succinate, *13*:476
di-*n*-butyltin oxide, *16*:578
dicapthon, *13*:439
dicarboximide, *13*:425
dichlorvos, *13*:442
dicofol, *13*:461
dicrotophos, *13*:446
dieldrin, *13*:435
di-2,3-epoxypropyl ether, *13*:268
diethyltin oxide, *16*:578
N,*N*-diethyl-*m*-toluamide, *13*:477
diflubenzuron, *13*:461
β-dihydroheptachlor, *13*:434
diisocyanates, *13*:267
dimefax, *13*:445
dimethoate, *13*:446
dimethrin, *13*:458
2-dimethylaminoethyl methacrylate, *15*:372
dimethyl carbate (*cis*-dimethylbicyclo-[2.2.1]-
5-heptene-2,3-dicarboxylate), *13*:477

dimethyl phthalate, *13*:477
dimetilan, *13*:455
dinitrocresol, *13*:428
dinitro-*o*-cresol, *13*:267
dinocap, *13*:429
dinoseb, *13*:429
di-*n*-octyltin oxide, *16*:578
dioxacarb, *13*:453
dioxane, *13*:267
dioxathion, *13*:443
di-*n*-propyl isocinchomeronate, *13*:477
disodium itaconate, *13*:870
disulfoton, *13*:447
endosulfan, *13*:435
endrin, *13*:435
epichlorohydrin, *13*:267
essential oil, *16*:319
ethion, *13*:445
2-ethyl-2-butyl-1,3-propanediol, *13*:477
ethyl cyanoacetate, *15*:908
ethylene dibromide, *13*:267, 467
ethylene dichloride, *13*:267, 467
ethylene oxide, *13*:268, 467
2-ethyl-1,3-hexanediol, *13*:477
ethyl methacrylate, *15*:372
ethyl *p*-nitrophenyl amylphosphonate, *13*:452
ethyl *p*-nitrophenyl *n*-butylphosphonate, *13*:
452
ethyl *p*-nitrophenyl *t*-butylphosphonate, *13*:
452
ethyl *p*-nitrophenyl ethylphosphonate, *13*:452
ethyl *p*-nitrophenyl isopropylphosphonate, *13*:
452
ethyl *p*-nitrophenyl methylphosphonate, *13*:
452
O-ethyl *O*-*p*-nitrophenyl
phenylphosphonothioate, *13*:443
ethyl *p*-nitrophenyl *n*-propylphosphonate, *13*:
452
2-ethyl-2-nitro-1,3-propanediol, *15*:915
eucalyptus oil, *16*:319
fenitrothion, *13*:439
fenson, *13*:463
fenthion, *13*:439
fenvalerate, *13*:458
fluorides, inorganic, *13*:268
fonofos, *13*:448
formaldehyde, *13*:268
formetanate, *13*:455
fospirate, *13*:440
fumaric acid, *14*:786
genite, *13*:463
geranium, algerian oil, *16*:320
geranium oil, *16*:319
glycidyl ethers, *13*:268
glycidyl methacrylate, *15*:372
HDI, *13*:812
HEMPA, *13*:473
heptachlor, *13*:434
heptachlor epoxide, *13*:434

2,2,5-*endo*-6-*exo*-8,9,10-heptachlorobornane, *13*:434
higher olefins, *16*:495
histamine, *15*:769
hydrazines, *13*:268
hydrogen cyanide, *13*:467
hydrogen fluoride, *13*:268
hydrogen sulfide, *13*:269
hydroprene, *13*:459
hydroquinone, *13*:269
of hydroquinone, *13*:55
p-hydroxybenzaldehyde, *13*:75
2-hydroxyethyl methacrylate, *15*:372
2-hydroxymethyl-2-nitro-1,3-propanediol, *15*: 915
3-hydroxy-2-naphthalenecarboxylic acid, *15*: 746
hydroxypropyl methacrylate, *15*:372
indium, *13*:209
iodine, *13*:660
iodofenfos, *13*:440
IPDI, *13*:812
iris, concrete oil, *16*:320
iron, *15*:595
iron(II) chloride tetrahydrate, *15*:574
iron(III) chloride heptahydrate, *15*:574
iron(III) oxide, *13*:787
iron(II) sulfate, *15*:574
iron(II) sulfate heptahydrate, *15*:574
iron oxide, *14*:749
isobornyl methacrylate, *15*:372
isobornyl thiocyanate, *13*:429
isocyanates, *13*:811
isodecyl methacrylate, *15*:372
D-*o*-isopentylphenyl *N*-methylcarbamate, *13*: 455
L-*o*-isopentylphenyl *N*-methylcarbamate, *13*: 455
isophthalonitrile, *15*:906
o-isopropoxyphenyl *N*-methylcarbamate, *13*: 452
isopropyl alcohol, *13*:269
m-isopropylphenyl *N*-methylcarbamate, *13*:455
p-isopropylphenyl *N*-methylcarbamate, *13*:455
itaconic acid, *13*:870
ketene, *13*:882
ketones, *13*:269, 901
lauryl methacrylate, *15*:372
lavender oil, *16*:319, 320
lead, *13*:269; *14*:197
lead acetate, *14*:198
lead arsenate, acid, *13*:421
lead arsenate, basic, *13*:421
lead salts, *14*:175
lemongrass oil, *16*:319
lemon oil, *16*:319
lindane, *13*:433
lube oils, *14*:523
mace, *16*:319
magnesium chloride, *15*:574

magnesium chloride hexahydrate, *15*:574
magnesium hexafluorosilicate hexahydrate, *15*: 574
magnesium sulfate, *15*:574
magnetic tape components, *14*:749
malathion, *13*:270, 441
maleic acid, *14*:786
maleic anhydride, *14*:786
malonic acid, *14*:796
malononitrile, *14*:806
manganese, *14*:841
manganese acetate, *15*:574
manganese acetate tetrahydrate, *15*:574
MDI, *13*:812
mercury, *13*:270; *15*:153
mercury compounds, *15*:166
mesityl oxide, *13*:918
metalkamate, *13*:456
methacrylic acid, *15*:372
methallyl chloride, *13*:467
methamidophos, *13*:443
methanol, *15*:411
methiocarb, *13*:455
methomyl, *13*:453
methoprene, *13*:459
methoxychlor, *13*:432
p-methoxylphenyl diethyl phosphate, *13*:451
m-methoxyphenyl *N*-methylcarbamate, *13*:452
o-methoxyphenyl *N*-methylcarbamate, *13*:452
methyl alcohol, *13*:270
methyl bromide, *13*:467
N-methylcarbamates, *13*:456
methyl chloromethyl ether, *15*:704
4,4′-methylene-bis(2-chloroaniline), *13*:270
methylene chloride, *13*:270; *16*:767
methylmercury, *15*:167
methyl methacrylate, *15*:372
1-methyl-1-nitro 2,2-bis(*p*-chlorophenyl)-
ethane, *13*:430
2-methyl-2-nitro-1,3-propanediol, *15*:915
2-methyl-2-nitro-1-propanol, *15*:915
methyl parathion, *13*:270, 438
p-methylphenyl diethyl phosphate, *13*:451
m-methylphenyl *N*-methylcarbamate, *13*:452
p-methylsulfonylphenyl diethyl phosphate,
13:451
p-methylsulfoxylphenyl diethyl phosphate, *13*: 451
p-methylthiophenyl diethyl phosphate, *13*:451
m-methylthiophenyl *N*-methylcarbamate, *13*: 452
o-methylthiophenyl *N*-methylcarbamate, *13*: 452
mevinphos, *13*:446
mexacarbate, *13*:455
mineral nutrients, *15*:598
mirex, *13*:436
molybdenum, *15*:679
molybdenum compounds, *15*:679
monocrotophos, *13*:447

Toxicity (*Continued*)
  mononitrotoluenes, *15*:930
  monosodium vanadate, *15*:574
  morphothion, *13*:447
  mycotoxins, *16*:263
  naled, *13*:442
  naphthalene, *15*:714
  1-naphthaleneamine, *15*:726
  2-naphthaleneamine, *15*:727
  2-naphthalenol, *15*:734
  2-naphthylamine, *15*:726, 728
  neotenin, *13*:459
  nickel, *13*:270
  nickel acetate, *15*:574
  nickel carbonyl, *13*:270; *15*:812
  niobium, *15*:830
  nitric acid, *15*:869
  nitric oxide, *15*:869
  nitrobenzene, *15*:922
  2-nitro-1-butanol, *15*:915
  nitrochlorobenzenes, *15*:924
  nitrogen dioxide, *15*:869
  nitrogen, oxides, *13*:271
  nitroglycerin, *13*:271
  nitroparaffins, *15*:981
  p-nitrophenyl diethyl phosphate, *13*:451
  nitrosamines, *15*:991
  N-nitrosodicyclohexylamine, *15*:991
  N-nitrosodiethanolamine, *15*:991
  N-nitrosodiethylamine, *15*:991
  N-nitrosodimethylamine, *15*:991
  N-nitrosodipropylamine, *15*:991
  N-nitrosomethyl(benzyl)amine, *15*:991
  N-nitrosomorpholine, *15*:991
  N-nitrosonornicotine, *15*:991
  N-nitrosopiperidine, *15*:991
  N-nitrosoproline, *15*:991
  N-nitrosopyrrolidine, *15*:991
  nutmeg, *16*:319
  nutmeg oil, *16*:319
  2,2,5-*endo*-6-*exo*-8,9,9,10-octachlorobornane,
    13:434
  octadecyl isocyanate, *13*:812
  orange oil, *16*:319
  organomercury compounds, *16*:587
  organotin compounds, *13*:271
  organotins, *16*:578
  ovex, *13*:463
  oxalic acid, *16*:630
  oxamyl, *13*:453
  oxydemeton methyl, *13*:446
  oxythioquinox, *13*:463
  ozone, *16*:699
  paramethadione, *13*:137
  parathion, *13*:271, 438
  pentac, *13*:463
  permethrin, *13*:458
  perthane, *13*:432
  phenol, *13*:271
  phenothrin, *13*:458

  2-phenylcyclohexanol, *13*:477
  phenyl diethyl phosphate, *13*:451
  phenyl diethyl phosphates, *13*:451
  phenylethylamine hydrochloride, *15*:775
  phenyl ethyl phosphonates, *13*:452
  phenyl isocyanate, *13*:812
  phenylmercury compounds, *15*:163
  phenyl *N*-methylcarbamate, *13*:452
  phorate, *13*:447
  phosgene, *13*:271
  phosmet, *13*:443
  phosphamidon, *13*:446
  phosphorus, *15*:2
  piperidine, *15*:776
  piperonyl butoxide, *13*:425
  piperonyl sulfoxide, *13*:425
  plutonium nitride, *15*:881
  polychlorinated biphenyls, *13*:271
  potassium acetate, *15*:574
  potassium carbonate, *15*:574
  potassium chloride, *15*:574
  potassium fluoride, *15*:574
  potassium hexafluorosilicate, *15*:574
  potassium iodide, *15*:574
  propargite, *13*:463
  prophos, *13*:443
  propoxur, *13*:453
  n-propyl *N*,*N*-diethylsuccinamate, *13*:477
  propyl isome, *13*:425
  pyrimicarb, *13*:455
  refined petroleum solvent, *13*:271
  repellents, *13*:476
  resmethrin, *13*:458
  of resorcinol, *13*:55
  ronnel, *13*:439
  rose oil, *16*:319
  rosewood oil, *16*:320
  ryania, *13*:427
  safrole, *16*:320
  salicylaldehyde, *13*:75
  sandalwood oil, *16*:319
  sassafras, *16*:320
  schradan, *13*:445
  selenium, *15*:591
  silica, crystalline, *13*:271
  sodium acetate, *15*:574
  sodium arsenite, *13*:421
  sodium bicarbonate, *15*:574
  sodium carbonate, *15*:574
  sodium chloride, *15*:574
  sodium fluoride, *13*:421; *15*:574
  sodium fluoroaluminate, *15*:574
  sodium fluorosilicate, *13*:421
  sodium hexafluorosilicate, *15*:574
  sodium hydrogen malonate, *14*:796
  sodium hydroxide, *13*:271
  sodium iodide, *15*:574
  sodium metasilicate, *15*:574
  sodium monohydrogen phosphate
    heptahydrate, *15*:574

sodium phosphomolybdate, *15*:574
sodium selenate, *15*:574
sodium selenite, *15*:574
sodium silicate (2:1), *15*:574
sodium silicate (3:1), *15*:574
stirofos, *13*:443
substituted phenyl *N*-methylcarbamates, *13*:452
sulfotepp, *13*:438
sulfur dioxide, *13*:271
sulfuric acid, *13*:271
sulphenone, *13*:463
2,4-TDI, *13*:812
telodrin, *13*:434
temephos, *13*:440
TEPA, *13*:473
terbufos, *13*:447
1,1,2,2-tetrachloroethylene, *13*:271
tetradifon, *13*:463
tetraethyllead, *16*:582
tetraethyl pyrophosphate, *13*:438
1,2,3,4-tetrahydronaphthalene, *15*:705
tetralin, *15*:705
tetramethyllead, *16*:582
tetramethylsuccinonitrile, *15*:904
thorium nitride, *15*:881
thuja oil, *16*:328
thujone, *16*:320
thyme oil, *16*:319
tin(II) chloride, *15*:574
toluene, *13*:273
toluene diisocyanate, *13*:273
toxaphene, *13*:433
1,1,1-tribromo 2,2-bis(*p*-chlorophenyl)ethane, *13*:430
1,1,1-trichloro 2,2-bis(*p*-bromophenyl)ethane, *13*:430
1,1,1-trichloro 2,2-bis(*p*-chlorophenyl)ethane, *13*:430
1,1,1-trichloro 2,2-bis(*p*-fluorophenyl)ethane, *13*:430
1,1,1-trichloro 2,2-bis(*p*-methylphenyl)ethane, *13*:430
1,1,1-trichloro 2,2-bis(*p*-phenylethane), *13*:430
1,1,1-trichloro 2-chloro-2,2-bis(*p*-chlorophenyl)-ethane, *13*:430
1,1,1-trichloro 2-deutero-2,2-bis(*p*-chlorophenyl)ethane, *13*:430
1,1,1-trichloroethane, *13*:273
trichloroethylene, *13*:273
triclorfon, *13*:441
tricyclohexyltin hydroxide, *13*:463
trimethadione, *13*:137
1,1,1-trimethyl 2,2-bis(*p*-chlorophenyl)ethane, *13*:430
trimethylolpropane trimethacrylate, *15*:372
triphenylphosphine, *16*:649
trisodium vanadate, *15*:574
tungsten, *13*:274

tungsten carbide, *13*:274
uranium nitride, *15*:881
vanadium, *13*:274
vanadyl chloride, *15*:574
VAZO 52, *15*:903
VAZO 64, *15*:903
vinyl chloride, *13*:274
vinyl halides, *13*:274
waste anesthetic gases and vapors, *13*:274
wintergreen oil, *16*:319
wormwood, *16*:320
xylene, *13*:274
zinc acetate dihydrate, *15*:574
zinc chloride, *15*:574
zinc fluorosilicate, *15*:574
zinc oxide, *13*:274
zinc sulfate, *15*:574
zinc sulfate heptahydrate, *15*:574
Toxicology.   See *Industrial hygiene and toxicology.*
Toxic Substances Control Act, *13*:184, 254
Toxilic acid [*110-16-7*], *14*:770.   (See also *Maleic acid.*)
Toxilic anhydride [*108-31-6*], *14*:770.   (See also *Maleic anhydride.*)
Toyobo Spun-Bond, *16*:78
TPE.   See *Thermoplastic elastomers.*
TPX [poly(4-methyl-1-pentene)], *16*:478
Traceability
   measurements to standards, *15*:38
Tracers
   isotopes as, *13*:859
Trachoma, *13*:413
Track dating, *16*:843
Track-etch detector, *16*:827
Track-etch detectors
   calibration of, *16*:833
Track-etch membranes, *16*:846
Track-registration threshold, *16*:831
Tractors
   noise source, *16*:13
Trademarks, *16*:851
Trade secrets, *16*:851
Tradlon, *13*:557
Traffic
   noise source, *16*:11
Traffic flow
   queuing theory applied to, *16*:520
Tramazoline hydrochloride [*3715-90-0*], *15*:727
Tranquilizer
   GHB as, *13*:98
Tranquilizers, *13*:122.   (See also *Hypnotics; sedatives.*)
Transalkylation
   aluminum alkyls, *16*:569
Transcobalamin II [*12651-28-4*], *15*:598
Transcyanohydrination, *15*:895

Transesterification
  in methacrylate mfg, 15:366
Transfer
  patent rights, 16:886
Transfer function, 13:485
Transfer molding
  for plastic composites, 13:976
Transferrin
  iron transport, 15:595
  in iron transport, 15:588
Transformer alloys, 14:668
Transformer oils, 13:551
Transformers
  magnetic materials for, 14:664
  nondestructive testing of, 16:57
Transglycosylation
  microbial, 15:467
Transistors
  field-effect, in electrodes, 13:729
  in ICS, 13:621
Transition-metal alkyls, 16:583
Transition metals
  metal π-complex formation, 16:592
Transitions
  electronic, 14:529
  radiative and nonradiative, 14:272
Transmethylase [9033-25-4], 15:597.   (See also
    Methyltransferase.)
Transmission loss
  sound, 13:516
Transmittance
  of optical filters, 16:522
Transpiration cooling, 13:598
Transportation
  noise source, 16:9
Transportation simplex method
  in operations planning, 16:510
Transuranic wastes, 16:206
Trapa bicornis, 16:251
Trapa bispinosa, 16:251
Trapa natans, 16:251
Travertine, 14:345
Tree extract
  as odor modifier, 16:301
Treeing, 13:543
Trees, electrochemical
  formation of in XLPE insulation, 13:569
Tremors
  (−) adrenaline induced, 15:758
Trevira 30/150, 16:76
Trevira-Spunbond, 16:79
TRH.   See Thyrotropin-releasing hormone.
sym-Triacetone dialcohol [3682-91-5], 13:914
Trialkyl aluminum
  in isoprene polymerization, 13:818
Triallyl cyanurate [101-37-1]
  in polymer curing, 13:567

Triamcinolone [124-94-7], 13:173
Triammonium hexakiscyanoferratae [14221-48-
    8], 13:768
Triammonium tris(oxalato)ferrate [14221-47-7],
    13:780
Triazines
  from nitriles, 15:892
Triazolam [28911-01-5]
  properties of, 13:122
Triazure, 13:172
Tribarium bis(hexacyanoferrate) [21729-04-4],
    13:769
Tribarium bis(hexakiscyanoferrate) [21729-04-4]
    (3:2)
  in ferricyanic acid synthesis, 13:768
Tribasic lead sulfate [12202-17-4], 14:167
Tribasic lead sulfate monohydrate [12397-06-7],
    14:161
Triboelectric effects
  in nondestructive tests, 16:53
Tribolium confusum, 13:465
Triboluminescence
  of phosphors, 14:527
Tribromoaniline [147-82-0]
  in bromine fixation, 16:283
1,1,1-Tribromo 2,2-bis(p-chlorophenyl)ethane
    [4399-08-0]
  as insecticide, 13:430
Tribromoborane [10294-33-4]
  IC dopant, 13:638
Tribromosalicylanilide [38848-67-8]
  applications of, 13:240
TRIBUT information system, 13:331
Tributylaluminum [1116-70-7]
  in LLDPE catalyst, 16:391
Tri-n-butylaluminum [1116-70-7]
  properties, 16:565
Tri-n-butylborane [122-56-5]
  properties, 16:562
Tri-sec-butylborane [1113-78-6]
  properties, 16:562
Tributyl citrate [77-94-1]
  paper coating defoamer, 16:791
Tributyl lead acetate [2587-82-8], 14:193
Tributyl[(1-oxo-2-propenyl)oxy]stannane,
    polymer with oxiranylmethyl-2-propenoate
    [65289-97-6], 15:205
Tributyl phosphate [126-73-8]
  degradation by radiation, 16:240
  indium extraction, 13:208
  in niobium processing, 15:825
  nuclear fuel reprocessing, 16:173
  in nuclear fuel reprocessing, 16:176
  paper coating defoamer, 16:791
Tri-n-butylphosphine-π-cyclopentadienylmethyl-
    nickel [7298-70-0], 15:809
3-Tributylphosphoranylidenedihydro-2,5-furan-
    dione [64750-72-7], 14:779

Tributyltin acrylate [*13331-52-7*]
 in antifouling paints, *16*:579
Tributyltin γ-amino-butyrate [*76082-06-9*], *15*:205
Tributyltin ω-amino-hexanoate [*76082-07-0*], *15*:205
Tributyltin ω-amino-undecanoate [*76082-08-1*], *15*:205
Tri-*n*-butyltin chloride [*1461-22-9*]
 immunodeficiency effect, *13*:170
 properties, *16*:573
Tributyltin fluoride [*1983-10-4*]
 applications of, *13*:236
 properties, *16*:573
Tributyltin oxide [*56-35-9*]
 as antifoulant, *13*:239
 applications of, *13*:236
 properties, *16*:573
Tributyltin sulfide [*4808-30-4*]
 as repellant, *16*:579
Tricalcium aluminate [*12042-78-3*], *14*:356
 in portland cement, *14*:376
Tricalcium phosphate [*7758-87-4*]
 from the ocean, *16*:289
Tricalcium silicate [*12168-85-3*]
 in portland cement, *14*:376
Tricaprylmethylammonium chloride [*5137-55-3*]
 in phase-transfer catalysts, *14*:862
Tricarboboxyl[(1,2,3,4,5-η)-1-vinyl-2,4-cyclopent-adien-1-yl]manganese, polymer with styrene [*35443-58-4*], *15*:189
Tricarbonylbis(tributylphosphine) cobalt [*41985-14-2*]
 oxo process catalyst, *16*:647
Tricarbonylbis(triphenylphosphine)ruthenium [*14741-36-7*]
 oxo process catalyst, *16*:639
Tricarbonyl[[(2,3,4,5-η)-2,4-hexadienyl]acrylate]-iron, polymer with methyl acrylate [*51796-23-7*], *15*:189
Tricarbonyl[[(2,3,4,5-η)-2,4-hexadienyl]2-propen-oate]iron [*51745-70-1*], *15*:187
Tricarbonyl[(1,2,3,4-η)-1,3,5-hexatriene]-iron, stereoisomer [*50277-87-7*], *15*:186
Tri-μ-carbonylnonacarbonyltetrarhodium [*19584-30-6*]
 oxo process catalyst, *16*:639
Tricarbonyl[2-(η⁶-phenyl)ethylmethacrylate]-chromium [*53761-68-5*], *15*:187, 189
Tricarbonyl[2-(η⁶-phenyl)ethyl 2-methyl-2-propenoate]chromium, polymer with methyl methacrylate [*53775-40-9*], *15*:189
Tricarbonyl[(η⁶-phenyl)methylacrylate]chrom-ium [*35004-37-6*], *15*:187
Tricarbonyl[(η⁶-phenylmethyl)acrylate]chrom-ium(2, *15*:189
Tricarbonyl(phosphorustrifluoride)nickel [*14264-32-5*], *15*:807

Tricarbonyl[(1,2,3,4,5,6-η-styrene]chromium, polymer with styrene [*35443-56-2*], *15*:189
Tricarbonyl(trimethylphosphite-*P*)nickel [*17099-58-0*], *15*:807
Tricarbonyl(triphenylarsine)nickel [*37757-32-7*], *15*:807
Tricarbonyl[(1,2,3,4,5,η)-vinylbenzene]chromium [*31870-79-8*], *15*:186
Tricarbonyl[(1,2,3,4,5-η)-1-vinyl-2,4-cyclopenta-diene-1-yl] manganese [*12116-27-7*], *15*:186, 189
Trichloracetaldehyde [*75-87-6*]
 carbonyl fluoride reaction, *13*:800
Trichlorfon [*52-68-6*], *13*:438, 441
Trichloroacetic acid [*76-83-9*]
 paint removers, *16*:763
Trichloroacetonitrile [*545-06-2*]
 as fumigant, *13*:467
Trichlorobenzene [*25323-29-2*]
 insulation properties of, *13*:551
1,2,3-Trichlorobenzene [*87-61-6*]
 wood preservative, *13*:480
1,2,4-Trichlorobenzene [*120-82-1*], *13*:432
 wood preservative, *13*:481
1,1,1-Trichloro 2,2-bis(*p*-bromophenyl)ethane [*2990-17-2*]
 as insecticide, *13*:430
1,1,1-Trichloro-2,2-bis(*o*-chlorophenyl)ethane [*50-29-3*]
 as insecticide, *13*:429
1,1,1-Trichloro-2,2-bis(*p*-chlorophenyl)ethane [*789-02-6*], *13*:429
 as insecticide, *13*:430
1,1,1-Trichloro 2,2-bis(*p*-fluorophenyl)ethane [*475-26-3*]
 as insecticide, *13*:430
1,1,1-Trichloro-2,2-bis-(*p*-methoxyphenyl)ethane [*72-43-5*]
 as insecticide, *13*:431
1,1,1-Trichloro 2,2-bis(*p*-methylphenyl)ethane [*4413-31-4*]
 as insecticide, *13*:430
1,1,1-Trichloro 2,2-bis(*p*-nitrophenyl)ethane [*2971-23-5*]
 as insecticide, *13*:430
1,1,1-Trichloro 2,2-bis(*p*-phenyl)ethane [*2971-22-4*]
 as insecticide, *13*:430
1,1,1-Trichloro-5-bromo-3-methyl-2-pentene [*74631-07-5*]
 isoprene reaction product, *13*:825
5,5,5-Trichloro-1-bromo-2-methyl-2-pentene [*74631-08-5*]
 isoprene reaction product, *13*:825
2,2,2-Trichloro-1-(butyryloxy)ethylphosphonate [*126-22-7*]
 as insecticide, *13*:442

Trichlorocarbanilide [101-20-2]
  applications of, 13:240
1,1,1-Trichloro 2-chloro-2,2-bis(p-chlorophenyl)-
    ethane [3563-45-9]
  as insecticide, 13:430
1,1,1-Trichloro-2-(o-chlorophenyl)-2-(p-chloro-
    phenyl)ethane [789-02-6]
  as insecticide, 13:429
1,1,1-Trichloro 2-deutero-2,2-bis(p-chlorophenyl)-
    ethane [13259-51-3]
  as insecticide, 13:430
Trichloroethane [71-55-6]
  exposure limit, 13:257
1,1,1-Trichloroethane [71-55-6]
  metal cleaning, 15:299
  stabilization of, 15:983
  toxicity, 13:273
Trichloroethanol [115-20-8]
  hypnotic action, 13:124
Trichloroethylene [79-01-6]
  in case hardening, 15:323
  as fumigant, 13:467
  metal cleaning, 15:299
  toxicity, 13:273
1,1,1-Trichloro 2-fluoro-2,2-bis(p-chlorophenyl)-
    ethane [1545-69-9]
  as insecticide, 13:430
Trichloroisocyanurate [87-90-1]
  applications of, 13:230
N-(Trichloromethylthio)-4-cyclohexene-1,2-di-
    carboximide [133-06-2]
  applications of, 13:245
N-(Trichloromethylthio)phthalimide [133-07-3]
  applications of, 13:245
Trichloronitromethane. See Chloropicrin.
Trichlorooxovanadium [7727-18-6]
  in nitride formation, 15:877
2,4,5-Trichlorophenol [95-95-4]
  applications of, 13:226
2,3,5-Trichloro-4-(propylsulfonyl)pyridine
    [38827-35-9]
  applications of, 13:244
Trichlorosilane [10025-78-2]
  in IC manufacture, 13:625
  isoprene reaction, 13:826
Trichlorotrifluoroethane [354-58-5]
  as insulator, 13:552
1,1,2-Trichloro-1,2,2-trifluoroethane [76-13-1]
  bromine/TEL reaction, 14:184
Trichoplusia, 13:471
Trichoplusia ni, 13:480
Tricresyl phosphate [1330-78-5], 14:496
  as antiwear agent, 14:493
  in TEL, 14:191
Tricyclohexylphosphine p-allylmethylnickel
    [76422-11-2], 15:809

Tricyclohexylphosphine triphenylboranenickel
    [76404-18-7]
  nickel arene from, 15:809
Tricyclohexylphosphine 1,3,5-
    triphenylbenzenenickel [76404-17-6], 15:809
Tricyclohexyltin hydroxide [13121-70-5], 13:463
  in insect control, 16:579
  solubility, 16:573
1-Tridecene [2437-56-1]
  properties, 16:481
Tridecyl alcohol [112-70-9]
  in metallic coating, 15:260
  paper coating defoamer, 16:791
Tridione, 13:135
Tridodecylethylammonium bromide [53290-56-5]
  in niobium analysis, 15:829
Triethanolamine [102-71-6]
  paint removers, 16:763
Triethyl 1-acetamido-1,1,3-propanetricarboxylate
    [6959-27-3]
  in L-glutamic acid synthesis, 15:767
Triethylaluminum [97-93-8]
  di-n-butylmagnesium complex, 16:560
  encapsulation co-catalyst, 15:482
  in ethylene oligomerization, 16:487
  in initiator systems, 13:367
  in isoprene polymerization, 13:829
  lead chloride reaction, 16:582
  properties, 16:565
  propylene polymerization cocatalyst, 16:458
Triethylaluminum–dipyridyl [97-93-8]
  in isoprene reaction, 13:822
Triethylammonium phosphate [35365-94-7]
  in paint and varnish removers, 16:763
Triethylborane [97-94-9]
  properties, 16:562
Triethylenediamine [280-57-9]
  isocyanate reaction catalyst, 13:793
Triethylene glycol [112-27-6]
  in inks, 13:382
  oxalic acid from, 16:622
Triethylene glycol methacrylate [109-16-0]
  properties, 15:354
Triethylene melamine [51-18-3], 13:162.   (See
    also 2,4,6-Tris(1-aziridinyl)-s-triazine.)
  in neoplastic therapy, 13:172
Triethylene phosphoramide [545-55-1], 13:
    148.   (See also Tris(1-aziridinyl)phosphine
    oxide.)
  in neoplastic therapy, 13:172
Triethylenetetramine [112-24-3]
  metallic copolymers from, 15:205
Triethyl(ethylsulfonyl)lead [13689-27-5], 14:183
Triethylhydroxylead [1529-35-7], 14:181

Triethyl indium [*923-34-2*], *13*:211

(Triethyllead)sodium [*6996-04-9*], *14*:181

Triethylmethyllead [*1762-28-3*], *14*:184

Triethyl phosphate [*78-40-0*]
  isocyanate reaction catalyst, *13*:797
  in TEPP synthesis, *13*:437

1-(Triethylsilyl)aziridine [*15000-97-2*], *13*:150

Triethylsilyl chloride [*994-30-9*]
  aziridine reaction, *13*:150

Trifluoroacetonitrile [*353-85-5*]
  isocyanic acid reaction, *13*:800

1,1,1-Trifluoro 2,2-bis(*p*-chlorophenyl)ethane
    [*361-67-9*]
  as insecticide, *13*:430

Trifluoroiodomethane [*2314-97-8*]
  in isotope separation, *13*:858

Trifluoromethane [*75-46-7*]
  in etch gas, *13*:639

Trifluoromethyldichlorocarbanilide [*369-77-7*]
  applications of, *13*:240

Trifluorophosphine [*7783-55-3*]
  nickel carbonyl reaction, *15*:807

TRIGA (training reactor, isotopes general atomic)
    reactors, *16*:141

Triglycerides
  in hyperlypidemic disease detection, *15*:82
  measurement, *15*:80

Trihalomethanes (THMs)
  in drinking water, *16*:708

Tri-*n*-hexylaluminum [*1116-73-0*]
  properties, *16*:565

Tri-*n*-hexylborane [*1188-92-7*]
  properties, *16*:562

Trihydrogen hexakiscyanoferrate [*17126-46-4*],
    13:768.   (See also *Ferricyanic acid.*)

2,3,4-Trihydroxybenzaldehyde [*2144-08-3*], *13*:70

2,3,5-Trihydroxybenzaldehyde [*74186-01-9*],
    13:70

2,3,6-Trihydroxybenzaldehyde [*64168-39-4*],
    13:70

2,4,5-Trihydroxybenzaldehyde [*35094-87-2*],
    13:70

2,4,6-Trihydroxybenzaldehyde [*487-70-7*], *13*:
    70.   (See also *Phloroglucinaldehyde.*)

3,4,5-Trihydroxybenzaldehyde [*13677-79-7*],
    13:70

11α,17,21-Trihydroxypregn-4-ene-3,20-dione
    [*566-35-8*], *16*:41

3,5,3'-Triiodothyronine [*6893-02-3*], *13*:673; *15*:
    593

Triiron tetroxide [*1317-61-9*], *13*:775.   (See also
    Magnetite.)

Triisobutylaluminum [*100-99-2*], *16*:567
  ethylene reaction, *16*:569

properties, *16*:565

Triisobutylborane [*1116-39-8*]
  properties, *16*:562

Trilead bis(hexakiscyanoferrate) [*28653-61-4*]
  in ferricyanic acid synthesis, *13*:768

Tri-med lure [*12002-53-6*]
  as insect attractant, *13*:481

Trimellitic anhydride [*552-30-7*]
  resin from in fluorescent product, *14*:561

Trimesoyl chloride [*3867-55-8*]
  aziridine reaction, *13*:143

Trimethadione [*127-48-0*]
  properties of, *13*:135

Trimethobenzamide [*138-56-7*], *13*:77

1,1,2-Trimethoxyethane [*149-73-5*]
  in paint and varnish removers, *16*:764

3-(Trimethoxysilyl)-propyldimethyloctadecylam-
    monium chloride [*27668-52-6*]
  applications of, *13*:234

Trimethylaluminum [*75-24-1*]
  in LED mfg, *14*:288
  properties, *16*:565
  zinc chloride reaction, *16*:586
  zinc reaction, *16*:586

Trimethylamine [*75-50-3*]
  in acetylcholine bromide synthesis, *15*:756
  in ion-exchange resin mfg, *13*:687
  odor-problem chemical, *16*:303
  polymer reaction, *14*:93

2-(*N,N,N*-Trimethylamino)ethyl methacrylate
    chloride [*33611-56-2*], *15*:356

2-(Trimethyl)amminium-ethyl methacrylate
    chloride [*5039-78-1*], *15*:352

1,1,1-Trimethyl 2,2-bis(*p*-chlorophenyl)ethane
    [*19685-37-1*]
  as insecticide, *13*:430

Trimethylborane [*593-90-8*], *16*:555
  properties, *16*:562

3,3,5-Trimethylcyclohexanol [*116-02-9*], *13*:919

3,3,5-Trimethylcyclohexanone [*73-94-9*], *13*:919

2,2,6-Trimethylcyclohex-5-ene-1,4-dione [*1125-
    21-9*], *13*:920

3,3,5-Trimethyl-2-cyclohexen-1-one [*78-59-1*],
    13:918.   (See also *Isophorone.*)

3,5,5-Trimethyl-3-cyclohexen-1-one [*471-01-2*],
    13:919

3,3,5-Trimethylcyclohexylamine [*15901-42-5*],
    13:919

3,3,5-Trimethylcyclohexyl methacrylate [*75673-
    26-6*]
  properties, *15*:353

2,2,6-Trimethyl-*m*-dioxin-4-one [*5394-63-8*], *13*:
    885

Trimethylenimines.   See *Azetidines.*

Trimethylgallium [*1445-79-0*], *16*:562
  in LED mfg, *14*:287

2,3,4-Tri-*O*-methyl-D-glucose [*4060-09-7*]
  in methylated dextran, *15*:442

3,5,5-Trimethylhexanal [*68039-71-4*]
　odor modification with, *16*:303
1,1,3-Trimethylindan [*2613-76-5*]
　from o-DIPB, *13*:50
Trimethyl indium [*3385-78-2*], *13*:211
2,3,3-Trimethylindolenine [*1640-39-7*], *13*:215
1,3,3-Trimethyl-2-methyleneindoline [*118-12-7*],
　13:215
2,6,8-Trimethyl-4-nonanone [*123-18-2*], *13*:908,
　934
Trimethylolmelamine [*1017-56-7*] (TMM)
　in papermaking, *16*:817
Trimethylolpropane [*77-99-6*]
　TDI reaction, *13*:815
Trimethylolpropane trimethacrylate [*3290-92-4*]
　properties, *15*:347, 354
Trimethyloxonium tetrafluoroborate [*420-37-1*]
　in N$_2$ fixation, *15*:958
2,4,4-Trimethyl-6-oxycyclohexane carbonitrile
　[*58114-70-8*], *13*:919
Trimethyl(phenyl)silane [*768-32-1*], *14*:471
Trimethyl phosphite [*121-45-9*]
　nickel carbonyl reaction, *15*:807
Trimethylsilylketene [*4071-85-6*]
　properties, *13*:875
Trimethylsulfonium iodide [*2181-42-2*], *13*:669
Trimethylsulfoxonium iodide, *13*:669
Trimethylvinylgermane [*753-97-9*], *15*:187
Trimethylvinylsilane [*754-05-2*], *15*:187
Trimofix [*28371-99-5*]
　in perfumes, *16*:968
Trinickel orthophosphate heptahydrate [*14396-
　43-1*], *15*:806
Triniding, *15*:320
1,3,5-Trinitrobenzene [*99-35-4*], *13*:10
1,1,1-Trinitroethane [*595-86-8*]
　properties, *15*:970
Trinitromethane [*517-25-9*], *15*:969
　properties, *15*:970
1,2,3-Trinitronaphthalene [*76530-13-7*], *15*:725
1,2,4-Trinitronaphthalene [*76530-14-8*], *15*:725
1,3,5-Trinitronaphthalene [*2243-94-9*], *15*:725
1,3,6-Trinitronaphthalene [*59054-75-0*], *15*:725
1,3,8-Trinitronaphthalene [*2364-46-7*], *15*:725
1,4,5-Trinitronaphthalene [*2243-95-0*], *15*:725
　nomenclature, *15*:702
2,4,6-Trinitroresorcinol [*82-71-3*]
　in explosive detonators, *13*:60
Trinitrotoluene
　manufacture, *15*:846
2,4,6-Trinitrotoluene [*118-96-7*], *13*:10
Trinitrotoluene [*118-96-7*] (tnt)
　isomer control in mfg, *15*:843
Tri-n-octylaluminum [*1070-00-4*]
　properties, *16*:565

Triode sputtering, *15*:268
Triolein [*122-32-7*]
　from almond oil, *16*:270
3,5,7-Trioxa-2,4,6,8,12-pentasilatridecane 12,12′-
　(1,1′-ferrocenediyl)bis-[2-chloro-
　2,4,4,6,6,8,8,12-octamethyl] [*1295-99-4*], *15*:
　200
Trioxymethylene [*110-88-3*]
　cashew oil reaction, *16*:272
Tri-n-pentylborane [*1883-38-1*]
　properties, *16*:562
Tri-n-pentyl phosphate [*10496-10-3*]
　as electrode solvent, *13*:724
sym-Triphenylbenzene [*612-71-5*]
　nickel complex reaction, *15*:809
Triphenyl indium [*3958-47-2*], *13*:211
Triphenylketenimine [*14181-84-1*], *13*:875
Triphenyl lead acetate [*1162-06-7*], *14*:193
Triphenylphosphine [*603-35-0*]
　in metal-nitrogen complexes, *15*:957
　oxo catalyst ligand, *16*:640
　ozonide reaction, *16*:686
Triphenylphosphine imines
　ketene reactions, *13*:882
Triphenylphosphine oxide [*791-28-6*]
　from iminophosphorane-ketene reaction, *13*:
　882
Triphenyl phosphite [*102-84-1*]
　iodine compounds formation, *13*:667
Triphenylphosphoranylideneketene [*15596-07-3*]
　properties, *13*:875
Triphenylstannic chloride [*639-58-7*]
　immunodeficiency effect, *13*:170
Triphenyltin hydroxide [*76-87-9*]
　properties, *16*:573
　solubility, *16*:573
2,4,6-Triphenyl-1,3,5-triazine [*493-77-6*], *15*:905
Triphos. See Bis[2-(diphenylphosphinyl)-
　ethyl]phenyl phosphine.
Triplet state
　phosphorescence from, *14*:557
Triplex ten-twenty, *13*:988
Tripotassium hexakiscyanoferrate [*13746-66-2*],
　13:769
　in Prussian Blue synthesis, *13*:771
Tripotassium tris(oxalato)ferrate trihydrate
　[*5936-11-8*], *13*:780
Tri-n-propylaluminum [*102-67-0*]
　properties, *16*:565
Tri-n-propylborane [*1116-61-6*]
　properties, *16*:562
Tripropylene glycol [*1638-16-0*]
　in inks, *13*:382
Tris(1-aziridinyl)phosphine oxide [*545-55-1*], *13*:
　148
　in neoplastic therapy, *13*:172
1-Tris(1-aziridinyl)phosphine oxide [*57-39-6*]
　in insect control, *13*:472

Tris(1-aziridinyl)-*s*-triazine [*51-18-3*], *13*:162
2,4,6-Tris(1-aziridinyl)-*s*-triazine [*51-18-3*]
  in neoplastic therapy, *13*:172
Tris(2,2′-bipyridine)iron(2, *13*:778
Tris(2,2′-bipyridine)iron(3, *13*:778
Tris(2,2′-bipyridine)iron(II) dibromide [*15388-40-6*], *13*:778
Tris(2,2′-bipyridine)iron(II) dichloride [*14751-83-8*], *13*:778
Tris(2,2′-bipyridine)iron(II) diperchlorate [*15388-48-4*], *13*:778
Tris(2,2′-bipyridine)iron(III) triperchlorate [*15388-50-8*], *13*:778
Tris(β-cyanoethyl)nitromethane [*1466-48-4*], *15*:977
Tris(4,7-diphenyl-1,10-phenanthroline)iron(2, *13*:780
Tris (*N,N′,N″*-ethyleneimino) phosphine oxide [*545-55-1*], *13*:148. (See also *Tris(1-aziridinyl)phosphine oxide.*)
Tris(hydroxymethyl)aminomethane [*77-86-1*]
  pH standard, *13*:3
Tris-(hydroxymethyl)nitromethane [*126-11-4*], *15*:914
  applications of, *13*:248
Tris(2-methacryloxyethyl)amine [*13884-43-0*]
  properties, *15*:354
Tris(2-methyl-1-aziridinyl)borane [*17862-61-2*], *13*:148
Tris nitro, *15*:914
Tris(nonylphenyl) phosphite [*26523-78-4*]
  in LLDPE, *16*:395
Trisodium hexafluoroaluminate [*13775-53-6*]
  in cementation coatings, *15*:248
Trisodium hexakiscyanoferrate [*14217-21-1*], *13*:769
Trisodium vanadate [*13721-39-6*]
  toxicity, *15*:574
Tris(2,4-pentanedionato)iron(III) [*14024-18-1*], *13*:778
Tris(1,10-phenanthroline)iron(2, *13*:780
Tris(1,10-phenanthroline)iron(3+) [*13479-49-7*], *13*:781
Tris(η³-2-propenyl) chromium [*12082-46-1*]
  in isoprene polymerization, *13*:829
Tris(2,2′,2″-terpyridine)iron(3+) [*47779-99-7*], *13*:781
Tris(trimethylphosphine)dimethylnickel [*42725-08-4*], *15*:809
Tris(tri-*o*-tolylphosphine)(3-pentenenitrile)nickel(0) [*41686-95-7*], *15*:808
Tritium [*10028-17-8*]
  disposal of, *16*:208
  in fuel reprocessing, *16*:175
  in nuclear reactors, *16*:141, 161
  radiation exposure from, *16*:217

in radioimmunoassay, *15*:86
  separation, *16*:162
Tri-*o*-tolylphosphine [*6163-58-2*]
  nickel complex formation, *15*:808
Trommsdorff effect, *15*:384
Trophicard, *15*:760
Tropical almond, *16*:250
Tropine [*120-29-6*]
  microbial synthesis of, *15*:462
Tropolone [*533-75-5*], *13*:876
Tropylium ions
  metal π-complex formation, *16*:592
Tropylium metal complexes, *16*:607
Troysan 174, *13*:242
Troysan 192, *13*:242
Troysan PMA, *13*:236
Trucks
  for LPG transportation, *14*:390
Trypanosoma cruzi, *13*:413
Trypanosomiasis, *13*:414
Tryptamine [*61-54-1*], *13*:220
  from aziridine, *13*:151
  as neuroregulator, *15*:782
  properties, *15*:782
Tryptamine hydrochloride [*343-94-2*], *15*:782
Tryptamine picrate [*6159-31-5*], *15*:782
Tryptophan [*73-22-3*], *13*:213; *14*:802
  microbial synthesis of, *15*:463
  in nuts, *16*:249, 252
  in tryptamine biosynthesis, *15*:782
L-Tryptophan [*73-22-3*]
  microbial syntheis of, *15*:464
DL-Tryptophan-hydantoin [*21753-16-2*]
  microbial transformation to L-tryptophan, *15*:465
Tryptophan hydroxylase [*9037-21-2*]
  in serotonin biosynthesis, *15*:770
TSCA. See *Toxic Substances Control Act.*
Tsetse fly, *13*:413
TSH. See *Thyroid-stimulating hormone.*
Tsuruga 1, *16*:160
Tuberculosis
  lactic acid elevation, *13*:80
  treatment, *13*:62, 178
Tuberculosis-causing bacteria
  disinfection by *o*-phenylphenol, *13*:226
Tuberculosis (cow), *15*:553
Tuberculosis (human), *15*:553
Tubeworms
  antimicrobial agents used against, *13*:238
Tub sizing, *16*:820
Tucum nut, *16*:251
Tufftriding, *15*:320
Tufperm YEP-H, *14*:657
Tufperm YEP-S, *14*:657
TULSA
  information service, *16*:903

Tumble mixers, *15*:629
Tumors
    remission of asparagine-dependent, *15*:490
Tung, *16*:248
Tung nut, *16*:251
Tung oil [*8001-20-5*]
    in fabric for insulation, *13*:578
    in insulation enamels, *13*:580
    in varnishes, *13*:554
Tungsten [*7440-33-7*]
    amalgamation with mercury, *15*:147
    in integrated circuits, *13*:641
    toxicity, *13*:274
    TZM alloy, *15*:679
Tungsten-180 [*14265-79-3*], *13*:842
Tungsten-182 [*14265-80-6*], *13*:842
Tungsten-183 [*14265-81-7*], *13*:842
Tungsten-184 [*14265-82-8*], *13*:842
Tungsten-186 [*14265-83-9*], *13*:842
Tungsten carbide [*12070-13-2*], *15*:263
    bearing coating, *14*:516
    in coatings, *15*:241
    nitride form, *15*:877
    toxicity, *13*:274
Tungsten carbonyl [*14040-11-0*]
    in chemical deposition coating, *15*:264
Tungsten diselenide [*12067-46-8*]
    as solid lubricant, *14*:508
Tungsten disulfide [*12138-09-9*]
    as solid lubricant, *14*:508
Tungsten fiber
    in metal laminates, *13*:945
Tungsten fluoride [*7783-82-6*]
    in chemical deposition coating, *15*:263
Tungsten nitride [*12058-38-7*] (1:1), *15*:872
Tungsten nitride [*12033-72-6*] (2:1), *15*:872
    properties, *15*:874
Tungsten oxide (1:3) [*1314-35-8*]
    recovery, *13*:61
Turbidity
    in water, *16*:706
Turbine blades
    coatings for, *15*:247
Turbines, *15*:609.   (See also *Impellers.*)
    as mixers, *15*:609
Turbine vanes
    coatings for, *15*:247
Turboexpander process
    for LPG, *14*:385
Turbulent flow
    mass transfer, *14*:950
Turkey point 3, *16*:155
Turkey x disease, *16*:263
Turmeric [*8024-37-1*], *16*:314
Turnbull's blue, *13*:769.   (See also *Prussian Blue.*)
Turpentine oil [*8006-64-2*], *16*:307, 328

in orange oil, *16*:325
TWA.   See *Time-weighted average.*
Tween 20 [*9005-64-5*], *15*:479
Tween 80 [*9005-65-6*], *13*:479; *15*:780
Twines
    fibers for, *16*:376
Twin-screw mixers, *15*:636
Twistor memories, *14*:659
Two-film theory
    in mass transfer, *14*:930
Two-step ethylene oligomerization process, *16*:488
Tymnet
    communications system, *16*:901
Typar 3351, *16*:76
Type-I censored life testing, *15*:20
Type-II censored life test, *15*:19
Type metals, *14*:156
Typhoid fever, *15*:553
    from water, *16*:705
Typhus, *13*:414
Tyramine [*51-67-2*]
    as neuroregulator, *15*:783
    in octopamine biosynthesis, *15*:774
    properties, *15*:783
Tyramine hydrochloride [*60-19-5*], *15*:783
Tyrosinase [*9002-10-2*]
    copper in, *15*:589
Tyrosine [*16-18-4*]
    in (−) adrenaline biosynthesis, *15*:758
    in nuts, *16*:249
    oxidation, *13*:219
    in tyramine biosynthesis, *15*:783
L-Tyrosine [*60-18-4*]
    microbial synthesis of, *15*:464
Tysine, *15*:737
Tysland-Hole process, *13*:745
Tyvek, *16*:94
Tyvek 1073-B, *16*:76
TZM [*11135-86-7*]
    molybdenum-base alloy, *15*:679

U

Udimet 200 [*12616-78-3*], *15*:789
Udimet 500 [*11068-87-4*], *15*:789
Udimet 700 [*11068-91-0*], *15*:789
U-gas process
    lignite gasification, *14*:338
UHMW–HDPE.   See *Ultrahigh molecular weight high density polyethylene.*
Uinta formation, *16*:335

Ulcer
  diagnosis, *15*:79
  from steroid treatment, *13*:174
Ulcers
  treatment, *15*:445
Ultrafiltration, *15*:106, 115
  for milk products, *15*:562
Ultralente iletin [*8049-62-5*], *13*:613
Ultralente insulin [*8049-62-5*], *13*:613
Ultramarine blue [*57455-37-5*]
  in inks, *13*:378
Ultraperm 10, *14*:659
Ultraperm F, *14*:659
Ultraperm Z, *14*:659
Ultrasonic cleaning
  metals, *15*:299
Ultrasonic detectors
  for level measurement, *14*:433
Ultrasonic inspection
  clad metals, *15*:285
Ultrasonics
  in nondestructive testing, *16*:65
Ultrasonic sealing, *16*:722
Ultrasonic testing, *16*:49
Ultrasonic tests, *16*:65
Ultrasonic welding, *15*:277
Ultrasuede, *14*:246
Ultraviolet cured inks, *13*:385
Ultraviolet curing
  of inks, *13*:375
Ultraviolet ozone generators, *16*:690
U-matic, *14*:750
Umbelliferone [*93-35-6*], *13*:46
γ-Undecalactone [*104-67-6*]
  in perfumes, *16*:967
1-Undecene [*821-95-4*]
  properties, *16*:481
Undecylenic acid [*112-38-9*]
  preservative, *13*:247
Undecylenic aldehyde [*112-45-8*]
  in perfumes, *16*:967
Undercoats, *16*:751
Underwater operations
  oxygen supply, *16*:671
Unicracking process, *15*:665
Unidak process, *15*:709
Union carbide process
  polyethylene, *16*:392
Union oil vertical-kiln retort, *16*:339
Unipen, *15*:735
Unipol process
  ethylene polymerization, *16*:392
Unisel, *16*:79
Unit absorbers
  acoustic performance of, *13*:518
United Nations
  information sources, *13*:307

Uniterm index
  information system, *16*:901
Upholstery
  fibers for, *16*:376
  leather substitutes for, *14*:233
  polypropylene usage, *16*:383
Uracil [*66-22-8*]
  in nucleic acid, *15*:587
Uraninite, *16*:144
Uranium [*7440-61-1*]
  fission efficiency, *16*:190
  in fission fuel, *16*:143
  HF reaction, *16*:181
  isotope separation by laser, *14*:76
  isotope separation process, *13*:846
  in lignite ash, *14*:329
  in nuclear reactor, *16*:162
  from the ocean, *16*:279
  oxidation, *16*:668
  prospecting methods, *16*:829
  recovery by ion exchange, *13*:702
  recovery by leaching, *16*:668
  recovery from wastewater, *15*:212
  resources and reserves, *16*:143
  from seawater, *16*:284
  from uranium tetrafluoride, *14*:573
Uranium-232 [*14158-29-3*], *16*:180
Uranium-233 [*32650-43-4*], *16*:188
  reprocessing of, *16*:173
  from thorium, *16*:169
  from thorium-232, *16*:187
Uranium-235 [*15117-96-1*], *13*:842; *16*:161, 188
  as fission fuel, *16*:140
  in fission fuel, *16*:143
  reprocessing of, *16*:173
  resources and reserves, *16*:143
Uranium-238 [*24678-82-8*], *13*:842
  in breeder reactors, *16*:168
  in FBRS, *16*:187
  radiation exposure from, *16*:217
  reprocessing of, *16*:173
Uranium-239 [*13982-01-9*], *16*:188
Uranium carbides
  in nuclear fuels, *16*:140
Uranium carbide [*12071-33-9*], UC$_2$
  nitrogen reaction, *15*:934
Uranium, cmpd with silicon and aluminum (3:1), *16*:140
Uranium dioxide [*1344-57-6*]
  as fission fuel, *16*:140
  fuel assembly for, *16*:165
Uranium fluoride [*10049-14-6*] (1:4)
  fluorination, *16*:181
Uranium hexafluoride [*7783-81-5*]
  in fuel reprocessing, *16*:173
  isotope separation of, *13*:845
  as lubricant, *14*:515
  NaF complex, *16*:181
  nuclear fuel from, *16*:165

Uranium hexafluoride (*Continued*)
  photodissociation by lasers, *14*:76
  U-235 from, *16*:162
Uranium mining
  radon dosimetry in, *16*:845
Uranium nitride [*25658-43-9*], *15*:934
  properties, *15*:874
Uranium nitride [*25658-43-9*] (1:1), *15*:872
Uranium nitride [*12033-85-1*] (2:3), *15*:872
β-Uranium nitride [*12033-85-1*], *15*:882
Uranium ores
  disposal of tailings, *16*:211
Uranium oxide [*1344-59-8*]
  in breeder fuel, *16*:169
  in nuclear reactors, *13*:859
Uranium prospecting
  using radon dosimetry, *16*:846
Uranium silicide [*12039-74-6*]
  in nuclear fuels, *16*:140
Uranium tetrafluoride [*10049-14-6*]
  magnesium reaction, *14*:573
Uranyl nitrate [*10102-06-4*]
  in lecithin analysis, *14*:264
  in purex process, *16*:176
  tributyl phosphate complex, *16*:176
Urate oxidase [*9002-12-4*]
  copper in, *15*:589
Urea [*57-13-6*]
  detection of using ISES, *13*:728
  diketene reaction, *13*:884
  in leather manufacture, *14*:216
  in metallo isocyanate synthesis, *13*:801
  in paper coatings, *16*:791
  in phenytoin synthesis, *13*:135
  removal from intestinal tract, *15*:490
Urea adhesives
  for plywood, *14*:14
Urea–formaldehyde [*9011-05-6*]
  as insulation, *13*:595
Urea–formaldehyde polymer [*9011-05-6*]
  in encapsulation, *15*:479
  in fluorescent product, *14*:560
Urea–formaldehyde resin [*9011-05-6*]
  in paper, *16*:793
Urea–formaldehyde resin modified with
    diethylenetriamine [*61472-52-4*], *16*:817
Urea–formaldehyde resins [*9011-05-6*]
  in papermaking, *16*:817
Urea–melamine adhesives
  for plywood, *14*:14
Urea peroxide [*35220-04-3*]
  in digoxin assay, *15*:87
Urea peroxyhydrate [*124-43-6*], *13*:15, 29
Urease [*9002-13-5*]
  BUN reagent, *15*:81
Ureides
  as grease thickeners, *14*:502

2-Ureidoethyl methacrylate [*4206-97-7*]
  properties, *15*:354
Uremia
  diagnosis, *15*:79
Urethane [*51-79-6*]
  in metallic coating, *15*:266
Urethane polymers
  as leather substitutes, *14*:231
  in leather substitutes, *14*:239
Urethanes
  as magnetic tape binders, *14*:741
  by nitro group carbonylation, *13*:808
Uretoneimines, *13*:798
7,7′-Ureylene-bis-4-hydroxy-2-naphthalenesul-
    fonic acid [*137-47-4*], *15*:742
Uric acid [*69-93-2*]
  from purines, *16*:260
  xanthine oxidase reaction, *15*:592
Urine
  analysis of lead content in, *14*:192
  analysis of with ISES, *13*:733
  oxalic acid in, *16*:618
U-ring packing, *16*:733
Urocanic acid [*104-98-3*]
  microbial synthesis of, *15*:465
USPSD information system, *13*:331
USP zinc insulin crystals reference standard, *13*:
    609
Ustilago sphaerogena, *13*:784
Uv absorbers
  from dihydroxybenzenes, *13*:58
Uv radiation
  as initiator, *13*:367
Uv stabilizers
  for olefin fibers, *16*:361

V

Vaccines, *13*:167
Vacodur 16, *14*:656
Vacoflex Z, *14*:659
Vacoperm 100, *14*:659
Vacozet 655, *14*:681
Vacreator
  for milk processing, *15*:538
Vacuum arc remelt process, *15*:799
Vacuum coatings, *15*:265
Vacuum-induction melting, *15*:799
Vacuum metallizing, *15*:266
  particles, *15*:474
Vacuum-tube spacers
  from sheet mica, *15*:436

Valence-bond theory (VBT), *16*:594

δ-Valerolactone [*108-29-2*], *13*:97

Valine [*72-18-4*]
  microbial synthesis of, *15*:463
  in nuts, *16*:249, 252

Valinomycin [*2001-95-8*]
  as ionophore, *13*:725

Valium, *13*:122, 127

Valproic acid [*99-66-1*]
  properties of, *13*:135

Valves
  in feedback control, *13*:489
  noise control, *16*:23

Vampire bats
  repellant for, *16*:579

Vanadic acid, ammonium salt [*7803-55-6*]
  nitride formation, *15*:877

Vanadium [*7440-62-2*]
  amalgamation with mercury, *15*:147
  in magnetic materials, *14*:663
  in metallic coatings, *15*:252
  as nutrient, *15*:598
  from the ocean, *16*:279, 292
  toxicity, *13*:274

Vanadium-50 [*14391-89-0*], *13*:840

Vanadium-51 [*7440-62-2*], *13*:840

Vanadium carbide [*12070-10-9*]
  in metallic coating, *15*:252
  in nitride pseudobinary system, *15*:873

Vanadium–cobalt–iron alloys
  as magnetic materials, *14*:679

Vanadium compounds
  recovery from fuels, *14*:642

Vanadium nitride [*24646-85-3*], *15*:877
  properties, *15*:874

Vanadium nitride [*24646-85-3*] (1:1), *15*:872

Vanadium nitride [*12209-81-3*] (2:1), *15*:872

Vanadium oxide [*12035-98-2*]
  in maleic anhydride synthesis, *14*:770
  in nitride formation, *15*:877

Vanadium oxychloride [*7727-18-6*]
  in LLDPE catalyst, *16*:391
  removal from silicon tetrachloride, *15*:906

Vanadium oxychloride [*7727-18-6*] (1:1:3)
  in isoprene polymerization, *13*:829

Vanadium pentoxide [*1314-62-1*]
  in *n*-butyllithium analysis, *14*:468

Vanadium tetrachloride [*7632-51-1*]
  in isoprene polymerization, *13*:829

Vanadium trichloride [*7718-98-1*]
  in isoprene polymerization, *13*:829

Vanadyl chloride [*7727-18-6*]
  toxicity, *15*:574

Van Arkel process, *15*:877

Vancide PA, *13*:244

Vancide 89RE, *13*:245

Vancide TH, *13*:242

Vancide 51Z, *13*:243

Van Dyke process, *16*:633

Vanillin [*121-33-5*], *13*:45
  economic aspects, *13*:53
  from lignin, *14*:296, 306
  in perfumes, *16*:949, 954

Vanilmandelic acid [*1321-73-9*]
  measurement, *15*:84

Vans
  laminated glass windows in, *13*:989

Vantel, *14*:248

Vanthoffite [*15557-33-2*], *14*:637

Van't Hoff relation
  for osmotic pressure, *15*:113

Vanway O ring, *16*:726

Vapor barriers
  use with thermal insulation, *13*:599

Vapor degreasing, *15*:298

Vapor-phase epitaxy (VPE)
  in LED manufacture, *14*:287

Vapor plating
  mechanical, *15*:260

VAR. See *Vacuum arc remelt.*

Vari-color filters, *16*:534

Varigam paper, *16*:544

Varilure papers, *16*:544

Varion, *13*:696

Varnishes, *16*:632

Varnishes, insulating
  properties of, *13*:555

Varnish removers, *16*:762. (See also *Paint and varnish removers.*)

Vasculitis
  induced by immune complexes, *13*:169

Vasoactive intestinal polypeptide [*40077-57-4*] (VIP)
  as neuroregulator, *15*:783

Vasoconstrictine, *15*:757

Vasodilation
  neurotensin induced, *15*:772

Vasopressin [*11000-17-2*]
  effect on learning and memory, *15*:132
  sulfur in, *15*:587

Vasotonin, *15*:757

VAZO 52, *15*:901

Vazo 64, *15*:888, 901

VBT. See *Valence-bond theory.*

Veba-Chemie process
  for MIBK, *13*:910

Vegetable amber, *16*:950

Vegetable oils, *16*:307. (See also *Oils, essential.*)
  in packings, *16*:738

Vegetable tanning, *14*:212

Vehicles
  noise source, *16*:9

Veneers
 laminated wood, *14*:39
Venom, snake
 lecithinase of, *14*:256
Ventilation
 in radioactive areas, *16*:242
Ventron process
 mercury recovery, *15*:170
Vera process, *16*:210
Veratraldehyde [*120-14-9*]
 papaverine from, *13*:63
Veratridine [*71-62-5*], *13*:427
Veratrole [*91-16-7*], *13*:43
Vermiculite [*1318-00-9*], *15*:416
 as thermal insulator, *13*:592
Vermilion. See *Mercuric sulfide.*
Vermillion [*19122-79-3*], *15*:143
Veronol [*141-13-9*]
 in perfumes, *16*:968
Vertofix coeur [*32388-55-9*]
 in perfumes, *16*:968
Verv. See *Calcium stearoyl-2-lactylate.*
Vespula, *13*:481
Vessels
 explosively clad, *15*:289
Vestas, *15*:4
Vetiveria zizanioides Stapf, *16*:328
Vetiveria zizanoides, *16*:953
Vetiver oil [*8016-96-4*], *16*:328
Vetiverone [*15764-04-2*], *16*:328
Vetivert acetate [*68917-34-0*]
 in perfumes, *16*:954
Vetivert oil [*8016-96-4*]
 in perfumes, *16*:953
VHS, *14*:750
Vibration isolation
 materials for, *13*:530
Vicalloy 1
 magnetic properties, *14*:671
Vicalloy II, *14*:694
Victoria Blue [*1325-93-5*], *15*:695
 in inks, *13*:378
Victor Meyer reaction, *15*:841
Video-disk systems
 lasers for, *14*:285
Video recorders, *14*:733
Video tape, *14*:701
Vidicon tube
 infrared sensitive, *13*:345
Vigoureux printing, *16*:631
Viledon-M, *16*:79
Vilsmeier-Haack formylation
 of indoles, *13*:216
VIM. See *Vacuum induction melting.*
Vinaccia, *13*:114. (See also *Cream of tartar.*)

Vinca, *15*:465
Vinca alkaloids
 as antimetabolites, *13*:172
Vindoline [*2182-14-1*]
 microbial demethylation, *15*:465
Vinegar
 microbial preparation, *15*:459
Vinyl acetate [*108-05-4*]
 copolymer, *15*:192
 copolymer with ethylene, *16*:419
 in encapsulation, *15*:481
 from ethylene, *16*:483
 health effect, *13*:274
 methacrylate copolymerization, *15*:384
 polymerization of, *13*:356
 solubility of, *14*:83
Vinyl acetate–vinyl chloride copolymer [*9003-22-9*]
 thickening agent, *16*:764
Vinyl acetylene [*108-05-4*]
 biacetyl from, *13*:925
Vinyl acrylics
 in paint, *16*:748
1-Vinylaziridine [*5628-99-9*], *13*:155
Vinylbenzyl chloride [*1592-20-7*]
 polymerization and quaternization, *14*:93
N-Vinylcarbazole [*1484-13-5*]
 copolymer, *15*:192
 polymerization of, *13*:356
Vinyl chloride [*74-01-4*]
 in emulsion polymerization, *14*:85
 kinetics of destruction, *13*:201
 maleic anhydride copolymers, *14*:780
 methacrylate copolymerization, *15*:384
 polymerization of, *13*:356
 solubility of, *14*:83
 toxicity, *13*:274
3-Vinylcyclohexyl methacrylate [*76392-15-9*]
 properties, *15*:353
Vinylene bisthiocyanate [*14150-71-1*]
 applications of, *13*:243
Vinyl fabrics
 expanded, *14*:233
 as leather substitutes, *14*:233
Vinylferrocene [*1271-51-8*], *15*:186
 polymerization of, *15*:185
Vinylferrocene, polymer with methyl acrylate [*35139-57-2*], *15*:189
Vinyl halides
 toxicity, *13*:274
Vinylidene chloride [*75-35-4*]
 methacrylate copolymerization, *15*:384
 solubility of, *14*:83
Vinyl isocyanate [*3555-94-0*]
 polymerization, *13*:796
Vinylketene [*50888-73-8*], *13*:882
Vinyl methacrylate [*4245-37-8*]
 properties, *15*:353

Vinyloxybenzene [766-94-9]
  isoprene reaction, 13:821
Vinyloxyethoxyethyl methacrylate [76392-22-8]
  properties, 15:353
Vinyl plastics
  colorants for, 14:568
Vinyl plastisol
  in leatherlike materials, 14:236
Vinyl polymers
  in leather coatings, 14:222
  as leather substitutes, 14:231
Vinylpyridine [1337-81-1]
  dyeing of in fibers, 16:362
2-Vinylpyridine–styrene–butadiene terpolymer
  [9019-71-0], 13:57
N-Vinylpyrrolidinone [88-12-0]
  copolymer, 15:192
N-Vinyl-2-pyrrolidinone [88-12-0]
  polymerization of, 13:356
Vinylruthenocene [76082-22-9], 15:186
Vinyls
  corrosion resistant paint, 16:755
Vinyltrimethylstannane [754-06-3], 15:187
Vinyon [9003-22-9]
  nonwovens from, 16:108
Vinyon HH, 16:109
Vinyzene BP-5, 13:236
Viola odorata, 16:953
Violet acid [578-85-8], 15:738. (See also 5-
  Hydroxy-2,7-naphthalenedisulfonic acid.)
Violet leaves absolute [8024-08-6]
  in perfumes, 16:953
VIP. See Vasoactive intestinal polypeptide.
Viral equation
  for polymer mol wt, 15:121
Viruses
  effect on cattlehides, 14:208
  liquid-crystal phases, 14:420
  sterilization by ozone, 16:703
Visadur films, 14:568
Viscolizer, 15:529
Viscometers
  for paint, 16:757
Viscose rayon
  nonwoven fabrics from, 16:108
Viscosity
  units, 14:485
Viscosity index, 14:489
Viscosity-index (VI) improvers, 14:495
Viscosity number
  of polymers, determination, 16:437
Viscosity temperature coefficient, 14:489
Viscous flow
  Darcy units, 15:224

Visine, 15:737
Vitamin [59-02-9]
  antiaging effects, 15:137
Vitamin A, 14:471
  added to milk, 15:554
Vitamin C [50-81-7]
  antiaging effects, 15:138
  encapsulation, 15:476
  in walnuts, 16:261
Vitamin content
  of milk, 15:524
Vitamin D
  added to milk, 15:554
Vitamin D₃, 15:586. (See also CC.)
Vitamin K
  preparation, 15:716
Vitamin K₃, 15:716
Vitamins
  fortification of milk, 15:554
  in nuts, 16:253
Vitellin
  animal phosphatides associated with, 14:252
Viton, 16:727
Vitrification
  nuclear wastes, 16:210
Vitrolite
  calibrated white block, 14:563
Vitrolube 1220, 14:510
Vivelle, 16:79
Voandzeia subterranea, 16:250
Volhard method
  iodine determination, 13:658
Voloxidation
  tritium, 16:211
  for tritium, 16:180
Voltex, 16:120
Von Baeyer, Adolf, 13:213. (See also Baeyer.)
Von Heyden process, 15:706
Vortex burners
  for incinerators, 13:198
Vortex incinerator, 13:190
VPE. See Vapor-phase epitaxy.
2V Permendur [37188-80-0], 14:651, 663
V-ring packing, 16:733
Vulcanized natural asphalts
  electric insulation filler, 13:565

W

Wacker acetaldehyde process, 16:584
Wacker process, 16:592
  with butenes, 13:906

Wacker process (*Continued*)
  catalysts, *16*:611
Wad
  manganese oxide mixture, *14*:825
Waferboard, *14*:21
Waiting-line models
  in operations planning, *16*:520
Wallace formula, *14*:747
Wallboard, *14*:22
Wall paints, *16*:751
Walnut, *16*:250
Walther equation, *14*:489
Wanderhaufen malt unit, *14*:816
Wankel engines
  linings for, *15*:883
WARC.   See *World Administrative Radio
    Conference.*
Warping
  of plywood, *14*:7
Washburn equation, *16*:808
Washer mica, *15*:421
Waspaloy [*11068-93-2*], *15*:789
Waste
  municipal, classification of, *13*:184
Waste management, *16*:206.   (See also *Nuclear
    reactors.*)
Wastes
  combustion of, *13*:188
  incineration of, *13*:188
  industrial, classification of, *13*:183
  nuclear, *16*:206
  from pulp mills, *16*:795
Wastewater
  ozone disinfection, *16*:711
Wastewaters
  oxygen treatment, *16*:669
Water [*7732-18-5*], *16*:282.   (See also *Ocean raw
    materials.*)
  in air, *16*:657
  emissivity, *13*:342
  hypochlorite purification, *15*:182
  iodine treatment, *13*:660
  lime in treatment of, *14*:379
  in neutron shielding, *16*:241
  nuclear reactor coolant, *16*:152
  packaging of, *16*:722
  radiolytic decomposition in nuclear reactors,
    16:155
  recovery from the ocean, *16*:282
  resins for desalination, *13*:688
Water-based inks, *13*:391
Water chestnut, *16*:249, 251
  structure, *16*:249
Water chinquapin nut, *16*:251
Water gas
  in ammonia mfg, *15*:946
Water glass.   See *Sodium silicate.*

Waterleaf paper, *16*:778
Watermelon, *16*:251
Water Pollution Control Act, *13*:254
Waterproofing
  matches, *15*:5
Water purification
  with iodine, *13*:674
  by magnetic separators, *14*:728
Water softening
  ion exchange in, *13*:698
  with lime, *14*:379
Water treatment
  magnetic, *14*:730
  with ozone, *16*:703
Watts plating bath, *15*:804
Waveguides, *15*:506
Wax
  case hardening, *15*:321
  odor compensation of, *16*:303
  in packings, *16*:734
  paper size, *16*:812
Wax cracking, *16*:487
  olefins from, *16*:485
Waxes
  in electrical insulation, *13*:565
  in inks, *13*:379
  from lignites, *14*:340
Weakness
  (−) adrenaline induced, *15*:758
Wear resistance
  enhancement of by ion implantation, *13*:717
Weathering
  of insulation materials, *13*:547
Web-fed lithographic process, *13*:385
Wedges, spectral, *16*:552
Weibull distribution, *15*:23
Weibull probability paper, *15*:23
Weissenberg rheogoniometer, *16*:757
Weissenstein process, *13*:23
WELDASEARCH
  information service, *16*:903
Welding, *15*:277.   (See also *Explosively clad
    metals.*)
  explosion-clad vessels, *15*:291
  lasers for, *14*:65
  lead, *14*:140
  of magnesium alloys, *14*:608
Welding flux
  lead bromide as, *14*:163
Welding solder
  lead-tin alloys, *14*:154
Wemco C
  dielectric strength, *13*:541
Werner complexes, *16*:596
Weston cell, *15*:40
Wetherill cross-belt separator,
    14:728
Wet look, *14*:231

Wet-strength additives
   for paper, *16*:816
Wet-web-strength
   in papermaking, *16*:807
Whipping cream, *15*:560
Whisker–ceramic fiber
   in metal laminates, *13*:945
White lead [*1344-36-1*]
   as solid lubricant, *14*:508
White mica, *15*:420
White phosphorus [*7723-14-0*]
   as insecticide, *13*:422
White rot, *13*:225
White wax [*8012-89-3*]
   repellent, *13*:479
Whiting, *14*:345
Wiberg-Soderfors process, *13*:745
Wiedemann-Franz relation, *15*:224
Wien's displacement law, *13*:338
Wigner effect, *16*:141
Willow
   catechol in, *13*:46
Wilson's disease
   nutrients associated with, *15*:573
Wilson's stain, *16*:794
Windows
   underwater, *13*:991
   water removal by molecular sieves, *15*:663
Windshields
   birdproof, *13*:990
   laminated glass in, *13*:978
Wine
   tartaric acid from, *13*:114
Winkler gasifier
   lignite, *14*:337
Winkler generator
   for lignite gasification, *14*:329
Wintergreen oil [*68917-75-9*], *16*:319, 329
Wipes
   nonwoven, *16*:123
Wire
   insulation coverings for, *13*:564
   melt spinning, *15*:229
   recording on, *14*:733
Wire-arc spraying, *15*:254
Wire coatings
   flame-retardant, *13*:402
Wire, enameled magnet, *13*:554
Wire, hook-up
   insulation of, *13*:554
Wire memories, *14*:693
Wires
   polyethylene, *16*:401
Wiring devices
   mercury, *15*:155
Wisconsin process
   nitrogen fixation, *15*:864

Wiswesser line notation, *13*:280
Wittig reaction, *13*:669
   organolithium compounds in, *14*:466
Wofatit, *13*:696
Wolff-Chaikoff effect, *13*:674
Wolff rearrangement, *13*:880
Wolfram [*7440-33-7*]
   from the ocean, *16*:280
Wolframite [*1332-08-7*]
   from beach sand, *16*:289
Wood
   emissivity, *13*:342
   as an engineering material, *14*:2
   fiber properties, *13*:970
   lignin from, *14*:304
   structure, *14*:2
Wood alcohol.   See *Methanol.*
Wood-based composites, laminated, *14*:1
Wood bleaching
   oxalic acid in, *16*:633
Wood fiber, *14*:17
Wood preservation
   organotin compounds, *16*:578
Wood pulp
   nonwoven fabrics from, *16*:107
Wood sugar
   by hydrolysis, *14*:309
Wood varieties
   lignin content, *14*:297
Wool
   moisture regain, *16*:358
   in packings, *16*:734
Wool grease
   in grease thickeners, *14*:502
Word processors
   in information retrieval, *13*:281
Work flow
   queuing theory applied to, *16*:520
World Administrative Radio Conference, *15*:496
World Aluminum Abstracts
   information service, *16*:903
World Patents Index
   information system, *16*:905
WORLD PETROCHEMICAL DATA BASE, *13*:327, 331
World Textiles
   information service, *16*:903
Worm gears
   lubricants for, *14*:488
Wrapper inks, *13*:382
Writing inks, *13*:395
Wrought iron, *13*:736
Wuchereria bancrofti, *13*:413
Wuestite
   iron(II) oxide as, *13*:775
Wulfenite [*14913-82-7*], *15*:670
Wurtz coupling, *16*:557

Wurtz-Fittig reaction, *13*:669
Wurtz reaction
  organotins by, *16*:573

X

Xanthan gum [*11138-66-2*], *15*:447
  from corn sugar, *15*:440
  properties, *15*:448
Xanthates
  from nitro alcohols, *15*:913
Xanthine [*69-89-6*]
  microbial synthesis of, *15*:465
  xanthine oxidase reaction, *15*:592
Xanthine oxidase [*9002-17-9*]
  Mo content, *15*:693
  molybdenum in, *15*:592
Xanthines
  CNS stimulants, *15*:135
Xanthomonas campestris, *15*:447, 449
Xanthone [*90-47-1*]
  as photoinitiator, *13*:368
Xanthosine-5′-monophosphate [*523-98-8*] (5′-
    XMP)
  microbial transformation to 5′-GMP, *15*:465
Xenon [*7440-63-3*]
  in air, *16*:657
  ion implantation, *13*:715
  in metallic coating, *15*:268
  from the ocean, *16*:280
Xenon-124 [*15687-60-2*], *13*:841
Xenon-126 [*27982-81-6*], *13*:841
Xenon-128 [*27818-78-6*], *13*:841
Xenon-129 [*13965-99-6*], *13*:841
Xenon-130 [*14808-51-6*], *13*:841
Xenon-131 [*14683-11-5*], *13*:841
Xenon-132 [*14155-79-4*], *13*:841
Xenon-134 [*15751-43-6*], *13*:841
Xenon-135 [*14995-62-1*]
  as fission-chain poison, *16*:139
Xenon-136 [*15751-79-8*], *13*:841
  in isotope separation, *13*:849
Xenopsylla cheopis, *13*:414
Xeroradiography, *16*:64
X-7287L, *13*:231
XLPE. See *Polyethylene, cross-linked*.
5′-Xmp. See *Xanthosine-5′-monophosphate*.
X-500 polymer [*54573-52-3*]
  liquid-crystal behavior, *14*:418
X rays
  radiation exposure from, *16*:217
X-ray screens
  phosphors for, *14*:544

X-ray testing methods
  nondestructive, *16*:49
Xylan [*9014-63-5*]
  biodegradation of, *13*:225
Xylene [*1330-20-7*]
  in isocyanate manufacture, *13*:803
  isomerization by zeolites, *15*:667
  in paint and varnish removers, *16*:764
  toxicity, *13*:274
m-Xylene [*108-38-3*]
  ammoxidation, *15*:906
o-Xylene [*95-47-6*]
  as phthalic anhydride feedstock, *15*:705
p-Xylene [*106-42-3*]
  microbial oxidation, *15*:460
p-Xylene diisocyanate [*25854-16-4*], *13*:806
Xylene Red B [*3520-42-1*]
  fluorescent dye, *14*:547
Xylenes
  separation by molecular sieves, *15*:664
3,5-Xylenol [*108-68-9*]
  from isophorone, *13*:920
Xylitone isomers, *13*:918
D(−) Xylonamide [*74421-67-3*], *13*:101
Xylonic acid [*526-91-0*], *13*:101
D(−) Xylonic acid, γ-lactone [*15384-37-9*], *13*:101
Xylonic acid, tetraacetate [*74497-94-2*], *13*:101
Xylose [*58-86-6*]
  in microbial synthesis, *15*:467

Y

Y-12
  isotope separation, *13*:844
YAG.   See *Yttrium aluminum garnet*.
YAG laser
  in Pt coating, *15*:256
Yankee dryer, *16*:785, 816
Yara yara, *15*:734
Yarns
  from olefin fibers, *16*:375
  in packings, *16*:735
Yeast
  chromium source, *15*:592
  lignite substrates, *14*:341
  zinc in, *15*:590
Yeasts
  malic acid in, *13*:106
  in polysaccharide preparation, *15*:439
Yellow-dwarf virus disease, *13*:415

Yellow prussiate of potash [*14459-95-1*], *13*: 767. (See also *Tetrapotassium hexakiscyanoferrate trihydrate.*)

Yield strength
of alloys, *14*:596

YIG. See *Yttrium iron garnet.*

Ylang-ylang [*8006-81-3*], *13*:63; *16*:329

Yogurt, *15*:564

Ytterbium-168 [*15743-54-1*], *13*:842

Ytterbium-170 [*13982-08-6*], *13*:842

Ytterbium-171 [*14041-50-0*], *13*:842

Ytterbium-172 [*14041-52-2*], *13*:842

Ytterbium-173 [*14041-51-1*], *13*:842

Ytterbium-174 [*14683-29-5*], *13*:842

Ytterbium-176 [*15751-45-8*], *13*:842

Ytterbium, compd with cobalt [*39416-96-1*] (1:5), *14*:676

Ytterbium, compd with cobalt [*12671-28-2*] (2: 17), *14*:676

Ytterbium nitride [*24600-77-9*], *15*:872

Yttrium [*7440-65-5*]
in lead-bath coatings, *15*:253
from the ocean, *16*:280, 292

Yttrium-89 [*7440-65-5*], *13*:840

Yttrium–aluminum garnet [*12005-21-9*]
optical-filter coating, *16*:526

Yttrium aluminum garnet (YAG)
lasers, *14*:56

Yttrium aluminum oxide [*12009-21-9*] (3:5:12)
in phosphors, *14*:540

Yttrium, compd with cobalt [*12017-71-9*] (1:5), *14*:676

Yttrium, compd with cobalt [*12017-78-6*] (2:17), *14*:676

Yttrium iron garnet (YIG) [*12063-56-8*]
as magnetic insulator, *14*:694
in microwave devices, *14*:667

Yttrium nitride [*25764-13-0*], *15*:872

Yttrium oxide, Eu and Tb doped [*68784-54-3*]
phosphor, *14*:527

Yttrium oxide, Eu doped [*68585-82-0*]
phosphor, *14*:544

Yttrium phosphate, Tb doped
phosphor, *14*:544

Yttrium phosphate vanadate
phosphor, *14*:527

Yttrium trioxide [*1314-36-9*]
phosphor host, *14*:541

Yttrium vanadate, Eu doped [*68585-81-9*]
phosphor, *14*:527

Yttrium ytterbium erbium fluoride (0.79:0.20:0.01:3)
phosphor, *14*:543

Yttrium ytterbium thulium fluoride (0.639:0.35:0.01:3)
phosphor, *14*:543

Z

Zanflo, *15*:451

Zarontin, *13*:135

Zea mays
radiobiological effects, *16*:845

Zein coating for nuts, *16*:263

Zeocarb, *13*:696

Zeolite [*1318-02-1*], *15*:638

Zeolite A, *15*:639

Zeolite 3A, *15*:660

Zeolite 4A, *15*:660

Zeolite 5A, *15*:660

Zeolite AW-300, *15*:660

Zeolite AW-500, *15*:660

Zeolite calcium A [*68989-22-0*], *15*:648

Zeolite calcium X, *15*:647

Zeolite F, *15*:659

Zeolite L, *15*:639

Zeolite LZY-52, *15*:660

Zeolite LZY-62, *15*:660

Zeolite LZY-72, *15*:660

Zeolite LZY-82, *15*:660

Zeolite NaX [*68989-73-1*], *15*:653

Zeolite omega, *15*:639

Zeolite potassium A [*68989-21-9*], *15*:647

Zeolites, *13*:678; *15*:638. (See also *Molecular sieves.*)
in ion exchange, *13*:678

Zeolite sodium X [*68989-23-1*], *15*:647

Zeolite T, *15*:651

Zeolite W, *15*:659

Zeolite X, *15*:639

Zeolite Y, *15*:639

Zeolon [*12173-98-7*], *15*:655

Zeolon 100, *15*:660

Zeolon 200, *15*:660

Zeolon 300, *15*:660

Zeolon 500, *15*:660

Zeolon 900, *15*:660

Zeolon H [*53569-61-2*], *15*:640

Zerolit, *13*:696

Ziegler catalyst, *16*:385
in isoprene polymerization, *13*:818

Ziegler catalysts
for hdpe, *16*:441
for polypropylene fiber, *16*:357

Ziegler-Natta catalysts, *16*:584
for polypropylene, *16*:458

Ziegler-Natta polymerization, *16*:592

Ziegler-Natta process
metal complex catalysts for, *16*:610

Ziegler oligomerization processes
ethylene, *16*:488

Ziegler polymerization
  of ethylene, *16*:433
Ziegler process
  one-step oligomerization of ethylene, *16*:489
  two-step oligomerization of ethylene, *16*:490
Ziese's salt [*12275-00-2*], *16*:592.   (See also
    η^2-*Ethyleneplatinum trichloride*.)
Zinc [*7440-66-6*]
  in cable sheathing, *13*:585
  in coatings, *15*:241
  HgO dry-cell reaction, *15*:162
  from lead refining, *14*:121
  in magnesium alloys, *14*:596
  metabolic functions, *15*:590
  nutrient properties, *15*:571
  from the ocean, *16*:279, 292
  in packings, *16*:734, 737
  surface treatment of, *15*:304
  wear coefficient, *14*:482
Zinc-64 [*14378-32-6*], *13*:840
Zinc-65 [*13982-39-3*]
  reactor corrosion product, *16*:155
Zinc-66 [*14378-33-7*], *13*:840
Zinc-67 [*14378-34-8*], *13*:840
Zinc-68 [*14378-35-9*], *13*:840
Zinc-70 [*14378-36-0*], *13*:840
Zinc acetate [*557-34-6*]
  dioctylphosphinic acid reaction, *15*:217
Zinc acetate dihydrate [*5970-45-6*]
  toxicity, *15*:573
Zinc alkyls, *16*:585
Zincalume, *15*:246
Zinc ammonium lactate [*74421-68-4*]
  in lactic acid resolution, *13*:80
Zinc beryllium silicate [*39413-47-3*]
  toxicity, *14*:545
Zinc bis(dioctylphosphinate) [*2915-86-8*], *15*:217
Zinc carbonate [*3486-35-9*]
  from diethylzinc, *16*:586
Zinc chloride [*7646-85-7*]
  Grignard reaction, *16*:586
  in lead dezincing, *14*:122
  toxicity, *15*:573
Zinc deficiency
  in nuts, *16*:253
Zinc dimethyldithiocarbamate [*137-30-4*]
  applications of, *13*:243
  in electrical insulation, *13*:565
Zinc diorganodithiophosphate
  as antiwear agent, *14*:493
Zinc dithiophosphate [*19210-06-1*]
  in hydraulic oils, *14*:488
  lubricant antioxidant, *14*:492
Zinc ferrite [*1317-55-1*], *13*:776
Zinc fluorosilicate [*16871-71-9*]
  toxicity, *15*:573

Zinc glycolate [*14008-38-9*], *13*:94
Zinc(II) dihydrogen phosphate [*13598-37-3*]
  in metal surface treatment, *15*:304
Zinc(II) phosphate tetrahydrate [*7543-51-3*]
  in metal treating, *15*:305
Zinc insulin [*8049-62-5*], *13*:609
Zinc lactate [*16039-53-5*]
  in lactic acid resolution, *13*:80
Zinc 2-mercaptobenzothiazole [*155-04-4*]
  in antimicrobial agent, *13*:246
Zinc metalloenzymes, *15*:590
Zinc molybdate [*13767-32-3*], *15*:687
Zinc naphthenate [*12001-85-3*], *15*:752
  in coatings, *13*:94
Zinc nitrate [*7779-88-6*]
  in anode coating mfg, *15*:178
Zinc nitride [*1313-49-1*], *15*:872
Zinc Omadine, *13*:245
Zinc oxide [*1314-13-2*]
  applications of, *13*:237
  in electrical insulation, *13*:565
  fungistat in paint, *16*:747
  as grease filler, *14*:503
  in inks, *13*:377
  in matches, *15*:5
  in metal treating, *15*:306
  paper pigment, *16*:789
  as solid lubricant, *14*:508
  toxicity, *13*:274
Zinc permanganate hexahydrate [*23414-72-4*],
    14:845
Zinc phosphide [*22569-71-7*]
  as insecticide, *13*:422
Zinc pyridinethione [*13463-41-7*]
  applications of, *13*:245
Zinc selenide [*1315-09-9*]
  optical-filter coating, *16*:526
Zinc silicate, Mn doped [*68611-47-2*]
  phosphor, *14*:527
Zinc sulfate [*7733-02-0*]
  toxicity, *15*:573
Zinc sulfate heptahydrate [*7446-20-0*]
  toxicity, *15*:573
Zinc sulfate monohydrate [*7446-19-7*]
  in electroplating solution, *14*:610
Zinc sulfide [*1314-98-3*]
  in inks, *13*:377
  optical filter coating, *16*:526
  in paper, *16*:777
  as semiconductor, *14*:286
Zineb, *13*:246.   (See also *Zinc
    ethylenebis(dithiocarbamate)*.)
Zingiber officiale Rosc., *16*:323
Zircaloy, *16*:165, 206
Zircaloy-2, *15*:830; *16*:156
Zircaloy 4, *16*:152
Zircon [*14940-68-2*]
  from beach sand, *16*:289

in mica books, *15*:420

Zirconia [*1314-23-4*]
  in insulation, *13*:598

Zirconium [*7440-67-7*]
  cladding of, *15*:281
  in coatings, *15*:241
  HF reaction, *16*:181
  in magnesium alloys, *14*:596
  in nuclear reactors, *16*:140
  from the ocean, *16*:280, 292
  tannages containing, *14*:214

Zirconium-90 [*13982-15-5*], *13*:840

Zirconium-91 [*14331-93-2*], *13*:840

Zirconium-92 [*14392-15-5*], *13*:840

Zirconium-94 [*14119-12-1*], *13*:840

Zirconium-96 [*15691-06-2*], *13*:840

Zirconium carbide [*12020-14-3*]
  in nitride pseudobinary system, *15*:873

Zirconium fluoride [*7783-64-4*] (1:4), *16*:181
  in fuel reprocessing, *16*:181

Zirconium hydride [*7704-99-6*]

in reactors, *16*:141

Zirconium lactate [*74421-69-5*], *13*:88

Zirconium–niobium [*11107-78-1*] (99:1), *15*:830

Zirconium–niobium [*11135-92-5*] (97.5:2.5), *15*:830

Zirconium nitrate [*13826-66-9*]
  in anode coating mfg, *15*:178

Zirconium nitride [*25658-42-8*], *15*:872
  properties, *15*:874

Zirconium phosphate [*13772-29-7*]
  as ion-exchange material, *13*:686

Zirconocene polymers, *15*:209

Zirconyl acetate peroxyhydrate, *13*:15

Zirconyl nitrate [*13826-66-9*]
  in anode coating mfg, *15*:177

Zsk extruders, *16*:396

ZSM-5 [*58339-99-4*], *15*:639

ZSM-11, *15*:640

ZSM-12, *15*:667

ZSM-21, *15*:667